STUDENT SOLUTIONS MANUAL

TO ACCOMPANY

Chang CHEMISTRY

FIFTH EDITION

Jerry Mills
TEXAS TECH UNIVERSITY

Raymond Chang
WILLIAMS COLLEGE

McGRAW-HILL, INC.

NEW YORK ST. LOUIS SAN FRANCISCO AUCKLAND BOGOTÁ
CARACAS LISBON LONDON MADRID MEXICO CITY
MILAN MONTREAL NEW DELHI SAN JUAN
SINGAPORE SYDNEY TOKYO TORONTO

STUDENT SOLUTIONS MANUAL

TO ACCOMPANY

Chang

CHEMISTRY

STUDENT SOLUTIONS MANUAL TO ACCOMPANY **Chang: Chemistry**

This book is printed on recycled paper containing a minimum of 50% total recycled fiber with 10% postconsumer de-inked fiber.

2 3 4 5 6 7 8 9 0 SEM SEM 9 0 9 8 7 6 5 4

ISBN 0-07-011004-2

The editors were Jennifer Speer and Maggie Lanzillo;
the production supervisor was Paula Flores.
Cover photograph: Ken Karp.
Semline, Inc., was printer and binder.

INTRODUCTION FOR THE STUDENT

Purpose and Plan of This Manual

"Chemistry is hard!" No doubt you have heard this from your friends. It is true that for most students chemistry is more difficult and requires more work than many other, if not most, college courses. You will in all likelihood confirm this fact for yourself as you progress through this course. What makes it hard? The simple fact is that all courses in the physical sciences, mathematics, and engineering share a common emphasis on *solving problems*, and it is this process that causes students so much difficulty.

Solving problems is an acquired skill like swimming or playing chess; to be good, you must practice a great deal. You cannot learn to swim fast or play chess well by watching somebody else do it; similarly, you cannot learn to solve chemistry problems well by passively watching your instructor or tutor work at the blackboard. Proficiency in solving problems is the result of experience accumulated through successfully working a large number of sample or practice problems. This type of skill cannot be acquired by intense study immediately before an examination (cramming).

The purpose of this manual is to help you learn to solve chemistry problems. You will discover that the more problems you work successfully, the better you will become at solving chemistry problems. You will also find that these same skills carry over to other disciplines. You will become better at solving mathematics, engineering, and physics problems because of your experience in chemistry.

Using this Manual

You will find detailed solutions and explanations of almost three-quarters of the end-of-chapter Problems in the fifth edition of Chemistry by Raymond Chang. (A few end-of-chapter Review Questions are also answered, but the emphasis is very much on the Problems.)

We present each solution in the same way. After the problem number, there is usually a brief introduction or commentary to focus on the essential features of the problem. This often involves a reference to an appropriate section in your text for idea-review purposes, or a question posed to concentrate your thinking. Similar problems in the text are referenced, as are other similar worked problems in this manual. The actual solution and problem analysis is set out in a box immediately following this commentary. At the end of the analysis we occasionally add follow-up questions to stimulate thought or discussion with other students or with your instructor; the answers to some of these may not be easy.

To use this manual to best effect, have it open beside your text as you work the problems in each chapter. Start to work each problem with the solution box covered. Use the introductory comments, and try to work out as much of the answer as you can without looking at the solution box. If you get stuck or need a hint, uncover just enough of the solution to see how to proceed further, and try to finish the problem. If you get stuck again, uncover more of the solution. Some problems will be easy, but a few will be tough, and you will have to uncover and study the entire solution. Don't be discouraged by this. Once you see the logic of the problem, go back and try to solve it again without looking at the answer.

A Caveat on Problem Solving

There is a major difference between mentally following how a problem is worked when it is presented to you step-by-step (as it is in this manual and the examples in the text) and actually reading a problem and solving it yourself without reference to the worked-out solution. Many students who have spent long hours studying do poorly on exams and are justifiably frustrated and discouraged. A very common cause for the poor grade, in spite of the hard work, is studying the wrong way by using worked-out solutions in the wrong way. *Do not* say "Ah yes, I see how this problem is worked and I can follow the solution." You as a student need to read the problem, think about it, refer to the text if necessary, and actually work the problem out! Only refer to the solution *after* you have an answer or are really lost. Why? Because if you work the problem yourself, you have to analyze the question and decide what type of problem it is before you can solve the mathematics. In other words, you are developing a concept of the material, and are not simply "plugging in" an equation. If you only use this solutions manual in an "I follow" method, you will not learn the material and you will not perform well on exams.

Working the problems out yourself is even more important when the homework problems are "cubby-holed" for you as they are in this text and most other modern textbooks. That is, you know that because of the method of organization that a certain problem is covered in the text on pages blank-to-blank. Try not to use this cubby-holing as too much of a crutch. If you do, you will not learn how to analyze problems. (You should try to work as many as possible of the "Miscellaneous Problems" at the last of the end-of-chapter problems because they are not pre-analyzed for you.) Remember, on an exam the questions will not be cubby-holed for you.

One becomes good at working problems by working many problems.

CONTENTS

Note: If you have not already done so, please read the "Introduction for the Student" at the beginning of this Student Solutions Manual.

Chapter 1

TOOLS OF CHEMISTRY

1.4 Read Section 1.2 to review the exact meanings of the terms *qualitative* and *quantitative*.

> ***Solution:*** (a) Quantitative. This is clearly a statement involving a measurable distance.
>
> (b) Qualitative. This is a value judgment. There is no numerical scale of measurement for artistic excellence.
>
> (c) Qualitative. If the numerical values of the densities of ice and water were given, it would be a quantitative statement.
>
> (d) Qualitative. Another value judgment.
>
> (e) Qualitative. Even though numbers are involved, they are not the result of exact measurements.
>
> What do you suppose is the difference between a qualitative chemical analysis and a quantitative chemical analysis?

> ***Solution:*** The answer to this question will depend upon which detective is chosen.

1.8 What is the difference between weight and mass? If necessary, review the definitions in Section 1.7.

> ***Solution:*** The correct statement is "The mass of the student is 56 kg." The weight of the student is dependent on the force of gravity.

1.11 Study the italicized definitions of these terms in Section 1.7. Which depends upon the amount of material being considered? Is weight an intensive or an extensive property?

> ***Solution:*** Mass and volume are both extensive properties. Density is the mass of an object divided by its volume (see Section 1.7). Is density an extensive property too?

1.13 Study the italicized definitions of these terms in Section 1.6. What is the important distinction between these two types of properties?

> ***Solution:*** (a) Chemical property. Oxygen gas is consumed in a combustion reaction; its composition and identity are changed.
>
> (b) Chemical property. The fertilizer is consumed by the growing plants; it is turned into vegetable matter (different composition).
>
> (c) Physical property. The measurement of the boiling point of water does not change its identity or composition.
>
> (d) Physical property. The measurement of the densities of lead and aluminum does not change their composition.
>
> (e) Chemical property. This may not be obvious, but the sensation of taste is the result of a chemical reaction between the sugar and sensory receptors in the taste buds.

1.14 This problem is similar to 1.13 above.

> ***Solution:*** (a) Physical change. The helium isn't changed in any way by leaking out of the balloon.
>
> (b) Chemical change in the battery.
>
> (c) Physical change. The orange juice concentrate can be regenerated by evaporation of the water.
>
> (d) Chemical change. Photosynthesis changes water, carbon dioxide, etc. into complex organic matter.
>
> (e) Physical change. The salt can be recovered unchanged by evaporation.

1.15 Study the italicized definitions of these terms in Section 1.6. What is the difference between them?

> ***Solution:*** (a) Extensive property. The length of an object depends upon how much there is of it.
>
> (b) Extensive property. Combining objects together changes their total area.
>
> (c) Extensive property. The volume of an object depends on how much of it there is.
>
> (d) Intensive property. Combining two glasses of water, both at the same temperature, doesn't change the temperature of the water.
>
> (e) Extensive property. Mass is a measure of the amount of matter in the system under study.
>
> Is density (mass divided by volume) an extensive or an intensive property? Refer to question 1.27. Would mass multiplied by volume be an extensive or an intensive property?

1.16 See Table 1.1 in your textbook.

Solution: Li, lithium; F, fluorine; P, phosphorus; Cu, copper; As, arsenic; Zn, zinc; Cl, chlorine; Pt, platinum; Mg, magnesium; U, uranium; Al, aluminum; Si, silicon; Ne, neon.

1.18 What is the difference between elements and compounds? Review Section 1.3 if necessary.

Solution: If you need additional help on this problem, see Problem 1.19 below.

(a) element (b) compound (c) element (d) compound

1.19 Find and carefully study the definitions of and distinctions between elements, compounds, homogeneous mixtures, and heterogeneous mixtures in Section 1.3. How can substances and mixtures be distinguished on the basis of their compositions? What is the key distinction between elements and compounds?

Solution: (a) Seawater is a mixture of salt, water, and other substances. Its composition can vary, so it can't be a compound or an element. A well-stirred sample of seawater is uniform throughout, and must therefore be a homogeneous mixture.

(b) Helium gas cannot be broken down or separated into simpler substances. It is an element.

(c) Sodium chloride is very different from sodium (a metal) and chlorine (a yellowish-green gas), so it can't be a mixture (the components don't retain their identities). It can be broken down into sodium and chlorine by chemical means. It is a compound.

(d) A soft drink contains dissolved carbon dioxide. Assuming that the cap is on the bottle, the liquid contents should be a homogeneous mixture. If the cap on the bottle is removed, then temporarily it may not be homogeneous. On standing, however, the liquid must be in equilibrium with atmospheric carbon dioxide, and it should again be a homogeneous mixture.

(e) A milk shake can be separated into finely crushed ice, water, sugar, fat globules, etc. It is a heterogeneous mixture.

(f) Clear air with no particulate pollutants is a homogeneous mixture (nitrogen, oxygen, carbon dioxide, water vapor, etc.).

(g) You can usually see distinct components of concrete without a magnifying glass. Heterogeneous mixture.

1.21 If you don't remember any of these, review Section 1.7.

Solution:

(a) length: m (b) area: m^2 (c) volume: m^3

(d) mass: kg (e) time: s (f) force: N or kg m/s^2

(g) energy: J or kg m^2/s^2 (h) temperature: K

Which of these are base units and which are derived units? Are all base units extensive quantities?

1.22 Review Table 1.3 for the definitions. If you need review on scientific notation, refer to Section 1.8.

Solution: (a) 10^6 (b) 10^3 (c) 10^{-1} (d) 10^{-2} (e) 10^{-3} (f) 10^{-6} (g) 10^{-9} (h) 10^{-12}

1.23 Density is a derived quantity that will be used constantly throughout your study of chemistry. Be sure you understand it completely. Review Section 1.7 if you are unsure of this important definition.

Solution: The density of an object is its mass divided by its volume. The units of g/mL or g/cm^3 are usually used. Why isn't the SI unit kg/m^3 the common expression for density?

Density is an intensive property. If both mass and volume are extensive properties, why should density be intensive? Would the product of mass and volume be intensive or extensive? Why?

1.24 The equations for interconversion between degrees Celcius and degrees Fahrenheit are give in Section 1.7.

Solution:

$$?°\text{C} = (°\text{F} - 32°\text{F}) \times \frac{5°\text{C}}{9°\text{F}}$$

$$?°\text{F} = \frac{9°\text{F}}{5°\text{C}} \times (°\text{C}) + 32°\text{F}$$

1.25 This is a straightforward problem if you remember the definition of density. If you don't, go back to Section 1.7 and relearn it. What is divided by what? What do the units become?

This problem is like Example 1.1. Find this example, cover the answer, and read the question. Try to set up and solve the example without looking at the covered answer. If you don't know how to proceed, uncover the solution one line at a time until you see how to work the problem. When you have worked through the example completely, cover the solution again and rework it. Don't forget the units!

The solution to Problem 1.25 is in the box below. To get the most out of working problems, cover the solution and try to work the problem as described above. Don't look at the answer until you are either baffled or finished with the problem.

Solution: The density of the sphere is given by

$$d = \frac{m}{V} = \frac{1.20 \times 10^4 \text{ g}}{1.05 \times 10^3 \text{ cm}^3} = 11.4 \text{ g/cm}^3$$

If you are not sure of how to work with numbers in exponential or scientific notation, carefully study Section 1.8 on this topic.

Would the density have been different if a cube of lead, rather than a sphere, had been used? What would be the density expressed in SI units? It is sometimes useful to realize that when the definition of density is stated as "mass

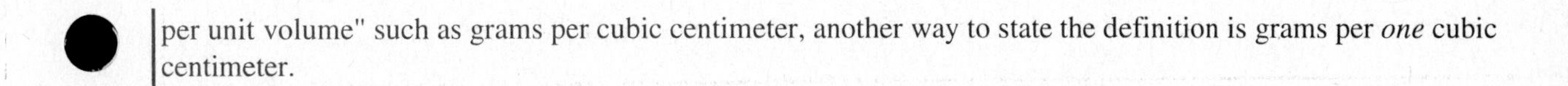
per unit volume" such as grams per cubic centimeter, another way to state the definition is grams per *one* cubic centimeter.

1.26 This problem is similar to Example 1.2. Cover the answer to this example, read the question, and try to solve it without looking at the answer. If you can't, study the solution and rework the problem with the answer covered. Notice how the solution is set up so the units work out to those of mass in Example 1.2.

Use Example 1.2 as a model for this problem.

Solution: The mass of the mercury is given by

$$m = dV = 13.6\ \text{g/mL} \times 95.8\ \text{mL} = 1.30 \times 10^3\ \text{g}$$

1.28 In this problem, the density and mass are provided and you are asked to solve for the volume.

Solution: The volume of the lithium is given by

$$V = m/d = (1.20 \times 10^3\ \text{g}) / (0.53\ \text{g/cm}^3) = 2.3 \times 10^3\ \text{cm}^3$$

1.30 The formulas for the interconversion of Fahrenheit and Celsius temperatures can be found at the end of Section 1.7. Example 1.3 is a model of this type of calculation. Cover the answer to Example 1.3 and try to work both parts before trying Problem 1.30. If you do this, you should have no trouble.

Solution:

(a) $?°\text{C} = (105°\text{F} - 32°\text{F})\left(\frac{5°\text{C}}{9°\text{F}}\right) = 41°\text{C}$

(b) $?°\text{F} = \left(\frac{9°\text{F}}{5°\text{C}}\right)(-11.5°\text{C}) + 32°\text{F} = 11.3°\text{F}$

(c) $?°\text{F} = \left(\frac{9°\text{F}}{5°\text{C}}\right)(6.3 \times 10^3°\text{C}) + 32°\text{F} = 1.1 \times 10^4°\text{F}$

1.31 You won't find the procedure for solving this question in Chapter 1, but you can find it with a little algebra. At the unknown temperature t, you have the situation that °F = °C. Use either of the Fahrenheit-Celsius formulas given at the end of Section 1.7, substitute t for both °C and °F in the equation, and find t. Cover the solution below with a piece of paper and try to work out the answer.

Solution: We write

$$?°\text{F} = \frac{9°\text{F}}{5°\text{C}} \times (°\text{C}) + 32°\text{F}$$

$$t = \frac{9}{5}t + 32°\text{F}$$

$$t - \frac{9}{5}t = 32°\text{F}$$

$$-\frac{4}{5}t = 32°F$$

$$t = -40°C = -40°F$$

1.33 The correct use of scientific notation is an essential skill in any science. If you have any difficulty with this problem, go back to the beginning of Section 1.8 and study the material carefully, then rework this example.

Solution: (a) 2.7×10^{-8} (b) 3.56×10^{2} (c) 4.7764×10^{4} (d) 9.6×10^{-2}

1.34 The correct use of scientific notation is an essential skill in any science. If you have any difficulty with this problem, go back to the beginning of Section 1.8 and study the material carefully, then rework this example.

Solution: (a) 0.0152 (b) 0.0000000778

1.35 Study Section 1.8 on handling numbers in scientific notation before attempting this problem. Can you add or subtract two numbers in scientific notation if they have different exponents? What about multiplication and division?

Solution: In addition and subtraction the exponents must be the same. In multiplication the exponents are added, and in division they are subtracted.

(a) $145.75 + (2.3 \times 10^{-1}) = 145.75 + 0.23 = 145.98$

(b) $\left(\frac{79500}{2.5 \times 10^{2}}\right) = \left(\frac{7.95 \times 10^{4}}{2.5 \times 10^{2}}\right) = 3.18 \times 10^{2}$

(c) $(7.0 \times 10^{-3}) - (8.0 \times 10^{-4}) = (7.0 \times 10^{-3}) - (0.80 \times 10^{-3}) = 6.2 \times 10^{-3}$

(d) $(1.0 \times 10^{4})(9.9 \times 10^{6}) = 9.9 \times 10^{10}$

(e) $0.0095 + (8.5 \times 10^{-3}) = (9.5 \times 10^{-3}) + (8.5 \times 10^{-3}) = 1.8 \times 10^{-2}$

(f) $\left(\frac{653}{5.75 \times 10^{-8}}\right) = \left(\frac{6.53 \times 10^{2}}{5.75 \times 10^{-8}}\right) = 1.14 \times 10^{10}$

(g) $850{,}000 - (9.0 \times 10^{5}) = (8.5 \times 10^{5}) - (9.0 \times 10^{5}) = -0.5 \times 10^{5} = -5 \times 10^{4}$

(h) $(3.6 \times 10^{-4})(3.6 \times 10^{6}) = 1.3 \times 10^{3}$

1.37 Before working this problem, go back to Section 1.8 and study the Guidelines for Using Significant Figures, then apply them to Example 1.4. Cover the answer and try to do all the parts without looking at the solution. After you can work all parts of Example 1.4, try this problem. Cover the solution below and work (a) through (h).

Solution: (a) four, (b) two, (c) five, (d) ambiguous (why?), (e) three, (f) one, (g) one, (h) two

1.38 This problem is very similar to Problem 1.37 above.

Solution: (a) one, (b) three, (c) three, (d) four, (e) two or three, (f) one, (g) one or two

1.39 Before attempting this problem, go back to Section 1.8 and study the rules and examples following Example 1.4 dealing with the handling of significant figures. Cover the solution to Example 1.5, read the problem, and try to work out the answers. If you are wrong on any of them, review the rules on significant figures again.

Cover the solution below and work (a) through (f).

Solution: (a) 10.6 m, (b) 0.79 g, (c) 16.5 cm^2, (d) 1.28 (why are there no units?), (e) 3.18×10^{-3} mg, (f) 8.14×10^{7} dm

1.40 Your success in this course depends in part on your complete mastery of the factor-label method of solving problems. Study the text of Section 1.9 thoroughly. What is a unit factor? Why is such an expression called a unit factor? Why is the reciprocal of a unit factor also a unit factor?

Try to apply the factor-label method to Examples 1.6 and 1.7. Cover the solutions with a piece of paper and try to work out the unit conversions. If you get stuck, uncover the answers, study the explanations, and rework the examples.

Problem 1.40 is similar to Example 1.6. Cover the solution below and write down the relationships between m and dm. Make a unit factor and set up the conversion. Do the units work out?

Solution: 22.6 m = ? dm. By definition 1 m = 10 dm. The unit factor is

$$\left(\frac{10\ \text{dm}}{1\ \text{m}}\right)$$

$$?\ \text{dm} = 22.6\ \text{m}\left(\frac{10\ \text{dm}}{1\ \text{m}}\right) = 226\ \text{dm or } 2.26 \times 10^{2}\ \text{dm}$$

1.41 Problem 1.41 is similar to Example 1.7. Cover the solution below and write down the relationships between mg and g and between g and kg. Make unit factors out of each and set up the conversion from mg to kg as a chain multiplication. Do the units work out?

Solution: 25.4 mg = ? kg. By definition 1 g = 1000 mg and 1 kg = 1000 g. The unit factors are

$$\left(\frac{1\ \text{g}}{1000\ \text{mg}}\right) \text{ and } \left(\frac{1\ \text{kg}}{1000\ \text{g}}\right)$$

$$? \text{ kg} = 25.4 \text{ mg} \left(\frac{1 \text{ g}}{1000 \text{ mg}}\right)\left(\frac{1 \text{ kg}}{1000 \text{ g}}\right) = 0.0000254 \text{ kg or } 2.54 \times 10^{-5} \text{ kg}$$

1.42 This problem is very similar to Example 1.7

Solution: By definition 1 lb = 453.6 g, so the unit factor is $\left(\frac{453.6 \text{ g}}{1 \text{ lb}}\right)$. It is also necessary to use the unit factor for converting g to mg.

$$? \text{ mg} = 242 \text{ lb} \left(\frac{453.6 \text{ g}}{1 \text{ lb}}\right)\left(\frac{1 \times 10^3 \text{ mg}}{1 \text{ g}}\right) = 1.10 \times 10^8 \text{ mg}$$

1.43 This problem is similar to Example 1.8. Does multiplication by unit factors ever change the actual magnitude of a quantity? Be sure in working this problem to cube the unit factor.

Solution: 68.3 cm^3 = ? m^3. By definition 1 m = 100 cm. The unit factor is

$$\left(\frac{1 \text{ m}}{100 \text{ cm}}\right)$$

$$? \text{ m}^3 = 68.3 \text{ cm}^3 \left(\frac{1 \text{ m}}{100 \text{ cm}}\right)\left(\frac{1 \text{ m}}{100 \text{ cm}}\right)\left(\frac{1 \text{ m}}{100 \text{ cm}}\right)$$

$$? \text{ m}^3 = 68.3 \text{ cm}^3 \left(\frac{1 \text{ m}}{100 \text{ cm}}\right)^3 = \left(\frac{68.3 \text{ m}^3}{1 \times 10^6}\right) = 6.83 \times 10^{-5} \text{ m}^3$$

1.44 If you have not already done so, study Section 1.9 thoroughly before trying this problem. Cover the solution below and try to set up the problem. What are the units of the answer? What is the relationship between the cost of gold in dollars and its mass in ounces? What is the relationship between mass in ounces and mass in grams? Set up the unit factors relating these quantities and try to solve the problem.

Solution: The unit factors are $\left(\frac{1 \text{ oz}}{28.4 \text{ g}}\right)$ and $\left(\frac{\$327}{1 \text{ oz}}\right)$

$$?\$ = 1.00 \text{ g} \left(\frac{1 \text{ oz}}{28.4 \text{ g}}\right)\left(\frac{\$327}{1 \text{ oz}}\right) = \$11.50$$

1.46 If you have not already done so, study Section 1.9 thoroughly before trying to solve this problem. Cover the solution below and try to set up the answer. What should be the units of the answer? What are the units of the quantity you are trying to convert? What is the relationship between distance and time in this problem? What is the relationship between minutes and seconds? Set up the unit factors relating these quantities and try to solve the problem. Do the units cancel correctly? Problem 1.47 gives the conversion between meters and miles.

Solution: Starting with speed = distance÷time, we have

$$\text{time} = \frac{\text{distance}}{\text{speed}} = \frac{(9.3 \times 10^7 \text{ mi})\left(\frac{1609 \text{ m}}{1 \text{ mi}}\right)}{\left(\frac{3.00 \times 10^8 \text{ m}}{1 \text{ s}}\right)\left(\frac{60 \text{ s}}{1 \text{ min}}\right)} = 8.3 \text{ min}$$

1.48 If you have studied Section 1.9 and worked through Examples 1.6, 1.7, and 1.8, parts (a) and (b) should present no real problem. Use the relationships given to set up unit factors and work the problem.

Part (c) involves changing cm/s to miles/hour. Go back to Example 1.9, cover the answer, and try to set up the solution to the double unit conversion. If you get stuck, uncover enough of the answer to get yourself going again, and try to finish the problem. If you can't, study the solution, cover it, and rework the problem. What is the relationship between miles and meters? (See Problem 1.47) What is the relationship between seconds and hours? Use these to make unit factors and try to work the problem. Problems 1.47, 1.50, 1.51, and 1.52 are similar.

In part (d) how many grams of lead are contained in one million grams of blood if the lead content is 0.62 ppm? What is the relationship between grams and pounds? Use these to make unit factors and try to work this problem.

Solution:

(a) $5.9 \text{ ft}\left(\frac{1 \text{ m}}{3.28 \text{ ft}}\right) = 1.8 \text{ m};$ $\quad 158 \text{ lb}\left(\frac{453.6 \text{ g}}{1 \text{ lb}}\right)\left(\frac{1 \text{ kg}}{1000 \text{ g}}\right) = 71.7 \text{ kg}$

(b) $\left(\frac{55 \text{ mi}}{1 \text{ hr}}\right)\left(\frac{1.609 \text{ km}}{1 \text{ mi}}\right) = 88 \text{ km/h}$

(c) $\left(\frac{3.0 \times 10^{10} \text{ cm}}{1 \text{ s}}\right)\left(\frac{3600 \text{ s}}{1 \text{ hr}}\right)\left(\frac{1 \text{ mi}}{1609 \text{ m}}\right)\left(\frac{1 \text{ m}}{100 \text{ cm}}\right) = 6.7 \times 10^8 \text{ mph}$

(d) $6.0 \times 10^3 \text{ g of blood}\left(\frac{0.62 \text{ g Pb}}{1 \times 10^6 \text{ g blood}}\right) = 3.7 \times 10^{-3} \text{ g Pb}$

1.50 If you have not already done so, study Section 1.9 on the use of the factor-label method. What is a unit factor? Why can a unit factor be used to multiply a quantity without changing the magnitude of that quantity? (a) How far does light travel in one second? How many seconds are there in one 365-day year? (b) and (c) What is the relationship between inches and centimeters? How many inches are in one yard? In one foot? (d) and (e) Review Temperature Scales in Section 1.7 and use Example 1.3 as a model for these problems. (f) Study Example 1.8. Why is the entire unit factor cubed? (g) From part (f) how many cm^3 are in one m^3? How many cm^3 are in one mL? How many mL in one liter?

Solution:

(a) $1.42 \text{ yr}\left(\frac{365 \text{ day}}{1 \text{ yr}}\right)\left(\frac{24 \text{ h}}{1 \text{ day}}\right)\left(\frac{3600 \text{ s}}{1 \text{ h}}\right)\left(\frac{3.00 \times 10^8 \text{ m}}{1 \text{ s}}\right)\left(\frac{1 \text{ mi}}{1609 \text{ m}}\right) = 8.35 \times 10^{12} \text{ mi}$

(b) $32.4 \text{ yd}\left(\frac{36 \text{ in}}{1 \text{ yd}}\right)\left(\frac{2.54 \text{ cm}}{1 \text{ in}}\right) = 2.96 \times 10^3 \text{ cm}$

(c) $\left(\frac{3.0 \times 10^{10} \text{ cm}}{1 \text{ s}}\right)\left(\frac{1 \text{ in}}{2.54 \text{ cm}}\right)\left(\frac{1 \text{ ft}}{12 \text{ in}}\right) = 9.8 \times 10^8 \text{ ft/s}$

(d) $?°C = (47.4°F - 32.0°F)\left(\frac{5°C}{9°F}\right) = 8.6°C$

(e) $?°F = \left(\frac{9°F}{5°C}\right)(-273.15°C) + 32.00°F = -459.67°F$

(f) $71.2\ cm^3 \left(\frac{1\ m}{10^2\ cm}\right)^3 = 7.12 \times 10^{-5}\ m^3$

(g) $7.2\ m^3 \left(\frac{10\ dm}{1\ m}\right)^3 \left(\frac{1\ L}{1\ dm^3}\right) = 7.2 \times 10^3\ L$

1.51 In this problem you need to convert g to kg and cm^3 to m^3. The unit factors can be chain multiplied. Example 1.9 is similar.

Solution: By definition 1 kg = 1000 g and 1 m = 100 cm. The unit factors are

$$\left(\frac{1\ kg}{1000\ g}\right) \text{ and } \left(\frac{100\ cm}{1\ m}\right)$$

$$?\ kg/m^3 = 2.70 \left(\frac{g}{cm^3}\right)\left(\frac{1\ kg}{1000\ g}\right)\left(\frac{100\ cm}{1\ m}\right)^3 = 2.70 \times 10^3\ kg/m^3$$

1.52 Example 1.9 is a similar problem.

Solution: The problem requires you to convert from units of g/L to units of g/cm^3. All that is required to solve the problem is to use the unit factor converting L to cm^3.

$$\text{density} = \left(\frac{0.625\ g}{1\ L}\right)\left(\frac{1\ L}{1000\ cm^3}\right) = 6.25 \times 10^{-4}\ g/cm^3$$

1.54 Review Section 1.7 if you need help in distinguishing between weight and mass.

Solution: The mass of the astronaut remains unchanged, but his (or her) weight (gravitational pull on the body) is balanced by the outward pull due to the orbital motion.

1.55 See Problem 1.13 and the discussion of these terms in Section 1.6.

Solution: (a) Chemical property. Iron has changed its composition and identity by chemically combining with oxygen and water.

(b) Chemical property. The water reacts with chemicals in the air (such as sulfur dioxide) to produce acids, thus changing the composition and identity of the water.

(c) Physical property. The color of the hemoglobin can be observed and mearsured without changing its composition or identity.

(d) Physical property. The evaporation of water does not change its chemical properties. Evaporation is a change in matter from the liquid state to the gaseous state.

(e) Chemical property. The carbon dioxide is chemically converted into other molecules.

Are changes in state chemical or physical properties? Refer to Section 1.5 for review.

1.58 In solving unfamiliar problems it is often helpful to draw some sort of figure to clarify the true relationships between the involved quantities. Make a drawing of the two temperature scales side-by-side and make sure that the 0 and +100 points on the "S" scale are exactly opposite the -117.3 and +78.3 points on the Celsius scale.

From your sketch determine the number of Celsius degrees between the 0 and +100 points on the "S" scale. This is the basis for your equivalency relationship. Covert this equality into a unit factor.

To convert an "S" temperature to a Celsius temperature, you must multiply by the above unit factor and then correct for the fact that the zeros on the two scales are at different points. How is this correction term applied? (Think about converting 0°S to a Celsius temperature.)

Use these hints to try to work out the equation without looking at the solution below.

Solution: There are 78.3 + 117.3 = 195.6 Celsius degrees between 0°S and 100°S. The unit factor is

$$\left(\frac{195.6°C}{100°S}\right).$$

Set up the equation like a Celsius to Fahrenheit conversion

$$?°C = \left(\frac{195.6°C}{100°S}\right)(?°S) - 117.3°C$$

Solving for ?°S gives

$$?°S = (?°C + 117.3°C)\left(\frac{100°S}{195.6°C}\right)$$

For 25°C we have

$$?°S = (25°C + 117.3°C)\left(\frac{100°S}{195.6°C}\right) = 73°S$$

Can you find the temperature at which the numerical readings on the two scales are equal?

1.59 Review Sections 1.7 and 1.8 and Example 1.4 if you are not sure of how to attack this problem.

Solution: The volume of a rectangular solid is obtained by multiplying the height times the width times the breadth.

$$\text{density} = \frac{m}{V} = \frac{52.7064 \text{ g}}{(8.53 \text{ cm})(2.4 \text{ cm})(1.0 \text{ cm})} = 2.6 \text{ g/cm}^3$$

What would the value of density be in units of g/mL?

1.60 Part (a) of this Problem is similar to Problem 1.25; Part (b) is similar to Problem 1.59 above; Part (c) is worked much like Problem 1.26.

Solution: In each case m = dV

(a) $m = (19.3 \text{ g/cm}^3)[\frac{4}{3}\pi(10.0 \text{ cm})^3] = 8.08 \times 10^4 \text{ g}$

(b) $m = (21.4 \text{ g/cm}^3)\left(0.040 \text{ mm} \times \frac{1 \text{ cm}}{10 \text{ mm}}\right)^3 = 1.4 \times 10^{-6} \text{ g}$

(c) $m = (0.798 \text{ g/mL})(50.0 \text{ mL}) = 39.9 \text{ g}$

1.61 This problem is similar to Problem 1.28. The strategy used in solving the problem is to first convert the mass of the mercury to volume by using the unit factor of the density of mercury. Then, knowing the volume of mercury, you can solve for the radius (and then the diameter) by knowing that the volume of a cylinder is $V = \pi r^2 \times \text{height}$.

Solution: Volume of mercury filling the cylinder $= \frac{m}{d} = \frac{105.5 \text{ g}}{13.6 \text{ g/cm}^3} = 7.76 \text{ cm}^3$

Volume of cylinder $= \pi r^2 h$, where r = inner radius of cylinder and h = height

Volume $= \pi r^2 h = 7.76 \text{ cm}^3$

$$r = \sqrt{\frac{7.76 \text{ cm}^3}{\pi \times 12.7 \text{ cm}}} = 0.441 \text{ cm; Cylinder diameter} = 2r = 0.882 \text{ cm}$$

1.62 In this problem the relationship between the mass of water and its volume is given by the density, i.e., 0.9976 g is equivalent to 1.000 cm^3. This can be used as a unit factor. How do you find the mass of the water in the flask?

Solution: Mass of water = 87.39 g - 56.12 g = 31.27 g

$$\text{Volume} = \frac{\text{mass}}{\text{density}} = \left(\frac{31.27 \text{ g}}{0.9976 \text{ g/cm}^3}\right) = 31.35 \text{ cm}^3$$

Is this method an effective way to measure the volumes of irregularly shaped containers?

1.64 In order to work this problem, you need to understand the physical prinicples involved in the experiment in Problem 1.63. The volume of the water displaced must equal the volume of the piece of silver. If the silver did not sink, would you have been able to determine the volume of the piece of silver?

> ***Solution:*** The liquid must be less dense than the ice in order for the ice to sink. The temperature of the experiment must be maintained at 0°C to prevent the ice from melting.

1.66 Percent error is a measure of accuracy that is used in scientific measurements. Review the difference between accuracy and precision in Section 1.8.

> ***Solution:*** For the Fahrenheit thermometer, we must convert the possible error of 0.1°F to °C, using the unit factor of $\frac{5°C}{9°F}$.
>
> $$?°C = (0.1°F) \times \frac{5°C}{9°F} = 0.056\ °C$$
>
> The percent error is the amount of uncertainty in a measurement divided by the value of the measurement, converted to percent by multiplication by 100.
>
> $$\text{Percent error} = \frac{\text{known error in a measurement}}{\text{value of the measurement}} \times 100\%$$
>
> $$\text{For the Fahrenheit thermometer, percent error} = \frac{0.056\ °C}{38.9\ °C} \times 100\% = 0.14\%$$
>
> $$\text{For the Celcius thermometer, percent error} = \frac{0.1\ °C}{38.9\ °C} \times 100\% = 0.26\%$$
>
> Which thermometer is more accurate?

1.67 To work this problem, we need to convert from cubic feet to L. Some tables will have this conversion factor of 28.3 L = 1 ft^3, but we can also calculate it using factor-label method described in Section 1.9.

> ***Solution:*** First, converting from cubic feet to liters
>
> $$(5.0 \times 10^7\ ft^3)\left(\frac{12\ in}{1\ ft}\right)^3\left(\frac{2.54\ cm}{1\ in}\right)^3\left(\frac{1\ mL}{1\ cm^3}\right)\left(\frac{1\ L}{1000\ mL}\right) = 1.4 \times 10^9\ L$$
>
> The g of vanillin is: $\left(\frac{2.0 \times 10^{-11}\text{ g vanillin}}{L}\right)(1.4 \times 10^9\ L) = 2.8 \times 10^{-2}$ g vanillin
>
> The cost is: $(2.8 \times 10^{-2}\text{ g vanillin})\backslash B(\backslash F(\$112\text{ g vanillin})) = \$0.063 = 6.3¢$

1.68 Make a guess at a reasonable answer to a problem before actually solving it. What is a good guess at the volume of air per breath?

> ***Solution:*** All of the oxygen needed must come from the 4% (that is, 20% - 16%) difference between inhaled and exhaled air. That is, the 4% of air that is oxygen must equal the 240 mL of pure oxygen that is required per minute.
>
> $$\text{Volume of air/min} = \frac{240\text{ mL of air/min}}{0.04} = 6000\text{ mL of air/min}$$

Since there are 12 breaths per min, $\left(\frac{6000 \text{ mL of air}}{\text{min}}\right)\left(\frac{\text{min}}{12 \text{ breaths}}\right) = 500 \text{ mL per breath}$

1.70 The italicized definitions of these terms in Section 1.4 give two differences. Can you think of any others? Are all metals solids? Are all nonmetals solids? Are any nonmetals liquids or gases?

Solution: Metals are good conductors of heat and electricity, whereas nonmetals are poor conductors.

1.75 Refer to Appendix 1.

Solution: (a) Berkelium (Berkeley, CA); Europium (Europe); Francium (France); Scandium (Scandinavia); Ytterbium (Ytterby, Sweden); Yttrium (Ytterby, Sweden).

(b) Einsteinium (Albert Einstein); Fermium (Enrico Fermi); Curium (Pierre and Marie Curie); Mendelevium (Dmitri Mendeleev); Lawrencium (Ernest Lawrence).

(c) Arsenic, Cesium, Chlorine, Chromium, Iodine.

1.76 If you don't remember which elements are nonmetals, refer to Figure 1.6

Solution: Helium and Selenium. (Tellurium is a metalloid whose name ends in *ium*.)

1.77 This problem requires you to use a supplemental source of information to Chapter 1. Many periodic tables list melting points and boiling points.

One element in the periodic table that wasn't discovered until 1875 has a melting point slightly above room temperature. What is that element? Refer to Table 8.2.

Solution: The two elements that are liquid at 25°C are mercury and bromine.

1.78 This problem, like problem 1.77, also requires a supplemental source of information. Refer to Table 5.1.

Solution: Hydrogen, nitrogen, oxygen, fluorine, chlorine, and the noble gases.

1.79 Study Section 1.4. How many elements are classified as nonmetals? As metalloids?

Solution: (a) Metallic character increases as you progress down a group in the periodic table. For example, moving down Group 4A, we move from the nonmetal carbon at the top to the metal lead at the bottom of the table.

(b) Metallic character decreases from the left side of the table (where the metals are situated) to the right side of the table (which is the location of the nonmetals).

1.81 Why are gold and platinum called "noble" metals?

Solution: The elements in Group 8A are called noble gases because for many years no compounds containing these elements had been made.

1.82 Study Section 1.4 and Figure 1.6.

Solution: F and Cl are Group 7A elements and should behave similarly. Na and K are both in Group 1A and should have similar chemical and physical properties. P and N are both Group 5A elements and should behave similarly.

Note: If you have not already done so, please read the "Introduction for the Student" at the beginning of this Student Solutions Manual.

Chapter 2

ATOMS, MOLECULES, AND IONS

2.7 This problem deals with the subsection on Radioactivity in Section 2.2. Read this material before trying to answer the problem; the loss of mass is not the result of any ordinary process like evaporation. Which type(s) of radioactivity would result in loss of mass?

> ***Solution:*** The radioisotope is probably emitting alpha particles (helium nuclei); every time this happens, the mass of the sample decreases by that of one helium nucleus.
>
> Would beta particle emission cause measurable mass loss? Would gamma ray emission? Is the law of conservation of mass (Section 2.1) violated by radioactive substances? Would John Dalton have had to modify his atomic theory if he had known about radioactivity?

2.9 This problem is really an exercise in unit conversions (Section 1.7) and the factor-label method (Section 1.9). You need the definitions of the centi- and pico- SI prefixes (Table 1.3). The essential problem is to find how many 100 pm diameter objects can be lined up in a 1 cm space. How can you find this?

> ***Solution:*** You must divide 1 cm by 100 pm. To convert centimeters to picometers we write
>
> $$1\text{ cm} = \left(\frac{1\text{ m}}{10^2\text{ cm}}\right)\left(\frac{10^{12}\text{ pm}}{1\text{ m}}\right) = 1\times 10^{10}\text{ pm}$$
>
> $$?\text{ atoms} = 1\times 10^{10}\text{ pm}\left(\frac{1\text{ atom}}{1\times 10^{2}\text{ pm}}\right) = 1\times 10^{8}\text{ atoms}$$

2.10 The calculation in this problem to Problem 2.9 above.

> ***Solution:*** Note that you are given the information to set up the unit factor relating meters and miles.
>
> $$r_{\text{atom}} = 10^4\, r_{\text{nucleus}} = 10^4\times 10\text{ cm}\left(\frac{1\text{ m}}{100\text{ cm}}\right)\left(\frac{1\text{ mi}}{1609\text{ m}}\right) = 0.62\text{ mi}$$

2.11 (a) What are the relative masses of the proton, neutron, and electron (see Table 2.1)? The picture here is of an "inside-out" atom with protons circling a negatively charged nucleus of roughly half the original mass (assume the neutrons are still in the nucleus). Bear in mind that scattering is caused by both the charges and the relative masses of the colliding objects. How will this compare with the results of the Rutherford experiment?

(b) The Thomson model pictured the atom as a fairly uniform, featureless mass of clustered positive and negative particles. The English physicists referred to it as the "plum pudding" model. What sort of scattering might be expected from this electrical "plum pudding"?

Solution: (a) The alpha particles would probably suffer much less scattering from collisions with individual orbiting protons; the charges and masses of the colliding particles are less than in the Rutherford experiment. There would still be some scattering resulting from collisions between the negative nucleus and the alpha particles, but this should also be less pronounced. The alpha particles (positive) would not be repelled by the nucleus (negative).

(b) The physicists believing in the Thomson model expected very little, if any, alpha particle scattering.

2.16 The information needed to work this problem is in the first part of Section 2.3 in the textbook. Read that section and work Example 2.1 before trying this problem.

Solution:

Isotope	$^{3}_{2}He$	$^{4}_{2}He$	$^{24}_{12}Mg$	$^{25}_{12}Mg$	$^{48}_{22}Ti$	$^{79}_{35}Br$	$^{195}_{78}Pt$
No. Protons	2	2	12	12	22	35	78
No. Neutrons	1	2	12	13	26	44	117

2.17 The information needed to work this problem is in the first part of Section 2.3 in the textbook. Read that section and work Example 2.1 before trying this problem. Problem 2.16 is similar.

Solution:

Isotope	$^{15}_{7}N$	$^{33}_{16}S$	$^{63}_{29}Cu$	$^{84}_{38}Sr$	$^{130}_{56}Ba$	$^{186}_{74}W$	$^{202}_{80}Hg$
No. Protons	7	16	29	38	56	74	80
No. Neutrons	8	17	34	46	74	112	122
No. Electrons	7	16	29	38	56	74	80

2.18 The problem is similar to 2.17 above. In part (b) of this problem, the ratio of neutrons to protons is simply the number obtained when the number of neutrons is divided by the number of protons. How does ths ratio seem to vary as the atomic number increases?

Solution:

	Isotope	$^{4}_{2}He$	$^{20}_{10}Ne$	$^{40}_{18}Ar$	$^{84}_{36}Kr$	$^{132}_{54}Xe$
(a)	No. Protons	2	10	18	36	54
	No. Neutrons	2	10	22	48	78
(b)	neutron/proton ratio	1.00	1.00	1.22	1.33	1.44

The neutron/proton ratio increases with increasing atomic number.

2.19 Read the first part of Section 2.3 to understand the meanings of the atomic number (A) and mass number (Z). Where are atomic numbers found on a periodic table of the elements? Are mass numbers shown on the periodic table? Why not? What is the relationship between the chemical identity of an atom (that is, which element) and the atomic number (A)?

Solution: (a) $^{23}_{11}Na$ (b) $^{64}_{28}Ni$ (c) $^{186}_{74}W$ (d) $^{201}_{80}Hg$

2.20 Go back to the first part of Section 2.3 and study the meanings of the sub- and superscripts in isotope symbols. What does each tell you? Does the atomic number identify the chemical element? Does it tell anything more? Does adding the atomic number to the element symbol tell anything more?

> ***Solution:*** The symbol ^{23}Na provides more information than $_{11}Na$. The mass number plus the chemical symbol identifies a specific isotope of Na (sodium) while combining the atomic number with the chemical symbol tells you nothing new. Can other isotopes of sodium have different atomic numbers?

2.28 Find and examine Table 2.1. What information does it supply about the electron? What are the units? How many electrons are in one mole of electrons?

> ***Solution:*** From Table 2.1 we find:
>
> Charge of one electron = 1.6022×10^{-19} Coulomb
> Mass of one electron = 9.1095×10^{-28} g
>
> Charge of one mole of electrons
>
> $$= \left(\frac{1.6022 \times 10^{-19} \text{ Coulomb}}{1 \text{ electron}}\right)\left(\frac{6.022 \times 10^{23} \text{ electrons}}{1 \text{ mol}}\right) = 9.648 \times 10^{4} \text{ Coulomb}$$
>
> $$\text{Mass of one mole of electrons} = (9.1095 \times 10^{-28} \text{ g})\left(\frac{6.022 \times 10^{23}}{1 \text{ mol}}\right) = 5.486 \times 10^{-4} \text{ g}$$

2.29 Does this statement refer to one mole of gold or one atom of gold? Does it refer to one specific isotope of gold (there are several), and, if so, which one? Why isn't the specific isotope symbol shown? Study the part of Section 2.3 on Atomic Masses to answer these questions.

> ***Solution:*** The phrase means that the average mass of one atom of the element gold, as found normally in Earth's crust is 197.0 amu. The amu is a unit of mass defined as exactly one twelfth the mass of one atom of carbon-12. Another way of looking at the problem is to say that the average gold atom has a mass 16.42 times as great as an atom of carbon-12 ($197.0 \div 12.00 = 16.42$).

2.30 You should review the part of Section 2.3 on Average Atomic Masses and Example 2.2 before working this problem. Why are the atomic masses not whole integers like the atomic numbers?

> ***Solution:*** $$34.968 \text{ amu} \left(\frac{75.53\%}{100\%}\right) + 36.956 \text{ amu} \left(\frac{24.47\%}{100\%}\right) = 35.45 \text{ amu}$$

2.31 This problem involves setting up a chain multiplication involving a lot of unit factors. The strategy is to first calculate how many particles 5.0 billion people could count in the space of one year at the rate of two per second per person. One then divides this number (what are the units?) into Avogadro's number. How many significant figures in the answer?

> ***Solution:*** In one year:

$$(5.0 \times 10^9 \text{ people})\left(\frac{365 \text{ days}}{1 \text{ year}}\right)\left(\frac{24 \text{ hours}}{1 \text{ day}}\right)\left(\frac{3600 \text{ s}}{1 \text{ hour}}\right)\left(\frac{2 \text{ particles}}{1 \text{ person}}\right)$$

$$= 3.2 \times 10^{17} \text{ particles/year}$$

$$\text{Total time} = \frac{6.022 \times 10^{23} \text{ particles}}{3.2 \times 10^{17} \text{ particles/year}} = 1.9 \times 10^6 \text{ years}$$

2.32 An alternate phrasing for this problem is as follows: how do you calculate the mass in grams of one fluorine atom (19.00 amu)? Study the last subsection of Section 2.3 on Molar Mass of an Element and Avogadro's Number to see how to attack this question.

Solution: From Section 2.3, 1 g = 6.022×10^{23} amu. To convert 19.00 amu to grams we have:

$$19.00 \text{ amu}\left(\frac{1 \text{ g}}{6.022 \times 10^{23} \text{ amu}}\right) = 3.155 \times 10^{-23} \text{ g}$$

The relationship is 1g/1 amu = 6.022×10^{23}. If you memorize Avogadro's number, do you also need to memorize the mass of one amu in grams? Alternatively, one can argue that one mole of fluorine atoms (19.00 g) contains 6.022×10^{23} F atoms, so

$$\left(\frac{19.00 \text{ amu}}{1 \text{ F atom}}\right) 6.022 \times 10^{23} \text{ F atoms} = 19.00 \text{ g}$$

$$\text{or } \left(\frac{1 \text{ g}}{1 \text{ amu}}\right) = 6.022 \times 10^{23}$$

2.34 Review Section 2.3 to find the relationship between grams and amu's. Remember that even though the mass units of amu's is unfamiliar, it is still simply a mass unit (albeit a very small unit) as is the pound. If necessary, study Problem 2.32.

Solution: The unit factor required is $\left(\frac{6.022 \text{ x } 10^{23} \text{ amu}}{1 \text{ g}}\right)$

$$? \text{ amu} = (8.4 \text{ g}) \text{ x } \left(\frac{6.022 \text{ x } 10^{23} \text{ amu}}{1 \text{ g}}\right) = 5.1 \text{ x } 10^{24} \text{ amu}$$

2.36 This problem is a variation of Example 2.6 in the textbook. How many cobalt atoms are in one mole of cobalt atoms? Study the last subsection of Section 2.3 if you can't answer that question. What is the unit factor connecting moles and number of particles?

Solution:

$$(6.00 \times 10^9 \text{ Co atoms})\left(\frac{1 \text{ mol}}{6.022 \times 10^{23} \text{ Co atoms}}\right) = 9.96 \times 10^{-15} \text{ mol}$$

2.37 This problem is similar to Example 2.3 in the textbook. How many moles of calcium atoms are in 40.08 g of calcium?

Solution: First we find the molar mass of Ca is 40.08 g: this can be expressed by the equality 1 mole of Ca = 40.08 g of Ca. To convert the amount of Ca in grams to Ca in moles, we write

$$(77.4 \text{ g of Ca})\left(\frac{1 \text{ mol of Ca}}{40.08 \text{ g of Ca}}\right) = 1.93 \text{ mol Ca}$$

2.38 This problem is a similar to Example 2.4 in the textbook. How many g of Au are in one mol of Au?

Solution: Since the molar mass of Au is 197.0 g, the mass of Au in grams is given by

$$(15.3 \text{ mol})\left(\frac{197.0 \text{ g of Au}}{1 \text{ mol of Au}}\right) = 3.01 \times 10^3 \text{ g Au}$$

2.39 This problem is a similar to Example 2.5 in the textbook. Should the answer that you get for the mass in grams of a single atom of any element be a very large or a very small number? Stated in another way, should the exponent in your answer be positive or negative?

Solution: Since the molar mass (in g) of each element is the mass of 6.022×10^{23} atoms of that element, then we know the mass of one atom of each element is simply the molar mass divided by 6.022×10^{23}.

(a) $$\frac{200.6 \text{ g Hg}}{6.022 \times 10^{23} \text{ Hg atoms}} = 3.331 \times 10^{-22} \text{ g/atom}$$

(b) $$\frac{20.18 \text{ g Ne}}{6.022 \times 10^{23} \text{ Ne atoms}} = 3.351 \times 10^{-23} \text{ g/atom}$$

(c) $$\frac{74.92 \text{ g As}}{6.022 \times 10^{23} \text{ As atoms}} = 1.244 \times 10^{-22} \text{ g/atom}$$

(d) $$\frac{207.2 \text{ g Pb}}{6.022 \times 10^{23} \text{ Pb atoms}} = 3.441 \times 10^{-22} \text{ g/atom}$$

2.40 This problem is very much like Example 2.6, except that the convesion is from grams to atoms rather than from atoms to grams.

Solution: In Problem 2.39(d), you found the mass of a single Pb atom. If you know the mass of a single Pb atom, it is simple to find the mass of 1.00×10^{12} lead atoms.

$$? \text{ g} = (1.00 \times 10^{12} \text{ Pb atoms})\left(\frac{207.2 \text{ g Pb}}{6.022 \times 10^{23} \text{ Pb atoms}}\right) = 3.44 \times 10^{-10} \text{ g}$$

2.41 This problem is a similar to Example 2.6 in the textbook. Also look at problem 2.36 above. How many atoms are present in 63.55 g of Cu?

Solution: The molar mass of Cu is 63.55 g, which is the mass of one mole of Cu atoms, which is 6.022×10^{23} atoms of Cu.

$$(3.14 \text{ g Cu})\left(\frac{1 \text{ mol Cu}}{63.55 \text{ g Cu}}\right)\left(\frac{6.022 \times 10^{23} \text{ Cu atoms}}{1 \text{ mol Cu}}\right) = 2.98 \times 10^{22} \text{ Cu atoms}$$

2.42 Conceptually this problem is identical to Problem 2.41 and Example 2.6. Try to work this problem without reference to them.

Solution: We find the number of atoms in each sample.

For 1.10 g of hydrogen:

$$1.10 \text{ g H}\left(\frac{1 \text{ mol}}{1.008 \text{ g H}}\right)\left(\frac{6.022 \times 10^{23} \text{ H atoms}}{1 \text{ mol}}\right) = 6.57 \times 10^{23} \text{ H atoms}$$

For 14.7 g of chromium:

$$14.7 \text{ g Cr}\left(\frac{1 \text{ mol}}{52.00 \text{ g Cr}}\right)\left(\frac{6.022 \times 10^{23} \text{ Cr atoms}}{1 \text{ mol}}\right) = 1.70 \times 10^{23} \text{ Cr atoms}$$

There are more hydrogen atoms than chromium atoms.

2.43 Review the subsection on Atomic Masses in Section 2.3. What is the reasoning that leads to the conclusion that the atomic mass of hydrogen is 1.008 amu? Try to apply the same method in this problem. What does the formula of carbon monoxide tell you about the relative numbers of carbon and oxygen atoms?

Solution: There is one oxygen atom for each carbon atom. For the mass of one oxygen atom we write

$$\left(\frac{12.01 \text{ amu}}{1 \text{ atom C}}\right)\left(\frac{1 \text{ atom C}}{1 \text{ atom O}}\right)\left(\frac{3.257 \text{ g}}{2.445 \text{ g}}\right) = 16.00 \text{ amu/O atom}$$

2.44 Review Section 2.4 and the law of definite proportions before you try to work this problem.

Solution: Since the formula for the compound is MO, then one molar mass of M will combine with one molar mass of O, (that is, 16.00 g). If we know how many grams of M will react with a certain number of grams of O, then we can use a ratio to determine the number of grams of M that will react with 16.00 g of O. That number is also the molar mass of M, so we can determine its identity by checking a table of atomic or molar masses.

The mass of oxygen in MO is 39.46 g – 31.70 g = 7.76 g. Therefore, for every 31.70 g of M, there is 7.76 g of O in the compound MO. A ratio can be used to determine how many grams of M will react with 16.00 g of O.

$$\frac{31.70 \text{ g M}}{7.76 \text{ g O}} = \frac{x \text{ g M}}{16.00 \text{ g O}} \qquad x = 65.4 \text{ g, which is the molar mass of M}$$

Therefore the metal is Zn.

2.50 In chemical formulas it matters where you put the numbers! If you aren't sure of the right answer here, go back and carefully study the first few paragraphs in Section 2.4.

Solution: The expression P_4 signifies one molecule containing four phosphorus atoms. The expression 4P signifies four unconnected, individual phosphorus atoms.

2.53 How does an empirical formula differ from a molecular formula? Which gives you more information? What information does an empirical formula convey? If you can't answer any of these questions, study the subsection on Empirical Formulas in Section 2.4 and work Example 2.7.

Solution: (a) CN, (b) CH, (c) C_9H_{20}, (d) P_2O_5, (e) BH_3, (f) $AlBr_3$, (g) $NaSO_2$, (h) N_2O_5, (i) $K_2Cr_2O_7$.

An empirical formula only tells the relative numbers of atoms of each type in a compound; the numbers are always the smallest whole numbers possible. If all the subscripts in a molecular formula are different prime numbers, is this also the empirical formula?

2.56 This problem is similar to Example 2.8 in the textbook.

Solution: Using the appropriate atomic masses,

(a) CH_4	12.01 + 4(1.008) = 16.04 amu
(b) H_2O	2(1.008) + 16.00 = 18.02 amu
(c) H_2O_2	2(1.008) + 2(16.00) = 34.02 amu
(d) C_6H_6	6(12.01) + 6(1.008) = 78.11 amu
(e) PCl_5	30.97 + 5(35.45) = 208.22 amu

2.58 Refer to Example 2.8 and Problem 2.56.

Solution: Using the appropriate atomic masses,

$C_{55}H_{72}MgN_4O_5$ 55(12.01) + 72(1.008) + 24.31 + 4(14.01) + 5(16.00) = 893.5 g

2.59 This is a problem of proportion. If 0.372 mole has a mass of 152 g, what is the mass of 1.0 mole?

Solution: We can write
$$\frac{152\text{ g}}{0.372\text{ mol}} = \frac{?\text{ g}}{1.0\text{ mol}}$$

$$\text{molar mass} = \frac{152\text{ g}}{0.372\text{ mol}} = 409\text{ g/mol}$$

2.60 What is the difference between the empirical formula and the molecular formula? Refer to Section 2.4 for help.

Solution: For B_2H_6 the molecular formula is double the empirical formula, BH_3. The molar mass of B_2H_6 is twice the molar mass of the empirical formula.

2.61 This problem involves the relationship between moles of a species and mass, and is similar to Example 2.9 in the textbook.

Solution: First we find the molar mass of C_2H_6 is 30.07 g: this can be expressed by the equality 1 mole of C_2H_6 = 30.07 g of C_2H_6. To convert the amount of C_2H_6 in grams to C_2H_6 in moles, we write

$$(0.334 \text{ g of } C_2H_6)\left(\frac{1 \text{ mol of } C_2H_6}{30.07 \text{ g of } C_2H_6}\right) = 0.0111 \text{ mol } C_2H_6 = 1.11 \times 10^{-2} \text{ mol } C_2H_6$$

2.62 This problem involves the use of density and the conversion of a mass to a number of molecules. If you don't remember how to convert volume to mass using density, study Example 1.2 in Section 1.7.

Solution:

$$\text{Mass of water} = (2.56 \text{ mL})\left(\frac{1.00 \text{ g}}{1.00 \text{ mL}}\right) = 2.56 \text{ g}$$

$$\text{Molar mass of } H_2O = (16.00) + 2(1.008) = 18.02 \text{ g/mol}$$

$$\text{Number of molecules} = 2.56 \text{ g}\left(\frac{1 \text{ mole}}{18.02 \text{ g}}\right)\left(\frac{6.022 \times 10^{23} \text{ molecules}}{1 \text{ mol}}\right) = 8.56 \times 10^{22} \text{ molecules}$$

Do you need to do a calculation to convert volume to mass when the density is 1.00 g/mL? How many hydrogen atoms are present in 2.56 g of water? See Example 2.10.

2.63 What is the ratio of the number of moles of chlorine to the number of moles of oxygen in the formula given?

Solution: The molecular formula for Cl_2O_7 means that there are 2 Cl atoms for every 7 O atoms (or 2 moles of Cl atoms for every 7 moles of O atoms). Therefore

$$\text{mole ratio} = 2 \text{ mol Cl per 7 mol O} = 1 \text{ mol } Cl_2 \text{ per } 3.5 \text{ mol } O_2 = \frac{1 \text{ mol } Cl_2}{3.5 \text{ mol } O_2}$$

2.64 A very similar problem is Example 2.10 in the text.

Solution: The molar mass of glucose $C_6H_{12}O_6$ is 180.2 g. There are six carbon atoms in every glucose molecule. Therefore, the total number of C atoms in 1.50 grams of glucose is

$$(1.50 \text{ g glucose})\left(\frac{6.022 \times 10^{23} \text{ molecules glucose}}{180.2 \text{ g glucose}}\right)\left(\frac{6 \text{ C atoms}}{1 \text{ molecule glucose}}\right) = 3.01 \times 10^{22} \text{ C atoms}$$

There is the same number of O atoms in glucose as C atoms, so the number of O atoms = 3.01×10^{22} O atoms.

There are twice as many H atoms in glucose as C atoms, so the number of H atoms = 6.02×10^{22} H atoms.

2.65 This is similar to Example 2.10. Go back and try to work this problem with the answer covered. If you get stuck, carefully study the solution and pay special attention to the use of the unit factors and the application of the shortcut in the explanation. The molar mass of urea is 60.06 g/mol.

Solution: Use Example 2.10 as a model for the calculations. We write

Carbon atoms

$$= (1.68 \times 10^4 \text{ g urea})\left(\frac{1 \text{ mol urea}}{60.06 \text{ g urea}}\right)\left(\frac{6.022 \times 10^{23} \text{ molecules urea}}{1 \text{ mol urea}}\right)\left(\frac{1 \text{ C atom}}{1 \text{ urea molecule}}\right)$$

$$= 1.68 \times 10^{26} \text{ C atoms}$$

For the other elements, we can use the same shortcut as Example 2.12. The hydrogen atom to carbon atom ratio in a urea molecule is 4/1, so for hydrogen, nitrogen, and oxygen we have:

$$\text{Hydrogen atoms} = (1.68 \times 10^{26} \text{ C atoms})\left(\frac{4 \text{ H atoms}}{1 \text{ C atom}}\right) = 6.72 \times 10^{26} \text{ H atoms}$$

$$\text{Nitrogen atoms} = (1.68 \times 10^{26} \text{ C atoms})\left(\frac{2 \text{ N atoms}}{1 \text{ C atom}}\right) = 3.36 \times 10^{26} \text{ N atoms}$$

$$\text{Oxygen atoms} = (1.68 \times 10^{26} \text{ C atoms})\left(\frac{1 \text{ O atom}}{1 \text{ C atom}}\right) = 1.68 \times 10^{26} \text{ O atoms}$$

2.66 This problem is similar to Problem 2.62. The problem is a straightforward one involving conversion of mass to numbers of molecules.

Solution: The molar mass of $C_{19}H_{38}O$ is 282.5 g.

$$1.0 \times 10^{-12} \text{ g}\left(\frac{1 \text{ mol}}{282.5 \text{ g}}\right)\left(\frac{6.022 \times 10^{23} \text{ molecules}}{1 \text{ mol}}\right) = 2.1 \times 10^9 \text{ molecules}$$

Notice that even though 1.0×10^{-12} g is an extremely small mass, it still is comprised of over a million billion pheromone molecules!

2.70 This is similar to Problem 2.17 with the exception that the atoms bear electrical charges. Does adding two electrons to an electrically neutral atom make its charge +2 or -2? Is it a cation or an anion? Study the first part of Section 2.5 if you can't answer these questions.

Solution:

Ion	Na^+	Ca^{2+}	Al^{3+}	Fe^{2+}	I^-	F^-	S^{2-}	O^{2-}	N^{3-}
No. protons	11	20	13	26	53	9	16	8	7
No. electrons	10	18	10	24	54	10	18	10	10

2.71 This problem is similar to Example 2.12 in the textbook. See if your can work that problem before trying this one.

Solution: The molar mass of KCN is 65.12 g. The number of KCN formula units in 1.00×10^{-3} g of KCN is

$$1.00 \times 10^{-3} \text{ g KCN}\left(\frac{1 \text{ mol KCN}}{65.12 \text{ g KCN}}\right)\left(\frac{6.022 \times 10^{23} \text{ formula units KCN}}{1 \text{ mol KCN}}\right) = 9.24 \times 10^{18} \text{ KCN units}$$

2.72 How can you tell whether a given formula represents an ionic or a molecular compound? Study Section 2.5 to find out!

Solution: Compounds of metals with nonmetals are usually ionic. Nonmetal-nonmetal compounds are usually molecular.

(a) $SiCl_4$ - molecular (b) LiF - ionic (c) $BaCl_2$ - ionic (d) B_2H_6 - molecular (e) KF - ionic (f) C_2H_4 - molecular

2.77 This is similar to Example 2.13. Study the first part of Section 2.7 and try to work this example with the answer covered. If you can't, uncover the solution and study it carefully. Use the example as a model for solving this problem.

Solution: Molar mass of SnO_2 = (118.7) + 2(16.00) = 150.7 g/mol.

$$\%Sn = \left(\frac{118.7 \text{ g/mol}}{150.7 \text{ g/mol}}\right) 100\% = 78.77\%$$

$$\%O = \left(\frac{2 \text{ x } 16.00 \text{ g/mol}}{150.7 \text{ g/mol}}\right) 100\% = 21.23\%$$

Why do we multiply the atomic mass of oxygen (16.00 amu) times 2?

2.79 This is similar to Example 2.13 in the textbook and Problem 2.77 above. Study the first part of Section 2.7 and try to work this example with the answer covered. If you can't, uncover the solution and study it carefully. Use the example as a model for solving this problem.

Solution: Molar mass of $CHCl_3$ = (12.01) + (1.008) + 3(35.45) = 119.4 g/mol.

$$\%C = \left(\frac{12.01 \text{ g/mol}}{119.4 \text{ g/mol}}\right) 100\% = 10.06\%$$

$$\%H = \left(\frac{1.008 \text{ g/mol}}{119.4 \text{ g/mol}}\right) 100\% = 0.8442\%$$

$$\%Cl = \left(\frac{3 \text{ x } 35.45 \text{ g/mol}}{119.4 \text{ g/mol}}\right) 100\% = 89.07\%$$

Why do we multiply the atomic mass of chlorine (35.45 amu) times 3?

2.81 The first part of this problem is a percent composition by mass problem like Example 2.13 and Problems 2.77 and 2.79. The second part of the problem is similar to 2.66.

Solution: The molar mass of cinnamic alcohol is 134.17 g.

(a)

$$\%C = \left(\frac{9 \times 12.01 \text{ g/mol}}{134.17 \text{ g/mol}}\right) 100\% = 80.56\%$$

$$\%H = \left(\frac{10 \times 1.008 \text{ g/mol}}{134.17 \text{ g/mol}}\right) 100\% = 7.51\%$$

$$\%O = \left(\frac{16.00 \text{ g/mol}}{134.17 \text{ g/mol}}\right) 100\% = 11.93\%$$

(b) $$0.469\text{ g}\left(\frac{1\text{ mol}}{134.17\text{ g}}\right)\left(\frac{6.022\times 10^{23}\text{ molecules}}{1\text{ mol}}\right) = 2.11\times 10^{21}\text{ molecules}$$

2.83 This problem can be solved by using the approach in Example 2.16. If you know the percent composition by mass for a compound, can you calculate the amounts of the component elements in a given mass of the compound? Cover the answer to Example 2.16 and try to solve it. If you can't, study the answer carefully and rework the problem. Apply the same technique to this problem.

Solution: (a) The mass of chlorine is 5.0 g.

(b) Molar mass of $NaClO_3$ = (22.99) + (35.45) + 3(16.00) = 106.44 g/mol.

$$\%\text{Cl} = \left(\frac{35.45\text{ g}}{106.44\text{ g}}\right)100\% = 33.31\%$$

$$\text{Mass of chlorine} = \left(\frac{33.31\%}{100.00\%}\right)60.0\text{ g} = 20.0\text{ g Cl}$$

(c) 0.10 mol KCl contains 0.10 mol chlorine.

$$\text{Mass of chlorine} = (0.10\text{ mol Cl})\left(\frac{35.45\text{ g}}{1\text{ mol}}\right) = 3.5\text{ g Cl}$$

(d) Molar mass of $MgCl_2$ = (24.31) + 2(35.45) = 95.21 g/mol

$$\%\text{Cl} = \left(\frac{70.90\text{ g}}{95.21\text{ g}}\right)100\% = 74.47\%$$

$$\text{Mass of chlorine} = \left(\frac{74.47\%}{100.00\%}\right)30.0\text{ g} = 22.3\text{ g Cl}$$

(e) Molar mass of Cl_2 = 2(35.45) = 70.90 g/mol

$$\text{Mass of chlorine} = (0.50\text{ mol})(70.90\text{ g/mol}) = 35.45\text{ g Cl}$$

The 0.50 mol Cl_2 has the greatest mass of chlorine.

2.84 This is similar to Example 2.16. Rather than asking how many moles of Fe_2O_3 are in 24.6 g of Fe_2O_3, (analogous to Problem 2.61 above), the problem asks how many moles of Fe are in 24.6 g of Fe_2O_3. How many moles of Fe are in one mole of Fe_2O_3?

Solution: The molar mass of Fe_2O_3 = 2(55.85) + 3(16.00) = 159.7 g/mol. Using the unit conversion factors

$$\frac{1\text{ mol Fe}_2\text{O}_3}{159.7\text{ g/mol}} \text{ and } \frac{2\text{ mol Fe}}{1\text{ mol Fe}_2\text{O}_3}$$

$$24.6\text{ g}\left(\frac{1\text{ mol Fe}_2\text{O}_3}{159.7\text{ g/mol}}\right)\left(\frac{2\text{ mol Fe}}{1\text{ mol Fe}_2\text{O}_3}\right) = 0.308\text{ mol Fe}$$

2.85 The key to working this stoichiometry problem is to realize that the formula HgS means that one mole of Hg is required for one mole of S. Moles of Hg and moles of S can be converted to mass using unit factors. This problem is similar to Example 2.15.

Solution: Using unit factors we convert: g of Hg → mol Hg → mol S → g S

$$246 \text{ g Hg} \left(\frac{1 \text{ mol Hg}}{200.6 \text{ g Hg}}\right)\left(\frac{1 \text{ mol S}}{1 \text{ mol Hg}}\right)\left(\frac{32.07 \text{ g S}}{1 \text{ mol S}}\right) = 39.3 \text{ g}$$

2.86 This problem is a somewhat more complicated extension of the type illustrated in Problem 2.85 above and in Example 2.16 in the textbook. The plan of attack is to convert the 20.4 g of Al to moles of Al, then the moles of Al to moles of I, then moles of I to moles of I_2, and then, finally, to convert moles of I_2 to g of I_2. How many moles of I are required per mole of Al? How many moles of I_2 are required per mole of Al?

Solution: The problem can be worked in step using unit conversions factors. Remember the sequence is

g of Al → mol of Al → mol of I → mol of I_2 → g of I_2

$$20.4 \text{ g Al} \left(\frac{1 \text{ mol Al}}{26.98 \text{ g Al}}\right)\left(\frac{3 \text{ mol I}}{1 \text{ mol Al}}\right)\left(\frac{1 \text{ mol } I_2}{2 \text{ mol I}}\right)\left(\frac{253.8 \text{ g } I_2}{1 \text{ mol } I_2}\right) = 288 \text{ g } I_2$$

Does the number of grams of iodine that are required for the reaction change because iodine is a dimer (occurs as I_2)?

2.87 Example 2.16 in the text is the same type of problem, as are Problems 2.83 and 2.84 above.

Solution: The molar mass of SnF_2 is 156.7 g.

$$24.6 \text{ g} \left(\frac{1 \text{ mol } SnF_2}{156.7 \text{ g}}\right)\left(\frac{2 \text{ mol F}}{1 \text{ mol } SnF_2}\right)\left(\frac{19.00 \text{ g F}}{1 \text{ mol F}}\right) = 5.97 \text{ g F}$$

2.88 Assume that the platinum makes up the balance of the percent composition in both compounds (i.e., the only elements present are Pt and Cl). This problem is similar to Example 2.14. Go to this example, cover the answer, and try to solve it. If you can't, study the solution thoroughly and rework the problem. Calculating an empirical formula is a common examination question.

Solution: The method demonstrated in Example 2.14 is general for this type of problem. It is always numerically convenient to assume you have exactly 100.0 g of each compound. The percents then translate directly to masses. For each compound find the number of moles of Pt and Cl.

Compound A:

$$\text{Moles Cl} = 26.7 \text{ g Cl} \left(\frac{1 \text{ mol Cl}}{35.45 \text{ g Cl}}\right) = 0.753 \text{ mol Cl}$$

$$\text{Moles Pt} = 73.3 \text{ g Pt} \left(\frac{1 \text{ mol Pt}}{195.1 \text{ g Pt}}\right) = 0.376 \text{ mol Pt}$$

The empirical formula is $Pt_{0.376}Cl_{0.753}$. Dividing the subscripts by 0.376 (Why?) gives $PtCl_2$.

Compound B:

$$\text{Moles Cl} = 42.1 \text{ g Cl} \left(\frac{1 \text{ mol Cl}}{35.45 \text{ g Cl}}\right) = 1.19 \text{ mol Cl}$$

$$\text{Moles Pt} = 57.9 \text{ g Pt} \left(\frac{1 \text{ mol Pt}}{195.1 \text{ g Pt}}\right) = 0.297 \text{ mol Pt}$$

The empirical formula is $Pt_{0.297}Cl_{1.19}$. Dividing by 0.297 gives $PtCl_4$.

What would happen if you divided by the larger number? Would you still arrive at the same formula? Try it and see. If more than two elements were present in these compounds, would it be possible to solve the problem with the data given?

2.89 This problem is a straightforward example of determining an empirical formula from percentage composition. The material can be reviewed in Section 2.7. Example 2.14 is very similar.

Solution: In each case assume 100 g of compound and convert element masses to numbers of moles:

(a) Hygrogen: $2.1 \text{ g} \left(\frac{1 \text{ mol}}{1.008 \text{ g}}\right) = 2.1 \text{ mol}$

Oxygen: $65.3 \text{ g} \left(\frac{1 \text{ mol}}{16.00 \text{ g}}\right) = 4.08 \text{ mol}$

Sulfur: $32.6 \text{ g} \left(\frac{1 \text{ mol}}{32.07 \text{ g}}\right) = 1.02 \text{ mol}$

The empirical formula is $H_{2.1}S_{1.02}O_{4.08}$. Dividing by 1.02 gives H_2SO_4.

(b) Aluminum: $20.2 \text{ g} \left(\frac{1 \text{ mol}}{26.98 \text{ g}}\right) = 0.749 \text{ mol}$

Chlorine: $79.8 \text{ g} \left(\frac{1 \text{ mol}}{35.45 \text{ g}}\right) = 2.25 \text{ mol}$

The empirical formula is $Al_{0.749}Cl_{2.25}$. Dividing by 0.749 gives $AlCl_3$.

(c) Carbon: $40.1 \text{ g} \left(\frac{1 \text{ mol}}{12.01 \text{ g}}\right) = 3.34 \text{ mol}$

Hydrogen: $6.6 \text{ g} \left(\frac{1 \text{ mol}}{1.008 \text{ g}}\right) = 6.5 \text{ mol}$

Oxygen: $53.3 \text{ g} \left(\frac{1 \text{ mol}}{16.00 \text{ g}}\right) = 3.33 \text{ mol}$

The empirical formula is $C_{3.34}H_{6.5}O_{3.33}$. Dividing by 3.33 gives CH_2O.

(d) Carbon: $18.4 \text{ g} \left(\frac{1 \text{ mol}}{12.01 \text{ g}}\right) = 1.53 \text{ mol}$

Nitrogen: $21.5 \text{ g} \left(\frac{1 \text{ mol}}{14.01 \text{ g}}\right) = 1.53 \text{ mol}$

Potassium: $60.1 \text{ g} \left(\frac{1 \text{ mol}}{39.10 \text{ g}}\right) = 1.54 \text{ mol}$

The empirical formula is $K_{1.54}C_{1.53}N_{1.53}$. Dividing by 1.53 gives KCN.

2.90 This problem is similar to Example 2.14 in the textbook. The percentage by mass of oxygen is obtained by subtracting the other percentages from one hundred.

Solution: It is convenient to assume that you have 100 g of sample.

The percentage of oxygen is found by difference: 100 % - (19.8% + 2.50% + 11.6%) = 66.1%

$$\text{Moles C} = 19.8 \text{ g C}\left(\frac{1 \text{ mol C}}{12.01 \text{ g C}}\right) = 1.65 \text{ mol C}$$

$$\text{Moles H} = 2.50 \text{ g H}\left(\frac{1 \text{ mol H}}{1.008 \text{ g H}}\right) = 2.48 \text{ mol H}$$

$$\text{Moles N} = 11.6 \text{ g N}\left(\frac{1 \text{ mol N}}{14.01 \text{ g N}}\right) = 0.828 \text{ mol N}$$

$$\text{Moles O} = 66.1 \text{ g O}\left(\frac{1 \text{ mol O}}{16.00 \text{ g O}}\right) = 4.13 \text{ mol O}$$

The empirical formula is $C_{1.65}H_{2.48}N_{0.828}O_{4.13}$. Dividing the subscripts by 0.828 (Why?) gives $C_2H_3NO_5$. Would you get a different answer if you had assumed an amount of sample different that 100 g?

2.91 If the molar mass of caffeine is 194.19 g, what is the molecular mass of caffeine? If you don't know, study the subsection on Molecular Mass in Section 2.4. This problem is similar to Example 2.17 in the textbook.

Solution: Find the molar mass corresponding to each formula.

For $C_4H_5N_2O$: 4(12.01) + 5(1.008) + 2(14.01) + (16.00) = 97.10 g/mol

For $C_8H_{10}N_4O_2$: 8(12.01) +10(1.008) + 4(14.01) + 2(16.00) = 194.20 g/mol

The molecular formula is $C_8H_{10}N_4O_2$.

2.92 The empirical formula of MSG can be found by the method shown in Problems 2.88 and 2.90 above. How can you find a molecular formula from an empirical formula if you know the molar or molecular mass? If you don't know, study the subsection on the Determination of Molecular Formulas at the end of Section 2.7. This problem is similar to Example 2.18. Try working it with the solution covered before trying this one.

Solution: Assuming 100.0 g of MSG, the amounts of each element are

$$\text{Carbon:} \quad 35.51 \text{ g}\left(\frac{1 \text{ mol}}{12.01 \text{ g}}\right) = 2.957 \text{ mol C}$$

$$\text{Hydrogen:} \quad 4.77 \text{ g}\left(\frac{1 \text{ mol}}{1.008 \text{ g}}\right) = 4.73 \text{ mol H}$$

$$\text{Oxygen:} \quad 37.85 \text{ g}\left(\frac{1 \text{ mol}}{16.00 \text{ g}}\right) = 2.366 \text{ mol O}$$

$$\text{Nitrogen:} \quad 8.29 \text{ g}\left(\frac{1 \text{ mol}}{14.01 \text{ g}}\right) = 0.592 \text{ mol N}$$

$$\text{Sodium:} \quad 13.60 \text{ g}\left(\frac{1 \text{ mol}}{22.99 \text{ g}}\right) = 0.5916 \text{ mol Na}$$

Dividing each numerical subscript in $C_{2.957}H_{4.73}O_{2.366}N_{0.592}Na_{0.5916}$ by the smallest number (0.5916) gives the empirical formula $C_5H_8O_4NNa$. The molar mass corresponding to this formula is

5(12.01) + 8(1.008) + 4(16.00) + (14.01) + (22.99) = 169.1 g/mol, so $C_5H_8O_4NNa$ is the molecular formula.

2.94 What is the law of multiple proportions? If you can't state it in your own words and explain its meaning, go back and study Section 2.4. This problem is similar to Example 2.11 in the textbook.

Solution: Calculate the number of grams of oxygen that combines with 1.00 g of nitrogen in each of these compounds.

$$1.00\text{ g N}\left(\frac{0.778\text{ g O}}{0.681\text{ g N}}\right) = 1.14\text{ g O}$$

$$1.00\text{ g N}\left(\frac{1.28\text{ g O}}{0.560\text{ g N}}\right) = 2.29\text{ g O}$$

For these two compounds the oxygen/oxygen ratio is 2.29/1.14 = 2.01. This is consistent with the law of multiple proportions; namely it is a small whole number ratio (2/1).

2.95 This problem, as is Problem 2.94, is an example of the law of multiple proportions.

Solution: Calculate the number of grams of chlorine that combines with 1.000 g of iron in each of these compounds.

$$1.000\text{ g Fe}\left(\frac{55.94\text{ g Cl}}{44.06\text{ g Fe}}\right) = 1.270\text{ g Cl}$$

$$1.000\text{ g Fe}\left(\frac{65.57\text{ g Cl}}{34.43\text{ g Fe}}\right) = 1.904\text{ g Cl}$$

The chlorine/chlorine ratio is 1.904/1.270 = 1.499, or 3/2. Since this is a ratio of small whole numbers, it is consistent with the law of multiple proportions.

2.98 With the isotopes given, how many different kinds of CF_4 molecules can you make? (Don't worry about the + charge.)

Solution: Since there are only two kinds of carbon, there are only two possibilities for CF_4

$${}^{12}_{6}\text{C}\,{}^{19}_{9}\text{F}_4^{+}\text{ (molecular mass 88 amu) and }{}^{13}_{6}\text{C}\,{}^{19}_{9}\text{F}_4^{+}\text{ (molecular mass 89 amu)}$$

There would be two peaks in the mass spectrum.

2.99 Given the building blocks (isotopes) in this problem, how many different H_2S molecules can you make? Are their molecular masses all different, or are some the same?

Solution: Since there are two hydrogen isotopes, they can be paired in three ways: ${}^1H$-${}^1H$, ${}^1H$-${}^2H$, and ${}^2H$-${}^2H$. There will then be three choices for each sulfur isotope. We can make a table showing all the possibilities (masses in amu):

	^{32}S	^{33}S	^{34}S	^{36}S
$^{1}H_2$:	34	35	36	38
$^{1}H^{2}H$:	35	36	37	39
$^{2}H_2$:	36	37	38	40

There will be seven peaks of the following mass numbers: 34, 35, 36, 37, 38, 39, and 40.

Very accurate (and expensive!) mass spectrometers can detect the mass difference between two hydrogen-1's and one hydrogen-2. How many peaks would be detected is such a "high resolution" mass spectrum?

2.100 This is a tricky problem. Assume that the only isotopes of C and H are hydrogen-1 and carbon-12. Given the molecular masses of 50 and 52 amu, is it possible for this molecule to contain more than one atom of chlorine? Try to come up with a combination of chlorine-37, carbon-12, and hydrogen-1 that has a molecular mass of 52. What will be the molecular mass of the same molecule built with chlorine-35? Can you explain the relative peak heights with the isotope abundance data?

Solution: There can only be one chlorine per molecule, since two chlorines have a combined mass in excess of 70 amu.

$$^{12}C^{1}H_3{}^{37}Cl \text{ molecular mass} = (12) + 3(1) + (37) = 52 \text{ amu}$$
$$^{12}C^{1}H_3{}^{35}Cl \text{ molecular mass} = (12) + 3(1) + (35) = 50 \text{ amu}$$

The abundance of chlorine-35 (75.5%) is three times that of chlorine-37 (24.5%), so the 50 amu peak will be three times a large as the peak at 52 amu.

2.108 Naming compounds is not the most exciting part of chemistry. It is a necessary evil. The rules are spelled out in Section 2.8. Cation and anion names can be found in Table 2.3. Examples 2.19, 2.21, and 2.23 will get you started; study the methods carefully.

Solution: (a) potassium dihydrogen phosphate; (b) potassium hydrogen phosphate; (c) hydrogen bromide (molecular compound); (d) hydrobromic acid (see Table 2.5); (e) lithium carbonate; (f) potassium dichromate; (g) ammonium nitrite; (h) phosphorus trifluoride (molecular compound. Why?); (i) phosphorus pentafluoride; (j) tetraphosphorus hexaoxide (or hexoxide); (k) cadmium iodide; (l) strontium sulfate; (m) aluminum hydroxide; (n) potassium hypochlorite; (o) silver carbonate; (p) iron(II) chloride (Why the (II)?); (q) potassium permanganate; (r) cesium chlorate; (s) potassium ammonium sulfate; (t) iron(II) oxide (Why the (II)?); (u) iron(III) oxide; (v) titanium(IV) chloride; (w) sodium hydride (Why not sodium hydrogen?); (x) lithium nitride; (y) sodium oxide; (z) sodium peroxide (How can you tell?).

2.109 Examples 2.20, 2.22, and 2.23 are models for this problem. Study the method carefully.

Solution: (a) $RbNO_2$; (b) K_2S; (c) NaHS; (d) $Mg_3(PO_4)_2$; (e) $CaHPO_4$; (f) KH_2PO_4; (g) IF_7; (h) $(NH_4)_2SO_4$; (i) $AgClO_4$; (j) $Fe_2(CrO_4)_3$; (k) CuCN; (l) $Sr(ClO_2)_2$; (m) $HBrO_4$; (n) HI (in water); (o) $Na_2(NH_4)PO_4$; (p) $PbCO_3$; (q) SnF_2; (r) P_4S_{10}; (s) HgO; (t) Hg_2I_2; (u) $CuSO_4{\cdot}5H_2O$

2.110 The answers to this problem are found in Table 2.7. Before looking at them, see how many you already know.

Solution: (a) CO_2 (b) $NaHCO_3$ (c) $Mg(OH)_2$ (d) N_2O (e) NaCl (f) $CaCO_3$ (g) $CaCO_3$ (h) $CaCO_3$ (i) CaO

(j) $Ca(OH)_2$

2.111 At first sight this may look like an impossible problem, but it is only a standard algebra word-problem in chemical terms. You can find the average atomic mass of lithium in the table inside the front cover of your textbook. There are only two isotopes. What must be the sum of their abundances? Can you express one abundance in terms of the other?

Solution: It would seem that there are two unknowns in this problem, the natural abundance of lithium-6 and the natural abundance of lithium-7. However, these two quantities are not independent of each other; they are related by the fact that they must sum to 1. Start by letting x be the natural abundance of lithium-6. Since the sum of the two abundances must be 1, we can write

Abundance lithium-7 = (1 - x)

We use the expression for average atomic mass given in Section 2.3

Average atomic mass of natural lithium = x(6.0151 amu) + (1 - x)(7.0160 amu) = 6.941 amu

Solving for x gives x = 0.07493 or 7.493% for the abundance of lithium-6. For lithium-7 the abundance is (1 - x) = 0.92507 or 92.507%

2.112 How many protons are in the nucleus of this element? See Section 2.3 if you can't figure this out. What is the atomic number of the element? What is the element? What is the ion charge?

Solution: Number of protons = (65 - 35) = 30 = atomic number. The element must be zinc (see the periodic table). Two electrons have been lost, so the charge is +2. It is Zn^{2+}.

2.113 Changing the electrical charge of an atom usually has a major effect on its chemical properties.

Solution: The two electrically neutral carbon isotopes should have nearly identical chemical properties.

2.114 Review Section 2.3 if necessary. Problem 2.17 is similar.

Solution: Recall that the total charge for a species is dependent on the relative number of protons and electrons. The identity of an element is determined by the number of neutrons, that is, the atomic number.

(a) A, F, and G are neutral. (b) B and E are negatively charged. (c) C and D are positively charged.

(d) A: ${}^{10}_{5}B$ B: ${}^{14}_{7}N^{3-}$ C: ${}^{39}_{19}K^{+}$ D: ${}^{66}_{30}Zn^{2+}$ E: ${}^{81}_{35}Br^{-}$ F: ${}^{11}_{5}B$ G: ${}^{19}_{9}F$

2.115 Part (a) is ambiguous. Part (b) uses incorrect terminology. Can you see the errors?

Solution: (a) Does this refer to hydrogen atoms or hydrogen molecules? One can't be sure.

(b) NaCl is an ionic compound; it doesn't form molecules.

116 This is easy if you review the material in Section 2.4. When is a molecule not a compound? How can you tell if a formula represents an ionic compound? (See Problem 2.72.)

> ***Solution:*** The species and their identification are as follows:
>
> (a) SO_2: molecule and compound
> (b) S_8: element and molecule
> (c) Cs: element
> (d) N_2O_5: molecule and compound
> (e) O: element
> (f) O_2: element and molecule
> (g) O_3: element and molecule
> (h) CH_4: molecule and compound
> (i) KBr: compound
> (j) S: element
> (k) P_4: element and molecule
> (l) LiF: compound

2.118 If you don't remember how to convert moles to atoms, review Problem 2.41.

> ***Solution:*** We calculate the number of atoms in each sample using the appropriate unit factors.
>
> (a) $2.5 \text{ mol } CH_4 \left(\frac{6.022 \times 10^{23} \text{ molecules}}{1 \text{ mol}}\right)\left(\frac{5 \text{ atoms}}{1 \text{ molecule}}\right) = 7.5 \times 10^{24} \text{ atoms}$
>
> (b) $10.0 \text{ moles He} \left(\frac{6.022 \times 10^{23} \text{ atoms}}{1 \text{ mol}}\right) = 6.02 \times 10^{24} \text{ atoms}$
>
> (c) $4.0 \text{ mol } SO_2 \left(\frac{6.022 \times 10^{23} \text{ molecules}}{1 \text{ mol}}\right)\left(\frac{3 \text{ atoms}}{1 \text{ molecule}}\right) = 7.2 \times 10^{24} \text{ atoms}$
>
> (d) $1.8 \text{ mol } S_8 \left(\frac{6.022 \times 10^{23} \text{ molecules}}{1 \text{ mol}}\right)\left(\frac{8 \text{ atoms}}{1 \text{ molecule}}\right) = 8.7 \times 10^{24} \text{ atoms}$
>
> (e) $3.0 \text{ mol } NH_3 \left(\frac{6.022 \times 10^{23} \text{ molecules}}{1 \text{ mol}}\right)\left(\frac{4 \text{ atoms}}{1 \text{ molecule}}\right) = 7.2 \times 10^{24} \text{ atoms}$
>
> The 1.8 mole sample of S_8 contains the most atoms.

2.120 Working this problem requires that you be able to calculate the molecular mass of a molecule (see Example 2.8) and convert mass expressed in amu to grams (see Section 2.3 for unit factor), to calculate the mass in grams of a number of atoms (see Problem 2.39), and to convert moles to grams (see Examples 2.3 and 2.4). If you have mastered the material in this chapter, you will have no trouble. If you have difficulties, go back and study the indicated sections thoroughly; these are basic skills in a chemistry course.

> ***Solution:*** (a) $16 \text{ molecules} \left(\frac{18.02 \text{ amu}}{1 \text{ molecule}}\right)\left(\frac{1 \text{ g}}{6.022 \times 10^{23} \text{ amu}}\right) = 4.788 \times 10^{-22} \text{ g}$
>
> (b) $2 \text{ atoms} \left(\frac{207.2 \text{ amu}}{1 \text{ atom}}\right)\left(\frac{1 \text{ g}}{6.022 \times 10^{23} \text{ amu}}\right) = 6.881 \times 10^{-22} \text{ g}$
>
> (c) $5.1 \times 10^{-23} \text{ mol} \left(\frac{4.003 \text{ g}}{1 \text{ mol}}\right) = 2.0 \times 10^{-22} \text{ g}$
>
> Two atoms of lead have the greatest mass. Would it have been faster to calculate all the masses in amu?

2.121 How many grams of element Y will combine with 33.42 g of element X? Why choose 33.42 g?

Solution: The mass of Y that will combine with 33.42 g of X (molar mass) is

$$\left(\frac{33.42 \text{ g X}}{27.22 \text{ g X}}\right) \times 84.10 \text{ g Y} = 103.3 \text{ g}$$

The atomic mass is 103.3 amu.

2.122 This problem is similar to, but easier than Problem 2.44.

Solution: If we assume 100 g of compound, the masses of Cl and X are 67.2 g and 32.8 g, respectively. Since we know there are three moles of Cl per mole of X (see formula), we can write

$$\left(\frac{32.8 \text{ g X}}{67.2 \text{ g Cl}}\right)\left(\frac{35.45 \text{ g Cl}}{1 \text{ mol Cl}}\right)\left(\frac{3 \text{ mol Cl}}{1 \text{ mol X}}\right) = 51.9 \text{ g/mol}$$

The element is chromium (atomic mass 52.00 amu)

2.123 If you can state and explain the law of multiple proportions in your own words, then this problem should be simple.

Solution: The law of multiple proportions requires that the masses of sulfur combining with phosphorus must be in the ratios of small whole numbers. For the three compounds shown, four phosphorus atoms combine with three, seven, and ten sulfur atoms, respectively. If the atom ratios are in small whole number ratios, then the mass ratios must also be in small whole number ratios.

2.124 Review Section 2.4 if you don't remember how to calculate molar mass. Part (b) is similar to Problem 2.41 except that you are asked to calculate the number of molecules rather than the number of atoms. Is the method different? How many significant figures should be in the answers?

Solution: (a) Molar mass $C_{2952}H_{4664}N_{812}O_{832}S_8Fe_4$

$$= 2952(12.01) + 4664(1.008) + 812(14.01) + 832(16.00) + 8(32.07) + 4(55.85)$$
$$= 6.532 \times 10^4 \text{ g/mol}$$

(b) $$(74.3 \text{ g hemoglobin})\left(\frac{1 \text{ mol hemoglobin}}{6.532 \times 10^4 \text{ g hemoglobin}}\right)\left(\frac{6.022 \times 10^{23} \text{ molecules hemoglobin}}{1 \text{ mol hemoglobin}}\right)$$
$$= 6.85 \times 10^{20} \text{ molecules hemoglobin}$$

2.125 Review Section 2.4 if you need refreshing on the concept of molar mass.

Solution: A 100 g sample of myoglobin (Mb) contains 0.34 g of iron (0.34 percent). The number of moles of iron therefore is

$$(0.34 \text{ g Fe})\left(\frac{1 \text{ mol}}{55.85 \text{ g Fe}}\right) = 6.1 \times 10^{-3} \text{ mol Fe}$$

The amount of Mb that contains one mole of iron must be

$$\left(\frac{100.00 \text{ g Mb}}{6.1 \times 10^{-3} \text{ mol Fe}}\right) = 1.6 \times 10^{4} \text{ g Mb/mol Fe}$$

2.126 The symbol "O" refers to moles of oxygen atoms, not oxygen molecules (O_2). Look at the molecular formulas given in parts (a) and (b). What do they tell you about the relative amounts of carbon and oxygen? Reread the first few paragraphs in Section 2.4 if you can't answer these questions.

Solution: (a) $0.212 \text{ mol C}\left(\frac{1 \text{ atom O}}{1 \text{ atom C}}\right) = 0.212 \text{ mol O}$

(b) $0.212 \text{ mol C}\left(\frac{2 \text{ atom O}}{1 \text{ atom C}}\right) = 0.424 \text{ mol O}$

2.127 This problem requires conversion of grams of a molecule to number of ions. If you have one molecular formula of KBr, then how many K^+ cations are there? How many Cl^- anions are there? If you have one molecular formula of Na_2SO_4, then how many Na^+ cations are there? How many SO_4^{2-} anions are there?

Solution: (a) The molar mass of KBr = 119.0 g/mol. We first convert g of KBr to moles of KBr. Using chain multiplication, we then convert moles of KBr to number of ions.

$$(8.38 \text{ g KBr})\left(\frac{1 \text{ mol KBr}}{119.0 \text{ g}}\right)\left(\frac{6.022 \times 10^{23} \text{ ions}}{1 \text{ mol KBr}}\right) = 4.24 \times 10^{22} \text{ ions of } K^+ \text{ and of } Br^-$$

(b) The molar mass of Na_2SO_4 = 142.1 g/mol. Using the same method as above, the number of Na^+ ions is

$$(5.40 \text{ g } Na_2SO_4)\left(\frac{1 \text{ mol } Na_2SO_4}{142.1 \text{ g } Na_2SO_4}\right)\left(\frac{6.022 \times 10^{23} \text{ } Na_2SO_4}{1 \text{ mol } Na_2SO_4}\right)\left(\frac{2 \text{ } Na^+ \text{ ions}}{1 \text{ } Na_2SO_4}\right) = 4.58 \times 10^{22} \text{ } Na^+ \text{ ions}$$

Since there are two Na^+ ions for every one SO_4^{2-} ion, then the number of SO_4^{2-} ions = 2.29×10^{22}

(c) The molar mass of $Ca_3(PO_4)_2$ = 310.2 g/mol. Using the same method as above

$$(7.45 \text{ g } Ca_3(PO_4)_2)\left(\frac{1 \text{ mol } Ca_3(PO_4)_2}{310.2 \text{ g } Ca_3(PO_4)_2}\right)\left(\frac{6.022 \times 10^{23} \text{ } Ca_3(PO_4)_2}{1 \text{ mol } Ca_3(PO_4)_2}\right)\left(\frac{3Ca^{2+} \text{ ions}}{1 \text{ } Ca_3(PO_4)_2}\right) = 4.34 \times 10^{22} \text{ } Ca^{2+} \text{ ions}$$

Since there are two PO_4^{3-} anions for every three Ca^{2+} ions, then the number of PO_4^{3-} anions is

$$\left(\frac{2 \text{ } PO_4^{3-} \text{ ions}}{3 \text{ } Ca^{2+} \text{ ions}}\right)(4.34 \times 10^{22} \text{ } Ca^{2+} \text{ ions}) = 2.89 \times 10^{22} \text{ } PO_4^{3-} \text{ anions}$$

2.128 This problem is a variation of Example 2.13 in the textbook and Problem 2.79 above.

Solution: The molar masses are: Al, 26.98 g/mol; $Al_2(SO_4)_3$, 342.2 g/mol; H_2O, 18.02 g/mol. Thus, using x as the number of H_2O molecules,

$$\left(\frac{2(\text{molar mass of Al})}{\text{molar mass of } Al_2(SO_4)_3 + x(\text{molar mass of } H_2O)}\right) \times 100\% = \text{mass \% of Al}$$

$$\left(\frac{2(26.98 \text{ g})}{342.2 \text{ g} + x(18.02 \text{ g})}\right) \text{x } 100\% = 8.20\%$$

Solving, we obtain x = 17.53; rounding off to a whole number of water molecules, x = 18. Therefore the formula is $Al_2(SO_4)_3 \cdot 18H_2O$.

2.129 Review Sections 2.3 and 2.4 if you have difficulty.

Solution: (a) Since the molar mass (g/mol) M is defined as the mass m (in grams) per mole n of molecules

$$\text{molar mass (g/mol)} = \frac{\text{mass (g)}}{\text{number of moles}}$$

Rearranging,

$$\text{number of moles} = \frac{\text{mass (g)}}{\text{molar mass (g/mol)}}$$

Using the appropriate abbreviations,

$$n = \frac{m}{M}$$

(b) We know that there are Avogadro's number N_A of molecules per one mole n of molecules. So,

$$\text{Avogadro's number} = \frac{\text{number of molecules}}{\text{one mole}}$$

Rearranging,

$$\text{number of molecules} = (\text{number of moles})(\text{Avogadro's number})$$

Using the appropriate abbreviations,

$$N = nN_A$$

2.131 The material on experimental determination of empirical formulas by combustion analysis is covered in Section 2.7.

Solution: (a) Using the information given in Experiment 1 to calculate the percentages of carbon and hydrogen,

% Carbon: $$\left(\frac{3.94 \text{ g } CO_2}{2.175 \text{ g}}\right)\left(\frac{12.01 \text{ g C}}{44.01 \text{ g } CO_2}\right) \text{x } 100\% = 49.4\% \text{ C}$$

% Hydrogen: $$\left(\frac{1.89 \text{ g } H_2O}{2.175 \text{ g}}\right)\left(\frac{2 \text{ x } 1.008 \text{ g H}}{18.02 \text{ g } H_2O}\right) \text{x } 100\% = 9.72\% \text{ H}$$

From experiment 2

% Nitrogen: $\left(\frac{0.436 \text{ g } NH_3}{1.873 \text{ g}}\right)\left(\frac{14.01 \text{ g N}}{17.03 \text{ g } NH_3}\right) \times 100\% = 19.2\% \text{ N}$

By difference, the percent of oxygen = 100% - (49.4% + 9.72% + 19.2 %) = 21.7% O

Assuming a 100 g sample,

$$\text{Moles C} = (49.4 \text{ g})\left(\frac{1 \text{ mol}}{12.01 \text{ g}}\right) \times 100\% = 4.11 \text{ mol C}$$

$$\text{Moles H} = (9.72 \text{ g})\left(\frac{1 \text{ mol}}{1.008 \text{ g}}\right) \times 100\% = 9.64 \text{ mol H}$$

$$\text{Moles N} = (19.2 \text{ g})\left(\frac{1 \text{ mol}}{14.01 \text{ g}}\right) \times 100\% = 1.37 \text{ mol N}$$

$$\text{Moles O} = (21.7 \text{ g})\left(\frac{1 \text{ mol}}{16.00 \text{ g}}\right) \times 100\% = 1.36 \text{ mol O}$$

The empirical formula is $C_{4.11}H_{9.64}N_{1.37}O_{1.36}$. Dividing the subscripts by 1.36 (why?) and simplifying gives C_3H_7NO for the empirical formula. The empirical formula molar mass is 73.10 g.

(b) Since the actual molar mass of lysine is 150 g, then $\frac{150 \text{ g}}{73.10 \text{ g}} \cong 2$. The molecular formula is $C_6H_{14}N_2O_2$.

2.133 Mass spectrometry is briefly covered in Section 2.6.

Solution: The question involves common, low molecular mass compounds, that is, compounds that frequently occur in air.

(a) H_2O (16 amu); NH_3 (17 amu); CH_4 (18 amu); SO_2 (64 amu).

(b) C_3H_8 might break off a fragment CH_3, which has a mass of 15 amu. No fragment of CO_2 can have a mass of 15 amu. (What is the structure of C_3H_8?)

(c) Calculating the exact masses for:

CO_2: 12.00000 amu + (2 x 15.99491 amu) = 43.98982 amu
C_3H_8: (3 x 12.00000 amu) + (8 x 1.00797 amu) = 44.06376 amu

These two masses differ by only 0.07394 amu. The measurements must be precise to ± 0.030 amu

43.98982 amu + 0.030 amu = 44.02 amu
44.06376 amu – 0.030 amu = 44.03 amu

2.135 What is the difference between hydrogen molecules and hydrogen atoms? Would the mass of 100 dimes be the same if they were stuck together in pairs rather than all being separated?

Solution: Yes, one gram of hydrogen molecules equals one gram of hydrogen atoms in the number of hydrogen atoms. There is no difference in the mass, only in the way that the particles are arranged.

What would the relative number of hydrogen molecules to hydrogen atoms be?

2.136 Section 2.6 discusses the use of mass spectrometry.

Solution: The observations mean that the amount of more abundant isotope was either increasing or the amount of the less abundant isotope was deceasing. One possible explanation is that the less abundant isotope was undergoing radioactive decay and was thus disappearing.

2.138 This is another straightforward determination of percent mass problem similar to Example 2.13.

Solution: Molar Mass of $C_4H_8Cl_2S$ = 4(12.01) + 8(1.008) + 2(35.45) + 32.07 = 159.1 g/mol

$$\%C = \left(\frac{48.04 \text{ g/mol}}{159.1 \text{ g/mol}}\right) 100\% = 30.19\%$$

$$\%H = \left(\frac{8.064 \text{ g/mol}}{159.1 \text{ g/mol}}\right) 100\% = 5.069\%$$

$$\%Cl = \left(\frac{70.90 \text{ g/mol}}{159.1 \text{ g/mol}}\right) 100\% = 44.56\%$$

$$\%S = \left(\frac{32.07 \text{ g/mol}}{159.1 \text{ g/mol}}\right) 100\% = 20.16\ \%$$

Why do the percentages not add to exactly 100%?

2.139 This problem is an interesting application of the factor-label method which demonstrates how large a number a mole is.

Solution: The number of dollars that would be given away would be:

$$(4.5 \times 10^9 \text{ yr})\left(\frac{365 \text{ days}}{\text{yr}}\right)\left(\frac{24 \text{ hr}}{\text{day}}\right)\left(\frac{60 \text{ min}}{\text{hr}}\right)\left(\frac{60 \text{ s}}{\text{min}}\right)\left(\frac{\$1.0 \times 10^6}{\text{s}}\right) = \$1.42 \times 10^{23}$$

The amount of $ remaining would be: $(\$6.02 \times 10^{23}) - (\$1.42 \times 10^{23}) = \$4.6 \times 10^{23}$

The number of moles remaining would be: $\left(\frac{\$4.6 \times 10^{23}}{\$6.02 \times 10^{23} \text{ per mole}}\right) = 0.76 \text{ mole}$

2.140 This problem is exactly analogous to Example 2.6 in the textbook.

Solution: $$(5.0 \text{ g Fe})\left(\frac{\text{mol Fe}}{55.85 \text{ g Fe}}\right)\left(\frac{6.022 \times 10^{23} \text{ atoms Fe}}{\text{mol Fe}}\right) = 5.4 \times 10^{22} \text{ atoms Fe}$$

2.142 The problem actually leads into the material in Chapters 3 and 4. To work the problem, you assume that all of the lost mass of Fe reacts to become Fe_2O_3. That is, every Fe atom that is lost actually reacts with 1.5 oxygen atoms (or 2 Fe atoms react with 3 oxygen atoms). Thus you can calculate what the mass of the "lost" pure Fe becomes when it forms Fe_2O_3. In Chapter 4 you will learn how to routinely use the factor-label method to work this type of problem.

Solution: The amount of Fe remaining is: $664\text{ g} - \left(\frac{1}{8}\right)(664\text{ g}) = 581\text{ g}$

The amount of Fe that reacted is: $\left(\frac{1}{8}\right)(664\text{ g}) = 83\text{ g}$

Therefore 83 g of Fe reacts to form the compound Fe_2O_3, where there are 1.5 oxygen atoms for every 1.0 Fe atom. That is, for every 1.0 mole of Fe atoms, 1.5 moles of O atoms will react to form Fe_2O_3 molecules. The amount of Fe that we have is:

$$(83\text{ g Fe})\left(\frac{1\text{ mol Fe}}{55.85\text{ g}}\right) = 1.49\text{ mol Fe}$$

which requires

$$(1.49\text{ mol Fe})\left(\frac{1.5\text{ mol O}}{\text{mol Fe}}\right) = 2.24\text{ mol O}$$

Converting moles O to g

$$(2.24\text{ mol O})\left(\frac{16.00\text{ g O}}{\text{mol O}}\right) = 35.8\text{ g of O} = 36\text{ g of O}$$

In summary, the 83 g of Fe that is missing reacted with 36 g of O to form 119 g of Fe_2O_3. The change in mass is due to the mass of the oxygen gained (36 g).

Could you have added the mass of the unreacted Fe to the mass of the Fe_2O_3 formed to obtain the new total mass?

2.144 The problem is a chain factor-label conversion problem.

Solution: The conversions are mL H_2O → g H_2O → mol H_2O → molecules H_2O → molecules D_2O

$$(400\text{ mL H}_2\text{O})\left(\frac{1\text{ g H}_2\text{O}}{\text{mL H}_2\text{O}}\right)\left(\frac{\text{mol H}_2\text{O}}{18.02\text{ g H}_2\text{O}}\right)\left(\frac{6.022\times 10^{23}\text{ molecules H}_2\text{O}}{\text{mol H}_2\text{O}}\right)\left(\frac{0.015\text{ \% molecules D}_2\text{O}}{100\text{ \% molecules H}_2\text{O}}\right)$$

$$= 2.01\times 10^{21}\text{ D}_2\text{O molecules}$$

In fact, should you expect to find very many D_2O molecules relative to the number of HDO molecules?

2.146 This problem is similar to Problem 2.17

Solution:

Symbol	$^{11}_{5}B$	$^{54}_{26}Fe^{2+}$	$^{31}_{15}P^{3-}$	$^{196}_{79}Au$	$^{222}_{86}Rn$
Protons	5	26	15	79	86
Neutrons	6	28	16	117	136
Electrons	5	24	18	79	86

Net Charge	0	+2	−3	0	0

2.148 What are the charges of the ions formed by the group 1A and 2A metals, aluminum, oxygen, nitrogen, and the group 7A elements? What criterion is used to determine the formula of a simple binary ionic compound? Review Section 2.5 if you can't answer these questions.

Solution: Group 1A metals form M^+ ions. Group 2A metals form Y^{2+} ions. Aluminum forms an Al^{3+} ion. Oxygen forms an O^{2-} ion. Nitrogen forms an N^{3-}, and the halogens form X^- ions. We make a table:

Nonmetals	1A Metals	2A Metals	Aluminum
Halogens	MX	YX_2	AlX_3
Oxygen	M_2O	YO	Al_2O_3
Nitrogen	M_3N	Y_3N_2	AlN

2.150 Refer to the Chemistry in Action on Allotropes.

Solution: Tin (Sn) and Sulfur (S) both have allotropes, as do a number of other elements.

Note: If you have not already done so, please read the "Introduction for the Student" at the beginning of this Student Solutions Manual.

Chapter 3

CHEMICAL REACTIONS I: CHEMICAL EQUATIONS AND REACTIONS IN AQUEOUS SOLUTION

3.5 This is similar to Example 3.1 and the information shown in Table 3.1. Study this material carefully if you can't see how to answer this problem.

> ***Solution:*** As in Example 3.1, several interpretations of the equation are possible. The "correct" interpretation would be apparent from the specific context.
>
> 1. Qualitative: Solid iron(II) sulfide (FeS) reacts with aqueous hydrochloric acid (HCl) to form iron(II) chloride ($FeCl_2$) and hydrogen sulfide gas (H_2S).
>
> 2. Molecular level: One iron(II) sulfide unit (ionic compound) combines with two units of hydrochloric acid to form one unit of iron(II) chloride (ionic compound) and one molecule of hydrogen sulfide.
>
> 3. Moles: One mole of iron(II) sulfide reacts with two moles of hydrochloric acid to form one mole of iron(II) chloride and one mole of hydrogen sulfide.
>
> 4. Quantitative: 87.92 g of iron(II) sulfide combines with 72.92 g of hydrochloric acid to form 126.75 g of iron(II) chloride and 34.09 g of hydrogen sulfide. Do the masses on each side sum to the same number?
>
> How would the equation be different if the problem involved iron(III) rather than iron(II)?

3.6 This problem is easy if you understand the material in Section 3.1.

> ***Solution:*** Without a good knowledge of chemistry, someone would not know whether the reaction was performed in the solid state or in solution. Additionally, they wouldn't know the quantities of materials to use, whether the reaction was fast or slow, or whether the reaction occurred spontaneously or had to be heated. The equation can be interpreted by a trained individual to provide a great deal of information. Even this particular equation is abbreviated because it doesn't provide any solution or solubility information such as Problem 3.5 above.

3.8 The subsection on Balancing Chemical Equations in Section 3.1 details the procedure for working these problems. Study it carefully. Note that you can always tell if your answer is correct by counting up the numbers of atoms on each side of the equation. The problems are similar to Example 3.2.

Solution: The balanced equations are as follows:

(a) $2KClO_3 \rightarrow 2KCl + 3O_2$

(b) $2KNO_3 \rightarrow 2KNO_2 + O_2$

(c) $NH_4NO_3 \rightarrow N_2O + 2H_2O$

(d) $NH_4NO_2 \rightarrow N_2 + 2H_2O$

(e) $2NaHCO_3 \rightarrow Na_2CO_3 + H_2O + CO_2$

(f) $P_4O_{10} + 6H_2O \rightarrow 4H_3PO_4$

(g) $2HCl + CaCO_3 \rightarrow CaCl_2 + H_2O + CO_2$

(h) $2Al + 3H_2SO_4 \rightarrow Al_2(SO_4)_3 + 3H_2$

(i) $CO_2 + 2KOH \rightarrow K_2CO_3 + H_2O$

(j) $CH_4 + 2O_2 \rightarrow CO_2 + 2H_2O$

(k) $Be_2C + 4H_2O \rightarrow 2Be(OH)_2 + CH_4$

(l) $3Cu + 8HNO_3 \rightarrow 3Cu(NO_3)_2 + 2NO + 4H_2O$

(m) $S + 6HNO_3 \rightarrow H_2SO_4 + 6NO_2 + 2H_2O$

(n) $2NH_3 + 3CuO \rightarrow 3Cu + N_2 + 3H_2O$

3.13 What is an electrolyte? What is a nonelectrolyte? How can you tell the difference by experiment? How can you tell from the formula of the compound? What are strong and weak electrolytes? If you can't answer these questions, you can't solve this problem. Go back and study Section 3.2.

Solution: Ionic compounds, strong acids, and strong bases (metal hydroxides) are strong electrolytes (completely broken up into ions in solution). Weak acids and weak bases are weak electrolytes. Molecular substances other than acids or bases are nonelectrolytes. Table 3.2 classifies some of the common solutes in water.

(a) very weak electrolyte; (b) strong electrolyte (ionic compound); (c) strong electrolyte (strong acid); (d) weak electrolyte (weak acid); (e) nonelectrolyte (molecular compound - neither acid nor base); (f) strong electrolyte (ionic); (g) nonelectrolyte (atomic); (h) weak electrolyte (weak base); (i) strong electrolyte (strong base).

3.14 The conduction of electric charge results from the free motion of charged particles. Review Figure 3.2

Solution: Since solutions must be electrically neutral, any flow of positive species (cations) must be balanced by the flow of negative species (anions). Therefore, the correct answer is (d).

3.15 The conduction of electric charge results from the free motion of charged particles. In which of the choices are the charged particles free to move around?

Solution: (a) Solid NaCl does not conduct. The ions are locked in a rigid lattice structure.

(b) Molten NaCl conducts. The ions can move around in the liquid state.

(c) Aqueous NaCl conducts.

3.16 What does it mean if a solution won't conduct electricity?

Solution: Since HCl dissolved in water conducts electricity, then we know that the HCl(*aq*) actually exists as

H^+*(aq)* cations and Cl^-*(aq)* anions. Since HCl dissolved in benzene solvent does not conduct electricity, then we must assume that the HCl molecules in benzene solvent do not ionize, but rather exist as un-ionized molecules.

3.17 How would you determine whether any species was an electrolyte or nonelectrolyte?

Solution: Measure the conductance to see if the solution carries an electrical current. If the solution is conducting, then you can determine whether the solution is a strong or weak electrolyte by comparing its conductance with that of a known strong electrolyte.

3.19 To answer this question, review the solubility rules given in Table 3.3 and Example 3.3.

Solution:

(a) Insoluble (rule 6)
(b) Insoluble (rule 4)
(c) Soluble (rule 3)
(d) Soluble (rule 1)
(e) Insoluble (rule 6)
(f) Soluble (rule 7)
(g) Soluble (rule 3)
(h) Insoluble (rule 7)
(i) Soluble (rules 2 and 3)

3.21 Notice that all of the compounds in (a) and (b) are ionic. Consider all the possible combinations of positive and negative ions; are any insoluble compounds? Study the solubility rules in Table 3.3 and Example 3.4.

Solution: (a) Both reactants are soluble ionic compounds. The other possible ion combinations, Na_2SO_4 and $Cu(NO_3)_2$, are also soluble (rules 1 and 3).

(b) Both reactants are soluble. Of the other two possible ion combinations, KCl is soluble (rules 1 and 5), but $BaSO_4$ is insoluble and will precipitate (rule 7).

3.22 What is the difference between an ionic equation and a net ionic equation? What are spectator ions? Review Section 3.3 and the solubility rules in Table 3.3 if you have trouble with this problem.

Solution: The equations are as follows:

(a) Ionic: $2Ag^+(aq) + 2NO_3^-(aq) + 2Na^+(aq) + SO_4^{2-}(aq) \rightarrow Ag_2SO_4(s) + 2Na^+(aq) + 2NO_3^-(aq)$

Net ionic: $2Ag^+(aq) + SO_4^{2-}(aq) \rightarrow Ag_2SO_4(s)$

(b) Ionic: $Ba^{2+}(aq) + 2Cl^-(aq) + Zn^{2+}(aq) + SO_4^{2-}(aq) \rightarrow BaSO_4(s) + Zn^{2+}(aq) + 2Cl^-(aq)$

Net ionic: $Ba^{2+}(aq) + SO_4^{2-}(aq) \rightarrow BaSO_4(s)$

(c) Ionic: $2NH_4^+(aq) + CO_3^{2-}(aq) + Ca^{2+}(aq) + 2Cl^-(aq) \rightarrow CaCO_3(s) + 2NH_4^+(aq) + 2Cl^-(aq)$

Net ionic: $Ca^{2+}(aq) + CO_3^{2-}(aq) \rightarrow CaCO_3(s)$

(d) Ionic: $2Na^+(aq) + S^{2-}(aq) + Zn^{2+}(aq) + 2Cl^-(aq) \rightarrow ZnS(s) + 2Na^+(aq) + 2Cl^-(aq)$

Net ionic: $Zn^{2+}(aq) + S^{2-}(aq) \rightarrow ZnS(s)$

(e) Ionic: $6K^{+}(aq) + 2PO_4^{3-}(aq) + 3Sr^{2+}(aq) + 6NO_3^{-}(aq) \rightarrow Sr_3(PO_4)_2(s) + 6K^{+}(aq) + 6NO_3^{-}(aq)$

Net ionic: $3Sr^{2+}(aq) + 2PO_4^{3-}(aq) \rightarrow Sr_3(PO_4)_2(s)$

3.30 This problem is similar to Example 3.5. Is it possible for a single species to be both a Brønsted acid and a Brønsted base?

Solution: (a) HI dissolves in water to produce H^+ and I^-, so HI is a Brønsted acid.

(b) CH_3COO^- can accept a proton to become acetic acid CH_3COOH, so it is a Brønsted base.

(c) $H_2PO_4^-$ can either accept a proton H^+ to become H_3PO_4 and thus behaves as a Brønsted base, or can donate a proton in water to yield H^+ and HPO_4^{2-}, thus behaving as a Brønsted acid.

(d) PO_4^{3-} in water can accept a proton to become HPO_4^{2-}, and is thus a Brønsted base.

(e) ClO_2^- in water can accept a proton to become $HClO_2$, and is thus a Brønsted base.

(f) NH_4^+ dissolved in water can donate a proton H^+, thus behaving as a Brønsted acid.

You should be able to name all of the above species. If you can't, review Section 2.8

3.35 Review the material in Section 3.5 of the textbook if you need help in working this problem. Example 3.6 is similar.

Solution: Even though the problem doesn't ask you to assign oxidation numbers, you need to be able to do so in order to determine what is being oxidized or reduced

	(i) Half Reactions	(ii) Oxidizing Agent	(iii) Reducing Agent
(a)	$Sr \rightarrow Sr^{2+} + 2e^-$ $O_2 + 4e^- \rightarrow 2O^{2-}$	O_2	Sr
(b)	$Li \rightarrow Li^+ + e^-$ $H_2 + 2e^- \rightarrow 2H^-$	H_2	Li
(c)	$Cs \rightarrow Cs^+ + e^-$ $Br_2 + 2e^- \rightarrow 2Br^-$	Br_2	Cs
(d)	$Mg \rightarrow Mg^{2+} + 2e^-$ $N_2 + 6e^- \rightarrow 2N^{3-}$	N_2	Mg
(e)	$Zn \rightarrow Zn^{2+} + 2e^-$ $I_2 + 2e^- \rightarrow 2I^-$	I_2	Zn

(f)	$C \rightarrow C^{2+} + 2e^-$ $O_2 + 4e^- \rightarrow 2O^{2-}$	O_2	C

Is it possible to have an oxidizing agent without a reducing agent? A reducing agent without an oxidizing agent?

3.36 Half-reactions specifically show what happens to each one of the reactants in an oxidation-reduction process. One half-reaction always has electrons on the left side (the reactant is the oxidizing agent); the other always has electrons on the right side (the reactant is the reducing agent). In writing half-reactions involving molecular, rather than ionic, compounds (e.g., parts (c) and (d)) treat the molecular compounds as though they were ionic for the purpose of working out electron balance.

Solution: In each part the reducing agent is the reactant in the first half-reaction and the oxidizing agent is the reactant in the second half-reaction. The coefficients in each half-reaction have been reduced to smallest whole numbers.

(a) The product is an ionic compound of Fe^{3+} and O^{2-}.

$$Fe \rightarrow Fe^{3+} + 3e^-$$

$$O_2 + 4e^- \rightarrow 2O^{2-}$$

(b) Na^+ does not change in this reaction. It is a "spectator ion."

$$2Br^- \rightarrow Br_2 + 2e^-$$

$$Cl_2 + 2e^- \rightarrow 2Cl^-$$

(c) Assume SiF_4 is made up of Si^{4+} and F^-.

$$Si \rightarrow Si^{4+} + 4e^-$$

$$F_2 + 2e^- \rightarrow 2F^-$$

(d) Assume HCl is made up of H^+ and Cl^-.

$$H_2 \rightarrow 2H^+ + 2e^-$$

$$Cl_2 + 2e^- \rightarrow 2Cl^-$$

3.40 This problem and problems 3.41 and 3.42 are all similar to Example 3.7. Review the guidelines for assigning oxidation numbers in Section 3.5.

Solution: The oxidation number for hydrogen is +1 (rule 4), and for oxygen is –2 (rule 3). The oxidation number for sulfur in S_8 is zero (rule 1). Remember that in a neutral molecule, the sum of the oxidation numbers of all the atoms must be zero, and in an ion the sum of oxidation numbers of all the elements in the ion must equal the net charge of the ion (rule 6).

H_2S (-2), S^{2-} (-2), HS^- (-2) < S_8 (0) < SO_2 (+4) < SO_3 (+6), H_2SO_4 (+6)

The number in parentheses denotes the oxidation number of sulfur.

3.41 Study the guidelines for assigning oxidation numbers in Section 3.5. This problem is similar to Example 3.7.

Solution: In each case the oxidation number of oxygen is -2 and that of hydrogen is +1 (rules 3 and 4, respectively). The numbers must sum to zero for neutral compounds (rule 6).

(a) +5 (b) +1 (c) +3 (d) +5 (e) +5 (f) +5

The molecules in parts (a), (e), and (f) can be made by strongly heating the compound in part (d). Are these oxidation-reduction reactions?

3.42 Study Example 3.7.

Solution: The oxidation numbers are assigned according to the guidelines provided in Section 3.5

(a) ClF: F -1 (rule 5), Cl +1 (rule 6)

(b) IF_7: F -1 (rule 5), I +7 (rules 5 and 6)

(c) CH_4: C -4 (rule 6), H +1 (rule 4)

(d) C_2H_2: C -1 (rule 6), H +1 (rule 4)

(e) C_2H_4: C -2 (rule 6), H +1 (rule 4)

(f) K_2CrO_4: K +1 (rule 2), Cr +6 (rule 6), O -2 (rule 3)

(g) $K_2Cr_2O_7$: K +1 (rule 2), Cr +6 (rule 6), O -2 (rule 3)

(h) $KMnO_4$: K +1 (rule 2), Mn +7 (rule 6), O -2 (rule 3)

(i) $NaHCO_3$: Na +1 (rule 2), H +1 (rule 4), C +4 (rule 6), O -2 (rule 3)

(j) Li_2: Li 0 (rule 1)

(k) $NaIO_3$: Na +1 (rule 2), I +5 (rules 5 and 6), O -2 (rule 3)

(l) KO_2: K +1 (rule 2), O -1/2 (rule 6)

(m) PF_6^-: P +5 (rule 6), F -1 (rule 5)

(n) $KAuCl_4$: K +1 (rule 2), Au +3 (rule 6), Cl -1 (rule 5)

3.44 For a review of assigning oxidation numbers, see Problem 3.42 above and Section 3.5 in the textbook.

Solution: (a) Mg: +2, N: -3; (b) Cs: +1, O: $-\frac{1}{2}$; (c) Ca: +2, C: -1; (d) C: +4, O: -2; (e) C: +3, O: -2; (f) Zn: +2, O: -2; (g) Na: +1, B: +3, H: -1; (h) W: +6, O: -2.

3.45 What happens to the oxidation number of nitrogen when nitric acid suffers reduction?

Solution: If nitric acid is a strong oxidizing agent and zinc is a strong reducing agent, then zinc metal will probably reduce nitric acid when the two react; that is, the oxidation number of N must decrease. Since the oxidation number of nitrogen in nitric acid is + 5 (verify!), then the nitrogen-containing product must have a smaller oxidation number for nitrogen. The only compound in the list that doesn't have a nitrogen oxidation number less than +5 is N_2O_5, (what is the oxidation number for N in N_2O_5?). This is never a product of the reduction of nitric acid.

3.46 Review Figure 3.8 in the textbook that shows the maximum and minimum oxidation numbers for the elements.

Solution: In order to work this problem, you need to assign the oxidation numbers to all the the elements in the compounds. In each case oxygen has an oxidation number of -2 (rule 3). These oxidation numbers should then be compared to the range of possible oxidation numbers that each element can have, as shown in Figure 3.8. Molecular oxygen is a powerful oxidizing agent. In SO_3 alone, the oxidation number of the element bound to oxygen (S) is at its maximum value (+6); the sulfur cannot be oxidized further. The other elements bound to oxygen in this problem have less than their maximum oxidation number and can undergo further oxidation.

3.51 What distinguishes a redox reaction from other types of chemical reactions?

Solution: Oxidation numbers must change in a redox reaction; in other words, electrons must be exchanged between reacting species. In both O_2 (molecular oxygen) and O_3 (ozone) the oxidation number of oxygen is zero. It is not a redox reaction.

3.52 The ammonium ion is a reducing agent and the nitrate ion is an oxidizing agent. With ammonium nitrate one product is water and the other is nitrous oxide (laughing gas) N_2O.

Solution: The ammonium nitrate decomposition reaction can be an explosion under the right conditions:

$$NH_4NO_3 \rightarrow N_2O + 2H_2O$$

Note that this reaction is a redox reaction. The nitrogen in the ammonium cation NH_4^+ has an oxidation number of −3; the nitrogen in the nitrate anion NO_3^- has an oxidation number of +5. The nitrogen in nitrous oxide N_2O has an oxidation number of +4. Therefore nitrogen is both oxidized and reduced. What type of redox reaction is this?

3.53 In other words, which metals can displace hydrogen from water? Study the discussion of Hydrogen Displacement in Section 3.5 and consult Figure 3.15.

Solution: Only (b) Li and (d) Ca will react with water.

3.54 Consult Figures 3.15 and 3.16 for the activity series that are needed to work this problem. How do net ionic equations differ from a "normal" equation?

Solution: (a) $Cu(s) + HCl(aq) \rightarrow$ no reaction, since $Cu(s)$ is less active that the hydrogen from acids.

(b) $I_2(s) + NaBr(aq) \rightarrow$ no reaction, since $I_2(s)$ is less reactive than $Br_2(l)$.

(c) $Mg(s) + CuSO_4(aq) \rightarrow MgSO_4(aq) + Cu(s)$, since $Mg(s)$ is more reactive than $Cu(s)$.

Net ionic equation: $Mg(s) + Cu^{2+}(aq) \rightarrow Mg^{2+}(aq) + Cu(s)$

(d) $Cl_2(g) + 2KBr(aq) \rightarrow Br_2(l) + 2KCl(aq)$, since $Cl_2(g)$ is more reactive than $Br_2(l)$

Net ionic equation: $Cl_2(g) + 2Br^-(aq) \rightarrow 2Cl^-(aq) + Br_2(l)$

What would the products be in (a) and (b) if the reactions did occur?

3.55 Section 3.6 discusses the *ion-electron method* for balancing redox equations. Review this Section before attempting to work this problem. Example 3.8 in the textbook is similar.

Solution: Step 1. We are given the unbalanced equation in the ionic form: $MnO_4^- + SO_2 \rightarrow Mn^{2+} + HSO_4^-$

Step 2. The equation for the two half reactions are:

Oxidation: $\overset{+4}{SO_2} \rightarrow \overset{+7}{HSO_4^-}$

Reduction: $\overset{+7}{MnO_4^-} \rightarrow \overset{+2}{Mn^{2+}}$

Step 3. We next balance the atoms other than H and O in each half reaction separately. In this case the oxidation half reaction already has one S atom on both sides of the equation and the reduction half reaction already is balanced in Mn atoms. This step then requires no additional work in this example.

Step 4. We need to independently balance the H and O atoms in each half-reaction. In acidic medium, we add H_2O molecules to balance the O atoms and H^+ to balance the H atoms. We balance the oxidation half-reaction for oxygen by adding two water molecules to the left side.

$$SO_2 + 2H_2O \rightarrow HSO_4^-$$

To achieve hydrogen balance we add $3H^+$ to the right side of the oxidation half-reaction. Notice there is already one hydrogen on the right in the HSO_4^-.

$$SO_2 + 2H_2O \rightarrow HSO_4^- + 3H^+$$

The oxidation half-reaction is now atom-balanced. We balance the reduction half-reaction for oxygen by adding 4 water molecules to the right side.

$$MnO_4^- \rightarrow Mn^{2+} + 4H_2O$$

The reduction half-reaction is balanced for hydrogen by adding $8H^+$ to the left side.

$$MnO_4^- + 8H^+ \rightarrow Mn^{2+} + 4H_2O$$

Step 5. Both half-reactions are now balanced for everything except charge. We balance charge by adding electrons.

$$SO_2 + 2H_2O \rightarrow HSO_4^- + 3H^+ + 2e^-$$

$$MnO_4^- + 8H^+ + 5e^- \rightarrow Mn^{2+} + 4H_2O$$

To equalize the exchange of electrons we multiply the oxidation half-reaction by 5 and the reduction half-reaction by 2.

$$5SO_2 + 10H_2O \rightarrow 5HSO_4^- + 15H^+ + 10e^-$$

$$2MnO_4^- + 16H^+ + 10e^- \rightarrow 2Mn^{2+} + 8H_2O$$

Step 6. Adding the two (now fully balanced) half-reactions and canceling electrons and extra H_2O and H^+ on each side gives the fully balanced equation

$$2MnO_4^- + 5SO_2 + 2H_2O + H^+ \rightarrow 2Mn^{2+} + 5HSO_4^-$$

Step 7. A final check shows that the resulting equation is "atomically" and "electrically" balanced.

3.56 Carefully study the steps of the ion-electron method in Section 3.6. Try to work Example 3.8 and Problem 3.55 above before trying any parts of this problem.

Solution: We follow the steps described in detail in Section 3.6 and Example 3.8 and Problem 3.55 above.

(a) The problem is given in ionic form, so combining Steps 1 and 2, the half-reactions are:

oxidation: $Fe^{2+} \rightarrow Fe^{3+}$
reduction: $H_2O_2 \rightarrow H_2O$

Combining Steps 3 and 4, balancing the second equation for O and H gives

$$2H^+ + H_2O_2 \rightarrow 2H_2O$$

Step 5 involves balancing both equations for charge with electrons:

$$Fe^{2+} \rightarrow Fe^{3+} + e^-$$
$$2H^+ + H_2O_2 + 2e^- \rightarrow 2H_2O$$

We multiply the first half-reaction by 2 to equalize charge, add, and cancel electrons to reach the balanced equation.

$$2Fe^{2+} + 2H^+ + H_2O_2 \rightarrow 2Fe^{3+} + 2H_2O$$

(b) The problem is given in ionic form, so combining Steps 1 and 2, the half-reactions are:

oxidation: $Cu \rightarrow Cu^{2+}$
reduction: $HNO_3 \rightarrow NO$

Combining Steps 3 and 4, balancing the reduction half reaction for O with H_2O and for H with H^+ gives

$$3H^+ + HNO_3 \rightarrow NO + 2H_2O$$

The oxidation half reaction is already mass balanced.

Step 5 involves balancing both equations for charge with electrons:

$$3e^- + 3H^+ + HNO_3 \rightarrow NO + 2H_2O$$
$$Cu \rightarrow Cu^{2+} + 2e^-$$

We multiply the first half-reaction by 2 and the second half reaction by 3 to equalize charge (6 e^- on each side of the equation, add, and cancel electrons to reach the balanced equation.

$$3Cu + 2HNO_3 + 6H^+ \rightarrow 3Cu^{2+} + 2NO + 4H_2O$$

(c) Half-reactions balanced for Cr and C:

oxidation: $C_2O_4^{2-} \rightarrow 2CO_2$
reduction: $Cr_2O_7^{2-} \rightarrow 2Cr^{3+}$

Balancing O and H with H_2O and H^+, respectively, gives

$$C_2O_4^{2-} \rightarrow 2CO_2 \text{ (no change)}$$
$$Cr_2O_7^{2-} + 14H^+ \rightarrow 2Cr^{3+} + 7H_2O$$

Balance charge with electrons

$$C_2O_4^{2-} \rightarrow 2CO_2 + 2e^-$$
$$Cr_2O_7^{2-} + 14H^+ + 6e^- \rightarrow 2Cr^{3+} + 7H_2O$$

Equalizing charges by multiplying the first equation by 3, adding, and canceling electrons gives the finished product

$$Cr_2O_7^{2-} + 14H^+ + 3C_2O_4^{2-} \rightarrow 2Cr^{3+} + 6CO_2 + 7H_2O$$

(d) Half-reactions balanced for Cl:

oxidation: $2Cl^- \rightarrow Cl_2$
reduction: $ClO_3^- \rightarrow ClO_2$

Balance O and H with H_2O and H^+, respectively

$$2Cl^- \rightarrow Cl_2 \text{ (no change)}$$
$$ClO_3^- + 2H^+ \rightarrow ClO_2 + H_2O$$

Balance charge with electrons

$$2Cl^- \rightarrow Cl_2 + 2e^-$$
$$ClO_3^- + 2H^+ + e^- \rightarrow ClO_2 + H_2O$$

Multiplying the second equation by 2, adding, and canceling electrons gives the fully balanced equation

$$2Cl^- + 2ClO_3^- + 4H^+ \rightarrow Cl_2 + 2ClO_2 + 2H_2O$$

Suppose the half-reactions were

$Cl^- \rightarrow ClO_2$
$ClO_3^- \rightarrow Cl_2$

Would the balanced equation be the same? Try it.

(e) Half-reactions balanced for S and I:

oxidation: $2S_2O_3^{2-} \rightarrow S_4O_6^{2-}$
reduction: $I_2 \rightarrow 2I^-$

Both half-reactions are already balanced for O, so we balance charge with electrons

$$2S_2O_3^{2-} \rightarrow S_4O_6^{2-} + 2e^-$$
$$I_2 + 2e^- \rightarrow 2I^-$$

The electron count is the same on both sides. We add and cancel electrons to obtain the finished equation

$$2S_2O_3^{2-} + I_2 \rightarrow S_4O_6^{2-} + 2I^-$$

(f) The two half-reactions are

oxidation: $Mn^{2+} \rightarrow MnO_2$
reduction: $H_2O_2 \rightarrow H_2O$

We balance oxygen with H_2O and hydrogen with H^+.

$$Mn^{2+} + 2H_2O \rightarrow MnO_2 + 4H^+$$
$$H_2O_2 + 2H^+ \rightarrow 2H_2O$$

Since this reaction is in basic solution, we add one OH^- to both sides for each H^+ and combine pairs of H^+ and OH^- on the same side of the arrow to form H_2O.

$$Mn^{2+} + 2H_2O + 4OH^- \rightarrow MnO_2 + 4H_2O$$
$$H_2O_2 + 2H_2O \rightarrow 2H_2O + 2OH^-$$

We remove extra H_2O and balance charge with electrons

$$Mn^{2+} + 4OH^- \rightarrow MnO_2 + 2H_2O + 2e^-$$
$$H_2O_2 + 2e^- \rightarrow 2OH^-$$

We can now add the reactions because they have the same number of electrons on opposite sides of the arrow. Canceling extra OH^- and electrons gives the completed balanced equation.

$$Mn^{2+} + H_2O_2 + 2OH^- \rightarrow MnO_2 + 2H_2O$$

Parts (g) through (i) can be solved by the same methods.

(g) $2Bi(OH)_3 + 3SnO_2^{2-} \rightarrow 2Bi + 3H_2O + 3SnO_3^{2-}$

(h) $3CN^- + 2MnO_4^- + H_2O \rightarrow 3CNO^- + 2MnO_2 + 2OH^-$

(i) $3Br_2 + 6OH^- \rightarrow BrO_3^- + 5Br^- + 3H_2O$ What type of redox reaction is this?

3.57 In the last step the products are manganese(II) and iron(III). Incidentally, the reason for converting manganese(II) to permanganate and then back to manganese(II) is that the permanganate ion is an intense purple color and it is easy to see exactly how much iron(II) it takes to reduce the permanganate.

Solution: The three redox equations are:

$$Mn + 2HNO_3 + 2H^+ \rightarrow Mn^{2+} + 2NO_2 + 2H_2O$$
$$2Mn^{2+} + 5IO_4^- + 3H_2O \rightarrow 2MnO_4^- + 5IO_3^- + 6H^+$$
$$MnO_4^- + 5Fe^{2+} + 8H^+ \rightarrow Mn^{2+} + 5Fe^{3+} + 4H_2O$$

3.59 Review the properties of aqueous solutions in Section 3.2. When an ionic solute dissolves, will the solution conduct electricity?

Solution: You would first accurately measure the electrical conductance of pure water. The conductance of a solution of the slightly soluble ionic compound X should be greater than that of pure water, indicating that some of the compound X had indeed dissolved.

3.60 How do you know whether a reaction is a redox reaction? See Sections 3.5 and 3.6.

Solution: In redox reactions the oxidation numbers of elements change. To test whether an equation represents a redox process, assign the oxidation numbers to each of the elements in the reactants and products. If they are different, it is a redox reaction. The rules for assigning oxidation numbers can be found in Section 3.5.

(a) On the left the oxidation number of chlorine in Cl_2 is zero (rule 1). On the right it is -1 in Cl^- (rule 2) and +1 in OCl^- (rules 3 and 5). Since chlorine is both oxidized and reduced, this is a disproportionation redox reaction.

(b) The oxidation numbers of calcium and carbon do not change. This is not a redox reaction; it is a precipitation reaction (see solubility rule 6 in Table 3.3).

(c) The oxidation numbers of nitrogen and hydrogen do not change. This is not a redox reaction; it is an acid-base reaction.

(d) The oxidation numbers of carbon, chlorine, chromium, and oxygen do not change. This is not a redox reaction; it doesn't fit easily into any catagory, but it could be considered as a type of combination reaction.

(e) The oxidation number of calcium changes from 0 to +2, and the oxidation number of fluorine changes from 0 to -1. This is a redox reaction and a combination reaction.

The remaining parts (f) through (j) can be worked in the same way.

(f) Redox (g) Precipitation (rule 7, Table 3.3) (h) Redox (i) Redox (j) Redox

3.61 Try to balance the equation by the ion-electron method. One half-reaction should show the oxidation of Au in the presence of HCl. The other should depict the reduction of nitric acid to nitrogen dioxide. The balanced equation will not have a 3:1 ratio of hydrochloric to nitric acid.

Solution: Notice that the oxidation number of chlorine doesn't change in this process. The gold oxidation half-reaction is

$$Au + 4HCl \rightarrow HAuCl_4 + 3H^+ + 3e^-$$

The nitrogen reduction half-reaction is

$$HNO_3 + H^+ + e^- \rightarrow NO_2 + H_2O$$

The balanced equation is

$$Au + 3HNO_3 + 4HCl \rightarrow HAuCl_4 + 3NO_2 + 3H_2O$$

3.62 Use Problem 3.60 (j) as a hint.

Solution: Iron(II) compounds can be oxidized easily to iron(III) compounds. If the sample were tested with a small amount of a strongly colored oxidizing agent like $KMnO_4$ solution (purple), the loss of color would imply the presence of an oxidizable substance like an iron(II) salt

3.63 What are the maximum and minimum possible oxidation numbers for Cl, S, and N? (See Section 3.5.)

Solution: In a redox reaction the oxidizing agent gains one or more electrons. In doing so, the oxidation number of the element gaining the electrons must become more negative. In the case of chlorine the -1 oxidation number (chloride ion) is already the lowest possible state, so chloride ion cannot accept any more electrons and hydrochloric acid is not an oxidizing agent. Is this the case with nitric and sulfuric acids?

3.64 The method for balancing these equations is discussed in detail in Section 3.6 and the Problems 3.55 and 3.56.

Solution: Remember that redox reactions often cannot be balanced by inspection. Be sure to work through these problems with the *ion-electron* method.

(a) $2I^-(aq) + 2H^+(aq) + H_2O_2(aq) \rightarrow I_2(s) + 2H_2O(l)$ [The brown color results from the iodine.]

(b) $Cr_2O_7^{2-}(aq) + 3SO_3^{2-}(aq) + 8H^+(aq) \rightarrow 2Cr^{3+}(aq) + 3SO_4^{2-}(aq) + 4H_2O(l)$ [$Cr_2O_7^{2-}(aq)$ is orange and $Cr^{3+}(aq)$ is green.]

(c) $MnO_2(s) + 2Cl^-(aq) + 4H^+(aq) \rightarrow Mn^{2+}(aq) + Cl_2(g) + 2H_2O(l)$

3.66 Try for an equation with NO_2, O_2, and H_2O. Which elements change oxidation state?

Solution: Notice that nitrogen is in its highest possible oxidation state (+5) in nitric acid (see Figure 3.8), so it cannot be both oxidized and reduced. The balanced equation is

$$4HNO_3 \rightarrow 4NO_2 + O_2 + 2H_2O$$

Oxygen changes oxidation state from -2 to 0, and nitrogen changes from +5 to +4. The yellow color of "old" nitric acid is caused by the small amounts of NO_2. This process is accelerated by light.

3.68 With what do you usually associate the word "fluoride"?

Solution: The following solubilities are in g/100 g H_2O

LiF	0.27	CaF_2	0.0016	AgF	182
LiCl	63.7	$CaCl_2$	74.5	AgCl	1.8×10^{-3}
LiBr	145	$CaBr_2$	142	AgBr	1.6×10^{-4}
LiI	165	CaI_2	209	AgI	2.1×10^{-6}

The fluorides do not fit in with the other halides.

3.70 Carefully study the steps of the ion-electron method in Section 3.6. Review Example 3.8 before trying any parts of this problem.

Solution: Shortening the steps described in detail in Section 3.6 and Example 3.8 and Problem 3.55 above:

The two half-reactions are:

oxidation: $S_2O_3^{2-} \rightarrow SO_4^{2-}$
reduction: $Br_2 \rightarrow Br^-$

Using H_2O to balance oxygen and H^+ to balance hydrogen.

$$5H_2O + S_2O_3^{2-} \rightarrow 2SO_4^{2-} + 10H^+ + 8e^-$$
$$Br_2 + 2e^- \rightarrow 2Br^-$$

Multiplying the reduction half reaction times 4 so that both the oxidation and reduction half reactions involve the same number of electrons (namely 8), and simplying:

$$5H_2O + S_2O_3^{2-} + 4Br_2 \rightarrow 2SO_4^{2-} + 10H^+ + 8Br^-$$

3.71 If you don't remember this type of equation, review Hydrogen Displacement in Section 3.5.

Solution: The reaction is too violent, possibly resulting in an explosion.

The explosion would involve the oxidation of hydrogen by oxygen. What is the product?

3.72 These reactions are classic laboratory methods for producing the desired products.

Solution: (a) $2KClO_3(s) \rightarrow 2KCl(s) + 3O_2(g)$

(b) $NH_4Cl(aq) + NaOH(aq) \rightarrow NH_3(g) + NaCl(aq) + H_2O(l)$

(c) $2HCl(aq) + MgCO_3(s) \rightarrow MgCl_2(aq) + CO_2(g) + H_2O(l)$

(d) $Zn(s) + 2HCl(aq) \rightarrow ZnCl_2(aq) + H_2(g)$

(e) $CaCO_3(s) \rightarrow CaO(s) + CO_2(g)$

3.73 Review Section 3.4 on Acid-Base Reactions if necessary to work the problem.

Solution: (a) An acid and a base react to form water and a salt. Potassium iodide is a salt. Therefore the reaction of the proper acid and base should yield postassium iodide

$$KOH(aq) + HI(aq) \rightarrow KI(aq) + H_2O(l)$$

The water could be evaporated to isolate the KI.

(b) Recalling that acids react with carbonates to form carbon dioxide,

$$2HI(aq) + K_2CO_3(aq) \rightarrow 2KI(aq) + CO_2(g) + H_2O(l)$$

3.74 All three desired products are water insoluble (review the solubility rules in Section 3.3). Use this fact in formulating your answer. Are the suggestions for the preparations below the only method of synthesis?

Solution: (a) $MgCl_2(aq) + 2NaOH(aq) \rightarrow Mg(OH)_2(s) + 2NaCl(s)$

(b) $AgNO_3(aq) + NaI(aq) \rightarrow AgI(s) + NaNO_3(aq)$

(c) $3Ba(OH)_2(aq) + 2H_3PO_4(aq) \rightarrow Ba_3(PO_4)_2(s) + 6H_2O(l)$

3.75 What type of compound is copper sulfate? Why can it be viewed as a salt? See Problem 3.73 for review.

Solution: The proper acid and base will undergo a neutralization reaction to form copper sulfate

$$Cu(OH)_2(s) + H_2SO_4(aq) \rightarrow CuSO_4(aq) + 2H_2O(l)$$

When the water is evaporated, the compound that remains is not the white, anhydrous $CuSO_4$, but rather the bright-blue hydrated compound $CuSO_4{\cdot}5H_2O$. Heating the compound will drive off the water, resulting in the desired anhydrous copper sulfate.

Why is the copper hydroxide used as a solid rather than in solution. Review the solubility rules in Table 3.3.

3.76 It is very important to be able to deduce all of the information that is possible from a chemical equation, as it is written. However, be careful not to make any conclusions that aren't justified. Refer to Section 3.2.

Solution: (a) No information is given about whether heat is needed to start the reaction. (b) The coefficients in a balanced equation can be interpreted as moles. (c) The equation gives no information as to the rate or speed of reaction. (d) Correct. (e) The equation gives no information as to how the reaction occurs (the reaction mechanism, see Chapter 13), but rather only the quantities of reactants and products. (f) The balanced equation does allow you to deduce that 800 molecules of SO_2 will reaction with 400 molecules of O_2, because the stoichiometry is two molecules of SO_2 to one of O_2.

3.77 If there is no light, what can be deduced?

Solution: Apparently both of the original solutions were strong electrolytes; the mixture of the two solutions would be expected also to conduct electricity. However, since the bulb did not light, then the mixture must not contain ions. Indeed the H^+ from the sulfuric acid reacts with the OH^- from the barium hydroxide to form water. The barium cations react with the sulfate anions to form the insoluble barium sulfate. Barium sulfate is an electrolyte, but is insoluble.

$$2H^+(aq) + SO_4^{2-}(aq) + Ba^{2+}(aq) + 2OH^-(aq) \rightarrow 2H_2O(l) + BaSO_4(s)$$

3.78 If we are trying to neutralize an acid, what type of species would we need?

Solution: The hydrogen carbonate ion HCO_3^- is behaving as a Brønsted base to accept a proton.

$$HCO_3^- + H^+ \rightarrow H_2CO_3 \rightarrow H_2O + CO_2$$

The heat generated during the reaction of the hydrogen carbonate base with the spilled acid forces the carbonic acid H_2CO_3 that was formed to decompose to water and gaseous carbon dioxide.

3.80 Refer to Figure 3.15 to see what metals are more reactive than aluminum.

Solution: Using the activity series for metals in Figure 3.15, we conclude that we could use Mg, Na, Ca, Ba , K or Li to reduce Al^{3+} to Al.

Could any of these metals be used to reduce Al^{3+} in water?

3.81 Use Table 3.3 in the text that gives the solubility rules.

Solution: (a) A soluble sulfate salt such as sodium sulfate or sulfuric acid could be added. Barium sulfate would precipitate leaving the sodium ions in solution. (b) Potassium carbonate, phosphate, or sulfide could be added which would precipitate the magnesium cations, leaving the potassium cations in solution.

3.82 This problem is similar to Problem 3.81 above. Refer to the solubility rules in Table 3.3

Solution: (a) Add Ag^+ ions to precipitate AgBr, leaving the nitrate ions in solution. (b) Add a cation other than ammonium or a Group 1A cation to precipitate the phosphate (see rule 6 in Table 3.3); the nitrate will remain in solution. (c) Same as (b), which will precipitate the carbonate ion, leaving the nitrate ion in solution.

3.83 Review Section 3.6 in the text and to Problems 3.55 and 3.56 above for a detailed explanation of balancing redox equations.

Solution: The balanced equation is: $Cl_2 + SO_2 + 2H_2O \rightarrow 2Cl^- + SO_4^{2-} + 4H^+$

How do you know that this reaction is an oxidation-reduction reaction?

3.87 Refer to Figure 3.8 to see the range of oxidation numbers for carbon.

Solution: Using the rules for assigning oxidation numbers presented in Section 3.5, we find that the oxidation number for C in CO_2 is +4, which is the maximum oxidation that carbon can have. Thus it cannot be further oxidized and CO_2 will not burn. In contrast, the oxidation number for C in CO is +2. The carbon in CO can be further oxidized (to what?) and therefore will burn.

3.89 Refer to the Chemistry in Action on Metal From the Sea.

Solution: (a) The preciptation reaction is: $Mg^{2+}(aq) + 2OH^-(aq) \rightarrow Mg(OH)_2(s)$

The acid-base reaction is: $Mg(OH)_2(s) + 2HCl(aq) \rightarrow MgCl_2(aq) + 2H_2O(l)$

The redox reaction are: $Mg^{2+} + 2e^- \rightarrow Mg(s)$

$2Cl^- \rightarrow Cl_2(g) + 2e^-$

(b) NaOH is much more expensive than CaO.

(c) Dolomite has the advantage of being an additional source of magnesium that can be recovered.

.90 Refer to Section 3.5 for the rules on assigning oxidation numbers.

Solution: Using the rules for assigning oxidation numbers given in Section 3.5, H is +1, F is -1, so O must be zero.

3.93 Review Section 3.6 in the text and to Problems 3.55 and 3.56 above for a detailed explanation of balancing redox equations. This particular problem is unlike most in the chapter because the question asks for the total molecular equation to be balanced rather than simply the net ionic equation. First the equation is broken into the simplest ionic form involving the redox changes and the ionic form is then balanced. The total neutral equation is then balanced by inspection.

Solution: The unbalanced ionic equation is: $MnF_6^{2-} + SbF_5 \rightarrow SbF_6^- + MnF_3 + F_2$

An even simpler form of the ionic equation is: $Mn^{4+} + F^- \rightarrow Mn^{3+} + F_2$

The half reactions are:

oxidation: $F^- \rightarrow F_2$

reduction: $Mn^{4+} \rightarrow Mn^{3+}$

Balanced for mass and charge:

$e^- + Mn^{4+} \rightarrow Mn^{3+}$

$2F^- \rightarrow F_2 + 2e^-$

The simplified net ionic equation is: $2Mn^{4+} + 2F^- \rightarrow 2Mn^{3+} + F_2$

The remainder of the problem involves the reconstruction of the balanced redox ionic equation to one of neutral molecules, because that is what the question asked for. Changing from the ionic equation to a molecular equation is done by inspection because no further redox reactions are involved.

The balanced equation is: $2K_2MnF_6 + 4SbF_5 \rightarrow 4KSbF_6 + 2MnF_3 + F_2$

3.94 Metal oxides plus water usually produce basic solutions while nonmetal oxides react with water to produce acids.

Solution: Sodium oxide is a metallic oxide: $Na_2O + H_2O \rightarrow 2NaOH$

Barium oxide is a metallic oxide: $BaO + H_2O \rightarrow Ba(OH)_2$

Carbon dioxide is a nonmetallic oxide: $CO_2 + H_2O \rightarrow H_2CO_3$

Dinitrogen pentaoxide is a nonmetallic oxide: $N_2O_5 + H_2O \rightarrow 2HNO_3$

Phosphorus Pentaoxide is a nonmetallic oxide: $P_4O_{10} + 6H_2O \rightarrow 4H_3PO_4$

Sulfur trioxide is a nonmetallic oxide: $SO_3 + H_2O \rightarrow H_2SO_4$

Note: P_4O_{10} is usually written as P_2O_5 and is commonly called phosphorus pentaoxide.

3.95 The oxides of nitrogen are briefly discussed in Chapter 8. To work the problem, simply assume the oxidation number of oxygen remains as -2.

Solution: Oxidation number of +1, N_2O; oxidation number of +2, NO; oxidation number of +3, N_2O_3; oxidation number of +4, NO_2 and N_2O_4; oxidation number of +5, N_2O_5.

Note: If you have not already done so, please read the "Introduction for the Student" at the beginning of this Student Solutions Manual.

Chapter 4

CHEMICAL REACTIONS II: MASS RELATIONSHIPS

4.5 Example 4.1 is a model for this type of calculation. Read the introductory paragraphs of Section 4.1 and then try to work the example with the answer covered. If you can't, uncover the solution one line at a time until you see how to proceed and then finish the computation. If you had any trouble, study the answer carefully. Notice that stoichiometric equivalencies are used to set up unit factors so that the mechanics of this type of problem is just like a unit conversion using the factor-label method (Section 1.9).

Solution: The balanced equation is given in the problem. We see that 2 moles of CO produce 2 moles of CO_2. We can write

$$\text{moles } CO_2 \text{ produced} = 3.60 \text{ mol CO}\left(\frac{2 \text{ mol } CO_2}{2 \text{ mol CO}}\right) = 3.60 \text{ mol } CO_2$$

This problem involved the conversion of moles to moles. Part (b) of Example 4.1 was a moles-to-grams conversion. Do you see how unit factors can be used for any such type of conversion?

4.6 This problem is also very similar to Example 4.1 in the text and to Problem 4.5 above.

Solution: From the balanced equation, one mole of $SiCl_4$ forms from two moles of chlorine. From this we can construct a unit factor.

$$\text{moles of } Cl_2 \text{ used} = 0.507 \text{ mol } SiCl_4\left(\frac{2 \text{ mol } Cl_2}{1 \text{ mol } SiCl_4}\right) = 1.01 \text{ mol } Cl_2$$

4.7 This problem is like Problem 4.5. Assume the answer is to be in tons of sulfur. Do you need to convert to grams?

Solution: If the mass of sulfur is to be found in tons, you don't need to convert from tons to grams and then back to tons. If 32.0 g of sulfur forms 64.0 g of sulfur dioxide, how many tons of sulfur dioxide come from 32.0 tons of sulfur? The balanced equation shows that one mole of SO_2 results from one mole of S. If we use ton-moles

$$\text{tons of S} = 2.6 \times 10^7 \text{ tons } SO_2\left(\frac{1 \text{ ton-mol } SO_2}{64.07 \text{ ton } SO_2}\right)\left(\frac{1 \text{ ton-mol S}}{1 \text{ ton-mol } SO_2}\right)\left(\frac{32.07 \text{ ton S}}{1 \text{ ton-mol S}}\right)$$

$$= 1.3 \times 10^7 \text{ tons of S}$$

Can you define a ton-mole in terms of carbon-12? What would be the magnitude of Avogadro's number in the ton-mole system?

4.8 This is a grams-to-grams type of calculation (see comment on Problem 4.5). Examples 4.2 and 4.3 illustrate this type of problem. Go back and try to solve those examples with the answers covered. If you have trouble, uncover the solutions one line at a time until you see how to finish the computation. Cover the answers and work the examples again; this is an important type of problem in chemistry.

Solution: Molar mass $NaHCO_3$ = (22.99) + (1.008) + (12.01) + 3(16.00) = 84.01 g/mol
Molar mass CO_2 = (12.01) + 2(16.00) = 44.01 g/mol

(a)
$$2\ NaHCO_3 \rightarrow Na_2CO_3 + H_2O + CO_2$$

(b) The balanced equation shows one mole of CO_2 comes from two moles of $NaHCO_3$

$$\text{Mass } NaHCO_3 = 20.5 \text{ g } CO_2 \left(\frac{1 \text{ mol } CO_2}{44.01 \text{ g } CO_2}\right)\left(\frac{2 \text{ mol } NaHCO_3}{1 \text{ mol } CO_2}\right)\left(\frac{84.01 \text{ g } NaHCO_3}{1 \text{ mol } NaHCO_3}\right)$$

$$= 78.3 \text{ g } NaHCO_3$$

4.10 Study Example 4.2 in the text before you try to work this problem.

Solution: From the balanced equation two moles of ethanol form from one mole of glucose.

Molar mass of glucose $C_6H_{12}O_6$ = 6(12.01) + 12(1.008) + 6(16.00) = 180.16 g/mol
Molar mass of ethanol C_2H_5OH = 2(12.01) + 5(1.008) + 1(16.00) + 1(1.008) = 46.07 g/mol

$$500.4 \text{ g glucose}\left(\frac{1 \text{ mol glucose}}{180.16 \text{ g glucose}}\right)\left(\frac{2 \text{ mol ethanol}}{1 \text{ mol glucose}}\right)\left(\frac{46.07 \text{ g ethanol}}{1 \text{ mol ethanol}}\right) = 255.9 \text{ g ethanol}$$

Since Volume = $\frac{\text{mass}}{\text{density}}$, the volume of ethanol produced is

$$V = \frac{255.9 \text{ g}}{0.789 \text{ g/mL}} = 324 \text{ mL} = 0.324 \text{ L}$$

4.11 What is the mass of the water lost in this process? Convert it to moles.

Solution: The mass of water lost is just the difference between the initial and final masses

$$\text{Mass } H_2O \text{ lost} = 15.01 \text{ g} - 9.60 \text{ g} = 5.41 \text{ g}$$

$$\text{Moles of } H_2O = 5.41 \text{ g } H_2O\left(\frac{1 \text{ mol } H_2O}{18.02 \text{ g } H_2O}\right) = 0.300 \text{ mol } H_2O$$

4.12 Example 4.3 is similar to this problem in that we combine several steps into one. If you have difficulty in following how this problem is worked, you should go back to Example 4.1 in the text.

Solution: The balanced equation shows that eight moles of KCN are needed to combine with four moles of Au.

$$29.0 \text{ g Au}\left(\frac{1 \text{ mol Au}}{197.0 \text{ g Au}}\right)\left(\frac{8 \text{ mol KCN}}{4 \text{ mol Au}}\right) = 0.294 \text{ mol KCN}$$

Could you have solved the problem if it asked for the amount of KCN required in grams rather than moles?

4.13 If you have difficulty with this problem, review Example 4.3 in the text. Several unit factors can be combined into one step.

Solution: The balanced equation is: $CaCO_3(s) \rightarrow CaO(s) + CO_2(g)$

$$1.0 \text{ kg } CaCO_3\left(\frac{1000 \text{ g}}{1 \text{ kg}}\right)\left(\frac{1 \text{ mol } CaCO_3}{100.1 \text{ g } CaCO_3}\right)\left(\frac{1 \text{ mol CaO}}{1 \text{ mol } CaCO_3}\right)\left(\frac{56.08 \text{ g CaO}}{1 \text{ mol CaO}}\right) = 5.6 \times 10^2 \text{ g CaO}$$

4.14 This problem is a mole-to-gram conversion problem.

Solution: (a) The balanced equation is $NH_4NO_3(s) \rightarrow N_2O(g) + 2H_2O(g)$

(b) $$0.46 \text{ mol } NH_4NO_3\left(\frac{1 \text{ mol } N_2O}{1 \text{ mol } NH_4NO_3}\right)\left(\frac{44.02 \text{ g } N_2O}{1 \text{ mol } N_2O}\right) = 2.0 \times 10 \text{ g } N_2O$$

4.18 The first step in solving this problem is finding the limiting reagent. Examples 4.4 and 4.5 serve as models for this procedure. Digest the first few paragraphs of Section 4.2, then study the steps in the solution of Example 4.4. Notice how the identity of the limiting reagent is discovered; both steps must always be applied. Once the limiting reagent has been identified, the computation is just like Examples 4.4 and 4.5

Solution: We are given the number of moles of each reactant. The balanced equation shows that 2 moles of NO react with 1 mole of O_2. The number of moles of NO needed to react with 0.503 moles of O_2 is

$$0.503 \text{ mol } O_2 \times \left(\frac{2 \text{ mol NO}}{1 \text{ mol } O_2}\right) = 1.01 \text{ mol NO}$$

We have less than this amount of NO (we have only 0.886 mol of NO), so NO is the limiting reagent.

The balanced equation shows 2 moles of NO produces 2 moles of NO_2, so

$$\text{Moles } NO_2 = 0.886 \text{ mol NO} \times \left(\frac{2 \text{ mol } NO_2}{2 \text{ mol NO}}\right) = 0.886 \text{ mol } NO_2$$

Why can't we use 0.503 mol of O_2 (rather than the 0.886 mol of NO) in the calculation of the yield of NO_2 ?

4.19 This problem is another another example of a limiting reagent problem. One reactant is present in excess; we must base all of our calculations on the other reactant, the *limiting* reagent. Examples 4.4 and 4.5 and Problem 4.18 above are similar.

Solution: First we calculate the number of moles of O_3 and NO. The molar mass of O_3 = 48.00 g/mol; the molar mass of NO = 30.01 g/mol.

$$\text{moles of } O_3 = 0.740 \text{ g } O_3 \left(\frac{1 \text{ mol } O_3}{48.00 \text{ g } O_3}\right) = 0.0154 \text{ mol } O_3$$

$$\text{moles of NO} = 0.670 \text{ g NO} \left(\frac{1 \text{ mol NO}}{30.01 \text{ g NO}}\right) = 0.0223 \text{ mol NO}$$

From the balanced equation we see that one mole of O_3 combines with one mole of NO. The number of moles of NO needed to combine with 0.0154 mol O_3 is

$$0.0154 \text{ mol } O_3 \left(\frac{1 \text{ mol NO}}{1 \text{ mol } O_3}\right) = 0.0154 \text{ mol NO}$$

Since there is 0.0223 mol NO present, more than what is needed to react with 0.0154 mol O_3, NO must be the excess reagent and O_3 the limiting reagent. We write

$$0.740 \text{ g } O_3 \left(\frac{1 \text{ mol } O_3}{48.00 \text{ g } O_3}\right)\left(\frac{1 \text{ mol } NO_2}{1 \text{ mol } O_3}\right)\left(\frac{46.01 \text{ g } NO_2}{1 \text{ mol } NO_2}\right) = 0.709 \text{ g } NO_2$$

The number of moles of NO (excess reagent) remaining at the end of the reaction is

$$0.0223 \text{ mol} - 0.0154 \text{ mol} = 6.9 \times 10^{-3} \text{ mol.}$$

4.20 You are to assume that propane is the limiting reagent in this problem.

Solution: (a) The balanced equation is: $C_3H_8(g) + 5O_2(g) \rightarrow 3CO_2(g) + 4H_2O(l)$

(b) From the balanced equation one mole of propane produces three moles of carbon dioxide. The molar mass of CO_2 is (12.01) + 2(16.00) = 44.01 g/mol.

$$3.65 \text{ mol } C_3H_8 \left(\frac{3 \text{ mol } CO_2}{1 \text{ mol } C_3H_8}\right)\left(\frac{44.01 \text{ g } CO_2}{1 \text{ mol } CO_2}\right) = 482 \text{ g } CO_2$$

Why do we assume that the limiting reagent is propane rather than oxygen in this problem?

4.21 This problem is another limiting reagent problem. Refer to Problem 4.18 for more details.

Solution: As in all limiting reagent problems, we first must ascertain which reagent is limiting.

(a) We first convert the mass of HCl given to moles

$$48.2 \text{ g HCl} \left(\frac{1 \text{ mol HCl}}{36.46 \text{ g HCl}}\right) = 1.32 \text{ mol HCl}$$

Since one mole of MnO_2 reacts with four moles of HCl, the number of moles of HCl needed to combine with 0.86 mole of MnO_2 is

$$0.86 \text{ mol } MnO_2 \left(\frac{4 \text{ mol HCl}}{1 \text{ mol } MnO_2}\right) = 3.4 \text{ mol HCl}$$

Since we do not have that much HCl present, HCl must be the limiting reagent and MnO_2 the excess reagent.

(b) The mass of Cl_2 formed is then:

$$1.32 \text{ mol HCl} \left(\frac{1 \text{ mol } Cl_2}{4 \text{ mol HCl}}\right)\left(\frac{70.90 \text{ g } Cl_2}{1 \text{ mol } Cl_2}\right) = 23.4 \text{ g } Cl_2$$

4.24 Review Section 4.3 before you work this problem. It is similar to Example 4.6 in the text.

Solution: To work the problem, we first work a grams-to-grams conversion such as Example 4.2 to determine how many kg of HF could be formed from 6.00 kg of CaF_2, that is, the *theoretical yield.* We then compare that theoretical yield to the actual yield of 2.86 kg of HF.

The balanced equation is: $CaF_2 + H_2SO_4 \rightarrow CaSO_4 + 2HF$

From the balanced equation, one mole of CaF_2 produces two moles of HF. The molar mass of CaF_2 is 78.08 g/mol, and the molar mass of HF is 20.01 g/mol.

$$\text{kg of HF} = 6.00 \text{ kg } CaF_2 \left(\frac{1 \text{ mol } CaF_2}{78.08 \text{ g } CaF_2}\right)\left(\frac{2 \text{ mol HF}}{1 \text{ mol } CaF_2}\right)\left(\frac{20.01 \text{ g HF}}{1 \text{ mol HF}}\right) = 3.08 \text{ kg HF}$$

The theoretical yield, as just calculated, is 3.08 kg HF. The actual yield is 2.86 kg HF.

$$\% \text{ yield} = \left(\frac{\text{actual yield}}{\text{theoretical yield}}\right)100\% = \left(\frac{2.86 \text{ g}}{3.08 \text{ g}}\right)100\% = 92.9\ \%$$

4.25 This problem is very similar in structure to Problem 4.24 above and to Example 4.6 in the text.

Solution: We use NG as an abbreviation for nitroglycerine. The molar mass of NG = 227.1 g/mol.

$$4\ C_3H_5N_3O_9 \rightarrow 6N_2 + 12CO_2 + 10H_2O + O_2$$

(a) First, calculating the theoretical yield of O_2.

$$2.00 \times 10^2 \text{ g NG} \left(\frac{1 \text{ mol NG}}{227.1 \text{ g NG}}\right)\left(\frac{1 \text{ mol } O_2}{4 \text{ mol NG}}\right)\left(\frac{32.00 \text{ g } O_2}{1 \text{ mol } O_2}\right) = 7.05 \text{ g } O_2$$

(b)Then, calculating the % yield

$$\% \text{ yield} = \left(\frac{\text{actual yield } O_2}{\text{theoretical yield } O_2}\right) \times 100\% = \left(\frac{6.55 \text{ g } O_2}{7.05 \text{ g } O_2}\right) \times 100\% = 92.9\%$$

4.26 This problem is very similar to Problem 4.24.

Solution: To work the problem, we first work a grams-to-grams conversion to determine how many kg of TiO_2 could be formed from 8.00 x 10^3 kg of $FeTiO_3$, that is, the *theoretical yield.* We then compare that theoretical yield to the actual yield of 3.67 x 10^3 kg of HF.

From the balanced equation, one mole of $FeTiO_3$ produces one mole of TiO_2. The molar mass of $FeTiO_3$ is 151.7 g/mol, and the molar mass of TiO_2 is 79.88 g/mol.

$$\text{kg of } TiO_2 = 8.00 \text{ x } 10^3 \text{ kg } FeTiO_3 \left(\frac{1 \text{ mol } FeTiO_3}{151.7 \text{ g } FeTiO_3}\right)\left(\frac{1 \text{ mol } TiO_2}{1 \text{ mol } FeTiO_3}\right)\left(\frac{79.88 \text{ g } TiO_2}{1 \text{ mol } TiO_2}\right) = 4.21 \text{ x } 10^3 \text{ kg } TiO_2$$

The theoretical yield, as just calculated, is 4.21 x 10^3 kg TiO_2 . The actual yield is 3.67 x 10^3 kg TiO_2.

$$\% \text{ yield} = \left(\frac{\text{actual yield}}{\text{theoretical yield}}\right) \text{ x } 100\% = \left(\frac{3.67 \text{ x } 10^3 \text{ kg}}{4.21 \text{ x } 10^3 \text{ kg}}\right) \text{ x } 100\% = 87.2\%$$

4.29 Study the definition of molarity in Section 4.4. This problem is like Example 4.7. Cover the answer and try to solve it.

Solution: Notice how the ratio of moles of solute to volume of solution is used as a unit factor.

$$\text{Moles NaOH} = \left(\frac{2.80 \text{ mol NaOH}}{1.00 \times 10^3 \text{ mL NaOH soln}}\right) 5.00 \times 10^2 \text{ mL NaOH soln} = 1.40 \text{ mol NaOH}$$

The molar mass of NaOH is 40.00 g/mol.

$$\text{Mass NaOH} = 1.40 \text{ mol NaOH} \left(\frac{40.00 \text{ g NaOH}}{1 \text{ mol NaOH}}\right) = 56.0 \text{ g NaOH}$$

4.30 This problem is a direct application of the definition of molarity. If you have trouble, study the answer and work the example again, reviewing Examples 4.7 and 4.8 in the text, and Problem 4.29 above.

Solution: The molar mass of NaOH is 40.00 g/mol.

$$\text{Moles NaOH} = 5.25 \text{ g NaOH} \left(\frac{1 \text{ mol NaOH}}{40.00 \text{ g NaOH}}\right) = 0.131 \text{ mol NaOH}$$

$$\text{Molarity} = \left(\frac{0.131 \text{ mol NaOH}}{1.00 \text{ L soln}}\right) = 0.131 \text{ M}$$

4.31 This is a slight variation of Problems 4.29 and 4.30. Notice how the molarity ratio is employed as a unit factor.

Solution: Moles $MgCl_2$ = 60.0 mL $MgCl_2$ soln $\left(\frac{0.100 \text{ mol } MgCl_2}{1000 \text{ mL } MgCl_2 \text{ soln}}\right) = 6.00 \times 10^{-3}$ mol $MgCl_2$

4.32 A similar problem involving this direct molarity calculation is Problem 3.30.

Solution: The molar mass of H_3PO_4 is 97.99 g/mol

$$1.50 \times 10^2 \text{ g } H_3PO_4 \left(\frac{1 \text{ mol } H_3PO_4}{97.99 \text{ g } H_3PO_4}\right)\left(\frac{1000 \text{ mL soln}}{1 \text{ L soln}}\right)\left(\frac{1}{750 \text{ mL soln}}\right) = 2.04 \text{ mol/L} = 2.04 \text{ M}$$

4.33 This problem is similar to Example 4.8 in the text and Problem 4.29 above.

Solution: The unit factor of 56.11 g KOH = molar mass of KOH can be included in the chain multiplication.

$$35.0 \text{ mL} \left(\frac{1 \text{ L}}{1000 \text{ mL}}\right)\left(\frac{5.50 \text{ mol KOH}}{1 \text{ L}}\right)\left(\frac{56.11 \text{ g KOH}}{1 \text{ mol KOH}}\right) = 10.8 \text{ g KOH}$$

4.34 Chain multiplcation using unit factors is useful in these calculations. It is not necessary to first convert grams to moles in a separate step such as in Problem 4.30.

Solution: Molar mass of C_2H_5OH = 46.07 g/mol; molar mass of $C_{12}H_{22}O_{11}$ = 342.3 g/mol; molar mass of NaCl = 58.44 g/mol.

(a) $$29.0 \text{ g } C_2H_5OH \left(\frac{1 \text{ mol } C_2H_5OH}{46.07 \text{ g } C_2H_5OH}\right)\left(\frac{1000 \text{ mL}}{1 \text{ L}}\right)\left(\frac{1}{545 \text{ mL}}\right) = 1.16 \text{ mol/L} = 1.16 \text{ M}$$

(b) $$15.4 \text{ g } C_{12}H_{22}O_{11} \left(\frac{1 \text{ mol } C_{12}H_{22}O_{11}}{342.3 \text{ g } C_{12}H_{22}O_{11}}\right)\left(\frac{1000 \text{ mL}}{1 \text{ L}}\right)\left(\frac{1}{74.0 \text{ mL}}\right) = 0.608 \text{ mol/L} = 0.608 \text{ M}$$

(c) $$9.00 \text{ g NaCl} \left(\frac{1 \text{ mol NaCl}}{58.44 \text{ g NaCl}}\right)\left(\frac{1000 \text{ mL}}{1 \text{ L}}\right)\left(\frac{1}{86.4 \text{ mL}}\right) = 1.78 \text{ mol/L} = 1.78 \text{ M}$$

4.36 Molarity calculations are extremely common and important. It is necessary that you be able to do them easily and quickly.

Solution: (a) $$6.57 \text{ g } CH_3OH \left(\frac{1 \text{ mol } CH_3OH}{32.04 \text{ g } CH_3OH}\right)\left(\frac{1000 \text{ mL}}{1 \text{ L}}\right)\left(\frac{1}{150 \text{ mL}}\right) = 1.37 \text{ M}$$

(b) $10.4\text{ g } CaCl_2 \left(\frac{1\text{ mol } CaCl_2}{111.0\text{ g } CaCl_2}\right)\left(\frac{1000\text{ mL}}{1\text{ L}}\right)\left(\frac{1}{2.20 \times 10^2\text{ mL}}\right) = 0.426\text{ M}$

(c) $7.82\text{ g } C_{10}H_8 \left(\frac{1\text{ mol } C_{10}H_8}{128.2\text{ g } C_{10}H_8}\right)\left(\frac{1000\text{ mL}}{1\text{ L}}\right)\left(\frac{1}{85.2\text{ mL}}\right) = 0.716\text{ M}$

4.37 In this problem the molarity ratio can be used as a unit factor to find volume when the amount of solute and the molarity are known. You must convert grams to moles in each part. The molecular formula of ethanol is C_2H_6O.

Solution: (a) Molar mass NaCl = 58.44 g/mol.

$$2.14\text{ g NaCl}\left(\frac{1\text{ mol NaCl}}{58.44\text{ g NaCl}}\right)\left(\frac{1000\text{ mL soln}}{0.270\text{ mol NaCl}}\right) = 136\text{ mL soln}$$

(b) Molar mass C_2H_6O = 46.07 g/mol.

$$4.30\text{ g ethanol}\left(\frac{1\text{ mol ethanol}}{46.07\text{ g ethanol}}\right)\left(\frac{1000\text{ mL soln}}{1.50\text{ mol ethanol}}\right) = 62.2\text{ mL soln}$$

(c) Molar mass acetic acid = 60.05 g/mol.

$$0.85\text{ g acetic acid}\left(\frac{1\text{ mol acetic acid}}{60.05\text{ g acetic acid}}\right)\left(\frac{1000\text{ mL soln}}{0.30\text{ mol acetic acid}}\right) = 47\text{ mL soln}$$

4.38 These problems are very similar to Example 4.7 and Problems 4.29 and 4.33 above. Consult them if you need these problems worked out in more detail.

Solution: A 250 mL sample of 0.100 M solution contains 0.0250 mol of solute (How do we know this?). The computation in each case is the same:

(a) $0.0250\text{ mol CsI}\left(\frac{259.8\text{ g CsI}}{1\text{ mol CsI}}\right) = 6.50\text{ g CsI}$

(b) $0.0250\text{ mol } H_2SO_4\left(\frac{98.09\text{ g } H_2SO_4}{1\text{ mol } H_2SO_4}\right) = 2.45\text{ g } H_2SO_4$

(c) $0.0250\text{ mol } Na_2CO_3\left(\frac{106.0\text{ g } Na_2CO_3}{1\text{ mol } Na_2CO_3}\right) = 2.65\text{ g } Na_2CO_3$

(d) $0.0250\text{ mol } K_2Cr_2O_7\left(\frac{294.2\text{ g } K_2Cr_2O_7}{1\text{ mol } K_2Cr_2O_7}\right) = 7.36\text{ g } K_2Cr_2O_7$

(e) $0.0250\text{ mol } KMnO_4\left(\frac{158.0\text{ g } KMnO_4}{1\text{ mol } KMnO_4}\right) = 3.95\text{ g } KMnO_4$

4.39 Study the first few paragraphs of Section 4.4. What are the molarities of chloride and iodide ions in 1.0 M solutions of sodium chloride and magnesium iodide? Why are they different? The molar mass of chloride ion Cl^- is the same as that of chlorine Cl (to three significant figures).

Solution: Calcium chloride is an ionic compound. (Why?) There are two moles of Cl^- per mole of $CaCl_2$.

(a) $$\text{Moles } Cl^- = 25.3 \text{ g } CaCl_2\left(\frac{1 \text{ mol } CaCl_2}{111.0 \text{ g } CaCl_2}\right)\left(\frac{2 \text{ mol } Cl^-}{1 \text{ mol } CaCl_2}\right) = 0.456 \text{ mol } Cl^-$$

$$\text{Concentration of } Cl^- = \left(\frac{0.456 \text{ mol } Cl^-}{0.325 \text{ L soln}}\right) = 1.40 \text{ mol/L} = 1.40 \text{ M}$$

(b) $$\text{Grams } Cl^- = 0.100 \text{ L soln}\left(\frac{1.40 \text{ mol } Cl^-}{1.00 \text{ L soln}}\right)\left(\frac{35.45 \text{ g } Cl^-}{1 \text{ mol } Cl^-}\right) = 4.96 \text{ g } Cl^-$$

4.43 Study the subsection on Dilution of Solutions in Section 4.4. What quantity remains unchanged when more solvent is added to a solution in the dilution process? How does this allow calculation of the molarity of the new, more dilute solution? This problem is similar to Example 4.9. Try to work the example with the answer covered. If you can't do it, study the solution and rework the problem. Notice how Equation (4.4) is used. What does the M × V product represent in that equation? Why are the two products equal?

Solution:

$M_{initial} = 2.00$ M $M_{final} = 0.646$ M

$V_{initial} = ?$ $V_{final} = 1.00$ L

From Equation (4.4): $$V_{initial} = \frac{(0.646 \text{ M})(1.00 \text{ L})}{(2.00 \text{ M})} = 0.323 \text{ L}$$

Dilute 0.323 L (323 mL) of 2.00 M HCl solution to a final volume of 1.00 L.

Can the same principle be applied in problems in which a solution is concentrated by removing some solvent? Why or why not?

4.44 Problem 4.43 above offers a more detailed explanation of dilution calculations.

Solution: $$M_{final} = \frac{M_{initial} \times V_{initial}}{V_{final}} = \frac{0.866 \text{ M} \times 25.0 \text{ mL}}{500 \text{ mL}} = 0.0433 \text{ M}$$

4.46 What would be the volume of the 0.100 M solution?

Solution:

$M_{initial} = 0.125$ M $M_{final} = 0.100$ M

$V_{initial} = 505$ mL $V_{final} = ?$

From Equation (4.4): $$V_{final} = \frac{(0.125 \text{ M})(505 \text{ mL})}{(0.100 \text{ M})} = 631 \text{ mL}$$

Add (631 - 505) mL = 126 mL of water.

4.48 Find the number of moles of calcium nitrate in each solution and assume the volumes are additive (a good assumption). Will the result be between 0.568 M and 1.396 M?

Solution: Moles of calcium nitrate in the first solution

$$\left(\frac{0.568 \text{ mol}}{1000 \text{ mL soln}}\right) 46.2 \text{ mL soln} = 0.0262 \text{ mol}$$

Moles of calcium nitrate in the second solution

$$\left(\frac{1.396 \text{ mol}}{1000 \text{ mL soln}}\right) 80.5 \text{ mL soln} = 0.112 \text{ mol}$$

The volume of the combined solutions = 46.2 mL + 80.5 mL = 126.7 mL. The concentration of the final solution is

$$\frac{(0.0262 + 0.112) \text{ mol}}{0.1267 \text{ L}} = 1.09 \text{ M}$$

Could you find the molarities of all the ions if the two solutions contained different compounds? Could you find the molarities of the ions if the two solutions had different solvents? What about if the volumes were not additive? (This sometimes happens.)

4.51 This problem combines several of the concepts in this chapter. You must be able to find the number of moles of a solute in a solution given the volume and the molarity (Section 4.4). You also need to be able to find the limiting reagent in a reaction (Section 4.2). What is the limiting reagent in this problem? What is the concentration of chloride ion in 0.150 M calcium chloride solution?

Solution: The balanced equation is: $CaCl_2(aq) + 2AgNO_3(aq) \rightarrow Ca(NO_3)_2(aq) + 2AgCl(s)$

$$\text{Moles Ag}^+ = \left(\frac{0.100 \text{ mol Ag}^+}{1000 \text{ mL soln}}\right) 15.0 \text{ mL soln} = 0.00150 \text{ mol Ag}^+$$

$$\text{Moles Cl}^- = \left(\frac{0.150 \text{ mol CaCl}_2}{1000 \text{ mL soln}}\right)\left(\frac{2 \text{ mol Cl}^-}{1 \text{ mol CaCl}_2}\right) 30.0 \text{ mL soln} = 0.00900 \text{ mol Cl}^-$$

$AgNO_3$ is the limiting reagent. Only 0.00150 mol AgCl can form

$$1.50 \times 10^{-3} \text{ mol AgCl} \left(\frac{143.4 \text{ g AgCl}}{1 \text{ mol AgCl}}\right) = 0.215 \text{ g AgCl}$$

In reconsidering this problem, could you tell which was the limiting reagent just by a careful look at the volumes and concentrations?

4.52 Review Example 4.10 before you try this problem.

Solution: We assume the precipitation is quantitative, that is, that all of the barium in the sample has been precipitated as barium sulfate.

The mass percent of barium in barium sulfate is given by

$$\% \text{ Ba} = \left(\frac{137.3 \text{ g Ba}}{233.4 \text{ g BaSO}_4}\right) \times 100\% = 58.83\%$$

The mass of barium in 0.4105 g of barium sulfate precipitate is

$$0.4105 \text{ g precipitate}\left(\frac{58.83\%}{100.0\%}\right) = 0.2415 \text{ g Ba}$$

Alternatively, we could have combined the two steps above into a single step, as in Example 4.10.

$$0.4105 \text{ g BaSO}_4\left(\frac{1 \text{ mol BaSO}_4}{233.4 \text{ g BaSO}_4}\right)\left(\frac{1 \text{ mol Ba}}{1 \text{ mol BaSO}_4}\right)\left(\frac{137.3 \text{ g Ba}}{1 \text{ mol Ba}}\right) = 0.2415 \text{ g Ba}$$

We find the percent by mass of Ba in the original compound

$$\left(\frac{0.2415 \text{ g Ba}}{0.6760 \text{ g compound}}\right) \times 100\% = 35.72\%$$

4.53 "Practically all" should be interpreted as "all" in this problem. Few reactions actually convert 100% of the reactants to products. How many moles of silver ion are in the sample? How many moles of chloride ion are needed?

Solution: The net ionic equation is

$$Ag^+(aq) + Cl^-(aq) \rightarrow AgCl(s)$$

One mole of Cl^- is required per mole of Ag^+. We find the number of moles of Ag^+

$$\text{Moles Ag}^+ = \left(\frac{0.0113 \text{ mol Ag}^+}{1000 \text{ mL soln}}\right) 2.50 \times 10^2 \text{ mL soln} = 2.83 \times 10^{-3} \text{ mol Ag}^+$$

$$\text{Mass NaCl} = 2.83 \times 10^{-3} \text{ mol Ag}^+\left(\frac{1 \text{ mol Cl}^-}{1 \text{ mol Ag}^+}\right)\left(\frac{1 \text{ mol NaCl}}{1 \text{ mol Cl}^-}\right)\left(\frac{58.44 \text{ g NaCl}}{1 \text{ mol NaCl}}\right) = 0.165 \text{ g NaCl}$$

4.54 This problem is similar to Example 4.11 in the textbook.

Solution: The net ionic equation is: $Cu^{2+}(aq) + S^{2-}(aq) \rightarrow CuS(s)$

The answer sought is the molar concentration of Cu^{2+}, that is, moles of Cu^{2+} ions per liter of solution. The factor-label method is used to convert, in order: g of CuS $\rightarrow$ moles CuS $\rightarrow$ moles Cu^{2+} $\rightarrow$ moles Cu^{2+} per liter soln.

$$[Cu^{2+}] = 0.0177 \text{ g CuS} \left(\frac{1 \text{ mol CuS}}{95.62 \text{ g CuS}}\right)\left(\frac{1 \text{ mol } Cu^{2+}}{1 \text{ mol CuS}}\right)\left(\frac{1}{0.800 \text{ L}}\right) = 2.31 \times 10^{-4} \text{ M}$$

4.58 In any acid-base titration one hydrogen ion always neutralizes one hydroxide ion. The only tricky thing to watch for is the presence of di- or triprotic acids or bases. Try to work Examples 4.12 and 4.13 with the answers covered. If you have trouble, uncover and study the answers. Pay special attention to the differences in the ways the monoprotic acid (HCl) and the diprotic acid (H_2SO_4) are handled with the stoichiometric equivalencies in these examples. Parts (a), (b), and (c) in this problem deal with mono-, di-, and triprotic acids, respectively.

Solution: The unit factors for the mono-, di-, and triprotic acids are set in boldface in each case. Study them carefully.

(a) $$\text{Moles HCl} = 25.00 \text{ mL soln} \left(\frac{2.430 \text{ mol HCl}}{1000 \text{ mL soln}}\right) = 6.075 \times 10^{-2} \text{ mol HCl}$$

$$\text{Volume NaOH soln} = 6.075 \times 10^{-2} \text{ mol HCl} \left(\frac{\mathbf{1\ mol\ NaOH}}{\mathbf{1\ mol\ HCl}}\right)\left(\frac{1000 \text{ mL soln}}{1.420 \text{ mol NaOH}}\right) = 42.78 \text{ mL soln}$$

(b) $$\text{Moles } H_2SO_4 = 25.00 \text{ mL soln} \left(\frac{4.500 \text{ mol } H_2SO_4}{1000 \text{ mL soln}}\right) = 0.1125 \text{ mol } H_2SO_4$$

$$\text{Volume NaOH soln} = 0.1125 \text{ mol } H_2SO_4 \left(\frac{\mathbf{2\ mol\ NaOH}}{\mathbf{1\ mol\ H_2SO_4}}\right)\left(\frac{1000 \text{ mL soln}}{1.420 \text{ mol NaOH}}\right) = 158.5 \text{ mL soln}$$

(c) $$\text{Moles } H_3PO_4 = 25.00 \text{ mL soln} \left(\frac{1.500 \text{ mol } H_3PO_4}{1000 \text{ mL soln}}\right) = 3.750 \times 10^{-2} \text{ mol } H_3PO_4$$

Volume NaOH soln

$$= 3.750 \times 10^{-2} \text{ mol } H_3PO_4 \left(\frac{\mathbf{3\ mol\ NaOH}}{\mathbf{1\ mol\ H_3PO_4}}\right)\left(\frac{1000 \text{ mL soln}}{1.420 \text{ mol NaOH}}\right) = 79.23 \text{ mL soln}$$

4.59 This problem is a straightforward example of a titration problem. You should always write a balanced equation for the neutralization reaction to ascertain how many acid formula units are required relative to base formula units. Review Examples 4.12 and 4.13 if you have trouble with this problem.

Solution: The neutralization reaction is

$$CH_3COOH(aq) + NaOH(aq) \rightarrow CH_3COONa(aq) + H_2O(l)$$

The balanced equation shows that one mole of acetic acid combines with one mole of sodium hydroxide. We find the number of moles of NaOH in the 5.75 mL sample

$$5.75 \text{ mL} \left(\frac{1 \text{ L}}{1000 \text{ mL}}\right)\left(\frac{1.00 \text{ mol NaOH}}{1 \text{ L soln}}\right) = 5.75 \times 10^{-3} \text{ mol NaOH}$$

$$M = \frac{\text{moles solute}}{\text{liters solution}} = 5.75 \times 10^{-3} \text{ mol} \left(\frac{1}{50.0 \text{ mL soln}}\right)\left(\frac{1000 \text{ mL soln}}{1 \text{ L soln}}\right) = 0.115 \text{ mol/L} = 0.115 \text{ M}$$

4.60 Refer to Problem 4.58 above and Example 4.13 in the text if you need help.

Solution: (a) The neutralization reaction is: $HCl(aq) + KOH(aq) \rightarrow KCl(aq) + H_2O(l)$

We see that one mole of KOH neutralizes one mole of HCl. The amount of HCl is:

$$10.0 \text{ mL}\left(\frac{1 \text{ L}}{1000 \text{ mL}}\right)\left(\frac{0.30 \text{ mol HCl}}{1 \text{ L}}\right) = 3.0 \times 10^{-3} \text{ mol HCl}$$

$$\text{Volume} = 3.0 \times 10^{-3} \text{ mol HCl}\left(\frac{1 \text{ mol KOH}}{1 \text{ mol HCl}}\right)\left(\frac{1 \text{ L soln}}{0.50 \text{ mol KOH}}\right) = 6.0 \times 10^{-3} \text{ L} = 6.0 \text{ mL}$$

(b) The neutralization reaction is: $H_2SO_4(aq) + 2KOH(aq) \rightarrow K_2SO_4(aq) + H_2O(l)$

Two moles of KOH neutralize ome mole of H_2SO_4. The amount of H_2SO_4 is

$$10.0 \text{ mL}\left(\frac{1 \text{ L}}{1000 \text{ mL}}\right)\left(\frac{0.20 \text{ mol } H_2SO_4}{1 \text{ L}}\right) = 2.0 \times 10^{-3} \text{ mol } H_2SO_4$$

$$\text{Volume} = 2.0 \times 10^{-3} \text{ mol } H_2SO_4\left(\frac{2 \text{ mol KOH}}{1 \text{ mol } H_2SO_4}\right)\left(\frac{1 \text{ L soln}}{0.50 \text{ mol KOH}}\right)$$

$$= 8.0 \times 10^{-3} \text{ L} = 8.0 \text{ mL}$$

(c) The neutralization reaction is: $H_3PO_4(aq) + 3KOH(aq) \rightarrow K_3PO_4(aq) + 3H_2O(l)$

Three moles of KOH neutralize one mole of H_3PO_4. The amount of H_3PO_4 is

$$15.0 \text{ mL}\left(\frac{1 \text{ L}}{1000 \text{ mL}}\right)\left(\frac{0.25 \text{ mol } H_3PO_4}{1 \text{ L}}\right) = 3.8 \times 10^{-3} \text{ mol } H_3PO_4$$

$$\text{Volume} = 3.8 \times 10^{-3} \text{ mol } H_3PO_4\left(\frac{3 \text{ mol KOH}}{1 \text{ mol } H_3PO_4}\right)\left(\frac{1 \text{ L soln}}{0.50 \text{ mol KOH}}\right) = 0.023 \text{ L} = 23 \text{ mL}$$

4.62 If you are unsure of how to balance a redox equation, study Section 3.6 Once the equation is balanced, this problem is worked like Example 4.14.

Solution: The balanced equation is (see Section 3.6):

$$Cr_2O_7^{2-} + 6Fe^{2+} + 14H^+ \rightarrow 2Cr^{3+} + 6Fe^{3+} + 7H_2O$$

The number of moles of potassium dichromate in 26.0 mL of the solution is

$$26.0 \text{ mL}\left(\frac{1 \text{ L}}{1000 \text{ mL}}\right)\left(\frac{0.0250 \text{ mol}}{1 \text{ L}}\right) = 6.50 \times 10^{-4} \text{ mol}$$

From the balanced equation it can be seen that in this particular reaction 1 mole of dichromate is stoichiometrically equivalent to 6 moles of iron(II). The number of moles of iron(II) oxidized is therefore

$$6.50 \times 10^{-4} \text{ mol } Cr_2O_7^{2-}\left(\frac{6 \text{ mol } Fe^{2+}}{1 \text{ mol } Cr_2O_7^{2-}}\right) = 3.90 \times 10^{-3} \text{ mol } Fe^{2+}$$

The molar concentration of iron(II) is

$$3.90 \times 10^{-3}\ \text{mol}\left(\frac{1}{25.0\ \text{mL}}\right)\left(\frac{1000\ \text{mL}}{1\ \text{L}}\right) = 0.156\ \text{mol/L} = 0.156\ \text{M}$$

4.64 If you are unsure of how to balance a redox equation, study Section 3.6 Once the equation is balanced, this problem is worked like Example 4.14.

Solution: The balanced equation is (see Section 3.6): $MnO_4^- + 5Fe^{2+} + 8H^+ \rightarrow Mn^{2+} + 5Fe^{3+} + 4H_2O$

$$\text{mass Fe} = 23.30\ \text{mL KMnO}_4\left(\frac{0.0194\ \text{mol KMnO}_4}{1000\ \text{mL KMnO}_4}\right)\left(\frac{5\ \text{mol Fe}}{1\ \text{mol KMnO}_4}\right)\left(\frac{55.85\ \text{g Fe}}{1\ \text{mol Fe}}\right)$$

$$= 0.126\ \text{g Fe}$$

$$\%\text{Fe} = \frac{0.126\ \text{g}}{0.2792\ \text{g}} \times 100\% = 45.1\ \%$$

4.65 This problem is similar to Problem 4.62 above and to Example 4.14 in the text.

Solution: (a) The balanced equation is (see Section 3.6):

$$5H_2O_2 + 2MnO_4^- + 6H^+ \rightarrow 2Mn^{2+} + 5O_2 + 8H_2O$$

(b) The number of moles of potassium permanganate in 36.44 mL of the solution is

$$36.44\ \text{mL}\left(\frac{1\ \text{L}}{1000\ \text{mL}}\right)\left(\frac{0.01652\ \text{mol}}{1\ \text{L}}\right) = 6.020 \times 10^{-4}\ \text{mol of KMnO}_4$$

From the balanced equation it can be seen that in this particular reaction 2 mole of permanganage is stoichiometrically equivalent to 5 moles of hydrogen peroxide The number of moles of H_2O_2 oxidized is therefore

$$6.020 \times 10^{-4}\ \text{mol MnO}_4^-\left(\frac{5\ \text{mol H}_2\text{O}_2}{2\ \text{mol MnO}_4^-}\right) = 1.505 \times 10^{-3}\ \text{mol H}_2\text{O}_2$$

The molar concentration of H_2O_2 is

$$[\text{H}_2\text{O}_2] = 1.505 \times 10^{-3}\ \text{mol}\left(\frac{1}{25.0\ \text{mL}}\right)\left(\frac{1000\ \text{mL}}{1\ \text{L}}\right) = 0.0602\ \text{mol/L} = 0.0602\ \text{M}$$

4.66 How many moles of electrons are lost by the sulfite ion? How many moles of potassium iodate are present? Compare these numbers.

Solution: The balanced half-reaction for the oxidation of sulfite to sulfate in acid solution is

$$SO_3^{2-} + H_2O \rightarrow SO_4^{2-} + 2H^+ + 2e^-$$

The number of moles of sulfite ion is

$$32.5 \text{ mL} \left(\frac{1 \text{ L}}{1000 \text{ mL}}\right)\left(\frac{0.500 \text{ mol}}{1 \text{ L}}\right) = 0.0163 \text{ mol}$$

The number of moles of electrons lost by the sulfite is

$$0.0163 \text{ mol } SO_3^{2-} \left(\frac{2 \text{ mol } e^-}{1 \text{ mol } SO_3^{2-}}\right) = 0.0326 \text{ mol } e^-$$

The number of moles of potassium iodate is

$$1.390 \text{ g} \left(\frac{1 \text{ mol } KIO_3}{214.0 \text{ g}}\right) = 0.006495 \text{ mol } KIO_3$$

The number of moles of electrons transferred per mole of iodate is

$$\frac{0.0326 \text{ mol } e^-}{0.006495 \text{ mol } KIO_3} = 5.02 \text{ mol}$$

Since the oxidation number of iodine in iodate is +5 and five moles of electrons are transferred per mole of iodate, the final oxidation number of iodine is +5 - 5 = 0. The iodine-containing product of the reaction is most probably elemental iodine, I_2.

4.68 Refer to Problem 4.58 for help.

Solution: Using the balanced equation

$$\text{moles of } K_2Cr_2O_7 = 25.00 \text{ mL } Na_2SO_3 \left(\frac{0.3143 \text{ mol } Na_2SO_3}{1000 \text{ mL } Na_2SO_3}\right)\left(\frac{1 \text{ mol } K_2Cr_2O_7}{3 \text{ mol } Na_2SO_3}\right) = 0.002619 \text{ mol } K_2Cr_2O_7$$

$$\text{Molarity of } K_2Cr_2O_7 = \left(\frac{0.002619 \text{ mol } K_2Cr_2O_7}{0.02842 \text{ L } K_2Cr_2O_7}\right) = 0.09216 \text{ M}$$

4.69 Example 4.14 uses the balanced chemical equations that apply to this problem. The first titration gives the amount of iron(II) in the mixture. The second titration gives the total amount of iron. The reduction by zinc metal is done upon the original solution, not the solution remaining after the first $KMnO_4$ reaction.

Solution: We use the balanced equation from Example 4.14 to compute the amount of iron(II) in the original solution

$$23.0 \text{ mL} \left(\frac{1 \text{ L}}{1000 \text{ mL}}\right)\left(\frac{0.0200 \text{ mol } KMnO_4}{1 \text{ L}}\right)\left(\frac{5 \text{ mol } Fe^{2+}}{1 \text{ mol } KMnO_4}\right) = 0.00230 \text{ mol } Fe^{2+}$$

The concentration of iron(II) must be

$$[Fe^{2+}] = \left(\frac{0.00230 \text{ mol}}{0.0250 \text{ L}}\right) = 0.0920 \text{ mol/L}$$

The total iron concentration can be found by simple proportion because the same sample volume (25.0 mL) and the same $KMnO_4$ solution were used

$$[Fe]_{total} = \frac{40.0 \text{ mL } KMnO_4}{23.0 \text{ mL } KMnO_4} \times 0.0920 \text{ M} = 0.160 \text{ M}$$

$$[Fe^{3+}] = [Fe]_{total} - [Fe^{2+}] = 0.0680 \text{ M}$$

Why are two titrations with permanganate necessary in this problem?

4.70 The relevant equation is given in Problem 4.67.

Solution: The balanced equation is: $2MnO_4^- + 5C_2O_4^{2-} + 16H^+ \rightarrow 2Mn^{2+} + 10CO_2 + 8H_2O$
Therefore, 2 mol MnO_4^- is equivalent to 5 mol $C_2O_4^{2-}$

$$\text{moles of } MnO_4^- \text{ reacted} = \left(\frac{9.56 \text{ x } 10^{-4} \text{ mol } MnO_4^-}{1 \text{ L}}\right)\left(\frac{1 \text{ L}}{1000\text{mL}}\right) \text{ x } (24.2 \text{ mL}) = 2.31 \text{ x } 10^{-5} \text{ mol } MnO_4^-$$

Since 1 mol of Ca^{2+} is equivalent of 1 mol of $C_2O_4^{2-}$, then the mass of Ca^{2+} in 10.0 mL is

$$(2.31 \text{ x } 10^{-5} \text{ mol } MnO_4^-) \text{ x } \left(\frac{5 \text{ mol } C_2O_4^{2-}}{2 \text{ mol } MnO_4^-}\right)\left(\frac{40.08 \text{ g Ca}}{1 \text{ mol Ca}}\right) = 2.31 \text{ x } 10^{-3} \text{ g}$$

$$\left(\frac{2.31 \text{ x } 10^{-3} \text{ g}}{10.0 \text{ mL}}\right)\left(\frac{1000 \text{ mg}}{1 \text{ g}}\right) = 0.231 \text{ mg/mL Blood}$$

4.71 What makes a solution of an electrolyte conduct electric current? Which of the compounds (a)-(d) are strong electrolytes? Which, if any, are weak electrolytes? What is the total concentration of all the ions present in (a)-(d)? See Section 3.2 and Table 3.2.

Solution: Choice (d), the 0.20 M $Mg(NO_3)_2$, should be the best conductor; the total ion concentration is 0.60 M. The total ion concentrations for (a) and (c) are 0.40 M and 0.50 M, respectively (why?). For (b), acetic acid is a weak electrolyte (see Section 3.2).

4.72 In this problem you are asked to find an empirical formula (see Section 2.4). What is the mass of the oxygen in the original metal oxide? You know the molar masses of the metal and of oxygen; find the empirical formula.

Solution: The mass of oxygen in the oxide must be 2.40 g - 1.68 g = 0.72 g oxygen. Next find the number of moles of the metal and of the oxygen

$$\text{Moles X} = 1.68 \text{ g} \left(\frac{1 \text{ mol}}{55.9 \text{ g}}\right) = 0.0301 \text{ mol}$$

$$\text{Moles O} = 0.72 \text{ g} \left(\frac{1 \text{ mol}}{16.0 \text{ g}}\right) = 0.0450 \text{ mol}$$

The empirical formula is $X_{0.0301}O_{0.0450}$. Dividing by 0.0301 gives $X_{1.00}O_{1.50}$ or X_2O_3. The balanced equation is

$$X_2O_3(s) + 3CO(g) \rightarrow 2X(s) + 3CO_2(g)$$

4.73 What is the theoretical yield of hydrogen in this problem (Section 4.3)? What is the actual yield?

Solution: (a) $Zn(s) + H_2SO_4(aq) \rightarrow ZnSO_4(aq) + H_2(g)$

(b) From the balanced equation we see 1 mol H_2 is produced by 1 mol Zn. The theoretical yield is

$$\text{Mass } H_2 = 3.86 \text{ g Zn} \left(\frac{1 \text{ mol Zn}}{65.39 \text{ g Zn}}\right)\left(\frac{1 \text{ mol } H_2}{1 \text{ mol Zn}}\right)\left(\frac{2.016 \text{ g } H_2}{1 \text{ mol } H_2}\right) = 0.119 \text{ g } H_2$$

$$\text{The percent purity} = \left(\frac{0.0764 \text{ g } H_2}{0.119 \text{ g } H_2}\right) \times 100\% = 64.2\%$$

(c) We assume that the impurities are inert and do not react with the sulfuric acid to produce hydrogen. Could you solve the problem if the impurities also produced hydrogen?

4.74 Compare this problem to Problem 4.64. The two problems are very similar, even though one is a titration problem and the other is an (indirect) percent yield problem.

Solution: The wording of the problem suggests that the actual yield is less than the theoretical yield. The percent yield will be equal to the percent purity of the iron(III) oxide. We find the theoretical yield:

$$2.62 \times 10^3 \text{ kg } Fe_2O_3 \left(\frac{1 \text{ mol } Fe_2O_3}{159.7 \text{ g } Fe_2O_3}\right)\left(\frac{2 \text{ mol Fe}}{1 \text{ mol } Fe_2O_3}\right)\left(\frac{55.85 \text{ g Fe}}{1 \text{ mol Fe}}\right) = 1.83 \times 10^3 \text{ kg Fe}$$

$$\frac{\text{actual yield}}{\text{theoretical yield}} \times 100\% = \text{percent yield} = \frac{1.64 \times 10^3 \text{ kg Fe}}{1.83 \times 10^3 \text{ kg Fe}} \times 100\% = 89.6\ \% = \text{purity of } Fe_2O_3$$

4.75 How many moles of Ti and of Cl were in the unknown sample?

Solution: First find the percent by mass of Ti in TiO_2 and of Cl in AgCl

$$\%\text{Ti} = \left(\frac{\text{molar mass Ti}}{\text{molar mass } TiO_2}\right) \times 100\% = \left(\frac{47.88 \text{ g/mol}}{79.88 \text{ g/mol}}\right) \times 100\% = 59.94\%$$

$$\%Cl = \left(\frac{\text{molar mass Cl}}{\text{molar mass AgCl}}\right) \times 100\% = \left(\frac{35.45\ g/mol}{143.4\ g/mol}\right) \times 100\% = 24.72\%$$

Next find the masses of Ti and Cl in the TiO_2 and AgCl samples

$$\text{Mass Ti} = 0.777\ g \times 0.5994 = 0.466\ g\ Ti$$
$$\text{Mass Cl} = 5.575\ g \times 0.2472 = 1.378\ g\ Cl$$

The number of moles of Ti and Cl are

$$\text{Moles Ti} = 0.466\ g\ Ti \left(\frac{1\ mol\ Ti}{47.88\ g\ Ti}\right) = 0.00973\ mol\ Ti$$
$$\text{Moles Cl} = 1.378\ g\ Cl \left(\frac{1\ mol\ Cl}{35.45\ g\ Cl}\right) = 0.0389\ mol\ Cl$$

The empirical formula is $Ti_{0.00973}Cl_{0.0389}$. Dividing by 0.00973 gives $TiCl_4$.

4.76 Review Section 2.4 on empirical formulas if you need to.

Solution: We assume the increase in mass results from the element nitrogen. The mass of nitrogen is (0.378 g - 0.273 g) = 0.105 g. We find the number of moles of Mg and N:

$$\text{Mg:} \quad 0.273\ g\ Mg \left(\frac{1\ mol\ Mg}{24.31\ g\ Mg}\right) = 0.0112\ mol\ Mg$$

$$\text{N:} \quad 0.105\ g\ N \left(\frac{1\ mol\ N}{14.01\ g\ N}\right) = 0.00749\ mol\ N$$

The empirical formula is $Mg_{0.0112}N_{0.00749}$. Dividing by 0.00749 gives $Mg_{1.5}N$ or Mg_3N_2.

4.78 Hint: What is the mass of sulfur in the daily mass of coal? Write out the balanced equation for the reaction of calcium oxide with sulfur dioxide.

Solution: The equations for the combustion of sulfur and the reaction of SO_2 with CaO are:

$$S(s) + O_2(g) \rightarrow SO_2(g) \qquad SO_2(g) + CaO(s) \rightarrow CaSO_3(s)$$

We find the sulfur present in the daily coal consumption:

$$6.60 \times 10^6\ kg\ coal \left(\frac{1.6\%\ S}{100\%}\right) = 1.06 \times 10^5\ kg\ S$$

The daily amount of CaO is then

$$1.06 \times 10^5\ kg\ S \left(\frac{1\ mol\ S}{32.07\ g\ S}\right)\left(\frac{1\ mol\ SO_2}{1\ mol\ S}\right)\left(\frac{1\ mol\ CaO}{1\ mol\ SO_2}\right)\left(\frac{56.08\ g\ CaO}{1\ mol\ CaO}\right) = 1.85 \times 10^5\ kg\ CaO$$

4.79 A discussion of the reaction of metals with hydrogen ion can be found at the end of the subsection on Metal Displacement in Section 3.5. How much hydrogen ion is left after all the magnesium is gone?

Solution: The balanced equation is

$$Mg(s) + 2H^+(aq) \rightarrow Mg^{2+}(aq) + H_2(g)$$

We see that 1 mol Mg combines with 2 mol H^+. We find the number of moles of H^+ that combined with Mg

$$\text{Moles } H^+ = 4.47 \text{ g Mg}\left(\frac{1 \text{ mol Mg}}{24.31 \text{ g Mg}}\right)\left(\frac{2 \text{ mol } H^+}{1 \text{ mol Mg}}\right) = 0.368 \text{ mol } H^+$$

The number of moles of H^+ in the original solution is

$$\text{Moles } H^+ = \left(\frac{2.00 \text{ mol HCl}}{1000 \text{ mL soln}}\right)\left(\frac{1 \text{ mol } H^+}{1 \text{ mol HCl}}\right) 5.00 \times 10^2 \text{ mL soln} = 1.00 \text{ mol } H^+$$

The number of moles of H^+ remaining is the difference

$$\text{Moles } H^+ = 1.00 \text{ mol} - 0.37 \text{ mol} = 0.63 \text{ mol } H^+$$

$$[H^+] = \frac{0.63 \text{ mol } H^+}{5.00 \times 10^2 \text{ mL}} \times \left(\frac{1000 \text{ mL}}{1 \text{ L}}\right) = 1.3 \text{ mol/L} = 1.3 \text{ M}$$

What is the limiting reagent in this problem? What is the concentration of magnesium ion?

4.80 This problem is actually just an empirical formula calculation (see Section 2.4). If heating the first manganese compound produces a new manganese compound plus oxygen, would you expect the new manganese compound to have a larger or smaller percent of manganese?

Solution: (a) For compound X:

Mn: $63.3 \text{ g Mn}\left(\frac{1 \text{ mol Mn}}{54.94 \text{ g Mn}}\right) = 1.15 \text{ mol Mn}$

O: $36.7 \text{ g O}\left(\frac{1 \text{ mol O}}{16.00 \text{ g O}}\right) = 2.29 \text{ mol O}$

The empirical formula of X is $Mn_{1.15}O_{2.29}$. Dividing by 1.15 gives MnO_2.

For compound Y:

Mn: $72.0 \text{ g Mn}\left(\frac{1 \text{ mol Mn}}{54.94 \text{ g Mn}}\right) = 1.31 \text{ mol M}$

O: $28.0 \text{ g O}\left(\frac{1 \text{ mol O}}{16.00 \text{ g O}}\right) = 1.75 \text{ mol O}$

The empirical formula is $Mn_{1.31}O_{1.75}$. Dividing by 1.31 gives $MnO_{1.33}$ or Mn_3O_4.

(b) The balanced equation is: $3MnO_2 \rightarrow Mn_3O_4 + O_2$

4.82 This is a mass-to-mass conversion problem. Problem 4.8 is similar.

Solution: (a) The balanced equation is: $C_3H_8(g) + 3H_2O(g) \rightarrow 3CO(g) + 7H_2(g)$

(b) $$2.84 \times 10^3 \text{ kg } C_3H_8 \left(\frac{1 \text{ mol } C_3H_8}{44.09 \text{ g } C_3H_8}\right)\left(\frac{7 \text{ mol } H_2}{1 \text{ mol } C_3H_8}\right)\left(\frac{2.016 \text{ g } H_2}{1 \text{ mol } H_2}\right) = 9.09 \times 10^2 \text{ kg } H_2$$

4.84 This is a a challenging problem. First, the total moles of NaCl plus KCl can be determined from the amount of AgCl formed. Knowing both the total moles of NaCl plus KCl and the total mass of NaCl plus KCl allows you to calculate the mass of each NaCl and KCl.

Solution: In this problem the relative amounts of NaCl and KCl are not known. However, the total amount of Cl can be determined from the AgCl data. We find the number of moles of AgCl

$$1.913 \text{ g AgCl} \left(\frac{1 \text{ mol AgCl}}{143.4 \text{ g AgCl}}\right) = 0.01334 \text{ mol AgCl}$$

The sum of the number of moles of NaCl + KCl must equal 0.01334 mole; furthermore, the sum of the NaCl + KCl masses must equal 0.8870 g. We let x be the number of moles of NaCl. The number of moles of KCl is then 0.01334 - x. We write an equation for the mass of the mixture

$$x \text{ moles NaCl}\left(\frac{58.44 \text{ g NaCl}}{1 \text{ mol NaCl}}\right) + (0.01334 - x) \text{ moles KCl} \left(\frac{74.55 \text{ g KCl}}{1 \text{ mol KCl}}\right) = 0.8870 \text{ g}$$

Solving gives $x = 6.673 \times 10^{-3}$ mol NaCl. Therefore $(0.01334 - 0.006673) = 6.667 \times 10^{-3}$ mol KCl

NaCl: $6.673 \times 10^{-3} \text{ mol NaCl} \left(\frac{58.44 \text{ g NaCl}}{1 \text{ mol NaCl}}\right) = 0.3900 \text{ g NaCl}$

KCl: Total mass - mass NaCl = 0.8870 g - 0.3900 g = 0.4970 g KCl

The percentages by mass for each compound are

NaCl: $\frac{0.3900 \text{ g}}{0.8870} \times 100\% = 43.97\%$ KCl: $\frac{0.4970 \text{ g}}{0.8870 \text{ g}} \times 100\% = 56.03\%$

4.86 This problem combines the simple mass relationship covered in Section 4.1 with concentration units of molarity covered in Section 4.4.

Solution: The balanced equation for the displacement reaction is

$$Zn(s) + CuSO_4(aq) \rightarrow ZnSO_4(aq) + Cu(s)$$

We find the solution volume

$$7.89 \text{ g Zn} \left(\frac{1 \text{ mol Zn}}{65.39 \text{ g Zn}}\right)\left(\frac{1 \text{ mol } CuSO_4}{1 \text{ mol Zn}}\right)\left(\frac{1 \text{ L soln}}{0.156 \text{ mol } CuSO_4}\right) = 0.773 \text{ L}$$

Would you expect Zn to displace Cu^{2+} from solution, as shown in the equation? Refer to Figure 3.15.

4.87 First write out the balanced equation. The products are carbon dioxide, water, and sodium chloride. The molarity calculation is similar to Problem 4.30.

Solution: The balanced equation is

$$2HCl(aq) + Na_2CO_3(s) \rightarrow CO_2(g) + H_2O(l) + 2NaCl(aq)$$

We find the molarity of the HCl solution. Note that one mole of Na_2CO_3 combines with <u>two</u> moles of HCl.

$$\text{Molarity HCl} = [\text{HCl}] = \left(\frac{\text{moles HCl}}{\text{L soln}}\right)$$

$$= \left(\frac{0.256\text{ g }Na_2CO_3}{28.3\text{ mL}}\right)\left(\frac{1000\text{ mL}}{1\text{ L}}\right)\left(\frac{1\text{ mol }Na_2CO_3}{106.0\text{ g }Na_2CO_3}\right)\left(\frac{2\text{ mol HCl}}{1\text{ mol }Na_2CO_3}\right)$$

$$= 0.171\text{ mol/L} = 0.171\text{ M}$$

4.88 What does "monoprotic" mean? Review Section 3.4 if you don't remember. How many moles of acid are represented by the 3.664 g sample? What is the mass of a 1.000 mol sample?

Solution: The neutralization reaction is

$$HX(aq) + NaOH(aq) \rightarrow NaX(aq) + H_2O(l)$$

The number of moles of acid is equal to the number of moles of NaOH. (Why?)

$$\text{Moles HX} = \text{Moles NaOH} = 20.27\text{ mL soln}\left(\frac{0.1578\text{ mol NaOH}}{1000\text{ mL soln}}\right) = 3.199 \times 10^{-3}\text{ mol}$$

The molar mass is the mass (in grams) of one mole of the acid. We can write

$$\text{Molar mass HX} = \left(\frac{?\text{ g HX}}{1\text{ mol}}\right) = \left(\frac{3.664\text{ g HX}}{3.199 \times 10^{-3}\text{ mol}}\right) = 1145\text{ g/mol}$$

4.90 This problem uses the volume and molarity to calculate moles; then moles are converted to mass using the balanced equation.

Solution: The number of moles of oxalic acid in 500 mL is

$$\left(\frac{0.100\text{ mol }H_2C_2O_4}{1\text{ L}}\right)\left(\frac{1\text{ L}}{1000\text{ mL}}\right) \times 500\text{ mL} = 0.0500\text{ mol}$$

Since 1 mole of Fe_2O_3 is equivalent to 6 moles of $H_2C_2O_4$ (from the balanced equation), then the mass of rust is

$$0.0500 \text{ mol } H_2C_2O_4\left(\frac{1 \text{ mol } Fe_2O_3}{6 \text{ mol } H_2C_2O_4}\right)\left(\frac{159.7 \text{ g } Fe_2O_3}{1 \text{ mol } Fe_2O_3}\right) = 1.33 \text{ g}$$

4.92 See the "Chemistry in Action" segment on breath analyzers for appropriate balanced equations. Find the percent alcohol by mass (note that you are given the mass of the blood sample).

Solution: The number of moles of potassium dichromate in 28.64 mL is

$$28.64 \text{ mL}\left(\frac{1 \text{ L}}{1000 \text{ mL}}\right)\left(\frac{0.07654 \text{ mol}}{1 \text{ L}}\right) = 2.192 \times 10^{-3} \text{ mol}$$

From the balanced equation two moles of potassium dichromate are equivalent to three moles of ethanol (C_2H_5OH). The mass of ethanol in the 60.00 g blood sample is:

$$2.192 \times 10^{-3} \text{ mol } K_2Cr_2O_7\left(\frac{3 \text{ mol } C_2H_5OH}{2 \text{ mol } K_2Cr_2O_7}\right)\left(\frac{46.07 \text{ g } C_2H_5OH}{1 \text{ mol } C_2H_5OH}\right) = 0.1515 \text{ g } C_2H_5OH$$

The percent ethanol by mass is:

$$\frac{0.1515 \text{ g ethanol}}{60.00 \text{ g blood}} \times 100\% = 0.2525\%$$

This exceeds the 0.1% legal limit. The driver is totally smashed.

4.93 Remember that for any type of stoichiometry problem, you must start with a balanced equation. If necessary, review Section 3.6 on Balancing Redox Equations.

Solution: The balanced net ionic equation is

$$5Cu^+ + 8H^+ + MnO_4^- \rightarrow 5Cu^{2+} + Mn^{2+} + 4H_2O$$

From this equation we see that five moles of copper(I) combine with one mole of permanganate ion. The mass of copper(I) present is

$$0.146 \text{ mol } MnO_4^-\left(\frac{36.0 \text{ mL}}{1000 \text{ mL}}\right)\left(\frac{5 \text{ mol } Cu^+}{1 \text{ mol } MnO_4^-}\right)\left(\frac{63.55 \text{ g Cu}}{1 \text{ mol Cu}}\right) = 1.67 \text{ g Cu(I)}$$

4.94 This problem is a gram-to-gram conversion problem. Would the answer be any different if the sodium carbonate were simply identified as "impurity"?

Solution: The balanced equation is: $2NaHCO_3(s) \rightarrow Na_2CO_3(s) + CO_2(g) + H_2O(g)$

The molar masses are: $NaHCO_3$ = 84.01 g/mol; CO_2 = 44.01 g/mol, H_2O =18.02. Note that for every two moles of sodium hydrogen carbonate (2 x 84.01 g = 168.0 g) that reacts, the mass of one mole of carbon dioxide and one mole of water (44.01 g + 18.02 g = 62.03 g) will be lost. The loss of 8.415 g of mass corresponds to an original mass of sodium hydrogen carbonate of

$$\left(\frac{8.415\ g}{62.03\ g}\right) \times 168.0\ g = 22.79\ g\ of\ NaHCO_3$$

$$Mass\ percent\ of\ NaHCO_3 = \left(\frac{22.79\ g}{60.42\ g}\right) \times 100\% = 37.72\%$$

4.96 Why do we need to know that "aspirin" in a monoprotic acid? If it were diprotic, would the calculation be any different?

Solution: Since aspirin $C_9H_8O_4$ is a monoprotic acid, then we know that 1 mole of it is equivalent to 1 mole of sodium hydroxide in a neutralization reaction. Including the conversion from grams to grains in the same chain multiplication:

$$12.25\ mL\ soln \left(\frac{1\ L\ soln}{1000mL}\right)\left(\frac{0.1466\ mol\ NaOH}{1\ L\ soln}\right)\left(\frac{1\ mol\ aspirin}{1\ mol\ NaOH}\right)\left(\frac{180.2\ g\ aspirin}{1\ mol\ aspirin}\right)\left(\frac{1\ grain}{0.0648\ g}\right)$$

$$= 4.99\ grains$$

4.98 This is a difficult problem involving primarily mass-to-mass conversions. If you can work this problem without help and without a mistake, you are doing very well.

Solution: The balanced equations are:

$$C_8H_{18} + \frac{25}{2}O_2 \rightarrow 8CO_2 + 9H_2O$$

$$C_8H_{18} + \frac{17}{2}O_2 \rightarrow 8CO + 9H_2O$$

The quantity of octane burned is 2,650 g (1 gallon with a denisty of 2.650 kg/gallon). Let x be the mass of octane converted to carbon dioxide; therefore (2650 g - x) g is the mass of octane converted to carbon monoxide. The molar mass of octane C_8H_{18} is 114.2 g/mol.

The amount of CO_2 and H_2O produced by x g of C_8H_{18} are:

$$x\ g\ C_8H_{18}\left(\frac{1\ mol\ C_8H_{18}}{114.2\ g\ C_8H_{18}}\right)\left(\frac{8\ mol\ CO_2}{1\ mol\ C_8H_{18}}\right)\left(\frac{44.01\ g\ CO_2}{1\ mol\ CO_2}\right) = 3.083x\ g\ CO_2$$

$$x\ g\ C_8H_{18}\left(\frac{1\ mol\ C_8H_{18}}{114.2\ g\ C_8H_{18}}\right)\left(\frac{9\ mol\ H_2O}{1\ mol\ C_8H_{18}}\right)\left(\frac{18.02\ g\ H_2O}{1\ mol\ H_2O}\right) = 1.420x\ g\ H_2O$$

The amount of CO and H_2O produced by (2650 g - x) g of C_8H_{18} are:

$$(2650\ g - x)\ g\ C_8H_{18}\left(\frac{1\ mol\ C_8H_{18}}{114.2\ g\ C_8H_{18}}\right)\left(\frac{8\ mol\ CO}{1\ mol\ C_8H_{18}}\right)\left(\frac{28.01\ g\ CO}{1\ mol\ CO}\right) = (5200 - 1.962x)\ g\ CO$$

$$(2650\ g - x)\ g\ C_8H_{18}\left(\frac{1\ mol\ C_8H_{18}}{114.2\ g\ C_8H_{18}}\right)\left(\frac{9\ mol\ H_2O}{1\ mol\ C_8H_{18}}\right)\left(\frac{18.02\ g\ H_2O}{1\ mol\ H_2O}\right) = (3763 - 1.420x)\ g\ H_2O$$

The total mass of CO_2 + CO + H_2O produced is 11,530 g. That is

$(3.083x + 1.420x + 5200 - 1.962x + 3763 - 1.420x) = 11{,}530$ g; solving, $x = 2{,}290$ g

Since x is the amount of octane converted to carbon dioxide, then

$$\text{the efficiency} = \frac{\text{g octane converted}}{\text{g octane total}} \times 100\% = \frac{2290 \text{ g}}{2650 \text{ g}} \times 100\% = 86.49\%$$

4.99 Review the part of Section 4.4 on Dilution of Solutions if you have trouble with this problem.

Solution: Let us call the original solution soln 1, the first dilution soln 2, and the second dilution soln 3. We will first work the problem in parts, and then show that we can also work the problem in a single step.

Starting with soln 3 (since we know its concentration): $V_{soln\ 3} = 1000$ mL, $M_{soln\ 3} = 0.00383$ M; $V_{soln\ 2} = 25.00$ mL, $M_{soln\ 2} = ?$

$$M_{soln\ 2} = \left(\frac{M_{soln\ 3} \times V_{soln\ 3}}{V_{soln\ 2}}\right) = \left(\frac{0.00383 \text{ M} \times 1000\text{mL}}{25.00 \text{ mL}}\right) = 0.153 \text{ M}$$

Working backwards for the dilution of soln 1 to soln 2: $V_{soln\ 2} = 125.00$ mL, $M_{soln\ 2} = 0.153$ M; $V_{soln\ 1} = 15.00$ mL, $M_{soln\ 1} = ?$

$$M_{soln\ 1} = \left(\frac{M_{soln\ 2} \times V_{soln\ 2}}{V_{soln\ 1}}\right) = \left(\frac{0.153 \text{ M} \times 125.00\text{mL}}{15.00 \text{ mL}}\right) = 1.28 \text{ M}$$

We could have worked the problem in a single step using the factor-label method as follows:

$$\left(\frac{0.00383 \text{ mol}}{1 \text{ L soln 3}}\right)\left(\frac{1000 \text{ mL soln 3}}{25.00 \text{ mL soln 2}}\right)\left(\frac{125.00 \text{ mL soln 2}}{15.00 \text{ mL soln 1}}\right) = 1.28 \text{ M}$$

4.100 This problem is a simple variation on Yields of Reactions in Section 4.3.

Solution: The molar mass of C_6H_{14} = 86.17 g/mol; the molar mass of C_2H_4 = 28.05 g/mol. Since 481 g of C_2H_4 is equal to 42.5% yield, we can use as a unit factor (481 g of C_2H_4)/(0.425).

$$\text{Grams of } C_6H_{14} = \left(\frac{481 \text{ g } C_2H_4}{0.425}\right)\left(\frac{1 \text{ mol } C_2H_4}{28.05 \text{ g } C_2H_4}\right)\left(\frac{1 \text{ mol } C_6H_{14}}{1 \text{ mol } C_2H_4}\right)\left(\frac{86.15 \text{ g } C_6H_{14}}{1 \text{ mol } C_6H_{14}}\right) = 3.48 \times 10^3 \text{ g}$$

4.102 Remember that with any stoichiometry problem, you must start with a balanced equation. If you don't remember the equation for the metabolism of glucose, refer to Example 4.2

Solution: The balanced equation is: $C_6H_{12}O_6 + 6O_2 \rightarrow 6CO_2 + 6H_2O$

$$\left(\frac{5.0 \times 10^2 \text{ g glucose/day}}{180.2 \text{ g/mol glucose}}\right)\left(\frac{6 \text{ mol } CO_2}{\text{mol glucose}}\right)\left(\frac{44.01 \text{ g } CO_2}{\text{mol } CO_2}\right)\left(\frac{365 \text{ days}}{\text{yr}}\right)(5.5 \times 10^9 \text{ persons}) = 1.5 \times 10^{15} \text{g } CO_2$$

03 What precipitate is formed when barium hydroxide is mixed with sodium sulfate?

Solution: The balanced equation is: $Ba(OH)_2(aq) + Na_2SO_4(aq) \rightarrow BaSO_4(s) + 2NaOH(aq)$

moles $Ba(OH)_2$: (2.27 L)(0.0820 *M*) = 0.186 mole
moles Na_2SO_4: (3.06 L)(0.0664 *M*) = 0.203 mole, therefore $Ba(OH)_2$ is limiting

$$(0.186 \text{ mol Ba(OH)}_2)\left(\frac{\text{mol BaSO}_4}{\text{mol Ba(OH}_2)}\right)\left(\frac{233.4 \text{ g BaSO}_4}{\text{mol BaSO}_4}\right) = 43.4 \text{ g BaSO}_4$$

4.104 In part (c) don't be confused by the fact that the problem contains five equations. We only need be concerned that the equations are balanced and with the mole equivalence.

Solution: (a) The balanced equations are:

(1)	$Cu + 4HNO_3 \rightarrow Cu(NO_3)_2 + 2H_2O + 2NO_2$	Redox
(2)	$Cu(NO_3)_2 + 2NaOH \rightarrow Cu(OH)_2 + 2NaNO_3$	Precipitation
(3)	$Cu(OH)_2 \rightarrow CuO + H_2O$	Decomposition
(4)	$CuO + H_2SO_4 \rightarrow CuSO_4 + H_2O$	Acid-base
(5)	$CuSO_4 + Zn \rightarrow ZnSO_4 + Cu$	Redox

(b) The theoretical yields based on beginning with 65.6 g Cu which is $\frac{65.6 \text{ g Cu}}{63.55 \text{ g/mol}} = 1.03$ mol Cu:

(1) 1.03 mol × (187.6 g $Cu(NO_3)_2$/mol) = 193 g
(2) 1.03 mol × (97.57 g $Cu(OH)_2$/mol) = 100 g
(3) 1.03 mol × (79.55 g CuO/mol) = 81.9 g
(4) 1.03 mol × (159.6 g $CuSO_4$/mol) = 164 g
(5) 1.03 mol × (63.55 g Cu/mol) = 65.5 g

(c) All of the reaction steps are clean and almost quantitative, so the recovery yield should be high.

Why is there a small discrepancy between the theoretical yield of 65.5 g of Cu recovered in step (b5) and the quantity of 65.6 g of Cu that we started with?

4.106 Remember that all stoichiometry problems require a balanced equation. The problem then becomes a simple mole conversion problem coupled with units conversion. If the mass of the SO_2 is given in tons, and the quantity of the H_2SO_4 is asked for in units of tons, is it necessary to convert to grams and then back to tons?

Solution: The equation is: $SO_2 \xrightarrow{H_2O} H_2SO_4$ One mole SO_2 yields one mole H_2SO_4.

$$(4.0 \times 10^5 \text{ ton SO}_2)\left(\frac{\text{ton-mol SO}_2}{64.07 \text{ ton SO}_2}\right)\left(\frac{\text{ton-mol H}_2\text{SO}_4}{\text{ton-mol SO}_2}\right)\left(\frac{98.09 \text{ ton H}_2\text{SO}_4}{\text{ton-mol H}_2\text{SO}_4}\right) = 6.1 \times 10^5 \text{ ton H}_2\text{SO}_4$$

Note that using "ton-moles" saves us several steps. It is unnecessary to convert from tons to pounds to grams and then back again to tons.

Note: If you have not already done so, please read the "Introduction for the Student" at the beginning of this Student Solutions Manual.

Chapter 5

THE GASEOUS STATE

5.11 Before trying this problem, study Figure 5.4 in your textbook and read the caption carefully. Is the right-hand column open to the atmosphere in each case? What is the pressure of the trapped gas when the two mercury columns are at the same height (Figure 5.4(a))? What is the pressure of the trapped gas when the mercury columns are not at the same height (Figure 5.4(b)-5.4(d))? Now look at the left-hand figure in Problem 5.11. What is the pressure of the trapped gas?

Solution: P = 76.0 cmHg + 20.0 cmHg = 96.0 cmHg = 960 mmHg

The pressure of the trapped gas is the sum of the atmospheric pressure (76.0 cm Hg) plus the difference in height between the two mercury columns.
Now look at the right-hand figure in Problem 5.11. How is it different? Is there a pressure caused by the atmosphere? What is the pressure of the trapped gas?

Solution: 25.0 cm Hg = 250 mmHg

There is no pressure from the atmosphere acting on the right-hand mercury column (vacuum!), so the pressure of the trapped gas is just equal to the difference in height between the two mercury columns. In the apparatus on the left, what would be the pressure of the gas if the right-hand mercury column were 20.0 cm below the left-hand column? Would this be possible in the apparatus on the right? If both pieces of apparatus were filled with water rather than mercury, would the differences in column heights be different? What would they be? (Read Example 5.3 in your textbook.)

5.12 This is a straightforward unit conversion problem like Example 5.2. Cover the answers and work Examples 5.1 and 5.2. Using Example 5.2 as a model, write out the chain multiplication involving the unit factors relating atm to mmHg and atm to kPa in the space below. When you have done this, compare your setup with the answer below the dashed line.

Solution:

$$562 \text{ mmHg}\left(\frac{1 \text{ atm}}{760 \text{ mmHg}}\right)\left(\frac{1.01325 \times 10^2 \text{ kPa}}{1 \text{ atm}}\right) = 74.9 \text{ kPa}$$

$$2.0 \text{ kPa}\left(\frac{1 \text{ atm}}{1.01325 \times 10^2 \text{ kPa}}\right)\left(\frac{760 \text{ mmHg}}{1 \text{ atm}}\right) = 15 \text{ mmHg}$$

If you had trouble with this problem, go back to Section 1.9 and review the use of the factor-label method. Rework the examples in Section 1.9.

5.20 This problem involves the use of Boyle's law. Study Figure 5.5 and the discussion of Equation (5.2) in your text. Cover the solution to Example 5.4 and try to solve the problem. If you get stuck, uncover the answer one line at a time until you see how to solve the problem. Work the problem again with the solution covered. Use this example as a model for Problem 5.20. What are the volume units in Problem 5.20?

Solution:

$P_1 = 0.970$ atm $\quad P_2 = 0.541$ atm
$V_1 = 725$ mL $\quad V_2 = ?$

From the equation $P_1V_1 = P_2V_2$, we can write $V_2 = \left(\frac{P_1 \times V_1}{P_2}\right) = \left(\frac{0.970 \times 725}{0.541}\right) = 1.30 \text{ x } 10^3$ mL

Would the answer have been different if the pressures were given in mmHg or kPa?

5.22 This problem involves the use of Boyle's law.

Solution: We substitute 760 mmHg for 1.00 atm

$$P_2 = P_1\left(\frac{V_1}{V_2}\right) = 760 \text{ mmHg}\left(\frac{5.80 \text{ L}}{9.65 \text{ L}}\right) = 457 \text{ mmHg}$$

Should we arrive at the same answer if we had first solved for the new pressure in units of atm, and then converted to units of mmHg?

5.24 This is a Boyle's law problem similar to 5.20-5.23. The pressure at the surface can be assumed to be 1.00 atm. Finding the pressure at a depth of 4.08 m is the hard part. The pressure at that depth is the sum of the atmospheric pressure on the surface plus the pressure resulting from the weight of the water. The latter (in N/m^2) is the product of the mass (in kg) of a column of seawater (density 1.03 g/cm^3) 4.08 m high and 1.00 m^2 in cross section area times the acceleration due to gravity (see Review Question 5.4). Remember that density is mass/volume; a density of 1.03 g/cm^3 is equal to 1.03×10^3 kg/m^3.

Solution: The mass of the seawater column is

$$\text{Mass} = (4.08 \text{ m})(1.00 \text{ m}^2)\left(\frac{1.03 \times 10^3 \text{ kg}}{1 \text{ m}^3}\right) = 4.20 \times 10^3 \text{ kg}$$

The pressure resulting from the seawater column of 1 m^2 cross section is

$$\text{Pressure} = \left(\frac{\text{force}}{\text{area}}\right) = \frac{(4.20 \times 10^3 \text{ kg})(9.8067 \text{ m/s}^2)}{1 \text{ m}^2} = 4.12 \times 10^4 \text{ N/m}^2$$

The combined atmosphere and water pressure is

$$\begin{aligned}\text{Pressure} &= 4.12 \times 10^4 \text{ N/m}^2 + 1.01 \times 10^5 \text{ N/m}^2 \\ &= 0.412 \times 10^5 \text{ N/m}^2 + 1.01 \times 10^5 \text{ N/m}^2 \\ &= 1.42 \times 10^5 \text{ N/m}^2\end{aligned}$$

The increase in volume is the ratio V_2/V_1. By Boyle's law $P_1V_1 = P_2V_2$, so we can write

$$\left(\frac{V_2}{V_1}\right) = \left(\frac{P_1}{P_2}\right) = \left(\frac{1.42 \times 10^5 \text{ N/m}^2}{1.01 \times 10^5 \text{ N/m}^2}\right) = 1.41$$

In other words the diver's lungs would expand by a factor of 1.41 before reaching the surface.

5.25 If you do not know how to work this problem, read the section of your textbook in which Equation (5.3) appears. How do you find t(°C) given T(K)?

Solution: (a) 0°C + 273°C = 273 K; 37°C + 273°C = 310 K; the other two are similar: 373 K; 48 K.

(b) 77 K - 273 K = -196°C; the other two are similar: –268.8°C; 5.7 x 10^3°C.

PV=const What are the unit factors relating degrees Fahrenheit to degrees Celsius and kelvins? What is absolute zero on the Fahrenheit scale?

5.26 Study the paragraphs in which Equations (5.4) and (5.5) appear. Cover the answer to Example 5.5, read the problem, and try to solve it. If you have trouble, uncover the solution one line at a time until you see how to work the problem. After you have solved it, try it again with the answer covered. Use this as a model for Problem 5.26. What are the volume units?

Solution: T_1 = 25°C + 273°C = 298 K T_2 = 88°C + 273°C = 361 K
V_1 = 36.4 L V_2 = ?

Using Equation (5.5) we can write $V_2 = V_1\left(\frac{T_2}{T_1}\right) = 36.4 \text{ L}\left(\frac{361 \text{ K}}{298 \text{ K}}\right) = 44.1 \text{ L}$

Why must the temperatures be converted to kelvin? What result do you get if you use the Celsius temperatures? What happens to the volume in this case if one of the Celsius temperatures is negative? Could you solve Problem 5.26 with Charles' law if the pressure were not constant? What else would you need to know? Would the answer be different if the gas were something other than methane?

5.28 What happens to the pressure of the gas in the ball?

Solution: The pressure inside the ball increases when it is heated. This pushes out the dented surface. What gas law is operating here?

5.29 Carefully read the two paragraphs in your text under the caption Volume-Amount Relationship: Avogadro's Law. What is the ratio of volumes between hydrogen and ammonia in that discussion? Problem 5.29 has two parts. The first requires you to write out the balanced chemical equation for the reaction of ammonia with molecular oxygen (if you don't know the correct formula, see Table 5.1) to form water and nitric oxide. If you can't do this, carefully reread the subsection on balancing chemical equations in Section 3.1 of your textbook. Use Example 3.2(b) as a model for balancing the ammonia-oxygen equation. The second step in Problem 5.29 is reasoning out the volume-volume relationship between nitric oxide and ammonia. This is done just like the nitrogen-hydrogen-ammonia example in your textbook.

Solution: The balanced equation is: $4NH_3(g) + 5O_2(g) \rightarrow 4NO(g) + 6H_2O(g)$

The ammonia and nitric oxide coefficients are the same, so one volume of nitric oxide must be obtained from one volume of ammonia.

Could you have reached the same conclusion if you had seen that nitric oxide is the only nitrogen-containing product and that ammonia is the only nitrogen-containing reactant?

5.30 This problem is similar to Problem 5.29. Study the subsection of Section 5.3 on Avogadro's law if you don't see what to do. To what are the volumes proportional under the stated conditions?

Solution: Equal volumes of different gases at the same temperature and pressure contain equal numbers of particles (Avogadro's law). From the information given, the overall equation is

$$Cl_2(g) + 3F_2(g) \rightarrow 2(\text{formula of unknown gas})$$

To balance, the formula for the unknown gas must contain 1 Cl atom and 3 F atoms. The formula is $ClF_3(g)$.

SOME COMMENTS ON GAS LAW PROBLEMS

When first encountered, there appears to be a hopelessly large number of different-looking gas laws: Boyle's, Charles', Avogadro's, ideal gas, and many others. In truth all these equations are contained in the ideal gas law, PV = nRT. Think about this equation for a few moments. There is one constant R and four quantities of variable magnitude, P, V, T, and n. If you know the values of any three of the variable quantities, you can solve PV = nRT algebraically for the fourth one and then calculate its value by substituting in the three known others. As a simple exercise, take the ideal gas equation and solve it algebraically for each of the four variables (P, V, T, or n).

What happens when you only know the values of two of the variables in the ideal gas equation? You get a relationship between the other two. Suppose that n and T are known to have some fixed values; what can you say about the relationship between P and V? Look at the ideal gas equation PV = nRT. All the factors on the right side are fixed, so you can write

$$PV = \text{constant}$$

This is Boyle's law. Now suppose n and P have fixed values; what can you say about V and T? If you solve algebraically for V, you have V = (constant)T. This is Charles' law. Does this method also work for Avogadro's law?

There are three other possible relationships between different pairs of the variables; work out each one for yourself. The two involving n don't have names. Name one for yourself and the other for a friend.

What if only one variable has a fixed value? Let's look at the case in which n is constant. This corresponds to a situation involving a fixed amount of a gas in which the pressure, volume, and temperature can change. If you rearrange the ideal gas equation so that the constants are all on one side, you get

$$\frac{PV}{T} = \text{constant}$$

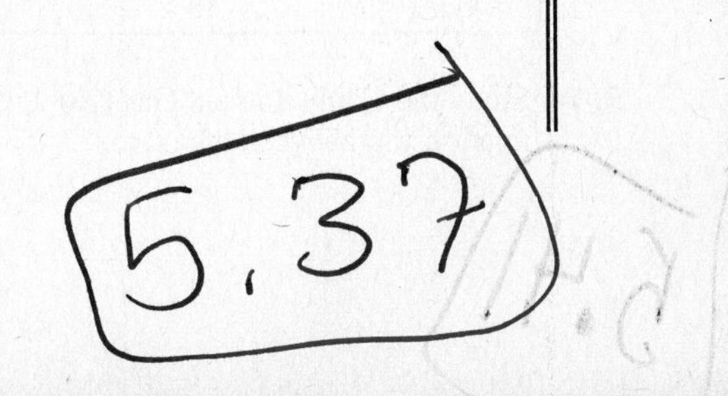

In other words, the expression PV/T must have a fixed value as long as the amount of gas doesn't change. Work out the three other relationships that can be obtained by letting P, V, or T be constant in value.

5.38 This problem is similar to Example 5.8 in the text. Study Section 5.4 on The Ideal Gas Law Equation. Using the ideal gas law equation (5.7), algebraically solve for P, for V, for T, and for n.

Solution: We are given n, V, and T. We solve the ideal gas equation for P

$$P = \frac{nRT}{V} = \frac{6.9 \text{ mol} \times 0.0821 \text{ L} \cdot \text{atm/K} \cdot \text{mol} \times (62 + 273) \text{ K}}{30.4 \text{ L}} = 6.2 \text{ atm}$$

5.40 This problem is similar to Problem 5.38. Study Section 5.4 on The Ideal Gas Law Equation. Using the ideal gas law equation (5.7), algebraically solve for P, for V, for T, and for n (Refer to the Comments on Gas Law Problems above).

Solution: (a) The balanced equation is: $NH_4NO_3(s) \rightarrow N_2O(g) + 2H_2O(l)$

(b) We convert to appropriate units:

$$P = 718 \text{ mmHg} \left(\frac{1 \text{ atm}}{760 \text{ mmHg}}\right) = 0.945 \text{ atm}$$

$$T = 24 + 273 = 297 \text{ K}$$

$$n = 0.580 \text{ g} \left(\frac{1 \text{ mol } N_2O}{44.02 \text{ g } N_2O}\right) = 0.0132 \text{ mol}$$

$$R = \frac{PV}{nT} = \frac{0.945 \text{ atm} \times 0.340 \text{ L}}{0.0132 \text{ mol} \times 297 \text{ K}} = 0.0820 \text{ L} \cdot \text{atm/K} \cdot \text{mol}$$

How does the answer correspond to known value to R? Notice that in this problem, we could have done all of the conversions in a single chain-multiplication step using unit factors in the ideal gas law equation.

5.42 Study the Comments on Gas Law Problems above before working this problem. Which variables do not change in this problem? What is the relationship between the others? This problem is similar to Example 5.8. Try to work that example with the answer covered before trying this one.

Solution: In this problem only n has a constant value. The relationship between the remaining three variables is the same as that in the Comments above. The method of solution is like that in Example 5.8.

Initial Conditions	Final Conditions
$P_1 = 1.2$ atm	$P_2 = 3.00 \times 10^{-3}$ atm
$V_1 = 2.50$ L	$V_2 = ?$
$T_1 = (25 + 273)$ K = 298 K	$T_2 = (-23 + 273)$ K = 250 K

$$V_2 = V_1 \times \frac{P_1}{P_2} \times \frac{T_2}{T_1} = 2.50 \text{ L} \times \frac{1.2 \text{ atm}}{3.00 \times 10^{-3} \text{ atm}} \times \frac{250 \text{ K}}{298 \text{ K}} = 8.4 \times 10^2 \text{ L}$$

5.44 Study the Comments on Gas Law Problems above. Which variables are constant in this problem? What is the relationship between the others?

5.41

Solution: We use Equation 5.8.

$$\frac{P_1V_1}{T_1} = \frac{P_2V_2}{T_2}$$

Note that the statement "...its absolute temperature is decreased by one-half" implies that $T_2/T_1 = 0.50$. Similarly, the statement "...pressure is decreased to one-third of its original pressure" indicates $P_2/P_1 = 1/3$.

$$V_2 = V_1\left(\frac{T_2}{T_1}\right)\left(\frac{P_1}{P_2}\right) = 6.0\text{ L} \times 0.50 \times 3.0 = 9.0\text{ L}$$

5.46 This problem is a direct substitution ("plug in") into the ideal gas law equation.

Solution: Since the quantity of gas is constant (that is, the number of moles), then we use the equation

$$\frac{P_1V_1}{T_1} = \frac{P_2V_2}{T_2}$$

$$V_1 = V_2\left(\frac{P_2}{P_1}\right)\left(\frac{T_1}{T_2}\right) = 94\text{ mL}\left(\frac{0.60\text{ atm}}{0.85\text{ atm}}\right)\left(\frac{(66 + 273)\text{ K}}{(45 + 273)\text{ K}}\right) = 71\text{ mL}$$

48 What does the term STP mean? This problem is the same type as Problem 5.46.

Solution: Since the quantity of gas is constant (that is, the number of moles), then we use the equation

$$\frac{P_1V_1}{T_1} = \frac{P_2V_2}{T_2}$$

$$V_2 = V_1\left(\frac{P_1}{P_2}\right)\left(\frac{T_2}{T_1}\right) = 6.85\text{ L}\left(\frac{772\text{ mmHg}}{760\text{ mmHg}}\right)\left(\frac{273\text{ K}}{(35 + 273)\text{ K}}\right) = 6.17\text{ L}$$

5.50 This problem involves direct substitution into the gas law equation. Refer to Example 5.8.

Solution: Since the quantity of gas is constant (that is, the number of moles), then we use the equation

$$V_2 = V_1\left(\frac{P_1}{P_2}\right)\left(\frac{T_2}{T_1}\right) = 566\text{ cm}^3\left(\frac{772\text{ mmHg}}{754\text{ mmHg}}\right)\left(\frac{(15 + 273)\text{ K}}{(10 + 273)\text{ K}}\right) = 5.90 \times 10^2\text{ cm}^3$$

5.52 What is the relationship between the number of moles of ozone molecules and the number of ozone molecules?

Solution: First we calculate the number of moles of O_3 present. Since $PV = nRT$, we write

$$n = \frac{PV}{RT} = \frac{(1.0 \times 10^{-3}\ \text{atm})(1.0\ \text{L})}{(0.0821\ \text{L·atm/K·mol})(250\ \text{K})} = 4.9 \times 10^{-5}\ \text{mol}$$

The number of O_3 molecules is given by

$$4.9 \times 10^{-5}\ \text{mol } O_3 \left(\frac{6.022 \times 10^{23}\ \text{molecules } O_3}{1\ \text{mol } O_3}\right) = 3.0 \times 10^{19}\ \text{molecules } O_3$$

5.54 Refer to Problem 5.52 if you need help.

Solution: First we determine the number of moles of gas under these conditions

$$n = \frac{PV}{RT} = \frac{(1.0 \times 10^{-6}\ \text{mmHg})\left(\frac{1\ \text{atm}}{760\ \text{mmHg}}\right)(1.0\ \text{L})}{(0.0821\ \text{L·atm/K·mol})(25 + 273)\ \text{K}} = 5.4 \times 10^{-11}\ \text{mol}$$

The number of molecules is

$$5.4 \times 10^{-11}\ \text{mol} \left(\frac{6.022 \times 10^{23}\ \text{molecules}}{1\ \text{mol}}\right) = 3.3 \times 10^{13}\ \text{molecules}$$

Even though a vacuum of 1×10^{-6} mmHg is considered to be a very good vacuum, an enormous number of molecules are still present in every liter!

5.56 Refer to Example 5.10 if you need help. Would you have to use equation (5.9) to solve this problem, or could you use the more direct form of the ideal gas law equation PV = nRT?

Solution: Using equation (5.9) for density

$$d = \frac{PM}{RT} = \frac{\left(\frac{733\ \text{mmHg}}{760\ \text{mmHg}}\right)\text{atm}\left(\frac{80.91\ \text{g}}{1\ \text{mol}}\right)}{(0.0821\ \text{L·atm/K·mol})(46 + 273)\ \text{K}} = 2.98\ \text{g/L}$$

An alternate approach to solving the problem is as follows. Realizing that the density is numerically the same as the number of grams in one liter, and that the number of moles n is numerically equal to grams of sample divided by molar mass M, then the answer above could have been solved by directly using the equation PV = nRT using a volume of 1.0 L

Substituting in the equation PV = nRT: $P(1.0\ \text{L}) = (x\ \text{g}/M)RT$

Substituting the data in for pressure, volume, molar mass, and R, solving for x also gives a value of = 2.98 g/L.

5.58 Example 5.11 is similar. The problem involves direct substitution into equation (5.10)

Solution: Using equation (5.10), and knowing that STP means 1.00 atm and 273 K,

$$M = \frac{mRT}{PV} = \frac{(0.400 \text{ g})(0.0821 \text{ L·atm/K·mol})(273 \text{ K})}{(1.00 \text{ atm})(0.280 \text{ L})} = 32.0 \text{ g/mol}$$

5.60 Example 5.11 is similar. The problem involves direct substitution into equation (5.10)

Solution: Using equation (5.10),

$$M = \frac{mRT}{PV} = \frac{(0.100 \text{ g})(0.0821 \text{ L·atm/K·mol})(293 \text{ K})}{(1.02 \text{ atm})(0.0221 \text{ L})} = 107 \text{ g/mol}$$

Since the molar mass of SF_4 is 108.1 g/mol, then the molecular formula is SF_4.

5.62 Examples 5.14 and 5.15 in the text are similar. The problem is worked in two parts. First, a gram-to-mole conversion is performed. Then the ideal gas law is used to convert the moles to gas volume.

Solution: First,using the balanced equation, we calculate the number of moles of H_2 produced.

$$n_{H_2} = 3.12 \text{ g Al} \left(\frac{1 \text{ mol Al}}{26.98 \text{ g Al}}\right)\left(\frac{3 \text{ mol } H_2}{2 \text{ mol Al}}\right) = 0.173 \text{ mol } H_2$$

Next we write:

$$V = \frac{nRT}{P} = \frac{(0.173 \text{ mol})(0.0821 \text{ L·atm/K·mol})(273 \text{ K})}{(1.00 \text{ atm})} = 3.88 \text{ L}$$

5.63 How do you find a molecular formula from an empirical formula? See 2.18. How do you find an empirical formula from percent composition by mass data? See Examples 2.15 and 2.16. How do you find the number of grams of fluorine in 0.2631 g of calcium fluoride? See Example 2.17. How do you find the molar mass of the gas in this problem? See Problem 5.52.

Solution: Let's first find the mass of F in the CaF_2

$$0.2631 \text{ g } CaF_2 \left(\frac{1 \text{ mol } CaF_2}{78.08 \text{ g } CaF_2}\right)\left(\frac{2 \text{ mol F}}{1 \text{ mol } CaF_2}\right)\left(\frac{19.00 \text{ g F}}{1 \text{ mol F}}\right) = 0.1280 \text{ g F}$$

Since the compound contains only P and F, the mass of P must be (0.2324 - 0.1280) g = 0.1044 g.

Next we find the empirical formula. Note that we have the actual masses of the elements, so we don't need to go through the "assume 100 g" procedure used for percent composition by mass problems. We directly convert the masses of F and P to moles:

$$n_P = 0.1044 \text{ g P} \left(\frac{1 \text{ mol P}}{30.97 \text{ g P}}\right) = 0.003371 \text{ mol P}$$

$$n_F = 0.1280 \text{ g F} \left(\frac{1 \text{ mol F}}{19.00 \text{ g F}}\right) = 0.006737 \text{ mol F}$$

The empirical formula is $P_{0.003371}F_{0.006737}$ or PF_2. The molar mass of PF_2 is 68.97 g.

Gas data: m = 0.2324 g; T = (77 + 273); K = 350 KV = 0.378 L; P = 97.3 mmHg $\left(\frac{1\ \text{atm}}{760\ \text{mmHg}}\right)$ = 0.128 atm

From the gas data and Equation (5.10) we find the molar mass of the gas.

$$\text{Molar mass} = \frac{dRT}{P} = \frac{mRT}{VP} = \frac{(0.2324\ \text{g})(0.0821\ \text{L·atm/K·mol})(350\ \text{K})}{(0.378\ \text{L})(0.128\ \text{atm})} = 138\ \text{g/mol}$$

Since this is twice the molar mass of PF_2, the molecular formula is P_2F_4.

5.64 How many moles of NaOH are needed to neutralize 1.00 mol of HCl? See Section 3.7. How many moles of HCl gas are used in this problem? See Problem 5.36. How do you calculate the molarity of a solution given the number of moles of solute and the solution volume? See Section 4.4 in the textbook

Solution: The number of moles of HCl gas is found from the ideal gas equation with V = 0.189 L, T = (25 + 273) K = 298 K, and P = 108 mmHg $\left(\frac{1\ \text{atm}}{760\ \text{mmHg}}\right)$ = 0.142 atm.

$$\text{Moles HCl} = \frac{PV}{RT} = \frac{(0.142\ \text{atm})(0.189\ \text{L})}{(0.0821\ \text{L·atm/K·mol})(298\ \text{K})} = 1.10 \times 10^{-3}\ \text{mol}$$

This will neutralize 1.10×10^{-3} mol of NaOH. The molarity of the NaOH solution is

$$\text{Molarity} = \left(\frac{1.10 \times 10^{-3}\ \text{mol}}{0.0157\ \text{L}}\right) = 0.0701\ \text{mol/L} = 0.0701\ \text{M}$$

What was the molarity of the hydrochloric acid in this problem? Would it have been any easier to calculate that number first?

5.65 How many moles of hydrogen gas are formed from how many moles of metal? Use your result to write a balanced equation for the reaction of the metal with HCl to form H_2 and the metal chloride. If chloride ion has a charge of -1, what is the charge of the metal ion? The formulas for the oxide and sulfate ions are in Table 2.3.

Solution: First find the number of moles of metal and of gas. For the gas V = 0.303 L, T = (17 + 273) K = 290 K, and P = 741 mmHg $\left(\frac{1\ \text{atm}}{760\ \text{mmHg}}\right)$ = 0.975 atm

Number of moles: $$n = \frac{PV}{RT} = \frac{(0.975\ \text{atm})(0.303\ \text{L})}{(0.0821\ \text{L·atm/mol·K})(290\ \text{K})} = 0.0124\ \text{mol}\ H_2$$

For the metal: $$0.225\ \text{g}\left(\frac{1\ \text{mol}}{27.0\ \text{g}}\right) = 0.00833\ \text{mol metal}$$

For each mole of metal 1.5 mole of H_2 forms. In other words, $\left(\frac{0.0124\ \text{mol}}{0.00833\ \text{mol}} = 1.5.\right)$

The balanced equation is therefore $M(s) + 3HCl(aq) \rightarrow 1.5H_2(g) + MCl_3(aq)$. The oxide and sulfate formulas are M_2O_3 and $M_2(SO_4)_3$.

5.67 How many moles of carbon dioxide gas could be produced from 3.00 g of pure calcium carbonate? How many moles of carbon dioxide gas are formed in this problem (see Problem 5.36)?

Solution: The balanced equation is $CaCO_3(s) + 2HCl(aq) \rightarrow CaCl_2(aq) + CO_2(g) + H_2O(l)$. We therefore know that 1 mol of $CaCO_3$ produces 1 mol of CO_2. For 3.00 g of pure $CaCO_3$, the yield of CO_2 would be

$$3.00\text{ g } CaCO_3\left(\frac{1\text{ mol } CaCO_3}{100.1\text{ g } CaCO_3}\right)\left(\frac{1\text{ mol } CO_2}{1\text{ mol } CaCO_3}\right) = 3.00 \times 10^{-2}\text{ mol } CO_2$$

The number of moles of gas formed is found from the ideal gas equation with V = 0.656 L, T = (20 + 273) K = 293 K, and $P = 792\text{ mmHg}\left(\frac{1\text{ atm}}{760\text{ mmHg}}\right) = 1.04\text{ atm}$

$$\text{Moles of gas} = n = \frac{PV}{RT} = \frac{(1.04\text{ atm})(0.656\text{ L})}{(0.0821\text{ L·atm/K·mol})(293\text{ K})} = 2.84 \times 10^{-2}\text{ mol}$$

The percent by mass of $CaCO_3$ is

$$\%CaCO_3 = \left(\frac{2.84 \times 10^{-2}\text{ mol}}{3.00 \times 10^{-2}\text{ mol}}\right)100\% = 94.7\%$$

Could this problem be solved if hydrochloric acid were the limiting reagent? Why or why not? Could this problem be solved if one of the impurities were another metal carbonate?

5.69 Review Section 5.5 on Stoichiometry Involving Gases and Examples 5.13 amd 5.14. The problem requires combining reaction stoichiometry and the ideal gas law.

Solution: The balanced equation is: $C_2H_5OH(l) + 3O_2(g) \rightarrow 2CO_2(g) + 3H_2O(l)$

The number of moles of C_2H_5OH burned is: $227\text{ g } C_2H_5OH\left(\frac{1\text{ mol } C_2H_5OH}{46.07\text{ g } C_2H_5OH}\right) = 4.93\text{ mol } C_2H_5OH$

From the balanced equation we see that 1 mol C_2H_5OH consumes 3 mol of O_2. The number of moles of O_2 needed is:

$$4.93\text{ mol } C_2H_5OH\left(\frac{3\text{ mol } O_2}{1\text{ mol } C_2H_5OH}\right) = 14.8\text{ mol } O_2$$

The volume of O_2 corresponding to 14.8 mol is:

$$V = \frac{nRT}{P} = \frac{(14.8\text{ mol})(0.0821\text{ L·atm/K·mol})(35 + 273)\text{ K}}{(790/760)\text{ atm}} = 3.60 \times 10^2\text{ L}$$

Since air is 21.0% O_2 by volume, we write:

$$V_{air} = V_{O_2}\left(\frac{100\%}{21.0\%}\right) = 3.60 \times 10^2\text{ L}\left(\frac{100}{21.0}\right) = 1.71 \times 10^3\text{ L}$$

5.75 If you know all the partial pressures, what is the total pressure? What is the total pressure without the nitrogen? The last part is similar to Problem 5.42.

Solution: Dalton's law guarantees that the total pressure of the mixture is the sum of the partial pressures.

(a) $P_{total} = 0.32 \text{ atm} + 0.15 \text{ atm} + 0.42 \text{ atm} = 0.89 \text{ atm}$

(b) We know

$P_1 = 0.15 \text{ atm} + 0.42 \text{ atm} = 0.57 \text{ atm}$ $\quad P_2 = 1.0 \text{ atm}$

$T_1 = (15 + 273) \text{ K} = 288 \text{ K}$ $\quad T_2 = 273 \text{ K}$

$V_1 = 2.5 \text{ L}$ $\quad V_2 = ?$

Using Equation (5.8) we have

$$V_2 = V_1 \times \frac{P_1}{P_2} \times \frac{T_2}{T_1} = 2.5 \text{ L} \times \frac{0.57 \text{ atm}}{1.0 \text{ atm}} \times \frac{273 \text{ K}}{288 \text{ K}} = 1.4 \text{ L at STP}$$

5.77 This is like Example 5.17. Read the short discussion on collecting gases over water in Section 5.6 and try to work Example 5.17 with the answer covered. If you have trouble, study the answer carefully.

Solution: The total pressure (745 mmHg) is the sum of the partial pressures of helium, neon, and water vapor. The neon pressure is therefore

$$P_{Ne} = (745 - 368 - 28.3) \text{ mmHg} = 349 \text{ mmHg}$$

Could this problem be solved if you didn't know the partial pressure of the helium?

5.79 This is like Example 5.17. Remember that when collecting a gas over water, you must make a pressure correction for the vapor pressure of the water.

Solution: The pressure of the hydrogen gas is

$$0.980 \text{ atm} - (23.8 \text{ mmHg})\left(\frac{1 \text{ atm}}{760 \text{ mmHg}}\right) = 0.949 \text{ atm}$$

The number of moles of hydrogen generated is

$$n = \frac{PV}{RT} = \frac{(0.949 \text{ atm})(7.80 \text{ L})}{(0.0821 \text{ L·atm/K·mol})(25 + 273) \text{ K}} = 0.303 \text{ mol}$$

Since one mole of Zn produces one mole of hydrogen, the amount of Zn consumed (in grams) is

$$0.303 \text{ mol Zn} \left(\frac{65.39 \text{ g Zn}}{1 \text{ mol Zn}}\right) = 19.8 \text{ g Zn}$$

5.81 This is like Example 5.17. Remember that when collecting a gas over water, you must make a pressure correction for the vapor pressure of the water.

Solution: The balanced equation is: $2NH_3(g) \rightarrow N_2(g) + 3H_2(g)$

$$\text{Mole fraction } X_{hydrogen} = \frac{\text{moles of } H_2}{\text{total moles}} = \frac{3}{3+1} = 0.750$$

$$\text{Mole fraction } X_{nitrogen} = \frac{\text{moles of } N_2}{\text{total moles}} = \frac{1}{3+1} = 0.250$$

Partial pressure of hydrogen $P_{hydrogen} = (X_{hydrogen})(\text{total pressure}) = (0.750)(866 \text{ mmHg}) = 650 \text{ mmHg}$

Partial pressure of nitrogen $P_{nitrogen} = (X_{nitrogen})(\text{total pressure}) = (0.250)(866 \text{ mmHg}) = 217 \text{ mmHg}$

Why doesn't the total pressure of 866 mmHg exactly equal the sum of the partial pressures of hydrogen and nitrogen?

5.90 Refer to Example 5.18 in the text.

Solution: Using Equation (5.14): $u_{rms} = \sqrt{\frac{3RT}{M}}$ with T = 273 + 65 = 338 K

$$O_2: \quad u_{rms} = \sqrt{\frac{3(8.314 \text{ J/K·mol})(338 \text{ K})}{32.00 \times 10^{-3}\text{kg/mol}}} = 513 \text{ m/s}$$

Similarly,
$$UF_6: \quad u_{rms} = \sqrt{\frac{3(8.314 \text{ J/K·mol})(338 \text{ K})}{352.00 \times 10^{-3}\text{kg/mol}}} = 155 \text{ m/s}$$

As should be the case, the heavier gas UF_6 has a lower average velocity than the lighter O_2.

5.91 Refer to Example 5.18 in the text. Do we need to know the pressure in order to calculate root-mean-square speeds?

Solution: Using Equation (5.14): $u_{rms} = \sqrt{\frac{3RT}{M}}$ with T = 273-23 = 250 K

$$N_2: \quad u_{rms} = \sqrt{\frac{3(8.314 \text{ J/K·mol})(250 \text{ K})}{28.02 \times 10^{-3}\text{kg/mol}}} = 471.7 \text{ m/s}$$

Similarly, O_2: $u_{rms} = 441.4$ m/s

O_3: $u_{rms} = 360.4$ m/s

5.92 What are the assumptions made in the kinetic molecular theory of gases? Refer to Section 5.7 if necessary.

Solution: The fourth assumption of the kinetic molecular theory of gases that is discussed in Section 5.7 of the text is that the average kinetic energy of any two gases at the same temperature are equal. More massive gases, in this case UF_6, will have a lower velocity than lighter gases such as He.

Do the calculations in Problem 5.90 confirm this?

5.93 Study the four assumptions of the kinetic molecular theory of gases in Section 5.7. What is the average kinetic energy of the neon atoms? If you know the average kinetic energy, how can you calculate the mean-square speed (not the average speed!)? To get the correct answer, m must be the mass of one neon atom in kilograms.

Solution: Since Ne is at the same temperature, it has the same average kinetic energy as the ammonia. We can write

$$\overline{KE} = \frac{1}{2} m \overline{u^2} \quad \text{and} \quad \overline{u^2} = \frac{2KE}{m}$$

We need the mass of one Ne atom in kilograms:

$$\text{Mass} = \left(\frac{20.18 \text{ g Ne}}{1 \text{ mol Ne}}\right)\left(\frac{1 \text{ kg}}{1000 \text{ g}}\right)\left(\frac{1 \text{ mol}}{6.022 \times 10^{23} \text{ atoms}}\right) = 3.351 \times 10^{-26} \text{ kg/atom}$$

$$\overline{u^2} = \frac{2\ \overline{KE}}{m} = \frac{2(7.1 \times 10^{-21} \text{ J})}{3.351 \times 10^{-26} \text{ kg/atom}} = 4.2 \times 10^5 \text{ m}^2/\text{s}^2$$

5.99 This is similar to Example 5.20. Review discussion of effusion in Section 5.8. Do heavier molecules have longer or shorter effusion times? Cover the answer and try to work Example 5.20. If your answer is wrong, carefully study the explanation and rework the example.

Solution: As in Example 5.20, $\frac{t_1}{t_2} = \sqrt{\frac{M_1}{M_2}}$

We write: $\frac{15.0 \text{ min}}{12.0 \text{ min}} = \sqrt{\frac{M_1}{28.02 \text{ g}}}$ Solving: $M_1 = \left(\frac{15.0}{12.0}\right)^2 (28.02 \text{ g}) = 43.8 \text{ g}$

The gas is CO_2 (molar mass = 44.01 g)

Why do lighter molecules have faster effusion rates? Why is the effusion time proportional to the square root of the molar mass?

5.100 Under conditions of identical pressure and temperature, what physical property of a molecule determines its diffusion rate? The latter half of this problem is like Example 5.19. After studying the discussion of diffusion in Section 5.8, try to solve Example 5.19 with the answer covered. Comparing the diffusion rates means taking their ratio. If you get stuck, study the answer. Try to understand the theory behind the formula.

Solution: The order of increasing diffusion rates will be the same as the order of decreasing molar masses (why?).

$$HI < Kr < ClO_2 < PH_3 < NH_3$$

increasing diffusion rate →

$$\frac{r_{NH_3}}{r_{HI}} = \sqrt{\frac{M_{HI}}{M_{NH_3}}} = \sqrt{\frac{127.9\ g/mol}{17.03\ g/mol}} = 2.74$$

Do you think diffusion rates would increase or decrease at higher temperatures?

5.102 This problem is an application of Graham's law as expressed in Equation (5.16), Example 5.19 in the text, and Problem 5.100 above.

Solution: Using Equation (5.16)

$$\frac{r_{CH_4}}{r_{Ni(CO)_x}} = 3.3 = \sqrt{\frac{\mathcal{M}_{Ni(CO)_x}}{\mathcal{M}_{CH_4}}}$$

$$\mathcal{M}_{Ni(CO)_x} = 174.7\ g/mol$$

(molar mass of Ni) + *x*(molar mass of CO) = 174.7

$x = 4.14 \cong 4$ CO units in $Ni(CO)_x$

5.108 This is like Example 5.21. Study Section 5.9 and try to work the example with the answer covered; it is a plug-in problem using the ideal gas equation and Equation (5.17). Study the answer if you have trouble.

Solution: This is a plug-in problem using the van der Waals equation. We find the pressure by first solving Equation (5.17) algebraically for P

$$P = \frac{nRT}{(V - nb)} - \frac{an^2}{V^2}$$

where n = 2.50 mol, V = 5.00 L, T = 450 K, a = 3.59 atm·L^2/mol^2, and b = 0.0427 L/mol

$$P = \frac{(2.50\ mol)(0.0821\ L\cdot atm/K\cdot mol)(450\ K)}{[(5.00\ L) - (2.50\ mol \times 0.0427\ L/mol)]} - \frac{(3.59\ atm\cdot L^2/mol^2)(2.50\ mol)^2}{(5.00\ L)^2} = 18.0\ atm$$

Using the ideal gas equation, $P = \frac{nRT}{V} = \frac{(2.50\ mol)(0.0821\ L\cdot atm/K\cdot mol)(450\ K)}{(5.00\ L)} = 18.5\ atm$

5.110 This problem simply requires you to use the ideal gas law equation to calculate the expected pressure, which is then compared to the actual pressure.

Solution: Using the ideal gas law equation: $P = \frac{nRT}{V}$

$$P = \frac{(10.0 \text{ mol})(0.0821 \text{ L·atm/K·mol})(300 \text{ K})}{1.50 \text{ L}} = 164 \text{ atm}$$

Since the actual, measured pressure is 130 atm, then this gas is not "ideal."

5.112 Even though these questions involve phenomena that you are familiar with, they all are direct applications of the gas laws.

Solution: (a) Neither the amount of gas in the tire nor its volume change appreciably. The pressure is proportional to the temperature (see Problem 5.41). Therefore, as the temperature rises, the pressure increases.

(b) As the paper bag is hit, its volume decreases so that its pressure increases. The popping sound occurs when the bag is broken.

(c) As the balloon rises, the pressure outside decreases steadily, and the balloon expands.

(d) The pressure inside the bulb is greater than 1 atm.

5.114 Hydrogen chloride gas is very soluble in water.

Solution: When the water enters the flask from the dropper, some hydrogen chloride dissolves, creating a partial vacuum. Pressure from the atmosphere forces more water up the vertical tube.

5.116 This is a good test of your understanding of Section 5.5.

Solution: (a) $NH_4NO_2(s) \rightarrow N_2(g) + 2H_2O(l)$

(b) From the balanced equation, 1 mol $NH_4NO_2(s)$ is equivalent to 1 mol $N_2(g)$

$$n_{N_2} = \frac{PV}{RT} = \frac{(1.20 \text{ atm})(0.0862 \text{ L})}{(0.0821 \text{ L·atm/K·mol})(295 \text{ K})} = 4.27 \times 10^{-3} \text{ mol}$$

The mass of NH_4NO_2 needed: $(64.06 \text{ g/mol})(4.27 \times 10^{-3} \text{ mol}) = 0.273 \text{ g}$

5.118 This problem is an excellent example of a situation where there are at least two methods to arrive at the answer: a long method and a short method. Obviously, both methods should produce the same answer. The first method is to calculate the number of moles of each gas using the equation PV = nRT and then add them to produce the total moles. The total moles can be used to calculate the total pressure. The mole fraction of each gas can be calculated and then be used to determine the partial pressures. The short method involves realizing that each gas behaves independently. Since the temperature is constant, the problem simply involves Boyle's law.

Solution: Using Equation (5.2): $P_1V_1 = P_2V_2$

Ne: $P_1 = 1.04$ atm; $V_1 = 502$ mL; $V_2 = 250$ mL

$$P_{Ne} = \frac{(1.04 \text{ atm})(502 \text{ mL})}{250 \text{ mL}} = 2.09 \text{ atm}$$

SF_6: $P_1 = 0.85$ atm; $V_1 = 364$ mL; $V_2 = 250$ mL $\qquad P_{SF_6} = \frac{(0.85 \text{ atm})(364 \text{ mL})}{250 \text{ mL}} = 1.24$ atm

5.120 The effusion information is used to calculate an average mass for the carbon monoxide/carbon dioxide mixture. The average molar mass can then be used to determine the quantity of each gas.

Solution: Calculating the rates of effusion

$$r_{He} = \frac{29.7 \text{ mL}}{2 \text{ min}} \qquad r_{CO/CO_2} = \frac{10.0 \text{ mL}}{2 \text{ min}}$$

$$\frac{r_{He}}{r_{CO/CO_2}} = \frac{29.7}{10} = \sqrt{\frac{M}{4.003}} \qquad M = 35.3 \text{ g/mol}$$

Let x be the fraction of CO in the mixture; therefore $(1 - x)$ is the fraction of CO_2. Molar mass of CO = 28.01 g/mol; molar mass of CO_2 = 44.01 g/mol.

$$x(28.01) + (1 - x)(44.01) = 35.5 \qquad \text{Solving, } x = 0.544$$

The mixture consists of 54.4% of CO and 45.6% of CO_2.

5.122 In order to work this problem, you must decide in each case how you can physically or chemically separate the two gases.

Solution: (a) The pressure P_{total} of the mixture of carbon dioxide and hydrogen must first be determined at some temperature in a container of known volume. The carbon dioxide can be removed by reaction with sodium hydroxide.

$$CO_2(g) + 2NaOH(aq) \rightarrow Na_2CO_3(aq) + H_2O(l)$$

The pressure $P_{hydrogen}$ is measured under the same conditions. Finally, $P_{carbon\ dioxide} = P_{total} - P_{hydrogen}$.

(b) Heliumm has a much lower boiling point than does nitrogen. Therefore, nitrogen can be removed by liquifaction. As above, the total pressure is measured first, and then the pressure of just the helium can be measured. The pressure of the nitrogen is simply the difference bewteen the total pressure and the pressure of helium.

5.124 This problem is a good test of your understanding of Section 5.5.

Solution: The reactions are: $\quad Na_2CO_3(s) + 2HCl(aq) \rightarrow 2NaCl(aq) + H_2O(l) + CO_2(g)$

$$MgCO_3(s) + 2HCl(aq) \rightarrow MgCl_2(aq) + H_2O(l) + CO_2(g)$$

The mass of carbon dioxide produced:

$$m = \frac{MPV}{RT} = \frac{(44.01 \text{ g/mol})(1.24 \text{ atm})(1.67 \text{ L})}{(0.0821 \text{ L·atm/K·mol})(299 \text{ K})} = 3.71 \text{ g}$$

Let x be the mass of Na_2CO_3 and $(7.63 - x)$ g is therefore the mass of $MgCO_3$.

$$\left(\frac{x \text{ g}}{106.0 \text{ g/mol}}\right)(44.01 \text{ g/mol}) + \left(\frac{(7.63 - x) \text{ g}}{84.32 \text{ g/mol}}\right)(44.01 \text{ g/mol}) = 3.71 \text{ g}$$

Solving, x = 2.55 g

$$\text{percent composition by mass of } Na_2CO_3\text{: } \frac{\text{mass of } Na_2CO_3}{\text{mass of } Na_2CO_3 + MgCO_3} \times 100\% = \frac{2.55 \text{ g}}{7.63 \text{ g}} \times 100\% = 33.4\%$$

5.126 Do you expect the solid calcium carbonate or the gas carbon dioxide to occupy more volume?

Solution: We need to determine the volume of an equal number of moles of $CaCO_3(s)$ and $CO_2(g)$. To simplify our calculations, we choose 1 cm^3 (that is, 1 mL) of $CaCO_3$ which must have a mass of 2.930 g/mol (why?)

$$\frac{2.930 \text{ g}}{100.1 \text{ g/mol}} = 0.02927 \text{ mol}$$

This is also the number of moles of CO_2 gas produced (why?)

$$V_{CO_2} = \frac{nRT}{P} = \frac{(0.02927 \text{ mol})(0.0821 \text{ L·atm/K·mol})(298 \text{ K})}{1.00 \text{ atm}} = 0.716 \text{ L} = 716 \text{ mL}$$

$$\text{Ratio of volumes: } \frac{V_{CO_2}}{V_{CaCO_3}} = \frac{716 \text{ mL}}{1 \text{ mL}} = 716$$

5.128 This is a very challenging problem A reasonable background in physics would also help. The basic chemistry question involves the use of the material in Section 5.7.

Solution: Circumference = $2\pi r = 2\pi\frac{15}{2}$ cm = 2π x 7.5 cm

$$\text{(a) Speed at which the target is moving} = \frac{2\pi \times 7.5 \text{ cm} \times 130/\text{s}}{100 \text{ cm/m}} = 61.3 \text{ m/s}$$

$$\text{(b) Time for target to travel 2.80 cm: } \frac{2.80 \times 10^{-2} \text{ m}}{61.3 \text{ m/s}} = 4.57 \times 10^{-4} \text{ s}$$

(c) The time in (b) is the time it takes for a Bi atom to travel 15.0 cm. Therefore, its speed is

$$\frac{15.0 \times 10^{-2} \text{ m}}{4.57 \times 10^{-4} \text{ s}} = 328 \text{ m/s}$$

$$u_{rms} = \sqrt{\frac{3RT}{M}} = \sqrt{\frac{3(8.314 \text{ J/K·mol})(1123 \text{ K})}{209.0 \times 10^{-3} \text{kg/mol}}} = 366.1 \text{ m/s}$$

The magnitudes of the speeds are comparable but not identical. This is not surprising since 328 m/s is the value for a particular Bi atom and u_{rms} is an average value.

5.130 Another way to ask the problem is "What mass of water vapor occupies 2.5 L at 65°C at a pressure of 187.5 mmHg?"

Solution: From the ideal gas law: $n = \frac{(187.5 / 760 \text{ atm})(2.500 \text{ L})}{(0.0821 \text{ L•atm/K•mol})(338 \text{ K})} = 0.0222 \text{ mol } H_2O$

Mass of vapor phase H_2O = (0.0222 mol H_2O)(18.02 g H_2O/mol) = 0.400 g H_2O

Mass of liquid phase H_2O remaining = 10.00 g - 0.40 g = 9.60 g H_2O

5.132 Refer to Example 5.20 in the text and Problem 5.99 above for a similar problem.

Solution: $\frac{t_1}{t_2} = \sqrt{\frac{M_1}{M_2}}$ $\quad \frac{1.62 \text{ min}}{2.43 \text{ min}} = \sqrt{\frac{\text{molar mass of unknown}}{\text{molar mass}_{\text{xenon}}}} = \sqrt{\frac{\text{molar mass of unknown}}{131.3}}$

Solving, the molar mass of the unknown = 58.3 g/mol. The molar mass of the empirical formula C_2H_5 = 29.06 g/mol. Since 58.3/29.1 ≈ 2, the formula is $(C_2H_5)_2 = C_4H_{10}$.

5.133 The first part of the question uses the ideal gas law. In the second part of the question, the temperature is constant, so this problem is a straightforward example of Boyle's Law. Refer to Example 5.4.

Solution: From the ideal gas law: $n = \frac{(132 \text{ atm})(120 \text{ L})}{(0.0821 \text{ L•atm/K•mol})(295 \text{ K})} = 654 \text{ mol } O_2$

Solving for the g of O_2 $\quad (654 \text{ mol } O_2) \times \frac{32.00 \text{ g } O_2}{\text{mol } O_2} = 2.09 \times 10^4 \text{ g } O_2$

Using equation (5.2) $\quad P_1V_1 = P_2V_2$; (132 atm)(120 L) = (1.00 atm)(V_2); $V_2 = 1.58 \times 10^4$ L

5.136 What process in Nature can remove carbon dioxide?

Solution: The partial pressure of carbon dioxide is higher in the winter because carbon dioxide is utilized less by photosynthesis in plants.

5.137 This is a simple problem that uses the ideal gas law and the mole concept.

Solution: From the ideal gas law: $n = \left(\frac{(1.1 \text{ atm})(0.500 \text{ L})}{(0.0821 \text{ L•atm/K•mol})(310 \text{ K})}\right) = 0.0216 \text{ mol gas}$

$$(0.0216 \text{ mol gas})\left(\frac{6.022 \times 10^{23} \text{ molecules}}{\text{mol gas}}\right) = 1.3 \times 10^{22} \text{ molecules}$$

The most common gases present in exhaled air are: CO_2, O_2, N_2, and H_2O.

5.138 Sodium bicarbonate is also called sodium hydrogen carbonate.

Solution: (a) The balanced equation is: $2NaHCO_3 \rightarrow Na_2CO_3 + CO_2 + H_2O$

$$(5.0\text{ g }NaHCO_3)\left(\frac{\text{mol }NaHCO_3}{84.01\text{ g }NaHCO_3}\right)\left(\frac{\text{mol }CO_2}{2\text{ mol }NaHCO_3}\right) = 0.030\text{ mol }CO_2$$

Using the ideal gas law: $V = \frac{nRT}{P} = \frac{(0.030\text{ mol})(0.0821\text{ L}\bullet\text{atm/K}\bullet\text{mol})(453\text{ K})}{1.3\text{ atm}} = 0.86\text{ L}$

(b) The balanced equation for the decomposition of NH_4HCO_3 is: $NH_4HCO_3 \rightarrow NH_3 + CO_2 + H_2O$
The advantage in using the ammonium salt is that you get twice as much gas for the money (both ammonia and carbon dioxide). The disadvantage is that the extra gas produced in the form of ammonia, because of the odor, would not be worth it!

5.140 In this problem, we first need to convert the ton of urea to grams of ammonia by use of the stoichiometry of the balanced equation. We then use the ideal gas law to solve for the volume of ammonia.

Solution: The balanced equation is: $CO_2 + 2NH_3 \rightarrow (NH_2)_2CO + H_2O$

$$(1.0\text{ ton urea})\left(\frac{2000\text{ lb}}{\text{ton}}\right)\left(\frac{453.6\text{ g}}{\text{lb}}\right)\left(\frac{\text{mol urea}}{60.06\text{ g urea}}\right)\left(\frac{2\text{ mol }NH_3}{\text{mol urea}}\right) = 3.0 \times 10^4\text{ mol }NH_3$$

Using the ideal gas law: $V = \frac{nRT}{P} = \frac{(3.0 \times 10^4\text{ mol})(0.0821\text{ L}\bullet\text{atm/K}\bullet\text{mol})(473\text{ K})}{150\text{ atm}} = 7.8 \times 10^3\text{ L }NH_3$

5.142 The method used to solve the problem will be to first calculate the mass on one square cm of the Earth's surface (caused by the atmosphere). We then multiply that mass times the number of square cm on the Earth's surface.

Solution: Mass of a single column of 1 cm^2 area = $76.0\text{ cm} \times 13.6\text{ g/cm}^3 = 1.03 \times 10^3\text{ g/cm}^2 = 1.03\text{ kg/cm}^2$

The surface area of the Earth in cm^2: $4\pi r^2 = 4\pi(6.371 \times 10^8\text{ cm})^2 = 5.10 \times 10^{18}\text{ cm}^2$

Mass of the Earth's atmosphere = (surface area of the earth in cm^2) × (mass per 1 cm^2 column)

Thus, the mass of the Earth's atmosphere: $(5.10 \times 10^{18}\text{ cm}^2)(1.03\text{ kg/cm}^2) = 5.25 \times 10^{18}\text{ kg}$

Note: If you have not already done so, please read the "Introduction for the Student" at the beginning of this Student Solutions Manual.

Chapter 6

THERMOCHEMISTRY

6.16 Carefully read and digest the five guidelines for the interpretation of thermochemical equations in Section 6.3. Which guidelines apply to each part of this problem?

Solution: (a) If all the coefficients in the equation are doubled, then the enthalpy change is also doubled. It becomes -2905.6 kJ. This is a consequence of the first and third guidelines. Since the equation should be read in terms of moles, doubling the amounts of the materials involved should double the amount of thermal energy released.

(b) The second guideline applies. Turning everything around means the whole process goes backwards. The 1452.8 kJ will be absorbed (endothermic, positive sign) rather than released (exothermic, negative sign).

(c) The fourth guideline applies. Water in the gas phase (water vapor) contains more thermal energy (44.0 kJ/mol) than liquid water. This energy is retained by the steam and is not released to the surroundings. The enthalpy change will be lessened by the thermal energy retained by the four moles of steam. It will be -1276.8 kJ. ($4 \times 44.0 - 1452.8 = -1276.8$)

6.19 What is the difference between specific heat and heat capacity (see Section 6.4)? How is the specific heat related to the heat capacity in this problem?

Solution: Solving Equation (6.2) for specific heat gives $s = \frac{C}{m}$.

$$\text{Specific heat} = \frac{C}{m} = \frac{85.7\ \text{J/°C}}{362\ \text{g}} = 0.237\ \text{J/g} \cdot \text{°C}$$

6.20 Is the evaporation of alcohol endo- or exothermic? Where does the thermal energy come from in this situation?

Solution: The evaporation of ethanol is an endothermic process with a fairly high $\Delta H°$ value. When the liquid evaporates, it absorbes heat from your body; hence the cooling effect.

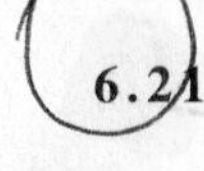

6.21 This is like Example 6.1. Cover the answer and try to work the example. If you can't, study the answer and try again with the answer covered. The specific heat of copper is in Table 6.1. Watch your sign conventions on this one!

Solution: From Equation (6.3):

$$q = ms\Delta t = (6.22 \times 10^3\ g)(0.385\ J/g \cdot °C)(324.3 - 20.5)°C = 728\ kJ$$

6.22 This problem is similar to Example 6.1 and Problem 6.21 above. Since the problem states that heat is "liberated," what is the sign of the answer?

Solution: The specific heat of mercury (Table 6.1) is 0.139 J/g · °C. We use Equation (6.3)

$$q = ms\Delta t = 366\ g \times 0.139\ J/g \cdot °C\ (12.0°C - 77.0°C) = -3.31 \times 10^3\ J = -3.31\ kJ$$

The result is negative because heat is lost from the system.

6.23 This type of problem is a classic. Equate the expression for the thermal energy lost by the iron with that gained by the gold. The final temperatures of the two metals will be the same. The specific heats of iron and gold are in Table 6.1.

Solution: Let t be the final temperature of the two metals.

$$\text{Heat gained by the gold} = ms\Delta t = (10.0\ g)(0.129\ J/g \cdot °C)(t - 18.0)°C$$

$$\text{Heat lost by the iron} = ms\Delta t = (20.0\ g)(0.444\ J/g \cdot °C)(t - 55.6)°C$$

Since one process is endothermic and the other is exothermic, the quantities of heat will be of opposite sign, and we must write $q_{Au} = -q_{Fe}$

$$(10.0\ g)(0.129\ J/g°C)(t - 18.0)°C = -(20.0\ g)(0.444\ J/g \cdot °C)(t - 55.6)°C$$

Solving for t, $t = 50.8°C$

Must the final temperature be between the two starting values? Do you need to know the amount of thermal energy transferred?

6.26 This is like Example 6.2. Study the subsection on constant volume calorimetry in Section 6.4 and try to work the example with the answer covered. If you can't, study the steps of the solution, cover it again, and rework the problem. Is this an exo- or an endothermic process?

Solution: The heat gained by the water and the calorimeter is

$$(1769\ J/°C)(1.126\ °C) + (300\ g)(4.184\ J/g \cdot °C)(1.126\ °C) = 3.41 \times 10^3\ J$$

The amount of heat released by one gram of Mg is

$$(3.41 \times 10^3\ J)\left(\frac{1\ kJ}{1000\ J}\right)\left(\frac{1}{0.1375\ g\ Mg}\right) = 24.8\ kJ/g\ Mg$$

The heat released by one mole of Mg is

$$\left(\frac{24.8 \text{ kJ}}{1 \text{ g Mg}}\right)\left(\frac{24.31 \text{ g Mg}}{1 \text{ mol Mg}}\right) = 603 \text{ kJ/mol Mg}$$

If the reaction were endothermic, what would happen to the temperature of the calorimeter and the water?

6.28 This problem is similar to Example 6.3. The given amount of HCl exactly neutralizes the given $Ba(OH)_2$. Also, the heat of neutralization is for the formation of *one mole* of water, and the final solution volume is 4.00×10^2 mL. We assume that the density and specific heat of the solution are the same as those of pure water.

Solution: The equation for the neutralization is: $2HCl(aq) + Ba(OH)_2(aq) \rightarrow 2H_2O(l) + BaCl_2(aq)$

One mole of water is formed from each mole of HCl. The number of moles of HCl is

$$200 \text{ mL}\left(\frac{1 \text{ L}}{1000 \text{ mL}}\right)\left(\frac{0.862 \text{ mol HCl}}{1 \text{ L}}\right) = 0.172 \text{ mol HCl}$$

The amount of heat liberated is

$$0.172 \text{ mol}\left(\frac{56.2 \times 10^3 \text{ J}}{1 \text{ mol}}\right) = 9.67 \times 10^3 \text{ J}$$

Assuming no heat loss to the surroundings, we write

$$q_{system} = q_{soln} + q_{calorimeter} + q_{rxn} = 0$$

$$q_{rxn} = -(q_{soln} + q_{calorimeter}) = -9.67 \times 10^3 \text{ J (exothermic)}$$

$$q_{soln} = (400 \text{ g})(4.184 \text{ J/g}\cdot{}^\circ\text{C})(t - 20.48^\circ\text{C}) \qquad q_{calorimeter} = (453 \text{ J/}^\circ\text{C})(t - 20.48^\circ\text{C})$$

$$(400 \text{ g})(4.184 \text{ J/g}\cdot{}^\circ\text{C})(t - 20.48^\circ\text{C}) + (453 \text{ J/}^\circ\text{C})(t - 20.48^\circ\text{C}) = 9.67 \times 10^3 \text{ J}$$

Solving for t: $t = 25.03^\circ\text{C}$

6.38 Study the discussion of Hess's law in Section 6.5. What happens to the enthalpy change in a thermochemical equation when the reaction is reversed (Section 6.3)?

Solution:

Reaction	ΔH° (kJ)
$S(rhombic) + O_2(g) \rightarrow SO_2(g)$	-296.06
$SO_2(g) \rightarrow S(monoclinic) + O_2(g)$	296.36
$S(rhombic) \rightarrow S(monoclinic)$	0.30 kJ

Which is the more stable allotropic form of sulfur?

6.39 What happens to the enthalpy change in a thermochemical equation when both sides are multiplied by a constant factor n (Section 6.3)? This problem is similar to Example 6.5. After studying the discussion of Hess's law in Section 6.5, cover the answer and try to solve the example. If you can't, study the solution carefully, cover the answer again, and rework the example. This is a very common type of exam question.

Solution:

Reaction	ΔH° (kJ)
$2C(graphite) + 2O_2(g) \rightarrow 2CO_2(g)$	-787.0
$3H_2(g) + \frac{3}{2}O_2(g) \rightarrow 3H_2O(l)$	-857.4
$2CO_2(g) + 3H_2O(l) \rightarrow C_2H_6(g) + \frac{7}{2}O_2(g)$	1559.8
$2C(graphite) + 3H_2(g) \rightarrow C_2H_6(g)$	-84.6 kJ

6.40 The three equations may be combined as shown to give the enthalpy of formation of methanol. Remember that when an equation is turned around, the sign of ΔH is reversed. Example 6.5 in the text is a similar problem.

Solution:

Reaction	ΔH° (kJ)
$CO_2(g) + 2H_2O(l) \rightarrow CH_3OH(l) + \frac{3}{2}O_2(g)$	726.4
$C(graphite) + O_2(g) \rightarrow CO_2(g)$	-393.5
$2[H_2(g) + \frac{1}{2}O_2(g) \rightarrow H_2O(l)]$	-571.6
$C(graphite) + 2H_2(g) + \frac{1}{2}O_2(g) \rightarrow CH_3OH(l)$	-238.7 kJ

We have just calculated an enthalpy at standard conditions, which we abbreviate ΔH°_{rxn}. In this case, the reaction in question was for the fomation of *one* mole of CH_3OH *from the elements* in their standard state. Therefore, the ΔH°_{rxn} that we calcuated is also, by definition, the standard heat of formation ΔH°_f of CH_3OH (-238.7 kJ/mol).

6.45 This is similar to Examples 6.5 and 6.2. Cover the answers and try to work both. If you get stuck, uncover the answer and study the method. Cover and rework the problem. Incidentally, the reaction in this problem is a decomposition, so the heat or enthalpy change is often called a heat or enthalpy of decomposition. Do you expect the enthalpy change to be positive or negative? See Problem 6.11 in the text.

Solution: $\Delta H^\circ = [\Delta H^\circ_f(CaO) + \Delta H^\circ_f(CO_2)] - \Delta H^\circ_f(CaCO_3)$

$= [(1\text{ mol})(-635.6\text{ kJ/mol}) + (1\text{ mol})(-393.5\text{ kJ/mol})] - (1\text{ mol})(-1206.9\text{ kJ/mol}) = 177.8\text{ kJ}$

6.46 In other words, you are to calculate the enthalpy change for the reaction of H^+ with OH^- to form water. The enthalpy of formation of water is in Table 6.4.

Solution: For $H^+(aq) + OH^-(aq) \rightarrow H_2O(l)$

$$\Delta H° = \Delta H_f^o(H_2O) - [\Delta H_f^o(H^+) + \Delta H_f^o(OH^-)]$$

$$\Delta H° = (1\text{ mol})(-285.8\text{ kJ/mol}) - [(1\text{ mol})(0\text{ kJ/mol}) + (1\text{ mol})(-229.6\text{ kJ/mol})] = -56.2\text{ kJ}$$

6.47 This is like Problem 6.45. All of the reactions shown are combustions (reactions with oxygen), so the standard enthalpies of reaction are called heats or enthalpies of combustion.

Solution: (a) $\Delta H° = 2\Delta H_f^o(H_2O) - 2\Delta H_f^o(H_2) - \Delta H_f^o(O_2)$
$= (2\text{ mol})(-285.8\text{ kJ/mol}) - (2\text{ mol})(0) - (1\text{ mol})(0) = -571.6\text{ kJ}$

(b) $\Delta H° = 4\Delta H_f^o(CO_2) + 2\Delta H_f^o(H_2O) - 2\Delta H_f^o(C_2H_2) - 5\Delta H_f^o(O_2)$
$= (4\text{ mol})(-393.5\text{ kJ/mol}) + (2\text{ mol})(-285.8\text{ kJ/mol}) - (2\text{ mol})(226.6\text{ kJ/mol}) - (5\text{ mol})(0) = -2599\text{ kJ}$

(c) $\Delta H° = 2\Delta H_f^o(CO_2) + 2\Delta H_f^o(H_2O) - \Delta H_f^o(C_2H_4) - 3\Delta H_f^o(O_2)$
$= (2\text{ mol})(-393.5\text{ kJ/mol}) + (2\text{ mol})(-285.8\text{ kJ/mol}) - (1\text{ mol})(52.3\text{ kJ/mol}) - (3\text{ mol})(0) = -1411\text{ kJ}$

(d) $\Delta H° = 2\Delta H_f^o(H_2O) + 2\Delta H_f^o(SO_2) - 2\Delta H_f^o(H_2S) - 3\Delta H_f^o(O_2)$
$= (2\text{ mol})(-285.8\text{ kJ/mol}) + (2\text{ mol})(-296.1\text{ kJ/mol}) - (2\text{ mol})(-20.15\text{ kJ/mol}) - (3\text{ mol})(0) = -1124\text{ kJ}$

6.48 How much heat is liberated when one molar mass of each of the alcohols burns? Should the enthalpies be positive or negative?

Solution: The given enthalpies are in units of kJ/g. We must convert them to units of kJ/mol.

(a) $$\left(\frac{-22.6\text{ kJ}}{1\text{ g}}\right)\left(\frac{32.04\text{ g}}{1\text{ mol}}\right) = -724\text{ kJ/mol}$$

(b) $$\left(\frac{-29.7\text{ kJ}}{1\text{ g}}\right)\left(\frac{46.07\text{ g}}{1\text{ mol}}\right) = -1.37 \times 10^3\text{ kJ/mol}$$

(c) $$\left(\frac{-33.4\text{ kJ}}{1\text{ g}}\right)\left(\frac{60.09\text{ g}}{1\text{ mol}}\right) = -2.01 \times 10^3\text{ kJ/mol}$$

6.49 Remember that when an enthalpy is written beside an equation, the amount of energy involved corresponds to the equation *as written*. How many grams of ammonia are produced when hydrogen and nitrogen react with the liberation of 92.6 kJ (as in the equation, as written). Example 6.4 in the text is a similar problem.

Solution: The amount of heat given off is

$$1.26 \times 10^4 \text{ g NH}_3 \left(\frac{1 \text{ mol NH}_3}{17.03 \text{ g NH}_3}\right)\left(\frac{-92.6 \text{ kJ}}{2 \text{ mol NH}_3}\right) = -3.43 \times 10^4 \text{ kJ}$$

6.51 The standard enthalpy of reaction in this problem also corresponds to the standard enthalpy of combustion of C_6H_{12}.

Solution: $\Delta H° = 6\Delta H_f^\circ(CO_2) + 6\Delta H_f^\circ(H_2O) - [\Delta H_f^\circ(C_6H_{12}) + 9\Delta H_f^\circ(O_2)]$

$$\Delta H° = (6 \text{ mol})(-393.5 \text{ kJ/mol}) + (6 \text{ mol})(-285.8 \text{ kJ/mol}) - (1 \text{ mol})(-151.9 \text{ kJ/mol}) - (9 \text{ mol})(0) = -3924 \text{ kJ}$$

Why is the standard heat of formation of oxygen zero?

6.53 How does this problem differ from Problem 6.46?

Solution: The balanced equation for the reaction is: $CaCO_3(s) \rightarrow CaO(s) + CO_2(g)$

From Problem 6.45 above: $\Delta H° = 177.8 \text{ kJ}$

The result above is for one mole of carbon dioxide produced. For 66.8 g of carbon dioxide, we write

$$66.8 \text{ g CO}_2 \left(\frac{1 \text{ mol CO}_2}{44.01 \text{ g CO}_2}\right)\left(\frac{177.8 \text{ kJ}}{1 \text{ mol CO}_2}\right) = 2.70 \times 10^2 \text{ kJ}$$

6.66 Study the discussion of enthalpy in Section 6.5 Under what conditions is the enthalpy change equal to the amount of heat absorbed or released in a process? How do you measure heat? See Section 6.4 if you don't remember.

Solution: Enthalpy change is the heat change measured at constant pressure.

(a) We need a constant-pressure calorimeter (Figure 6.10). From the amount and specific heat of sulfuric acid, the heat capacity of the calorimeter, the amount of magnesium reacted, and the initial and final temperatures, the heat evolved at constant pressue (or the enthalpy change) can be determined using the procedure shown in Section 6.4.

(b) Since the heat given off by molten naphthalene as it freezes is equal in magnitude (but opposite in sign) to the heat required to melt solid naphthalene at the same temperature (i.e., the melting point) and pressure, we can carry out the following process:

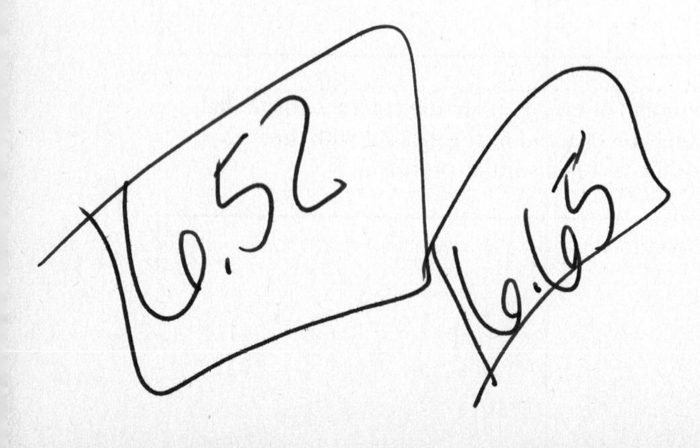

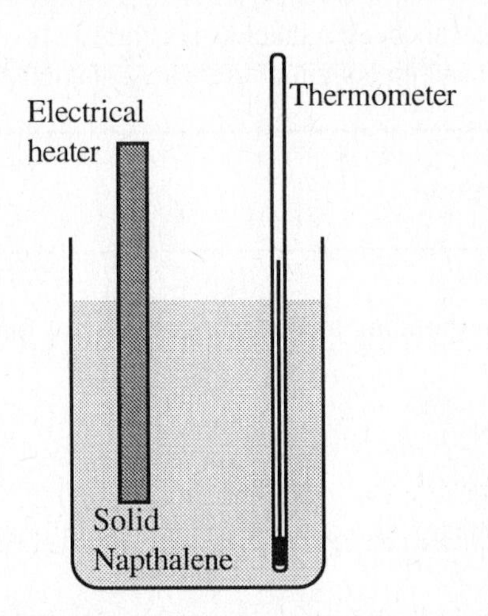

We apply a constant curent through the heater and plot temperature versus time as follows:

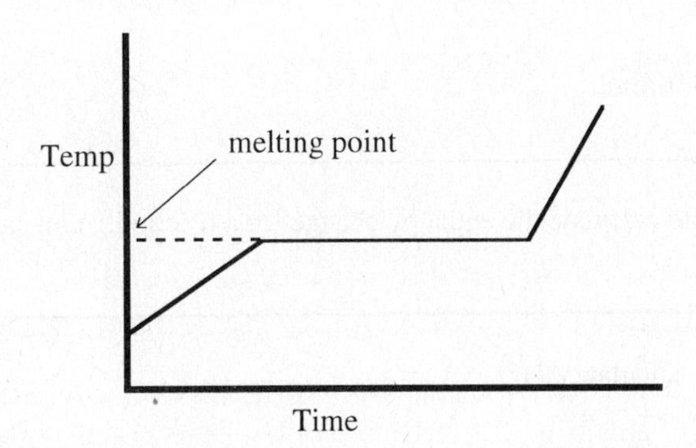

From the time it takes for solid naphthalene to completely turn into liquid we can calculate the heat supplied to the system (given by I^2R where I is the current and R is the resistance of the heater).

6.68 Review Section 6.7 and Example 6.6 if you need help. Before you can calculate the work involved, you must first use the ideal gas law (Section 5.4) to calculate the volume of the gas.

Solution: $V = \frac{nRT}{P} = \frac{(1 \text{ mol})(0.0821 \text{ L·atm/K·mol})(373 \text{ K})}{(1.0 \text{ atm})} = 31 \text{ L}$

$$w = -P\Delta V = -(1 \text{ atm})(31 \text{ L}) = -31 \text{ L atm} \left(\frac{101.3 \text{ J}}{1 \text{ L atm}}\right) = -3.1 \times 10^3 \text{ J} = -3.1 \text{ kJ}$$

Why is the sign negative?

6.69 In a chemical reaction are the same elements always on both sides of the equation? Is this true in nuclear reactions?

Solution: In a chemical reaction the same elements in the same numbers are always on both sides of the equation. This provides a consistent reference which allows the energy change in the reaction to be interpreted in terms of the chemical or physical changes that have occurred. In a nuclear reaction the same elements are not always on both sides of the equation and no common reference point exists.

6.70 This is like Problem 6.71 below.

Solution: The reaction corresponding to standard enthalpy of formation ΔH_f° of $AgNO_2(s)$ is

$$Ag(s) + \frac{1}{2}N_2(g) + O_2(g) \rightarrow AgNO_2(s)$$

Rather than measuring the enthalpy directly, we can use the information provided and apply Hess's law.

$$\Delta H_{rxn}^\circ = \Delta H_f^\circ(AgNO_2) + \frac{1}{2}\Delta H_f^\circ(O_2) - \Delta H_f^\circ(AgNO_3)$$

$$78.67 \text{ kJ} = (1 \text{ mol})\Delta H_f^\circ(AgNO_2) + 0 - (1 \text{ mol})(-123.02 \text{ kJ/mol})$$

$$\Delta H_f^\circ(AgNO_2) = -44.35 \text{ kJ/mol}$$

6.71 Review Section 6.5 for examples of how to calculate the enthalpy change for a reaction. You can find standard enthalpies of formation in Appendix 3.

Solution: (a) We use Equation (6.9) to calculate ΔH°

$$\Delta H_{rxn}^\circ = 4\Delta H_f^\circ(NH_3) + \Delta H_f^\circ(N_2) - 3\Delta H_f^\circ(N_2H_4)$$

$$= (4 \text{ mol})(-46.3 \text{ kJ/mol}) + (0) - (3 \text{ mol})(50.42 \text{ kJ/mol}) = -336.5 \text{ kJ}$$

(b)(i) The balanced equations are

$$N_2H_4(l) + O_2(g) \rightarrow N_2(g) + 2H_2O(l)$$

$$4NH_3(g) + 3O_2(g) \rightarrow 2N_2(g) + 6H_2O(l)$$

The standard enthalpy changes are

$$\Delta H_{rxn}^\circ = \Delta H_f^\circ(N_2) + 2\Delta H_f^\circ(H_2O) - \Delta H_f^\circ(N_2H_4) - \Delta H_f^\circ(O_2)$$
$$= (1 \text{ mol})(0) + (2 \text{ mol})(-285.8 \text{ kJ/mol}) - (1 \text{ mol})(50.42 \text{ kJ/mol}) - (1 \text{ mol})(0) = -622.0 \text{ kJ}$$

$$\Delta H_{rxn}^\circ = 2\Delta H_f^\circ(N_2) + 6\Delta H_f^\circ(H_2O) - 4\Delta H_f^\circ(NH_3) - 3\Delta H_f^\circ(O_2)$$
$$= (2 \text{ mol})(0) + (6 \text{ mol})(-285.8 \text{ kJ/mol}) - (4 \text{ mol})(-46.3 \text{ kJ/mol}) - (3 \text{ mol})(0) = -1529.6 \text{ kJ}$$

(b)(ii) We calculate the enthalpy change per kilogram of each substance

$N_2H_4(l)$: $\Delta H^\circ_{rxn} = \left(\frac{-622.0 \text{ kJ}}{1 \text{ mol } N_2H_4}\right)\left(\frac{1 \text{ mol } N_2H_4}{32.05 \text{ g } N_2H_4}\right)\left(\frac{1000 \text{ g}}{1 \text{ kg}}\right) = -1.941 \times 10^4 \text{ kJ/kg } N_2H_4$

$NH_3(g)$: $\Delta H^\circ_{rxn} = \left(\frac{-1529.6 \text{ kJ}}{4 \text{ mol } NH_3}\right)\left(\frac{1 \text{ mol } NH_3}{17.03 \text{ g } NH_3}\right)\left(\frac{1000 \text{ g}}{1 \text{ kg}}\right) = -2.245 \times 10^4 \text{ kJ/kg } NH_3$

Ammonia would be a better fuel.

6.72 If two moles of nitrogen react with six moles of hydrogen, how many moles of ammonia are formed? This problem is similar to Example 6.8

Solution: We initially have 8 moles of gas (2 of nitrogen and 6 of hydrogen). Since our product is 4 moles of ammonia, there is a net loss of 4 moles of gas (8 reactant → 4 product). The corresponding volume loss is

$$V = \frac{nRT}{P} = \frac{(4.0 \text{ mol})(0.0821 \text{ L·atm/K·mol})(298 \text{ K})}{1 \text{ atm}} = 98 \text{ L}$$

Using Equation (6.11): $w = -P\Delta V = -[1 \text{ atm} \times (-98 \text{ L})] = 98 \text{ L·atm } \frac{101.3 \text{ J}}{1 \text{ L·atm}} = 9.9 \times 10^3 \text{ J} = 9.9 \text{ kJ}$

Using Equation (6.12): $\Delta H = \Delta E + P\Delta V$ or $\Delta E = \Delta H - P\Delta V$

Using ΔH as -185.2 kJ = (2×-92.6 kJ), (because the question involves the formation of 4 moles of ammonia, not 2 moles of ammonia for which the standard enthalpy is given in the question), and $-P\Delta V$ as 9.9 kJ (for which we just solved):

$$\Delta E = -185.2 \text{ kJ} + 9.9 \text{ kJ} = -175.3 \text{ kJ}$$

6.74 You can use Appendix 3 for data to calculate the enthalpies. Also refer to Problem 6.47 and 6.48.

Solution: The combustion reactions are: $H_2(g) + \frac{1}{2}O_2(g) \rightarrow H_2O(l)$ $\qquad \Delta H^\circ_{comb} = -285.8 \text{ kJ}$

$CH_4(g) + O_2(g) \rightarrow CO_2(g) + 2H_2O(l)$ $\qquad \Delta H^\circ_{comb} = -890.4 \text{ kJ}$

Advantages: The product is water, which is nonpolluting. No carbon dioxide is produced; thus using hydrogen does not contribute to the greenhouse effect. The supply of hydrogen is essentially unlimited (in the form of water in the oceans).

Disadvantages: Hydrogen produces less heat on a molar basis than methane. At this stage hydrogen is less economical that methane which is the major component of natural gas. Because hydrogen requires a very high pressure to liquify, containers must be very large and strong to store useable quantities of hydrogen gas (see Problem 6.88).

6.76 What is the initial state of this system? What is the final state of this system?

Solution: The initial and final states of this system are identical. Since enthalpy is a state function, its value depends only upon the state of the system. The enthalpy change is zero.

Can you say the same for the change in internal energy? What about w and q?

6.77 The work is the pressure-volume work done in the reaction. What is the volume of the reactants under the conditions given? What is the volume of the products? This problem is exactly analogous to Problem 6.72 and Example 6.8.

Solution: We initially have 6 moles of gas (3 of chlorine and 3 of hydrogen). Since our product is 6 moles of hydrogen chloride, there is no change in the number of moles of gas. Therefore there is no volume change; $\Delta V = 0$.

Using Equation (6.11): $w = -P\Delta V = [-1 \text{ atm} \times (0 \text{ L})] = 0$

Using Equation (6.12): $\Delta H = \Delta E + P\Delta V$ or (since in this case $P\Delta V = 0$), $\Delta H = \Delta E$

Using ΔH as -553.8 kJ = (3 x -184.6 kJ), because the question involves the formation of 6 moles of hydrogen chloride, not 2 moles of hydrogen chloride for which the standard enthalpy is given in the question.

$$\Delta H = \Delta E = -553.8 \text{ kJ}$$

6.78 This question combines the ideal gas law, stoichiometry, and energy calculations (Section 6.7), demonstrating how cumulative the study of chemistry is.

Solution: From the balanced equation we see that two moles of sodium are stoichiometrically equivalent to one mole of hydrogen. The number of moles of hydrogen produced is

$$0.34 \text{ g Na}\left(\frac{1 \text{ mol Na}}{22.99 \text{ g Na}}\right)\left(\frac{1 \text{ mol } H_2}{2 \text{ mol Na}}\right) = 7.4 \times 10^{-3} \text{ mol } H_2$$

Using the ideal gas equation, we write

$$V = \frac{nRT}{P} = \frac{(7.4 \times 10^{-3} \text{ mol})(0.0821 \text{ L·atm/K·mol})(273 \text{ K})}{(1 \text{ atm})} = 0.17 \text{ L}$$

$$\Delta V = 0.17 \text{ L} \qquad w = -P\Delta V = -(1.0 \text{ atm})(0.17 \text{ L}) = -0.17 \text{ L atm}\left(\frac{101.3 \text{ J}}{1 \text{ L atm}}\right) = -17 \text{ J}$$

6.80 This is a Hess's law problem. We simply need to manipulate the given equations so that when summed, they yield the desired equation.

Solution:

Reaction	$\Delta H°$ (kJ)
$2H(g) \rightarrow H_2(g)$	-436.4
$2Br(g) \rightarrow Br_2(g)$	-192.5
$H_2(g) + Br_2(g) \rightarrow 2HBr(g)$	-104
$2H(g) + 2Br(g) \rightarrow 2HBr(g)$	-733 kJ

However, we want the enthalpy for the reaction: $H(g) + Br(g) \rightarrow HBr(g)$ $\Delta H° = \frac{-733\text{ kJ}}{2} = -367\text{ kJ}$

6.82 What kind of reaction is this? The required standard heats of formation are listed in Table 6.4. Will it make a difference if you use the standard enthalpy of formation for gas-phase water rather than liquid-phase?

Solution: Using the balanced equation, we can write

$$\Delta H^\circ_{rxn} = [2\Delta H^\circ_f(CO_2) + 4\Delta H^\circ_f(H_2O)] - [2\Delta H^\circ_f(CH_3OH) + 3\Delta H^\circ_f(O_2)]$$

$$-1452.8\text{ kJ} = (2\text{ mol})(-393.5\text{ kJ/mol}) + (4\text{ mol})(-285.8\text{ kJ/mol}) - (2\text{ mol})\Delta H^\circ_f(CH_3OH) - (3\text{ mol})(0\text{ kJ})$$

$$\Delta H^\circ_f(CH_3OH) = \frac{-477.4\text{ kJ}}{2} = -238.7\text{ kJ/mol}$$

6.84 This problem is a very thorough one involving the use of a bomb calorimeter. Review Section 6.4 before you attempt this problem.

Solution: (a) First solving for the heat capacity of the calorimeter:

$$q_{system} = q_{water} + q_{bomb} + q_{rxn} = 0$$

$$q_{system} = (1000\text{ g})(4.184\text{ J/g}\cdot°C)(3.60°C) + C \times 3.60°C + (-26.43\text{ kJ/g})(0.7521\text{ g}) = 0$$

Solving: $C = 1.34\text{ kJ/°C}$

(b) The combined heat capacity of the water and calorimeter is: $1.34\text{ kJ/°C} + (1000\text{ g})(4.184\text{ kJ/g°C}) = 5.52\text{ kJ/°C}$

$$q_{rxn} = -(q_{water} + q_{bomb}) = -(5.52\text{ kJ/°C})(3.52°C)$$

The molar mass of glucose is 180.2 g/mol, so the heat of combustion of 1 mole of glucose is

$$\Delta H^\circ_{comb}\text{ of glucose} = \left(\frac{-(5.52\text{ kJ/°C})(3.52°C)}{1.251\text{ g glucose}}\right)\left(\frac{180.2\text{ g glucose}}{1\text{ mol glucose}}\right) = -2799\text{ kJ/mol}$$

The reaction corresponding to this enthalpy is: $C_6H_{12}O_6(s) + 6O_2(g) \rightarrow 6CO_2(g) + 6H_2O(l)$

$$\Delta H^\circ_{rxn} = 6\Delta H^\circ_f(CO_2) + 6\Delta H^\circ_f(H_2O) - \Delta H^\circ_f(C_6H_{12}O_6) - 6\Delta H^\circ_f(O_2)$$

$$-2799\text{ kJ} = (6\text{ mol})(-393.5\text{ kJ/mol}) + (6\text{ mol})(-285.8\text{ kJ}) - (1\text{ mol})\Delta H^\circ_f(C_6H_{12}O_6) - (6\text{ mol})(0)$$

$\Delta H^\circ_f(C_6H_{12}O_6) = -1277\text{ kJ/mol}$ (The answer is slightly different from that in Appendix 3.)

(c) $q_{rxn} = -(q_{water} + q_{bomb}) = -(5.52\text{ kJ/°C})(2.57°C)$

The molar mass of lactic acid is 90.08 g/mol, so the heat of combustion of 1 mole of lactic acid is

$$\Delta H^\circ_{comb} \text{ of lactic acid} = \left(\frac{-(5.52\text{ kJ/°C})(2.57\text{°C})}{0.9485\text{ g lactic acid}}\right)\left(\frac{90.08\text{ g lactic acid}}{1\text{ mol lactic acid}}\right) = -1347\text{ kJ/mol}$$

The reaction corresponding to this enthalpy is: $C_3H_6O_3(s) + 3O_2(g) \rightarrow 3CO_2(g) + 3H_2O(l)$

$$\Delta H^\circ_{rxn} = 3\Delta H^\circ_f(CO_2) + 3\Delta H^\circ_f(H_2O) - \Delta H^\circ_f(C_3H_6O_3) - 3\Delta H^\circ_f(O_2)$$

$$-1347\text{ kJ} = (3\text{ mol})(-393.5\text{ kJ/mol}) + (3\text{ mol})(-285.8\text{ kJ/mol}) - (1\text{ mol})\Delta H^\circ_f(C_3H_6O_3) - (3\text{ mol})(0)$$

$$\Delta H^\circ_f(C_3H_6O_3) = -690.7\text{ kJ/mol}$$

(d) The reaction is: $C_6H_{12}O_6(s) \rightarrow 2C_3H_6O_3(s)$

$$\Delta H^\circ_{rxn} = 2[\Delta H^\circ_f(C_3H_6O_3)] - \Delta H^\circ_f(C_6H_{12}O_6)$$

$$\Delta H^\circ_{rxn} = (2\text{ mol})(-690.7\text{ kJ/mol}) - (1\text{ mol})(-1{,}277\text{ kJ/mol}) = -104.4\text{ kJ}$$

6.86 This problem is a straightforward one using Hess's law and Table 6.4.

Solution: Calculating the standard enthalpy for the reaction: $C(s) + \frac{1}{2}O_2(g) \rightarrow CO(g)$

This reaction corresponds to the standard enthalpy of formation for CO, so we use the Table 6.4 value of –110.5 kJ.

Calculating the standard enthalpy for the reaction: $C(s) + H_2O(g) \rightarrow CO(g) + H_2(g)$

$$\Delta H^\circ_{rxn} = [\Delta H^\circ_f(CO) + \Delta H^\circ_f(H_2)] - [\Delta H^\circ_f(C) + \Delta H^\circ_f(H_2O)]$$

$$\Delta H^\circ_{rxn} = = [(1\text{ mol})(-110.5\text{ kJ/mol}) + (1\text{ mol})(0)] - [(1\text{ mol})(0) + (1\text{ mol})(-241.8\text{ kJ/ mol})] = 131.3\text{ kJ}$$

The first reaction, which is exothermic, can be used to promote the second reaction, which is endothermic. Thus the two gases are produced altermately.

6.88 This problem is an example of a real and valuable type of calculation that you can perform using enthalpies and the ideal gas law. The thermodynamic data that you need is in Table 6.4.

Solution: First, calculating the energy produced by one gallon of octane C_8H_{18}

$$C_8H_{18}(l) + \frac{25}{2}O_2(g) \rightarrow 8CO_2(g) + 9H_2O(l)$$

$$\Delta H^\circ_{rxn} = 8\Delta H^\circ_f(CO_2) + 9\Delta H^\circ_f(H_2O) - [\Delta H^\circ_f(C_8H_{18}) + \frac{25}{2}\Delta H^\circ_f(O_2)]$$

$$\Delta H^\circ_{rxn} = (8\text{ mol})(-393.5\text{ kJ/mol}) + (9\text{ mol})(-285.8\text{ kJ/mol}) - [(1\text{ mol})(-249.9\text{ kJ/mol}) + (\frac{25}{2}\text{ mol})(0)] = -5470\text{ kJ}$$

The heat of combustion for one gallon of octane is

$$\left(\frac{2660\ g}{1\ gal}\right)\left(\frac{1\ mol\ octane}{114.2\ g\ octane}\right)\left(\frac{-5470\ kJ}{1\ mol\ octane}\right) = -1.274 \times 10^5\ kJ$$

The combustion of hydrogen corresponds to the standard heat of formation of water: $H_2(g) + \frac{1}{2}O_2(g) \rightarrow H_2O(g)$

Thus ΔH°_{rxn} is the same as ΔH°_f for $H_2O(g)$, which has a value of –285.8 kJ (from Table 6.4). The number of moles of hydrogen required to produce 1.274×10^5 kJ of heat is

$$n_{H_2} = \frac{1.274 \times 10^5\ kJ}{285.8\ kJ/mol} = 445.8\ mol\ H_2$$

Using the ideal gas law to calculate the volume of gas corresponding to 445.8 moles at 1 atm at 25°C.

$$V = \frac{nRT}{P} = \frac{(445.8\ mol)(0.0821\ L{\cdot}atm/K{\cdot}mol)(298\ K)}{1\ atm} = 1.09 \times 10^4\ L$$

That is, the volume of hydrogen that is energy-equivalent to one gallon of gasoline is over ten thousand liters!

6.90 To work this problem, we first must calculate the amount of ethane that would be required to heat the water. Knowing the amount of gas (in moles), we then can calculate the volume using the ideal gas law. The thermodynamic data for ethane is in Appendix 3. The problem is also similar to 6.90 above.

Solution: The combustion reaction is: $C_2H_6(l) + \frac{7}{2}O_2(g) \rightarrow 2CO_2(g) + 3H_2O(l)$

$$\Delta H^{\circ}_{rxn} = 2\Delta H^{\circ}_f(CO_2) + 3\Delta H^{\circ}_f(H_2O) - [\Delta H^{\circ}_f(C_2H_6) + \frac{7}{2}\Delta H^{\circ}_f(O_2)]$$

$$\Delta H^{\circ}_{rxn} = (2\ mol)(-393.5\ kJ/mol) + (3\ mol)(-285.8\ kJ/mol) - [(1\ mol)(-84.7\ kJ/mol) + (\frac{7}{2}\ mol)(0)] = -1560\ kJ$$

Heat required to heat the water = $ms\Delta t$ = (855 g)(4.184 J/g·°C)(98.0 – 25)°C = 2.61×10^5J = 261 kJ

The combusion of 1 mole of ethane produces 1560 kJ; the number of moles required to produce 261 kJ is

$$\frac{261\ kJ}{1560\ kJ} = 0.167\ mol$$

The volume of ethane is:

$$V = \frac{nRT}{P} = \frac{(0.167\ mol)(0.0821\ L{\cdot}atm/K{\cdot}mol)(296\ K)}{(752/760)(atm)} = 4.10\ L$$

6.91 This problem is a very simple application of Hess's Law. Review Section 6.5 if you need help.

Solution: Rearranging the equations as necessary so they can be added to yield the desired equation:

$$2B \rightarrow A \qquad -\Delta H_1$$
$$A \rightarrow C \qquad \Delta H_2$$
$$2B \rightarrow C \qquad \Delta H_2 - \Delta H_1$$

6.93 In order to work this problem, the assumption must be made that all of the heat required to vaporize the liquid nitrogen has come from cooling the water.

Solution: The heat lost by the water is calculated by using Equation (6.3):

$$q = ms\Delta t = (2.00 \times 10^2 \text{ g})(4.184 \text{ J/g}\bullet{}^\circ\text{C})(41.0^\circ\text{C} - 55.3^\circ\text{C}) = 1.20 \times 10^4 \text{ J}$$

The heat gained by the nitrogen is 1.20×10^4 J which is used to calculate the molar heat of vaporization ΔH_{vap}.

$$1.20 \times 10^4 \text{ J} = (60.0 \text{ g } N_2)\left(\frac{\text{mol } N_2}{28.02 \text{ g } N_2}\right)\Delta H_{vap}; \qquad \Delta H_{vap} = 5.60 \times 10^3 \text{ J/mol} = 5.60 \text{ kJ/mol}$$

6.94 What is the definition of standard enthalpy of formation? Refer to Section 6.5 for review.

Solution: Equation (a) corresponds to standard enthalpy of formation because the heat change that is measured is for the formation of one mole of compound from the elements in their standard states. Equations (b), (c) and (d) do not meet this criterion, so the standard enthapies of reaction do not correspond to standard enthalpies of formation. In Equation (b), diamond is not the stable allotrope of carbon; the stable allotrope is graphite.

6.95 This problem is similar to Example 6.3 in text.

Solution: The heat produced by the reaction heats the solution and the calorimeter: $q_{rxn} = -(q_{soln} + q_{cal})$

$$q_{soln} = ms\Delta t = (50.0 \text{ g})(4.184 \text{ J/g}\bullet{}^\circ\text{C})(22.17^\circ\text{C} - 19.25^\circ\text{C}) = 611 \text{ J}$$

$$q_{cal} = C\Delta t = (98.6 \text{ J/}^\circ\text{C})(22.17^\circ\text{C} - 19.25^\circ\text{C}) = 288 \text{ J}$$

$$-q_{rxn} = (q_{soln} + q_{cal}) = (611 + 288) \text{ J} = 899 \text{ J}$$

The 899 J produced was for 50.0 mL of a 0.100 *M* $AgNO_3$ solution.

$$0.0500 \text{ L} \times 0.100\ M = 5.00 \times 10^{-3} \text{ mol of } Ag^+.$$

On a molar basis the heat produced was:

$$(899 \text{ J})/(5.00 \times 10^{-3} \text{ mol } Ag^+) = 1.80 \times 10^5 \text{ J/mol } Ag^+ = 180 \text{ kJ/mol } Ag^+$$

The balanced equation involves 2 moles of Ag^+, so the heat produced is 2×180 kJ/mol = 360 kJ

Since the reaction produces heat (or by noting the sign convention above), then:

q_{rxn} = -360 kJ/mol Zn (or -360 kJ/2mol Ag^+)

6.96 What is the definition of work in this case?

Solution: Using equation (6.11) $w = -P\Delta V = -(1.0\text{ atm})(0.0196\text{ L} - 0.0180\text{ L}) = -1.6 \times 10^{-3}\text{ L}\bullet\text{atm}$

Using the conversion factor 1 L•atm = 101.3 J: $w = (-1.6 \times 10^{-3}\text{ L}\bullet\text{atm})\left(\frac{101.3\text{ J}}{\text{L}\bullet\text{atm}}\right) = -0.16\text{ J}$

6.98 This problem requires the simple use of Hess's law in addition to knowing the equation relating to the standard enthalpy of formation for diamond.

Solution: The equation corresponding to the standard enthalpy of formation of diamond is:

$$C(graphite) \rightarrow C(diamond)$$

Adding the equations:

$$C(graphite) + O_2(g) \rightarrow CO_2(g) \qquad \Delta H° = -393.5\text{ kJ}$$
$$CO_2(g) \rightarrow C(diamond) + O_2(g) \qquad \Delta H° = 395.4\text{ kJ}$$

$$C(graphite) \rightarrow C(diamond) \qquad \Delta H° = 1.9\text{ kJ}$$

Since the reverse reaction of changing diamond to graphite is exothermic, need you worry about any diamonds that you might have changing to graphite?

6.100 The equation is given in Problem 4.10.

Solution: The balanced equation is: $C_6H_{12}O_6 \rightarrow 2C_2H_5OH + 2CO_2$

Using the thermodynamic data in Appendix 3 and Equation (6.7),

$$\Delta H° = [2\Delta H_f^\circ(C_2H_5OH) + 2\Delta H_f^\circ(CO_2)] - \Delta H_f^\circ(C_6H_{12}O_6)$$
$$\Delta H° = (2\text{ mol})(-276.98\text{ kJ/mol}) + (2\text{ mol})(-393.5\text{ kJ/mol}) - (1\text{ mol})(-1274.5\text{ kJ/mol}) = -66.5\text{ kJ}$$

6.101 From a thermochemistry viewpoint, does the equation represent any type of special enthalpy?

Solution: The equation represents twice the standard enthalpy of formation of Fe_2O_3. From Appendix 3, the standard enthalpy of formation of Fe_2O_3 = -822.2 kJ, which is also for 2 moles of Fe.

$$\left(\frac{-822.2\text{ kJ}}{2\text{ mol Fe}}\right)\left(\frac{\text{mol Fe}}{55.85\text{ g Fe}}\right)(250\text{ g Fe}) = -1.84 \times 10^3\text{ kJ}$$

Note: If you have not already done so, please read the "Introduction for the Student" at the beginning of this Student Solutions Manual.

Chapter 7

QUANTUM THEORY AND THE ELECTRONIC STRUCTURE OF ATOMS

7.8 This is basically a factor-label method problem in which the speed of light functions as one of the unit factors (the distance traveled by any light wave in 1.00 s). Use the value for the speed of light given in the subsection on Electromagnetic Radiation in Section 7.1.

Solution: Since the speed of light is 3.00 x 10^8 m/s, we can write

$$1.3 \times 10^8 \text{ mi}\left(\frac{1.61 \text{ km}}{1 \text{ mi}}\right)\left(\frac{1000 \text{ m}}{1 \text{ km}}\right)\left(\frac{1}{3.00 \times 10^8 \text{m/s}}\right) = 7.0 \times 10^2 \text{ s}$$

Would the time be different for other types of electromagnetic radiation?

7.10 Part (a) is similar to Example 7.2. Study the subsection of Section 7.1 on the Properties of Waves and try to work the example with the answer covered. If you can't, go through the material again with greater care and study the answer. Rework the example. Part (b) is the reverse of Example 7.2. Follow the same directions as for part (a).

Solution: (a) Since $\lambda\nu = c$,:

$$\nu = \frac{c}{\lambda} = \left(\frac{3.00 \times 10^8 \text{ m/s}}{456 \text{ nm}}\right)\left(\frac{10^9 \text{ nm}}{1 \text{ m}}\right) = 6.58 \times 10^{14} \text{ s}^{-1} \text{ or Hz}$$

(b) We can write:

$$\lambda = \frac{c}{\nu} = \left(\frac{3.00 \times 10^8 \text{ m/s}}{2.20 \times 10^9 \text{ s}^{-1}}\right)\left(\frac{10^9 \text{ nm}}{1 \text{ m}}\right) = 1.36 \times 10^8 \text{ nm}$$

7.12 This problem is similar to Problem 7.10(b). Refer to Figure 7.4 to identify the color corresponding to the wavelength that you calculate.

Solution: $$\lambda = \frac{c}{\nu} = \frac{3.00 \times 10^8 \text{ m/s}}{9192631770 \text{ /s}} = 3.26 \times 10^{-2} \text{ m} = 3.26 \times 10^7 \text{ nm}$$

This radiation falls in the microwave region of the spectrum.

7.16 This problem is similar to Example 7.3. Try working that problem before you attempt this one. Be careful with your units in this calculation.

Solution: $$E = h\nu = \frac{hc}{\lambda} = \frac{(6.63 \times 10^{-34}\ \text{J s})(3.00 \times 10^{8}\ \text{m/s})}{624 \times 10^{-9}\ \text{m}} = 3.19 \times 10^{-19}\ \text{J}$$

7.18 Part (a) of this problem is simply a frequency-to-wavelength calculation like Problem 7.12. Part (b) is a frequency-to-energy conversion like Problem 7.16, and part (c) involves the conversion of the energy of *one* photon to the energy of *one mole* of photons.

Solution: (a) $$\lambda = \frac{c}{\nu} = \frac{3.00 \times 10^{8}\ \text{m/s}}{6.0 \times 10^{4}\ \text{/s}} = 5.0 \times 10^{3}\ \text{m} = 5.0 \times 10^{12}\ \text{nm}$$

The radiation does not fall in the visible region; it is radio radiation.

(b) Using Equation (7.2): $E = h\nu = (6.63 \times 10^{-34}\ \text{J s})(6.0 \times 10^{4}\ \text{/s}) = 4.0 \times 10^{-29}\ \text{J}$

(c) Converting to J/mol: $E = (4.0 \times 10^{-29}\ \text{J})(6.022 \times 10^{23}) = 2.4 \times 10^{-5}\ \text{J}$

7.19 In this problem you are to find a photon wavelength from the energy. Use the combined form of Equations (7.1) and (7.2) employed in Example 7.3. What does E represent in that example? Is that the same as the energy given in this problem? When you have found the wavelength, locate the region of the spectrum in Figure 7.4.

Solution: The energy given in this problem is for *one mole* of photons. To apply E = hν we must divide by Avogadro's number. The energy of one photon is

$$E = \left(\frac{1.0 \times 10^{3}\ \text{kJ}}{1\ \text{mol}}\right)\left(\frac{1\ \text{mol}}{6.022 \times 10^{23}}\right)\left(\frac{1000\ \text{J}}{1\ \text{kJ}}\right) = 1.7 \times 10^{-18}\ \text{J}$$

The wavelength of this photon can be found using the relationship E = hc/λ.

$$\lambda = \frac{hc}{E} = \frac{(6.63 \times 10^{-34}\ \text{J·s})(3.00 \times 10^{8}\text{m/s})}{1.7 \times 10^{-18}\ \text{J}}\left(\frac{10^{9}\ \text{nm}}{1\ \text{m}}\right) = 1.2 \times 10^{2}\ \text{nm}$$

The radiation is in the ultraviolet region.

7.20 This problem is a straightforward example of the application of Equation (7.2).

Solution: $$E = h\nu = \frac{hc}{\lambda} = \frac{(6.63 \times 10^{-34}\ \text{J s})(3.00 \times 10^{8}\ \text{m/s})}{(0.154 \times 10^{-9}\ \text{m})} = 1.29 \times 10^{-15}\ \text{J}$$

7.22 Study Section 7.2 on the photoelectric effect. What particle is ejected from the surface of a metal in this phenomenon? What is the electric charge of the particle? What would be the effect on the metal surface of losing these particles? Over a long period of time, might this effect have some influence on the ejected particles?

$E = \frac{hc}{\lambda} = h\nu$

> ***Solution:*** In the photoelectric effect, light of sufficient energy shining on a metal surface causes electrons to be ejected (photoelectrons). Since the electrons are charged particles, the metal surface becomes positively charged as more electrons are lost. After a long enough period of time, the positive surface charge becomes large enough to start attracting the ejected electrons back toward the metal with the result that the kinetic energy of the departing electrons becomes smaller.

7.30 Study Section 7.3. What is an emission spectrum? What is a line spectrum? Are the emission line spectra of different elements different from each other? Why?

> ***Solution:*** The arrangement of energy levels for each element is unique. The frequencies of light emitted by an element are characteristic of that element. Even the frequencies emitted by isotopes of the same element are very slightly different.

7.31 How is light broken up into its components?

> ***Solution:*** The emitted light could be analyzed by passing it through a prism.

7.32 Study the "Chemistry in Action" piece on fluorescent lights. Does ultraviolet light have lower or higher energy than visible light?

> ***Solution:*** Light emitted by fluorescent materials always has lower energy than the light striking the fluorescent substance. Absorption of visible light could not give rise to emitted ultraviolet light because the latter has higher energy.
>
> The reverse process, ultraviolet light producing visible light by fluorescence, is very common. Certain brands of laundry detergents contain materials called "optical brighteners" which, for example, can make a white shirt look much whiter and brighter than a similar shirt washed in ordinary detergent. How do you think "optical brighteners" might work?

7.33 This problem is closely related to the discussion in Problem 7.30.

> ***Solution:*** Excited atoms of the chemical elements emit the same characteristic frequencies or lines in a terrestrial laboratory, in the sun, or in a star many light years distant from Earth.

7.34 This problem is similar in part to Example 7.4 which involves the use of Equation (7.5) to calculate the difference in energy between two states of the hydrogen atom. Study Section 7.3, especially the part on the Emission Spectrum of the Hydrogen Atom. Try to work Example 7.4 with the answer covered. Use Equation (7.5) to calculate the energy difference, then convert to the frequency and wavelength. If you have trouble, study the answer carefully.

In this problem you do not need to use Equation (7.5) since you are given the energies of the states. In part (a) you are to calculate the photon wavelength corresponding to the indicated energy difference. This conversion is like Problem 7.19. In part (b) you must find the energy difference between the two states. In (c) you must again find a photon wavelength from a computed energy difference.

Solution: (a) The energy difference between states E_1 and E_4 is

$$E_4 - E_1 = (-1.0 \times 10^{-19})\text{J} - (-15 \times 10^{-19})\text{J} = 14 \times 10^{-19}\ \text{J}$$

We calculate the wavelength as in Example 7.3

$$\lambda = \frac{hc}{\Delta E} = \frac{(6.63 \times 10^{-34}\ \text{J·s})(3.00 \times 10^{8}\text{m/s})}{14 \times 10^{-19}\ \text{J}} = 1.4 \times 10^{-7}\ \text{m or } 1.4 \times 10^{2}\ \text{nm}$$

(b) The energy difference between states E_2 and E_3

$$E_3 - E_2 = (-5.0 \times 10^{-19}\ \text{J}) - (-10 \times 10^{-19}\ \text{J}) = 5 \times 10^{-19}\ \text{J}$$

(c) The energy difference between states E_1 and E_3 is

$$E_1 - E_3 = (-15 \times 10^{-19}\ \text{J}) - (-5.0 \times 10^{-19}\ \text{J}) = -10 \times 10^{-19}\ \text{J}$$

Ignoring the negative sign for ΔE, the wavelength is found as in part (a)

$$\lambda = \frac{hc}{\Delta E} = \frac{(6.63 \times 10^{-34}\ \text{J·s})(3.00 \times 10^{8}\text{m/s})}{10 \times 10^{-19}\ \text{J}} = 2.0 \times 10^{-7}\ \text{m or } 2.0 \times 10^{2}\ \text{nm}$$

$\Delta E = \frac{hc}{\lambda}$

7.35 This is the same as Example 7.4 which involves the use of Equation (7.5) to calculate the difference in energy between two states of the hydrogen atom. Study Section 7.3, especially the part on the Emission Spectrum of the Hydrogen Atom. Try to work Example 7.4 with the answer covered. Use Equation (7.5) to calculate the energy difference, then convert that to the frequency and wavelength. If you have trouble, study the answer carefully.

Solution: In this problem $n_i = 5$ and $n_f = 3$. The photon energy can be found using Equation (7.5)

$$\Delta E = R_H\left(\frac{1}{n_i^2} - \frac{1}{n_f^2}\right) = (2.18 \times 10^{-18}\ \text{J})\left(\frac{1}{5^2} - \frac{1}{3^2}\right) = -1.55 \times 10^{-19}\ \text{J}$$

The sign of ΔE means that this is energy associated with an emission process. As in the above problem we have

$$\lambda = \frac{hc}{\Delta E} = \frac{(6.63 \times 10^{-34}\ \text{J·s})(3.00 \times 10^{8}\text{m/s})}{1.55 \times 10^{-19}\ \text{J}} = 1.28 \times 10^{-6}\ \text{m or } 1.28 \times 10^{3}\ \text{nm}$$

Is the sign of the energy change consistent with the sign conventions for exo- and endothermic processes (Section 7.3)?

7.36 The Balmer series involves transitions to the $n_f = 2$ energy level. However, to work this problem, this piece of information is not needed.

Solution: We use more accurate values of h and c for this problem.

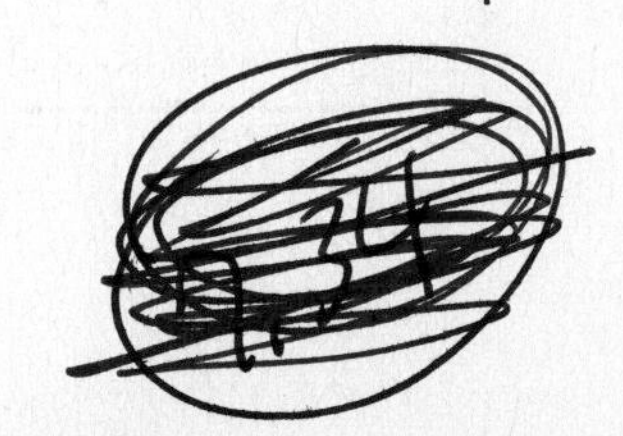

$$E = h\nu = \frac{hc}{\lambda} = \frac{(6.6256 \times 10^{-34}\ J\ s)(2.998 \times 10^{8}\ m/s)}{656.3 \times 10^{-9}\ m} = 3.027 \times 10^{-19}\ J$$

7.38 In this problem you must compute the two energies (Problem 7.16) and take the difference. Look at the two wavelengths. Do you think the energy difference will be large or small? Calculate it for one photon and for one mole of photons. For comparison, the energy needed to separate the two atoms in the hydrogen molecule (H_2) is 436,400 J/mol.

Solution: This problem must be worked to four-significant-figure accuracy. We use 6.6256×10^{-34} J·s for Planck's constant. First calculate the energy of each of the photons

$$E = \frac{hc}{\lambda} = \frac{(6.6256 \times 10^{-34}\ J{\cdot}s)(2.998 \times 10^{8} m/s)}{589.0 \times 10^{-9}\ m} = 3.372 \times 10^{-19}\ J$$

$$E = \frac{hc}{\lambda} = \frac{(6.6256 \times 10^{-34}\ J{\cdot}s)(2.998 \times 10^{8} m/s)}{589.6 \times 10^{-9}\ m} = 3.369 \times 10^{-19}\ J$$

For *one* photon the energy difference is

$$\Delta E = (3.372 \times 10^{-19}\ J) - (3.369 \times 10^{-19}\ J) = 3 \times 10^{-22}\ J \quad \text{(Why only one significant figure?)}$$

For *one mole* of photons the energy difference is

$$(3 \times 10^{-22}\ J)\left(\frac{6.022 \times 10^{23}}{1\ mol}\right) = 2 \times 10^{2}\ J/mol$$

7.39 In this problem you must first calculate the photon energy from the wavelength (Problem 7.16), then use Equation (7.5) to find n_i. Will the energy change (ΔE) for the hydrogen atom be positive or negative (see Problem 7.35)?

Solution: The photon energy is

$$E = \frac{hc}{\lambda} = \frac{(6.63 \times 10^{-34}\ J{\cdot}s)(3.00 \times 10^{8} m/s)}{434 \times 10^{-9}\ m} = 4.58 \times 10^{-19}\ J$$

Since this is an emission process, the energy change ΔE must be negative, or -4.58×10^{-19} J. From Equation (7.5) we have

$$\Delta E = -4.58 \times 10^{-19}\ J = R_H\left(\frac{1}{n_i^2} - \frac{1}{n_f^2}\right) = (2.18 \times 10^{-18}\ J)\left(\frac{1}{n_i^2} - \frac{1}{2^2}\right)$$

$$\frac{1}{n_i^2} = \left(\frac{-4.58 \times 10^{-19}\ J}{2.18 \times 10^{-18}\ J}\right) + \frac{1}{2^2} = -0.210 + 0.250 = 0.040$$

$$n_i = \frac{1}{\sqrt{0.040}} = 5$$

7.44 This is like Example 7.5. Read Section 7.4 and try the example with the answer covered. Remember that using the value 6.63 x 10^{-34} Js for Planck's constant requires that masses be in kg and speeds in m/s. Wavelength will be in meters. If you have trouble, study the answer and rework the example before doing this problem.

Solution: Using Equation (7.7) we have

$$\lambda = \frac{h}{mu} = \frac{6.63 \times 10^{-34}\ \text{J}\cdot\text{s}}{(1.675 \times 10^{-27}\ \text{kg})(4.00 \times 10^{1}\ \text{m/s})} = 9.90 \times 10^{-9}\ \text{m} = 9.90\ \text{nm}$$

How does this wavelength compare with the diameter of an atom (0.1 - 0.4 nm)? How does this wavelength compare with the diameter of a neutron (0.01 pm)? Can a neutron pass through a hole with a diameter equal to this wavelength?

7.46 This problem is analogous to Problem 7.44 above and to Example 7.5 in the text.

Solution: We first convert mph to m/s:

$$\left(\frac{35\ \text{mi}}{1\ \text{h}}\right)\left(\frac{1.61\ \text{km}}{1\ \text{mi}}\right)\left(\frac{1000\ \text{m}}{1\ \text{km}}\right)\left(\frac{1\ \text{h}}{3600\ \text{s}}\right) = 16\ \text{m/s}$$

We then use Equation (7.7)

$$\lambda = \frac{h}{mu} = \frac{6.63 \times 10^{-34}\ \text{J s}}{(2.5 \times 10^{-3}\ \text{kg})(16\ \text{m/s})} = 1.7 \times 10^{-32}\ \text{m} = 1.7 \times 10^{-23}\ \text{nm}$$

7.58 Study the discussion of quantum numbers in Section 7.7. Try to focus on the interdependence of the allowed values of the quantum numbers. If the principal quantum number n has the value $n = 2$, what values can the angular momentum quantum number l have? What about the magnetic quantum number m_l?

Solution: The angular momentum quantum number l can have integral (i.e. whole number) values from 0 to n-1. In this case $n = 2$, so the allowed values of the angular momentum quantum number are 0 and 1. If you don't know what angular momentum is, don't worry about it.

Each allowed value of the angular momentum quantum number labels a subshell. Within a given subshell (label l) there are $2l$+1 allowed energy states (orbitals) each labeled by a different value of the magnetic quantum number. The allowed values run from -l through 0 to +l (whole numbers only). For the subshell labeled by the angular momentum quantum number $l = 1$, the allowed values of the magnetic quantum number are -1, 0, and 1. For the other subshell in this problem labeled by the angular momentum quantum number $l = 0$, the allowed value of the magnetic quantum number is 0. Don't worry about why it's called the magnetic quantum number.

If the allowed whole number values run from -l to +l, are there always $2l$+1 values? Why?

7.60 In the symbol "4p" what do the "4" and the "p" mean? Study Sections 7.7, 7.8, and Table 7.2. See Examples 7.6 and 7.8 in the text.

Solution: (a) 2p: $n = 2$, $l = 1$, $m_l = 1, 0,$ or -1

(b) 3s: $n = 3$, $l = 0$, $m_l = 0$ (only allowed value)
(c) 5d: $n = 5$, $l = 2$, $m_l = 2, 1, 0, -1$, or -2

An orbital in a subshell can have any of the allowed values of the magnetic quantum number for that subshell. All the orbitals in a subshell have exactly the same energy.

7.62 Read the discussion of p orbitals in Section 7.8.

Solution: The two are identical in size, shape, and energy. They differ only in their orientation with respect to each other.

Can you assign a specific value of the magnetic quantum number to these orbitals? What are the allowed values of the magnetic quantum number for the 2p subshell?

7.64 This is like Problems 7.58 and 7.60 above, and to Examples 7.6 and 7.8 in the text.

Solution:

(a) 4p: $n = 4$, $l = 1$, $m_l = 1, 0$, or -1 (3 orbitals)

(b) 3d: $n = 3$, $l = 2$, $m_l = 2, 1, 0, -1$, or -2 (5 orbitals)

(c) 3s: $n = 3$, $l = 0$, $m_l = 0$ (1 orbital)

(d) 5f: $n = 5$, $l = 3$, $m_l = 3, 2, 1, 0, -1, -2$, or -3 (7 orbitals)

7.65 This is similar to Example 7.7, but you must also enumerate and label the subshells. Study Sections 7.7 and 7.8. What are the allowed values of the angular momentum quantum number when $n = 6$? What are the assigned "names" of these subshells (6s, 6p, etc.)? How many orbitals are in each subshell?

Solution: The allowed values of l are 0, 1, 2, 3, 4, and 5 (what formula tells you this?). These correspond to the 6s, 6p, 6d, 6f, 6g, and 6h subshells (what would come after h?). These subshells each have one, three, five, seven, nine, and eleven orbitals, respectively. In general, how many orbitals are in a subshell labeled by angular momentum quantum number l?

7.66 How many electrons can occupy one orbital? Study Section 7.9 and Example 7.9 in the text if you don't know.

Solution: There can be a maximum of two electrons occupying one orbital. (a) two; (b) six; (c) ten; (d) fourteen.

What rule of nature demands a maximum of two electrons per orbital? Do they have the same energy? How are they different? Would five 4d orbitals hold as many electrons as five 3d orbitals? In other words, does the principal quantum number n affect the number of electrons in a given subshell?

7.67 On the surface this may seem like an impenetrable problem. Often with such cases it is good problem solving technique to start by writing down the first few numbers in hope of seeing a pattern. Make a table in which one column lists the values of n (1, 2, 3,...) and a second column lists the number of orbitals (see Example 7.9). Do this up to $n = 5$ or 6. Do you see a pattern? How many electrons per orbital?

Solution: The table would look like this:

n value	orbital sum	total
1	1	1
2	1 + 3	4
3	1 + 3 + 5	9
4	1 + 3 + 5 + 7	16
5	1 + 3 + 5 + 7 + 9	25
6	1 + 3 + 5 + 7 + 9 + 11	36

In each case the total number of orbitals is just the square of the value of n. The total number of electrons is $2n^2$.

If you are mathematically inclined, try to prove the validity of the formula by mathematical induction. If you don't know what that is, don't bother to try.

7.68 This is similar to Problem 7.66 above.

Solution: 3s: two; 3d: ten; 4p: six; 4f: fourteen; 5f: fourteen

EXAM 1 ENDS HERE

7.69 Work out the electron configurations of these elements (Sections 7.9 and 7.10) and count the appropriate types of electrons.

Solution: The electron configuration for the elements are

(a) N: $1s^2 2s^2 2p^3$. There are three p-type electrons.
(b) Si: $1s^2 2s^2 2p^6 3s^2 3p^2$. There are six s-type electrons.
(c) S: $1s^2 2s^2 2p^6 3s^2 3p^4$. There are no d-type electrons.

7.70 In other words make a diagram showing all the orbitals allowed for n = 1 to 4.

Solution: See Figure 7.27 in your textbook.

7.71 See the discussion of shielding in Section 7.9.

Solution: The 3p orbital electrons are more effectively shielded by the inner electrons of the atom (that is, the 1s, 2s, and 2p electrons) than the 3s electrons. The 3s orbital is said to be more "penetrating" than the 3p and 3d orbitals. In the hydrogen atom there is only one electron so the 3s, 3p, and 3d orbitals have the same energy.

7.72 Equation (7.4) allows the calculation of the energies of the different orbitals of the hydrogen atom. On which quantum number(s) does the energy depend? On which, if any, doesn't it depend?

Solution: Equation (7.4) gives the orbital energy in terms of the principal quantum number n alone (for the hydrogen atom). The energy does not depend on any of the other quantum numbers. If two orbitals in the hydrogen atom have the same value of n, they have equal energy. (a) 2s > 1s; (b) 3p > 2p; (c) equal; (d) equal; (e) 5s > 4f.

7.83 Review the rules for the allowed values of the angular momentum, the magnetic, and the electron spin quantum numbers in Section 7.7. Remember that when the quantum numbers for a single electron are tabulated, they are in the order (principal quantum number, angular momentum number, magnetic quantum number, electron spin quantum number) $=(n, l, m_l, m_s)$.

Solution: (a) is wrong because the magnetic quantum number m_l can have only whole number values.

(c) is wrong because the maximum value of the angular momentum quantum number l is $n - 1$.

(e) is wrong because the electron spin quantum number m_l can have only half-integral values.

7.84 What does the Stern-Gerlach experiment detect? See Section 7.7.

Solution: Helium atoms would produce only one spot. Both electrons are paired and the atoms have no net electron spin or magnetic moment.

Lithium atoms would produce two spots just like the silver atoms in the original Stern-Gerlach experiment. There is one unpaired electron which can have either of two values for the electron spin quantum number.

Fluorine atoms would produce two spots. There is only one unpaired electron.

Neon atoms would produce only one spot for the same reasons as helium.

7.86 Find the elements in Table 7.3 and apply Hund's rule (Section 7.9) when appropriate. When in doubt, make an orbital box diagram as in Problem 7.87 below.

Solution: B(1), Ne(0), P(3), Sc(1), Mn(5), Se(2), Zr(2), Ru(4), Cd(0), I(1), W(4), Pb(2), Ce(2), Ho(3).

7.87 Make orbital box diagrams for the three species and apply Hund's rule (Section 7.9).

Solution:

↑↓	↑ | ↑ | ↑	↑↓	↑↓ | ↑ | ↑	↑↓	↑↓ | ↑↓ | ↑
$3s^2$	$3p^3$	$3s^2$	$3p^4$	$3s^2$	$3p^5$

S^+ (5 valence electrons)
3 unpaired electrons

S (6 valence electrons)
2 unpaired electrons

S^- (7 valence electrons)
1 unpaired electron

7.88 This is like Example 7.11.

Solution:

B:	$1s^22s^22p^1$	As:	$[Ar]4s^23d^{10}4p^3$
V:	$[Ar]4s^23d^3$	I:	$[Kr]5s^24d^{10}5p^5$
Ni:	$[Ar]4s^23d^8$	Au:	$[Xe]6s^14f^{14}5d^{10}$

What is the meaning of "[Ar]"? of "[Kr]"? of "[Xe]"?

7.90 These should be ground state electron configurations. Each subshell should be filled to the maximum extent possible. The Pauli exclusion principle must be obeyed. The numbers of electrons must match the atomic numbers of the elements.

Solution: For aluminum there are not enough electrons in the 2p subshell (how many should there be?). The number of electrons (13) is correct. The electron configuration should be $1s^22s^22p^63s^23p^1$. The configuration shown might be an excited state of an aluminum atom.

For boron there are too many electrons. The electron configuration should be $1s^22s^22p^1$. What would be the electric charge of a boron atom with the electron arrangement given in the problem?

For fluorine there are also too many electrons. The configuration shown is that of the F^- ion. The correct electron configuration is $1s^22s^22p^5$.

.92 Study Section 7.9.

Solution: Applying the Pauli exclusion principle and Hund's rule

(a) [↑↓] [↑↓] [↑↓][↑↓][↑]

$1s^2$ $2s^2$ $2p^5$

(b) [↑↓] [↑↓] [↑↓][↑↓][↑↓] [↑↓] [↑][↑][↑]

$1s^2$ $2s^2$ $2p^6$ $3s^2$ $3p^3$

(c) [↑↓] [↑↓] [↑↓][↑↓][↑↓] [↑↓] [↑↓][↑↓][↑↓] [↑↓] [↑↓][↑↓][↑][↑][↑]

$1s^2$ $2s^2$ $2p^6$ $3s^2$ $3p^6$ $4s^2$ $3d^7$

7.93 Is the number of electrons in an excited *neutral* atom the same as when the atom is in its ground electronic state? See Problems 7.88 and 7.89 for ground-state electron configurations.

Solution: The excited atoms are still neutral so the number of electrons is the same as the atomic number of the element.

(a) N (7 electrons) $1s^22s^22p^3$
(b) He (2 electrons) $1s^2$
(c) Na (11 electrons) $1s^22s^22p^63s^1$

7.91

(d) As (33 electrons) $[Ar]4s^2 3d^{10} 4p^3$
(e) Cl (17 electrons) $[Ne]3s^2 3p^5$

Could you still write ground-state electron configurations if (a)-(e) were ions of unknown charge? Could you identify the elements?

7.94 This is like Problem 7.19.

Solution: We first calculate the wavelength, then we find the color using Figure 7.4.

$$\lambda = \frac{hc}{E} = \frac{(6.63 \times 10^{-34}\ \text{Js})(3.00 \times 10^{8}\ \text{m/s})}{4.30 \times 10^{-19}\ \text{J}} = 4.63 \times 10^{-7}\ \text{m} = 463\ \text{nm, which is blue.}$$

7.95 These concepts are discussed in Sections 7.1-7.3.

Solution: (a) Wavelength and frequency are reciprocally related properties of any wave. The two are connected through Equation (7.1). See Example 7.1 for a simple application of the relationship to a light wave.

(b) Typical wave properties: wavelength, frequency, characteristic wave speed (sound, light, etc.). Typical particle properties: mass, speed or velocity, momentum (mass × velocity), kinetic energy. For phenomena that we normally perceive in everyday life (macroscopic world) these properties are mutually exclusive. At the atomic level (microscopic world) objects can exhibit characteristic properties of both particles and waves. This is completely outside the realm of our everyday common sense experience and is extremely difficult to visualize.

(c) Quantization of energy means that emission or absorption of only discrete energies is allowed (e.g., atomic line spectra). Continuous variation in energy means that all energy changes are allowed (e.g., continuous spectra).

7.98 Refer to Problem 7.35 and Example 7.4 for similar problems. The Lyman series refers to transitions to the $n = 1$ level and the Balmer series refers to transitions to the $n = 2$ level.

Solution: For the Lyman series, we want the longest wavelength (smallest energy), with $n_i = 2$ and $n_f = 1$. Using Equation (7.5)

$$\Delta E = R_H\left(\frac{1}{n_i^2} - \frac{1}{n_f^2}\right) = (2.18 \times 10^{-18}\ \text{J})\left(\frac{1}{2^2} - \frac{1}{1^2}\right) = -1.64 \times 10^{-18}\ \text{J}$$

$$\lambda = \frac{hc}{\Delta E} = \frac{(6.63 \times 10^{-34}\ \text{J·s})(3.00 \times 10^{8}\text{m/s})}{1.64 \times 10^{-18}\ \text{J}} = 1.21 \times 10^{-7}\ \text{m} = 121\ \text{nm}$$

For the Balmer series, we want the shortest wavelength (highest energy), with $n_i = \infty$ and $n_f = 2$.

$$\Delta E = R_H\left(\frac{1}{n_i^2} - \frac{1}{n_f^2}\right) = (2.18 \times 10^{-18}\ \text{J})\left(\frac{1}{\infty^2} - \frac{1}{2^2}\right) = -5.45 \times 10^{-19}\ \text{J}$$

$$\lambda = \frac{hc}{\Delta E} = \frac{(6.63 \times 10^{-34}\ \text{J·s})(3.00 \times 10^{8}\text{m/s})}{5.45 \times 10^{-19}\ \text{J}} = 3.65 \times 10^{-7}\ \text{m} = 365\ \text{nm}$$

Therefore the two series do not overlap.

7.100 This problem is similar to Example 7.3

Solution: $\lambda = \frac{hc}{\Delta E} = \frac{(6.63 \times 10^{-34}\ J{\cdot}s)(3.00 \times 10^8 m/s)}{(2.858 \times 10^5\ J)/(6.022 \times 10^{23})} = 4.19 \times 10^{-7}$ m or 419 nm

This wavelength is in the visible region (Refer to Figure 7.4). Since water is continuously being struck by visible radiation without decomposition, it seems unlikely that photodissociation of water by this method is feasible.

Why do we divide the energy by Avogado's number in the calculation above?

7.104 The first part of this problem involves the straightforward conversion of photon wavelength to photon energy. We can then determine the number of photons per pulse.

Solution: (a) $E = \frac{hc}{\lambda} = \frac{(6.63 \times 10^{-34}\ J{\cdot}s)(3.00 \times 10^8 m/s)}{633 \times 10^{-9}\ m} = 3.14 \times 10^{-19}$ J

The number of photons: $\frac{0.376\ J}{3.14 \times 10^{-19}\ J/photon} = 1.20 \times 10^{18}$ photons in each pulse

(b) Since a watt is a Js^{-1}, then the power per pulse is

$$\frac{0.376\ J}{1.00 \times 10^{-9}\ s} = 3.76 \times 10^8\ J/s = 3.76 \times 10^8\ W$$

Compare this with the power delivered by a 100-W light bulb!

7.108 The calculation in this problem is very similar to Problem 7.100.

Solution: $\lambda = \frac{hc}{E} = \frac{(6.63 \times 10^{-34}\ J{\cdot}s)(3.00 \times 10^8 m/s)}{(248 \times 10^3\ J)/(6.022 \times 10^{23})} = 4.83 \times 10^{-7}$ m or 483 nm

Any wavelength shorter than 483 nm will also promote this reaction. Once a person goes indoors, the reverse reaction Ag + Cl $\rightarrow$ AgCl takes place.

7.110 This problem is a straightforward application of Equation (7.1).

Solution: $\lambda = \frac{c}{\nu} = \frac{3.00 \times 10^8\ m/s}{2.45 \times 10^9/s} = 1.22 \times 10^{-1}\ m = 1.22 \times 10^8$ nm

7.111 Are wavelengths additive? Frequencies?

Solution: Since the energy corresponding to a photon of wavelength λ_1 equals the energy of photon of wavelength λ_2 plus the energy of photon of wavelength λ_3, then the equation must relate the wavelength to energy.

energy of photon 1 = (energy of photon 2 + energy of photon 3)

Since $E = \frac{hc}{\lambda}$, then: $\frac{hc}{\lambda_1} = \frac{hc}{\lambda_2} + \frac{hc}{\lambda_3}$ Dividing by hc, $\frac{1}{\lambda_1} = \frac{1}{\lambda_2} + \frac{1}{\lambda_3}$

7.112 This problem is similar in part to Example 7.4 which involves the use of Equation (7.5) to calculate the difference in energy between two states of the hydrogen atom.

Solution: In this problem $n_i = 110$ and $n_f = 109$. The photon energy can be found using Equation (7.5)

$$\Delta E = R_H\left(\frac{1}{n_i^2} - \frac{1}{n_f^2}\right) = (2.18 \times 10^{-18}\ \text{J})\left(\frac{1}{110^2} - \frac{1}{109^2}\right) = -3.32 \times 10^{-24}\ \text{J}$$

The sign of ΔE means that this is energy associated with an emission process. As in the above problem we have

$$\lambda = \frac{hc}{\Delta E} = \frac{(6.63 \times 10^{-34}\ \text{J·s})(3.00 \times 10^8 \text{m/s})}{3.32 \times 10^{-24}\ \text{J}} = 0.0599\ \text{m or } 5.99 \times 10^7\ \text{nm}$$

This wavelength is in the microwave region of the electromagnetic radiation spectrum.

7.113 This problem is more tedious than difficult. The calculations are a repetition of Example 7.4 and Problem 7.37 above.

Solution: The Balmer series corresponds to transitions to the n = 2 level.

For He^+

$$\Delta E = R_{He^+}\left(\frac{1}{n_i^2} - \frac{1}{n_f^2}\right) \qquad \lambda = \frac{hc}{\Delta E} = \frac{(6.63 \times 10^{-34}\ \text{Js})(3.00 \times 10^8\ \text{m/s})}{\Delta E}$$

For the transition $n = 3 \rightarrow 2$

$$\Delta E = (8.72 \times 10^{-18})\left(\frac{1}{3^2} - \frac{1}{2^2}\right) = 1.21 \times 10^{-18}\ \text{J} \quad \lambda = \frac{1.99 \times 10^{-25}\ \text{J m}}{1.21 \times 10^{-18}\ \text{J}} = 1.64 \times 10^{-7}\ \text{m} = 164\ \text{nm}$$

For the transition $n = 4 \rightarrow 2$, $\Delta E = 1.64 \times 10^{-18}$ J $\lambda = 121$ nm

For the transition $n = 5 \rightarrow 2$, $\Delta E = 1.83 \times 10^{-18}$ J $\lambda = 109$ nm

For the transition $n = 6 \rightarrow 2$, $\Delta E = 1.94 \times 10^{-18}$ J $\lambda = 103$ nm

For H, the calculations are identical except for the Rydberg constant. See Example 7.4 and Problem 7.37.

For the transition $n = 3 \rightarrow 2$, $\Delta E = 3.03 \times 10^{-19}$ J $\lambda = 657$ nm

For the transition $n = 4 \rightarrow 2$, $\Delta E = 4.09 \times 10^{-19}$ J $\lambda = 486$ nm

For the transition $n = 5 \rightarrow 2$, $\Delta E = 4.58 \times 10^{-19}$ J $\lambda = 434$ nm

For the transition $n = 6 \rightarrow 2$, $\Delta E = 4.84 \times 10^{-19}$ J $\lambda = 411$ nm

All of the Balmer transitions for He^+ are in the ultraviolet, and the transitions for H are all in the visible. Note that the negative sign for the energy, indicating that a photon has been emitted, has been omitted for simplicity.

7.115 At first glance this problem looks very difficult. However it simply involves converting the wavelength of the CO_2 laser to energy per photon, followed by calculating how many photons are required to produce 500 millijoules.

Solution: The energy of a photon of wavelength 10060 nm is:

$$E = h\nu = \frac{hc}{\lambda} = \frac{(6.63 \times 10^{-34}\ \text{J·s})(3.00 \times 10^{8}\ \text{m/s})}{10060 \times 10^{-9}\ \text{m}} = 1.98 \times 10^{-20}\ \text{J/photon}$$

The number of photons required: $\dfrac{500 \times 10^{-3}\ \text{J}}{1.98 \times 10^{-20}\ \text{J/photon}} = 2.53 \times 10^{19}$ photons

Note: If you have not already done so, please read the "Introduction for the Student" at the beginning of this Student Solutions Manual.

Chapter 8

PERIODIC RELATIONSHIPS AMONG THE ELEMENTS

8.18 What common features in electron configurations imply similar chemical properties? If you don't know, study the first few paragraphs of Section 8.2.

Solution: If two elements have similar outer (valence) electron configurations, they will resemble each other in chemical behavior.

(a) and (h) are similar; both have [core]np^5 configurations.

(b) and (e) are similar; both have [core]ns^2 configurations.

(c) and (g) are similar; both have [core]ns^2np^3 configurations.

(d) and (f) are similar; both have noble gas configurations.

8.19

8.20 This is similar to Problems 8.18.

Solution: (a) Group 1A, (b) Group 5A, (c) Group 8A, (d) Group 8B

Identify the elements.

8.21 Before trying this problem, be sure to study the subsection on Cations Derived from Transition Metals at the end of Section 8.2. Does the ion in this problem have any electrons in the 4s subshell? How many valence electrons does the neutral atom have?

Solution: There are no electrons in the 4s subshell because transition metals lose electrons from the ns valence subshell before they are lost from the (n-1)d subshell (Section 8.2). For the neutral atom there are only six valence electrons. The element can be identified as Cr (chromium) simply by counting six across starting with potassium (K, atomic number 19).

What is the electron configuration of neutral chromium? Are you sure? Check Figure 8.3.

.26 Are electrons gained or lost when a negative ion forms? When a positive ion is formed?

> ***Solution:*** Determine the number of electrons, and then "fill in" the electrons as you learned (Figure 7.29 and Table 7.3).
>
> (a) $1s^2$
>
> (b) $1s^2$
>
> (c) $1s^22s^22p^6$
>
> (d) $1s^22s^22p^6$
>
> (e) $[Ne]3s^23p^6$
>
> (f) [Ne]
>
> (g) $[Ar]4s^23d^{10}4p^6$
>
> (h) $[Ar]4s^23d^{10}4p^6$
>
> (i) [Kr]
>
> (j) [Kr]
>
> (k) $[Kr]5s^24d^{10}$
>
> (l) $[Kr]5s^24d^{10}5p^6$
>
> (m) [Xe]
>
> (n) $[Xe]6s^24f^{14}5d^{10}$
>
> (o) $[Kr]5d^{10}$
>
> (p) $[Xe]6s^24f^{14}5d^{10}$
>
> (q) $[Xe]4f^{14}5d^{10}$

8.27 This problem is similar to Problem 8.26.

> ***Solution:*** Refer to Figure 7.29 and Table 7.3 if you need help in writing electron configurations.
>
> (a) [Ne]
>
> (b) Same as (a). Do you see why?
>
> (c) [Ar]
>
> (d) Same as (c). Do you see why?
>
> (e) Same as (c)
>
> (f) $[Ar]3d^6$. Why isn't it $[Ar]4s^23d^4$?
>
> (g) $[Ar]3d^9$. Why not $[Ar]4s^23d^7$?
>
> (h) $[Ar]3d^{10}$. Why not $[Ar]4s^23d^8$?

8.28 This is like Problems 8.26 and 8.27 above.

> ***Solution:*** This exercise simply depends on determining the total number of electrons and using Figure 7.29 and Table 7.3.
>
> (a [Ar]
>
> (b) [Ar]
>
> (c) [Ar]
>
> (d) $[Ar]3d^3$
>
> (e) $[Ar]3d^5$
>
> (f) $[Ar]3d^6$
>
> (g) $[Ar]3d^5$
>
> (h) $[Ar]3d^7$
>
> (i) $[Ar]3d^8$
>
> (j) $[Ar]3d^{10}$
>
> (k) $[Ar]3d^9$
>
> (l) $[Kr]4d^{10}$
>
> (m) $[Xe]4f^{14}5d^{10}$
>
> (n) $[Xe]4f^{14}5d^8$
>
> (o) $[Xe]4f^{14}5d^8$

8.30 What does isoelectronic mean? See the last part of Section 7.3. Are electrons gained or lost when a negative ion forms from a neutral atom?

Solution: Two species are isoelectronic if they have the same number of electrons. Can two neutral atoms of different elements be isoelectronic (see Problem 8.24)?

(a) C and B^- are isoelectronic. (b) Mn^{2+} and Fe^{3+} are isoelectronic.

(c) Ar and Cl^- are isoelectronic. (d) Zn and Ge^{2+} are isoelectronic.

With which neutral atom are the positive ions in (b) isoelectronic?

8.34 Locate these elements in the periodic table and review the discussion of the trends in atomic radius in Section 8.3. How does atomic radius change going from left to right across a period? How does it change going down a column in a group? This problem is similar to Example 8.2.

Solution: (a) Cs is larger. It is below Na in Group 1A.

(b) Ba is larger. It is below Be in Group 2A.

(c) Sb is larger. It is below N in Group 5A.

(d) Br is larger. It is below F in Group 7A.

(e) Xe is larger. It is below Ne in Group 8A.

8.36 Review the subsection on Atomic Radius in Section 8.3.

Solution: The electron configuration of lithium is $1s^22s^1$. The two 1s electrons shield the 2s electron effectively from the nucleus. Consequently, the lithium atom is considerably larger than the hydrogen atom.

8.38 Review the subsection on Atomic Radius in Section 8.3.

Solution: Fluorine is the smallest atom in Group 7A. Atomic radius increases with increasing atomic number within a group.

8.40 Study the discussion of ionic radius in Section 8.3. What happens to the radius of an atom when an electron is removed? What happens when an electron is added? What is the trend in radii in an isoelectronic series of ions? See Example 8.3 in the text.

Solution: (a) Cl is smaller than Cl^-. An atom gets bigger when more electrons are added.

(b) Na^+ is smaller than Na. An atom gets smaller when electrons are removed.

(c) O^{2-} is smaller than S^{2-}. Both elements belong to the same group, and ionic radius increases going down a group.

(d) Al^{3+} is smaller than Mg^{2+}. The two ions are isoelectronic (What does that mean? See Section 8.2.) and in such cases the radius gets smaller as the charge becomes more positive.

(e) Au^{3+} is smaller than Au^+ for the same reason as part (b).

In each of the above cases from which atom would it be harder to remove an electron?

8.42 This problem is similar to Example 8.3.

Solution: The Cu^{+} ion is larger than Cu^{2+} because it has one more electron.

8.44 To solve this problem, you need to find a trend. What do O, S, Se, and Te have in common? What do their +2 cations have in common?

Solution: The four elements all belong to Group 6A. Since atomic radius increases going down a column in the periodic table, it follows that the order of increasing size is

$$O^{2-} < S^{2-} < Se^{2-} < Te^{2-}$$

8.45 What is the trend in radius in an isoelectronic series?

Solution: H^{-} and He are isoelectronic, and the radius becomes smaller in an isoelectronic series as the nuclear charge becomes more positive. H^{-} is larger.

8.46 How many elements are liquids or gases under the specified conditions? See the "Chemistry in Action" insert Section 8.3.

Solution: Bromine is liquid; all the others are solids.

8.47 This is like Example 8.4.

Solution: As in Example 8.4, we assume the approximate boiling point of argon to be the mean of the boiling points of neon and krypton

$$\text{bp} = \frac{-245.9°\text{C} + (-152.9°\text{C})}{2} = -199.4°\text{C}$$

The actual boiling point of argon is -185.7°C.

8.51 The ionization energies of these atoms can be found in Table 8.4. The conversion from energy of a photon to wavelength of a photon was the subject of Problem 7.19. Do the ionization energies in Table 8.4 refer to one single atom of an element?

Solution: In Table 8.4 the ionization energies are given in kJ/mol. In order to find wavelength and frequency for one photon, the energies must be divided by Avogadro's number.

(a) Ionization energy of a lithium atom

$$E = \left(\frac{5.20 \times 10^5 \text{ J}}{1 \text{ mol}}\right)\left(\frac{1 \text{ mol}}{6.022 \times 10^{23} \text{ atoms}}\right) = 8.635 \times 10^{-19} \text{ J/atom}$$

The frequency and wavelength can be found combining the equations $E = h\nu$ and $\lambda = c/\nu$.

$$\lambda = \frac{hc}{E} = \frac{(6.63 \times 10^{-34} \text{ J}\cdot\text{s})(3.00 \times 10^8 \text{ m/s})}{8.635 \times 10^{-19} \text{ J}} = 2.30 \times 10^{-7} \text{ m} = 230 \text{ nm}$$

(b) Ionization energy of a chlorine atom

$$E = \left(\frac{1251 \times 10^3 \text{ J}}{1 \text{ mol}}\right)\left(\frac{1 \text{ mol}}{6.022 \times 10^{23} \text{ atoms}}\right) = 2.077 \times 10^{-18} \text{ J}$$

$$\nu = \frac{E}{h} = \left(\frac{2.077 \times 10^{-18} \text{ J}}{6.63 \times 10^{-34} \text{ J}\cdot\text{s}}\right) = 3.13 \times 10^{15} \text{ s}^{-1} \text{ or Hz}$$

$$\lambda = \frac{c}{\nu} = \left(\frac{3.00 \times 10^8 \text{ m/s}}{3.13 \times 10^{15} \text{ s}^{-1}}\right) = 9.58 \times 10^{-8} \text{ m} = 95.8 \text{ nm}$$

(c) Ionization energy of a helium atom

$$E = \left(\frac{2373 \times 10^3 \text{ J}}{1 \text{ mol}}\right)\left(\frac{1 \text{ mol}}{6.022 \times 10^{23} \text{ atoms}}\right) = 3.941 \times 10^{-18} \text{ J}$$

$$\nu = \frac{E}{h} = \left(\frac{3.941 \times 10^{-18} \text{ J}}{6.63 \times 10^{-34} \text{ J}\cdot\text{s}}\right) = 5.94 \times 10^{15} \text{ s}^{-1} \text{ or Hz}$$

$$\lambda = \frac{c}{\nu} = \left(\frac{3.00 \times 10^8 \text{ m/s}}{5.94 \times 10^{15} \text{ s}^{-1}}\right) = 5.05 \times 10^{-8} \text{ m} = 50.5 \text{ nm}$$

Both photoionization and the photoelectric effect involve the light-induced liberation of electrons. In the photoelectric effect the light energy removes the electron from the surface of the metal; this is not the same as the ionization energy of the atom because the current-conducting electrons in a metal are already partially separated from their parent atoms.

8.52 See the discussion in Section 8.4 and look at Figure 8.14.

Solution: Apart from small irregularities, the ionization energies of elements in a period increase with increasing atomic number. We can explain this trend by referring to the increase in effective nuclear charge from left to right. A larger effective nuclear charge means a more tightly held outer electron, and hence a higher first ionization energy. Thus, in the third period, sodium has the lowest and neon has the highest first ionization energy.

8.54 Why is the second ionization energy of K so much larger than that of Ca? Write out the electron configurations of the +1 and +2 ions of K and Ca. This problem is similar to Example 8.5.

> ***Solution:*** To form the +2 ion of calcium, it is only necessary to remove two valence electrons. For potassium, however, the second electron must come from the atom's noble gas core which accounts for the much higher second ionization energy. Would you expect a similar effect if you tried to form the +3 ion of calcium?

8.55 To which periodic groups would each of the elements belong? What is the general trend in the magnitudes of first ionization energies across a period in the table? Study Section 8.4 and Figure 8.14.

> ***Solution:*** The ionization energy of 2080 kJ/mol would be associated with the element having the $1s^22s^22p^6$ electron configuration. This is a noble gas electron configuration. To justify your answer read Section 8.4.

8.56 This is like Problem 8.30.

> ***Solution:*** N and O^+ (7 electrons); Ne and N^{3-} (10 electrons); Ar and S^{2-} (18 electrons); Zn and As^{3+} (30 electrons); Cs^+ and Xe (54 electrons).

8.57

8.58 Believe it or not, you have seen the equation before. It is a slightly modified version of Equation (7.5), which is Bohr's formula for calculating the energies of the allowed electron orbits in the hydrogen atom. When Z=1, the two equations are identical. The equation works because He^+ is isoelectronic with the hydrogen atom (right?), and the only difference is the fact that the nuclear charge is +2 rather than +1. Assume that $n = 1$.

> ***Solution:*** The ionization energy is the difference between the $n = \infty$ state (final) and the $n = 1$ state (initial).

$$\Delta E = E_{\infty} - E_1 = (-2.18 \times 10^{-18}\ \text{J})(2)^2\left(\frac{1}{\infty}\right)^2 - (-2.18 \times 10^{-18}\ \text{J})(2)^2\left(\frac{1}{1}\right)^2$$

$$= 0 + (2.18 \times 10^{-18}\ \text{J})(2)^2\left(\frac{1}{1}\right)^2 = 8.72 \times 10^{-18}\ \text{J}$$

In units of kJ/mol:

$$(8.72 \times 10^{-18}\ \text{J})\left(\frac{1\ \text{kJ}}{1000\ \text{J}}\right)\left(\frac{6.022 \times 10^{23}}{1\ \text{mol}}\right) = 5.25 \times 10^3\ \text{kJ/mol}$$

Should this be larger than the ionization energy of helium (see Table 8.4)?

8.64 Are the values positive or negative? Take a guess at the answer. You might wish to consult Example 8.6 before attempting this problem.

> ***Solution:*** Based on electron affinity values, we would not expect the alkali metals to form anions. A few years ago most chemists would have answered this question with a loud "No!" In the early seventies a chemist named J. L. Dye at Michigan State University discovered that under very special circumstances alkali metals could be coaxed into accepting an electron to form negative ions! These ions are called alkalide ions.

8.66 Study the discussion of electron affinity in Section 8.5. Which has the greater electron affinity, O or O^-? Why?

Solution: Alkali metals have an electron configuration of ns^1 so they can accept another electron in the ns orbital. On the other hand, alkaline earth metals are ns^2 (see Example 8.6) and they have little tendency to accept another electron, as it would have to go into a higher energy orbital.

8.70 See the "Chemistry in Action: The Third Liquid Element" insert in Section 8.3. What other properties might you predict for Fr?

Solution: Since ionization energies decrease going down a column in the periodic table, francium should have the lowest first ionization energy of all the alkali metals. As a result, Fr should be the most reactive of all the Group 1A elements toward water, oxygen, etc. The reaction with oxygen would probably be similar to that of K, Rb, or Cs.

What would you expect the formula of the oxide to be? The chloride?

8.72 See the subsection on the Comparison of Group 1A and Group 1B Elements in Section 8.6.

Solution: The Group 1B elements are much less reactive than the Group 1A elements because the 1B elements have much higher ionization energies resulting from incomplete shielding of the nucleus by the inner d electrons (compared to the more effective shielding of the completely filled noble gas cores). Consequently, the outer s electrons of these elements are more strongly attracted by the nucleus.

8.74 This is like Example 8.7.

Solution: (a) Lithium oxide is a basic oxide. It reacts with water to form the metal hydroxide:

$$Li_2O(s) + H_2O(l) \rightarrow 2LiOH(aq)$$

(b) Calcium oxide is a basic oxide. It reacts with water to form the metal hydroxide:

$$CaO(s) + H_2O(l) \rightarrow Ca(OH)_2(aq)$$

(c) Sulfur dioxide is an acidic oxide. It reacts with water to form sulfurous acid:

$$SO_2(g) + H_2O(l) \rightarrow H_2SO_3(aq)$$

8.76 Review Section 8.6 and Section 2.8 before trying to work this problem.

Solution: LiH (lithium hydride): ionic compound; BeH_2 (beryllium hydride): covalent compound; B_2H_6 (diborane...you aren't expected to know that name): molecular compound; CH_4 (methane...do you know that one?): molecular compound; NH_3 (ammonia...you should know that one): molecular compound; H_2O (water...if you didn't know that one, you should be ashamed): molecular compound; HF (hydrogen fluoride): molecular compound. LiH and BeH_2 are solids, B_2H_6, CH_4, NH_3, and HF are gases, and H_2O is a liquid.

8.79 What other atomic properties correlate with high ionization energies?

Solution: Both ionization energy and electron affinity are affected by atomic size - the smaller the atom, the greater the attraction between the electrons and the nucleus. If it is difficult to remove an electron from an atom (that is, high ionization energy), then it follows that it would also be favorable to add an electron to the atom (large electron affinity).

8.80 This problem is not as formidable as it might seem. All you are asked to do is to find the value of the shielding constant σ when $(Z\text{-}\sigma)$ is substituted into the equation for the electron energy of a hydrogen-like atom. In this problem you are given the energy. What are the values of *Z* and *n*? See Problem 8.58.

Solution: For helium $n = 1$ and $Z = 2$ (why?). We have the equation

$$E_1 = (2.18 \times 10^{-18}\ \text{J})(2 - \sigma)^2\left(\frac{1}{1}\right)^2 = 3.94 \times 10^{-18}\ \text{J}$$

$$(2 - \sigma)^2 = \left(\frac{3.94 \times 10^{-18}\ \text{J}}{2.18 \times 10^{-18}\ \text{J}}\right) = 1.81$$

$$2 - \sigma = \sqrt{1.81} = 1.35$$

$$\sigma = 2 - 1.35 = 0.65$$

8.82 How many electrons does each of these species have?

Solution: This is an isoelectronic series with 10 electrons in each species. The nuclear charge interacting with these 10 electrons ranges from +8 for oxygen to +12 for magnesium. Therefore the +12 charge in Mg^{2+} will draw in the 10 electrons more tightly than the +11 charge in Na^+, than the +9 charge in F^-, and than the +8 charge in O^{2-}. Recall that the largest species will be the *easist* to ionize.

(a) increasing ionic radius: $Mg^{2+} < Na^+ < F^- < O^{2-}$

(b) increasing ionization energy: $O^{2-} < F^- < Na^+ < Mg^{2+}$

8.84 Refer to Problem 8.83 in the text for a hint.

Solution: The percent change in volume from F to F^- is

$$\frac{\text{volume of F}^-\text{ ion}}{\text{volume of F atom}} \times 100\% = \frac{\frac{4}{3}\pi\,(136\ \text{pm})^3}{\frac{4}{3}\pi\,(72\ \text{pm})^3} \times 100\% = 674\%$$

Therefore, there is an increase in volume of (674 - 100)% or 574% as a result of the formation of the F^- ion.

8.85 In other words what are the expected formulas of the products when these elements combine with O_2 or Cl_2? This is similar to Problem 8.76. How can you tell if a compound is ionic or molecular?

Solution: Ionic compounds are combinations of a metal and a nonmetal. Molecular compounds are usually nonmetal-nonmetal combinations.

(a) Na_2O (ionic); MgO (ionic); Al_2O_3 (ionic); SiO_2 (molecular); P_4O_6 and P_4O_{10} (both molecular); SO_2 or SO_3 (molecular); Cl_2O and several others (all molecular).

(b) NaCl (ionic); $MgCl_2$ (ionic); $AlCl_3$ (ionic); $SiCl_4$ (molecular); PCl_3 and PCl_5 (both molecular); SCl_2 (molecular).

8.86 Check the melting points of the alkali metals and the halogens in the Handbook of Chemistry and Physics or other data source containing this type of information. Ask your instructor if you don't know where to look.

Solution: According to the Handbook of Chemistry and Physics, 1966-67 Edition, potassium metal has a melting point of 63.6°C, bromine is a reddish brown liquid with melting point -7.2°C, and potassium bromide (KBr) is a colorless solid with melting point 730°C. M is potassium (K) and X is bromine (Br).

8.88 How common and how reactive are the noble gases?

Solution: Elements revealed themselves because of their abundances and/or their reactivities. Nobel gases are neither abundant nor reactive. Chemical periodicity was well established by the time argon was discovered; therefore, its existence suggested a missing group of atoms.

8.89 This problem requires you to use some of the information that you learned in Chapter 7.

Solution: Using the equation

$$IE = h\nu - \frac{1}{2}mv^2$$

First solving for hν, given the information that the wavelenght $\lambda = 162 \text{ nm} = 1.62 \times 10^{-9}$ m

$$c = \lambda\nu \quad \text{or, rearranging} \quad \nu = c/\lambda$$

$$h\nu = \frac{hc}{\lambda} = \frac{(6.63 \times 10^{-34} \text{ J}\bullet\text{s})(3.00 \times 10^{8} \text{ m/s})}{162 \times 10^{-9} \text{ m}} = 1.23 \times 10^{-18} \text{ J}$$

We are give the kinetic energy $\frac{1}{2}mv^2$ of the ejected electron as 5.34×10^{-19} J.

Substituting: $$IE = h\nu - \frac{1}{2}mv^2 = (1.23 \times 10^{-18} \text{ J}) - (5.34 \times 10^{-19} \text{ J}) = 5.30 \times 10^{-19} \text{ J}$$

Converting from J/photon to kJ/mol of photons

$$\frac{(5.30 \times 10^{-19}\text{ J})(6.022 \times 10^{23})}{1000} = 3.2 \times 10^{2}\text{ kJ/mol}$$

To make sure that the ejected electron is indeed the valence electron, use UV light of the *longest* wavelength (least energy) that can stilll eject electrons.

Given the mass of the electron as 9.1095×10^{-31} kg, could you determine the velocity of the ejected electron?

8.90 Review Section 8.3.

Solution: Only (b) is listed in order of decreasing radius. Review Problem 8.82. Answer (a) is listed in increasing size because the radius increases down a group. Answer (c) is listed in increasing size because the number of electrons is increasing.

8.92 Review the appropriate subsections in Section 8.6.

Solution: (a)(i) Both react with water to produce hydrogen; (ii) Their oxides are basic; (iii) Their halides are ionic. (b)(i) Both are strong oxidizing agents; (ii) both react with hydrogen to form HX (where X is Cl or Br); (iii) Both form halide ions (Cl^- or Br^-) when combined with electropositive metals (Na, K, Ca, Ba).

8.94 Carbon dioxide is the oxide of what catagory of elements (metal, nonmetal, or metalloid)?

Solution: The equation is: $CO_2(g) + Ca(OH)_2(aq) \rightarrow CaCO_3(s) + H_2O(l)$

The milky white color is due to calcium carbonate. Calcium hydroxide is a base and carbon dioxide is an acidic oxide. The products are a salt and water.

8.95 You need the information from Problem 8.58.

Solution:

$Li \rightarrow Li^{+} + e^-$ $\quad I_1 = 520$ kJ/mol

$Li^{2+} \rightarrow Li^{3+} + e^-$ $\quad I_3 = ?$

From Problem 8.58: $E = -(2.18 \times 10^{-18}\text{ J})(3^2)(\frac{1}{1^2}) = -1.96 \times 10^{-17}\text{ J}$

$$I_3 = \frac{(-1.96 \times 10^{-17}\text{ J})(6.022 \times 10^{23})}{1000} = 11.8 \times 10^{3}\text{ kJ/mol}$$

Given: $I_1 + I_2 + I_3 = 1.96 \times 10^{4}$ kJ/mol

$I_2 = 1.96 \times 10^{4}\text{ kJ/mol} - 520\text{ kJ/mol} - 11.8 \times 10^{3}\text{ kJ/mol} = 7.28 \times 10^{3}\text{ kJ/mol}$

8.96 This is a chemical detective problem using several concepts that you have acquired including the gas laws, stoichiometry, and mass percentages.

Solution: The first equation is: $X + H_2 \rightarrow Y$. We are given sufficient information from the decomposition reaction (the reverse reaction) to calculate the relative number of moles of X and H.

$$0.559 \text{ L} \times \frac{1 \text{ mol}}{22.4 \text{ L}} = 0.0250 \text{ mol } H_2 \quad \text{or} \quad 2 \times 0.0250 = 0.0500 \text{ mol H}$$

Let M be the molar mass of X. If we assume that the formula for Y is either MH or MH_2 or MH_3, then

$$\text{if Y = MH, then } \frac{\text{mol H}}{\text{mol X}} = \frac{0.0500 \text{ mol}}{(1.00 \text{ g})/(M \text{ g/mol})} = 1, \text{ then } M = 20.0 \text{ g/mol} = \text{the element Ne (closest mass)}$$

$$\text{if Y} = MH_2\text{, then } \frac{\text{mol H}}{\text{mol X}} = \frac{0.0500 \text{ mol}}{(1.00 \text{ g})/(M \text{ g/mol})} = 2, \text{ then } M = 40.0 \text{ g/mol} = \text{the element Ca (closest mass)}$$

$$\text{if Y} = MH_3\text{, then } \frac{\text{mol H}}{\text{mol X}} = \frac{0.0500 \text{ mol}}{(1.00 \text{ g})/(M \text{ g/mol})} = 3, \text{ then } M = 60.0 \text{ g/mol} = \text{? (no element of close mass)}$$

If we deduce that the element X = Ca, then the formula for the chloride Z is $CaCl_2$ (why?). (Why couldn't X be Ne?) Calculating the mass percent of chlorine in $CaCl_2$ to compare with the known results

$$\text{Cl:} \quad \frac{2 \times 35.45}{[40.08 + (2 \times 35.45)]} \times 100\% = 63.89\% \quad \text{Therefore X is calcium.}$$

8.98 This is another chemical detective problem. It is considerably easier than 8.96.

Solution: X must belong to Group 4A; it is probably Sn or Pb because it behaves as not-very-reactive metal (certainly not like an alkali metal). Y is a nonmetal since it doesn't conduct electricity. Since it is a light yellow solid, it is probably phosphorus (Group 5A). Z is an alkali metal since it reacts with air to form a basic oxide or peroxide.

8.100 Review Section 8.5 on Electron Affinity if you need help.

Solution: Noble gases have filled shells or subshells. Therefore, they have little tendency to accept electrons (endothermic).

8.102 Review Section 8.5 on Electron Affinity if you need help.

Solution: Z_{eff} increases from left to right across the table, so electrons are held more tightly. (This explains the electron affinity values of C and O.) Nitrogen has a zero value of electron affinity because of the stability of the half-filled 2p subshell (that is, N has little tendency to accept another electron).

8.103 How is electron affinity related to ionization energy?

Solution: The reaction corresponding to the electron affinity of Na^+ is: $Na^+ + e^- \rightarrow Na$
It is the reverse of the ionization energy for Na.

8.104 How is the ionization energy related to the ability of an oxidizing agent to remove an electron from another species?

Solution: Since PtF_6 is able to oxidize O_2 to form $O_2^+PtF_6^-$ (which requires 1176 kJ/mol in the gas phase), then it can be argued that PtF_6 should be able to oxidize any element that has an ionization energy of 1176 kJ/mol or less. Thus Xe with an ionization energy of 1167 kJ/mol should be oxidized by PtF_6.

This reaction is of historical interest. It produced the first noble gas compound originally formulated as $XePtF_6$.

8.105 This problem is very similar to Problem 7.19.

Solution:

Since for the reaction: $Cl + e^- \rightarrow Cl^-$ $\Delta H° = -348$ kJ,
then for the reaction: $Cl^- \rightarrow Cl + e^-$ $\Delta H° = 348$ kJ

The energy given in this problem is for *one mole* of photons. To apply $E = h\nu$ we must divide by Avogadro's number. The energy of one photon is

$$E = \left(\frac{3.48 \times 10^2 \text{ kJ}}{1 \text{ mol}}\right)\left(\frac{1 \text{ mol}}{6.022 \times 10^{23}}\right)\left(\frac{1000 \text{ J}}{1 \text{ kJ}}\right) = 5.78 \times 10^{-19} \text{ J}$$

The wavelength of this photon can be found using the relationship $E = hc/\lambda$.

$$\lambda = \frac{hc}{E} = \frac{(6.63 \times 10^{-34} \text{ J·s})(3.00 \times 10^8 \text{ m/s})}{5.78 \times 10^{-19} \text{ J}}\left(\frac{10^9 \text{ nm}}{1 \text{ m}}\right) = 3.44 \times 10^2 \text{ nm}$$

The radiation is in the ultraviolet region.

8.106 This problem is a straightforward application of Hess's law. Review Section 6.5.

Solution:

Standard enthalpy of atomization:	$Na(s) \rightarrow Na(g)$	$\Delta H° = 108.4$ kJ
Ionization energy:	$Na(g) \rightarrow Na^+(g)$	$\Delta H° = 495.9$ kJ
	$Na(s) \rightarrow Na^+(g)$	$\Delta H° = 604.3$ kJ

8.108 When do we write hydrogen first in the formula and when do we write it last?

Solution: The hydrides are: LiH (lithium hydride), CH_4 (methane), NH_3 (ammonia), H_2O (water), and HF (hydrogen fluoride).

The reactions with water: $LiH + H_2O \rightarrow LiOH + H_2$
$CH_4 + H_2O \rightarrow$ no reaction at room temperature
$NH_3 + H_2O \rightarrow NH_4^+ + OH^-$
$H_2O + H_2O \rightarrow H_3O^+ + OH^-$
$HF + H_2O \rightarrow H_3O^+ + F^-$

The last three reactions involve *equilibria* that will be discussed in later chapters.

8.110 Could you make partial assignments by know nothing more that the physical state of the elements? Figure 8.23 contains photographs of some of the halogens.

Solution: Fluorine is a yellow-green gas that attacks glass; chlorine is a yellow gas; bromine is a red liquid; and iodine is a dark, metallic-appearing solid.

8.111 What are the electron configurations of these species?

Solution: Considering electron configuration, Fe^{2+} [Ar]$3d^6 \rightarrow Fe^{3+}$ [Ar]$3d^5$
Mn^{2+} [Ar]$3d^5 \rightarrow Mn^{3+}$ [Ar]$3d^4$

A half-filled shell has extra stability. In oxidizing Fe^{2+} the product is a d^5-half-filled shell. In oxidizing Mn^{2+}, a d^5-half-filled shell is being lost, which requires more energy.

8.112 In order to work the problem, we need to determine the oxidation state of Ti in each compound. Section 3.7 can be used for review.

Solution: The compounds followed by the oxidation state of titanium are: K_3TiF_6, +3; $K_2Ti_2O_5$, +4, $TiCl_3$, +3; K_2TiO_4, +6; K_2TiF_6, +4. The compound of Ti that is unlikely to exist is K_2TiO_4, because of the oxidation state of +6. Titanium in an oxidation state greater than +4 (which is a noble gas configuration) is unlikely because of the high ionization energies.

8.114 We usually think of the radius of an atom or ion as a measure of the size of the cloud of electron density around the nucleus.

Solution: Since H^+ is only a proton and no electrons, it does not have a radius like other ions (in the sense that the radius is a measure of the electron distribution around the nucleus).

8.116 In order to answer this question, you will probably need some supplemental information.

Solution:
(a) Mg in $Mg(OH)_2$
(b) Na, liquid
(c) Mg in $MgSO_4{\cdot}7H_2O$
(d) Na in $NaHCO_3$
(e) K in KNO_3
(f) Mg
(g) Ca in CaO
(h) Ca
(i) Na in NaCl; Ca in $CaCl_2$

Note: If you have not already done so, please read the "Introduction for the Student" at the beginning of this Student Solutions Manual.

Chapter 9

CHEMICAL BONDING I: BASIC CONCEPTS

9.16 Study Section 9.2 and Equation (9.2). Assuming parts (a) through (d) are completely unrelated to each other, what predictions can you make based on the mathematical form of Coulomb's law? In other words, if the distance between the ions r becomes twice as large, what happens to the magnitude of E if nothing else changes? One of the best ways to familiarize yourself with a mathematical equation like Coulomb's law is to make up a few values for the variable r and calculate E for each case. Try it in this case.

> ***Solution:*** We use Coulomb's law to answer the question: $E = k\frac{Q_{cation}Q_{anion}}{r}$
>
> (a) Doubling the radius of the cation would increase the distance r between the centers of the ions. A larger value of r results in a smaller energy E of the ionic bond. Is it possible to say how much smaller E will be?
>
> (b) Tripling the charge on the cation will result in tripling of the energy E of the ionic bond, since the energy of the bond is directly proportional to the charge on the cation Q_{cation}.
>
> (c) Doubling the charge on both the cation and anion will result in quadrupling (or four-fold) the energy E of the ionic bond.
>
> (d) Decreasing the radius of both the cation and the anion to half of their original values is the same as halving the distance r between the centers of the ions. Halving the distance results in doubling the energy.

9.17 If you don't remember the guidelines for writing the formulas of ionic compounds, review Section 2.8. The rules for naming ionic compounds can be found in Section 2.8.

> ***Solution:*** (a) RbI, rubidium iodide (b) Cs_2SO_4, cesium sulfate (c) Sr_3N_2, strontium nitride
> (d) Al_2S_3, aluminum sulfide

9.18 This problem is similar to Example 9.1.

> ***Solution:*** Lewis representations for the ionic reactions are as follows.
>
> (a) $Na\cdot + :\ddot{\dot{F}}: \rightarrow Na^{+} :\ddot{\underset{..}{F}}:^{-}$ (b) $2K\cdot + :\ddot{S}: \rightarrow 2K^{+} :\ddot{\underset{..}{S}}:^{2-}$
>
> (c) $\cdot Ba\cdot + :\ddot{O}: \rightarrow Ba^{2+} :\ddot{\underset{..}{O}}:^{2-}$ (d) $\cdot\dot{Al}\cdot + :\dot{\underset{\cdot}{N}}\cdot \rightarrow Al^{3+} :\ddot{\underset{..}{N}}:^{3-}$

9.20 How can you tell from the formula whether a compound is ionic or molecular? See Section 2.5.

Solution: (a) I and Cl should form a molecular compound; both elements are nonmetals. One possibility would be ICl, iodine chloride.

(b) K and Br should form an ionic compound; K is a metal while Br is a nonmetal. The compound would be KBr, potassium bromide.

(c) Mg and F will form an ionic compound for the same reason as part (b). The substance will be MgF_2, magnesium fluoride.

(d) Al and F will also form the ionic compound AlF_3, aluminum fluoride.

9.25 Use the five steps in the LiF example as a model to work this problem. Needed data are either given in the problem or referenced in the LiF example (Section 9.3). Slightly more detail is provided for Problem 9.26 below.

Solution:

(1) $Na(s) \rightarrow Na(g)$	$\Delta H_1^o = 108$ kJ
(2) $\frac{1}{2}Cl_2(g) \rightarrow Cl(g)$	$\Delta H_2^o = 121.4$ kJ
(3) $Na(g) \rightarrow Na^+(g) + e^-$	$\Delta H_3^o = 495.9$ kJ
(4) $Cl(g) + e^- \rightarrow Cl^-(g)$	$\Delta H_4^o = -348$ kJ
(5) $Na^+(g) + Cl^-(g) \rightarrow NaCl(s)$	$\Delta H_5^o = ?$
$Na(s) + \frac{1}{2}Cl_2(g) \rightarrow NaCl(s)$	$\Delta H^o_{overall} = -411$ kJ

$$\Delta H_5^o = \Delta H^o_{overall} - \Delta H_1^o - \Delta H_2^o - \Delta H_3^o - \Delta H_4^o = (-411) - (108) - (121.4) - (495.9) - (-348) = -788 \text{ kJ}$$

The lattice energy of NaCl is 788 kJ/mol

9.26 This is like Problem 9.25 above.

Solution:

(1)	$Ca(s) \rightarrow Ca(g)$	$\Delta H_1^o = 121$ kJ
(2)	$Cl_2(g) \rightarrow 2Cl(g)$	$\Delta H_2^o = 242.8$ kJ
(3)	$Ca(g) \rightarrow Ca^+(g) + e^-$	$\Delta H_3^o{}' = 589.5$ kJ
	$Ca^+(g) \rightarrow Ca^{2+}(g) + e^-$	$\Delta H_3^o{}'' = 1145$ kJ
(4)	$2[Cl(g) + e^- \rightarrow Cl^-(g)]$	$\Delta H_4^o = 2(-348 \text{ kJ}) = -696$ kJ
(5)	$Ca^{2+}(g) + 2Cl^-(g) \rightarrow CaCl_2(s)$	$\Delta H_5^o = ?$
	$Ca(s) + Cl_2(g) \rightarrow CaCl_2(s)$	$\Delta H^o_{overall} = -795$ kJ

Thus we write $\Delta H^o_{overall} = \Delta H_1^o + \Delta H_2^o + \Delta H_3^o{}' + \Delta H_3^o{}'' + \Delta H_4^o + \Delta H_5^o$

$$\Delta H_5^o = (-795 - 121 - 242.7 - 589.5 - 1145 + 696) \text{ kJ} = -2197 \text{ kJ}$$

The lattice energy is represented by the reverse of equation (5); it is +2197 kJ/mol.

9.37 Study the discussion of electronegativity in Section 9.5. What is the definition of electronegativity? How can the difference in electronegativity between two elements be used to predict bond type? Where can electronegativity data be found in the textbook? Example 9.2 in the text is a similar problem.

Solution: The degree of ionic character in a bond is a function of the difference in electronegativity between the two bonded atoms. Figure 9.9 lists electronegativity values of the elements. The bonds in order of increasing ionic character are: N-N (zero difference in electronegativity) < S-O (difference 1.0) = Cl-F (difference 1.0) < K-O (difference 2.7) < Li-F (difference 3.0).

In the entire periodic table what pair of elements has the largest electronegativity difference?

9.38 This problem is similar to Problem 9.37 above.

Solution: C-H < Br-H < Na-I < Li-Cl < K-F

9.40 This problem is similar to Example 9.2 and Problem 9.37.

Solution: The order of increasing ionic character is Cl-Cl (zero difference in electronegativity) < Br-Cl difference 0.2) < Si-C (difference 0.7) < Cs-F (difference 3.3).

9.45 Some comments on the atomic arrangements in this problem: (e) the two hydrogens are connected to each nitrogen; (f) the three oxygens are connected to chlorine and the hydrogen is connected to an oxygen. Refer to Examples 9.3 and 9.4 in the text.

Solution: The Lewis dot structures are:

```
                                          ..
                                          P
     ..  ..              ..              /|\                      ..
(a) :I—Cl:       (b) H—P—H       (c) :P—+—P:          (d) H—S—H
     ..  ..              |               \|/                      ..
                         H                P
                                          ..

      ..  ..                ..   ..  ..                  ..       ..
(e) H—N—N—H       (f) H—O—Cl—O:            (g) :Br—C—Br:
      |   |                 ..   |   ..                  ..  ||   ..
      H   H                     :O:                         :O:
                                 ..
```

9.46 Don't forget to adjust the electron count for the ion charges. Does a negative charge mean there are more or fewer electrons? Refer to Examples 9.5 and 9.6 in the text for similar problems.

Solution: The appropriate Lewis dot structures are

(a) $\left[:\ddot{O}-\ddot{O}:\right]^{2-}$ (b) $\left[:C\equiv C:\right]^{2-}$ (c) $\left[:N\equiv O:\right]^{+}$ (d) $\left[H-\underset{H}{\overset{H}{N}}-H\right]^{+}$

9.47 For each bond in this molecule, verify whether the octet rule is obeyed.

Solution: (a) Neither oxygen atom has a complete octet. The leftmost hydrogen atom is forming two bonds.

(b) Correct structure:

```
   H  :O:
   |   ||   ..
H—C—C—O—H
   |        ..
   H
```

Do the two structures have the same number of electrons?

9.48 This is similar to Problem 9.47 above and the Example 9.7 in the text.

Solution: (a) Too many electrons. The correct structure is

H— C≡N :

(b) Hydrogen atoms do not form double bonds. The correct structure is

H— C≡C— H

(c) Too few electrons.

$:\ddot{O}{=}Sn{=}\ddot{O}:$

(d) Too many electrons. The correct structure is

```
 ..     ..
:F:    :F:
   \  /
    B
    |
   :F:
    ..
```

(e) Fluorine has more than an octet. The correct structure is

$H-\ddot{\underset{..}{O}}-\ddot{\underset{..}{F}}:$

(f) Oxygen does not have an octet. The correct structure is

```
     :O:
      ‖    ..
H— C— F:
           ..
```

(g) Too few electrons. The correct structure is

```
 ..    ..    ..
:F— N— F:
 ..    |     ..
      :F:
       ..
```

9.55 This problem is similar to Example 9.8. If you have trouble with writing Lewis structures, resonance forms, or formal charges, review the appropriate parts of Chapter 9 (Sections 9.6, 9.7, and 9.8).

Solution: The resonance structures are

```
          ..  −                     :O:
         :O:                       //
        /                ⟷   H—C
  H—C                              \
        \\                          :O: −
         :O:                         ..

H        :O: −        H        :O: −        H         :O:
 \     + /             \ −   + /             \ −   + //
  C=N                   C—N                    C—N
 /     \                /  ..   \\            /  ..   \
H        :O: −        H        :O:          H         :O: −
         ..                                            ..
      ⟷                     ⟷
```

9.56 Problem 9.55 above is similar to this one.

Solution: The resonance structures are as follows:

```
    .. −                :O:                  .. −
   :O:                   ‖                  :O:
    |                 − ..    .. −            |
..      .. −         :O—Cl—O:          − ..       ..
:O=Cl—O:      ⟷      ..     ..   ⟷    :O—Cl=O:
   .. +  ..               .. +          ..   .. +
```

9.58 This problem is like Problem 9.55 above and Example 9.8.

Solution: The structures of the most important resonance forms are

```
 ..   +   .. -          ..-  +                 +   +  ..2-
H—N=N=N:       ↔      H—N—N≡N:        ↔      H—N≡N—N:
                         ..                           ..

           H                          H
           |  +    ..-               -|   +
        H—C=N=N:          ↔        H—C—N≡N:
                                      ..
```

9.59 In Section 9.7 study the guidelines for choosing the most plausible Lewis structures of a molecule. Try to apply the guidelines to this problem.

Solution: (a) is a very good resonance form; there are no formal charges and each atom satisfies the octet rule.

(b) is a second choice after (a) because of the positive formal charge on the oxygen (high electronegativity).

(c) is a poor choice for several reasons. The formal charges are placed counter to the electronegativities of C and O, the oxygen atom does not have an octet, and there is no bond between that oxygen and carbon!

(d) is a mediocre choice because of the large formal charge and lack of an octet on carbon.

9.60 This is similar to Example 9.8 and Problem 9.55. Is NNO isoelectronic with the subject ion in Problem 9.57?

Solution: Three reasonable resonance structures with the formal charges are

```
  ..  +  ..            +  ..               ..  +
 ⁻:N=N=O:     ↔     :N≡N–O:⁻     ↔     2⁻:N–N≡O:⁺
                         ..                ..
```

9.65 Review the guidelines for identifying good Lewis structures on the basis of formal charge arrangement in Section 9.7. A similar question is Example 9.9.

Solution: An octet on Be can only be formed by making two double bonds as shown below:

```
   ..   2-    ..
 ⁺Cl=Be=Cl⁺
   ..         ..
```

This places a high negative charge on Be and distributes the charges counter to the electronegativities of the elements. It is not a plausible Lewis structure.

9.66 Refer to Example 9.9.

Solution: The resonance structures are

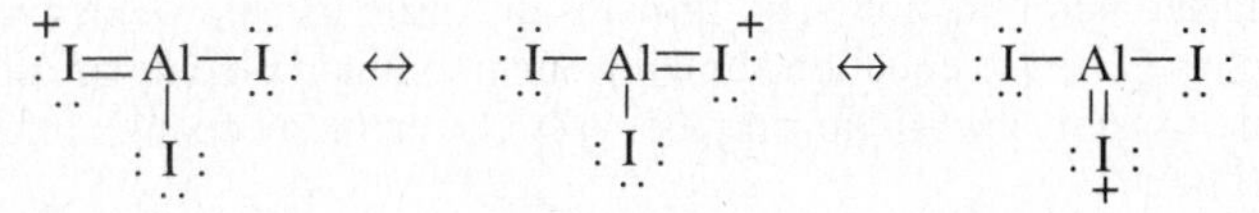

$^{+}$:I=Al–I: ↔ :I–Al=I:$^{+}$ ↔ :I–Al–I: (with :I: bonded to Al; in the third structure Al=I^{+})

9.68 Since Xe is in the fifth period, it can exceed the octet. In (d) and (e) all atoms are connected to Xe.

Solution: For simplicity, the three, nonbonding pairs of electrons around the fluorine atoms are omitted.

(a) F–Xe–F (b) F–Xe(F)(F)–F (c) Xe(F)$_6$ (d) F$_4$Xe=O (e) F–Xe(=O)(=O)–F

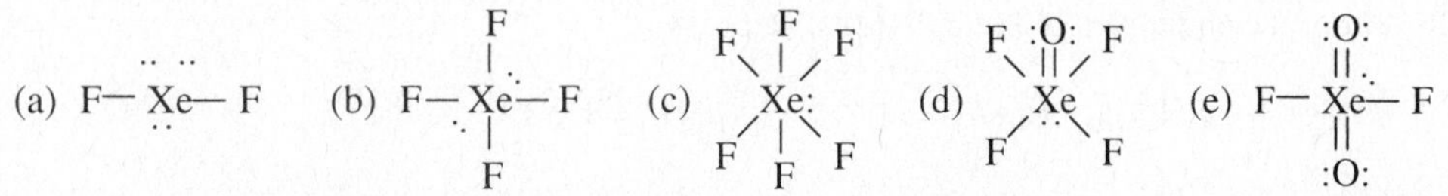

The octet rule is exceeded in each case.

9.70 This is like Problem 9.68 above and to Example 9.11 in the text.

Solution: For simplicity, the three, nonbonding pairs of electrons around the fluorine atoms are omitted.

F–Se–F (with two more F bonded to Se) Se(F)$_6$

The octet rule is not satisfied for Se in both compounds (why not?).

9.71 In part this is like Example 9.1. The new bond formed is discussed in Section 9.9.

Solution: The reaction can be represented as

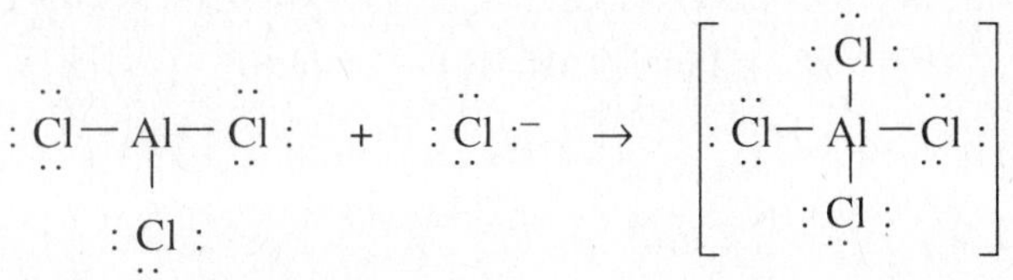

:Cl–Al(–Cl:)–Cl: + :Cl:$^{-}$ → [:Cl–Al(–Cl:)(–Cl:)–Cl:]$^{-}$

The new bond formed is called a coordinate covalent bond.

9.72 This problem is similar to Example 9.11.

Solution: A reasonable Lewis structure is

:O=N–O–H

9.76 This problem requires an understanding of Section 9.10. What is the difference between average bond energy and bond dissociation energy? Write a balanced equation showing an ammonia molecule breaking up into a nitrogen atom and three hydrogen atoms. What is the relationship between the enthalpy change for this reaction and the average bond energy of the N-H bond?

Solution: The enthalpy change for the equation showing ammonia dissociating into a nitrogen atom and three hydrogen atoms is equal to three times the average bond energy of the N-H bond (Why three?).

$$NH_3(g) \rightarrow N(g) + 3H(g) \qquad \Delta H° = 3\Delta H°(\text{N-H})$$

This equation is the sum of the three equations given in the problem, and by Hess's law (Section 6.3) the enthalpy change is just the sum of the enthalpies of the individual steps.

$$NH_3(g) \rightarrow NH_2(g) + H(g) \qquad \Delta H° = 435 \text{ kJ}$$

$$NH_2(g) \rightarrow NH(g) + H(g) \qquad \Delta H° = 381 \text{ kJ}$$

$$NH(g) \rightarrow N(g) + H(g) \qquad \Delta H° = 360 \text{ kJ}$$

$$NH_3(g) \rightarrow N(g) + 3H(g) \qquad \Delta H° = 1176 \text{ kJ}$$

$$\Delta H°(\text{N-H}) = \frac{1176 \text{ kJ}}{3} = 392 \text{ kJ}$$

Are the energies given in the problem average bond energies or bond dissociation energies?

9.77 Write the balanced equation for the dissociation of molecular fluorine into two fluorine atoms. What is the enthalpy change for this reaction? What is the relationship between that enthalpy change and the enthalpy of formation of atomic fluorine? What are the units for the enthalpy of formation of atomic fluorine?

Solution: When molecular fluorine dissociates, two fluorine atoms are produced. Since the enthalpy of formation of atomic fluorine is in units of kJ/mol, this number is half the bond dissociation energy of the fluorine molecule.

$$F_2(g) \rightarrow 2F(g) \qquad \Delta H° = 156.9 \text{ kJ}$$

$$\Delta H° = 2\Delta H_f^o(F) - \Delta H_f^o(F_2)$$

$$156.9 \text{ kJ} = (2 \text{ mol})[(\Delta H_f^o(F)] - (1 \text{ mol})(0)$$

$$\Delta H_f^o(F) = \frac{156.9 \text{ kJ}}{2 \text{ mol}} = 78.5 \text{ kJ/mol}$$

9.78 This problem is like Example 9.13.

Solution: (a)

Bonds Broken	*Number Broken*	*Bond Energy*	*Energy Change*
O=O (O_2)	1	498.7 kJ/mol	498.7 kJ/mol
Bonds Formed	*Number Formed*	*Bond Energy*	*Energy Change*
O-O (O_3)	2	*x* kJ/mol	2*x* kJ/mol

Total energy input = 498.7 kJ
Total energy released = $2x$ kJ

$\Delta H° = \Sigma BE(reactants) - \Sigma BE(products)$

$-107.2 \text{ kJ} = 498.7 \text{ kJ} - 2x \text{ kJ}$

$x = 303.0$ kJ = average O-O bond energy in O_3.

Considering the resonance structures for ozone (see Section 9.8), is it expected that the O-O bond energy in ozone is between the single O-O bond energy (142 kJ) and the double O=O bond energy (498.7 kJ)?

9.80 Check the definitions of these quantities in appropriate chapters.

Solution: (a) electron affinity of fluorine, (b) bond dissociation energy of molecular fluorine, (c) ionization energy of sodium, (d) lattice energy of sodium fluoride, (e) standard enthalpy of formation of sodium fluoride

9.82 The energy changes for the given reactions can be calculated easily using Hess's law (Section 6.3). Review the equations defining ionization energy [Equation (8.1)] and electron affinity (Section 8.5). What is the sum of the two equations?

Solution: By Hess's law, the overall enthalpy (energy) change in a reaction is equal to the sum of the enthalpy (energy) changes for the individual steps. The reactions shown in the problem are just the sums of the ionization energy of the alkali metal and the electron affinity of the halogen.

(a) Taking data from the referenced figures we have

$Li(g) \rightarrow Li^+(g) + e^-$ $\qquad \Delta H° = 520$ kJ
$I(g) + e^- \rightarrow I^-(g)$ $\qquad \Delta H° = -295$ kJ

$Li(g) + I(g) \rightarrow Li^+(g) + I^-(g)$ $\qquad \Delta H° = 225$ kJ

Parts (b) and (c) are solved in an exactly analogous manner.

(b) $Na(g) + F(g) \rightarrow Na^+(g) + F^-(g)$ $\qquad \Delta H° = 163$ kJ

(c) $K(g) + Cl(g) \rightarrow K^+(g) + Cl^-(g)$ $\qquad \Delta H° = 71$ kJ

9.84 Compounds composed of two different halogens are referred to as interhalogen compounds.

Solution: The three pairs of nonbonding electrons around each flourine have been omitted for simplicity.

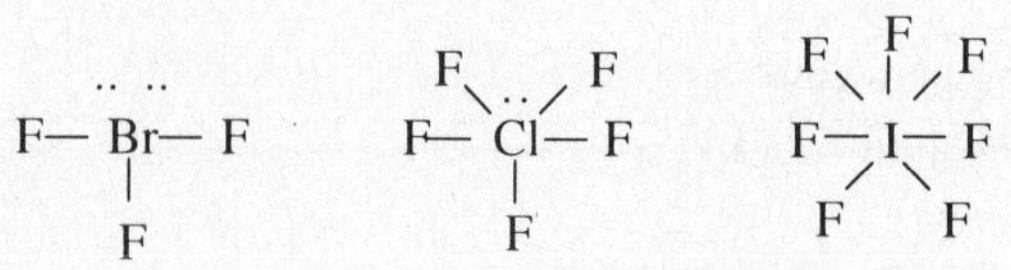

F— Br— F (with F below Br)   F— Cl— F (with two F above and one F below Cl)   F—I—F (with three F above and two F below I)

The octet rule is not obeyed in any of the compounds. In order for the octet rule to be obeyed, what would the value of n in the compound ICl_n have to be? [Hint: see Problem 9.45 (a)]

9.86 There are obviously many possible (correct) answers for this problem.

Solution: (a) An example of an aluminum species that satisfies the octet rule is the anion $AlCl_4^-$. The dot structure is drawn in Problem 9.71.

(b) An example of an aluminum species containing an expanded octet is anion AlF_6^{3-}. (How many pairs of electrons surround the central atom?)

(c) An aluminum species that has an incomplete octet is the compound $AlCl_3$. The dot structure is given in Problem 9.71.

9.88 You are given the structural framework, so all that needs to be done is to work out the dot structures. Of the elements phosphorus, fluorine, and oxygen, which can exceed an octet of electrons?

Solution: Four resonance structures together with the formal charges are

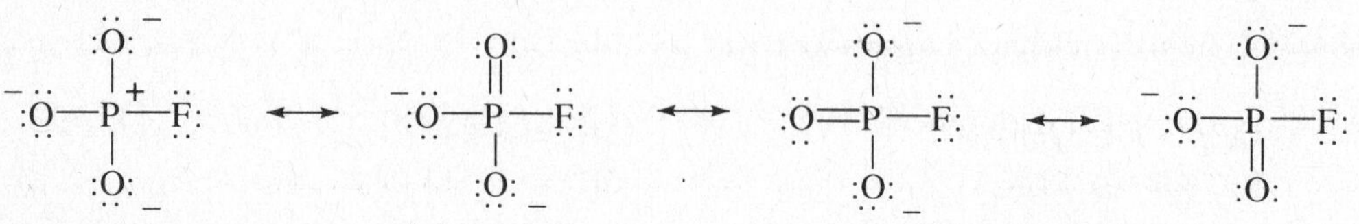

$^-$:O—P$^+$—F: (O$^-$ above and O$^-$ below P) ⟷ $^-$:O—P—F: (O double-bonded above, O$^-$ below P) ⟷ :O=P—F: (O$^-$ above and O$^-$ below P) ⟷ $^-$:O—P—F: (O$^-$ above, O double-bonded below P)

9.90 You should draw Lewis dot structures for the proposed compounds to help answer this question.

Solution: CF_2 would be very unstable because carbon does not have an octet. (How many electrons does it have?) LiO_2 would not be stable because the lattice energy would be too low to stabilize the solid. $CsCl_2$ requires a Cs^{2+} cation. The second ionization potential is too large to be compensated for by the lattice energy, so the compound is energetically unfavorable. PI_5 appears to be a reasonable species (compare to PF_5 in Example 9.10). However, the iodine atoms are too large to have five of them "fit" around a single phosphorus atom.

9.92 Rather than simply answer each question as true or false, you should also be able to explain why.

Solution: (a) false; (b) true; (c) false; (d) false.

For question (c), what is an example of a second-period species that violates the octet rule?

9.94 Review Section 9.10 and use the information in Table 9.4.

Solution: (a) Bond broken: C–H $\Delta H° = 414$ kJ/mol
Bond made: C–Cl $\Delta H° = -338$ kJ/mol

$$\Delta H^{\circ}_{rxn} = 414 - 338 = 76 \text{ kJ}$$

(b) Bond broken: C–H $\Delta H° = 414$ kJ/mol
Bond made: H–Cl $\Delta H° = -431.9$ kJ/mol

$$\Delta H^{\circ}_{rxn} = 414 - 431.9 = -18 \text{ kJ}$$

Based on energy considerations, reaction (b) will occur more readily since it is exothermic. Reaction (a) is endothermic.

9.96 Acetic acid, a typical organic acid, was the subject of Problem 9.47.

Solution: The rest of the molecule (in this problem, unidentified) would be attached at the end of the free bond.

(a) —C(=O)—O—H (b) —C(=O)—O:$^-$ ↔ —C(—O:$^-$)=O:

9.98 What is the definition of isoelectronic?

Solution: NH_4^+ and CH_4 are isoelectronic, as are CO and N_2, as are $B_3N_3H_6$ and C_6H_6.

9.100 This problem is exactly analogous to Problem 9.25.

Solution:

(1) $K(s) \rightarrow K(g)$	$\Delta H^{\circ}_1 = 90.00$ kJ
(2) $\frac{1}{2}Br_2(l) \rightarrow \frac{1}{2}Br_2(g)$	$\Delta H^{\circ}_2 = \frac{30.71 \text{ kJ}}{2}$
(3) $\frac{1}{2}Br_2(g) \rightarrow Br(g)$	$\Delta H^{\circ}_3 = \frac{192.5 \text{ kJ}}{2}$
(4) $K(g) \rightarrow K^+(g) + e^-$	$\Delta H^{\circ}_4 = 418.7$ kJ
(5) $Br(g) + e^- \rightarrow Br^-(g)$	$\Delta H^{\circ}_5 = -324$ kJ
(6) $K^+(g) + Br^-(g) \rightarrow KBr(s)$	$\Delta H^{\circ}_6 = ?$
$K(s) + \frac{1}{2}Br_2(l) \rightarrow KBr(s)$	$\Delta H^{\circ}_{overall} = -392.2$ kJ

$$\Delta H^{\circ}_6 = \Delta H^{\circ}_{overall} - \Delta H^{\circ}_1 - \Delta H^{\circ}_2 - \Delta H^{\circ}_3 - \Delta H^{\circ}_4 - \Delta H^{\circ}_5 = (-392.2) - (90.00) - (15.36) - (96.3) - (418.7) - (-324)$$

ΔH_6^o = -689 kJ   The lattice energy of KBr is 689 kJ/mol.

9.102 This problem is exactly analogous to Problem 9.25 and 9.100 above.

Solution:

(1) $Mg(s) \rightarrow Mg(g)$	ΔH_1^o = 150.2 kJ
(2) $\frac{1}{2}O_2(g) \rightarrow O(g)$	ΔH_2^o = 249.4 kJ
(3) $Mg(g) \rightarrow Mg^{2+}(g) + 2e^-$	ΔH_3^o = 2188 kJ
(4) $O(g) + e^- \rightarrow O^-(g)$	ΔH_4^o = -142 kJ
(5) $O^-(g) + e^- \rightarrow O^{2-}(g)$	ΔH_5^o = ?
(6) $Mg^{2+}(g) + O^{2-}(g) \rightarrow MgO(s)$	ΔH_6^o = -3890 kJ
$Mg(s) + \frac{1}{2}O_2(g) \rightarrow MgO(s)$	ΔH_f^o = -601.8 kJ

$$\Delta H_5^o = \Delta H_f^o - \Delta H_1^o - \Delta H_2^o - \Delta H_3^o - \Delta H_4^o - \Delta H_6^o = (-601.8) - (150.2) - (249.4) - (2188) - (-142) - (-3890) = 842.6 \text{ kJ}$$

The electron affinity for O^- is 842.6 kJ/mol.

9.104 We need to consider the difference between the bond dissociation energy and the process that produces F^+ and F^- from F_2.

Solution: The bond dissociation energy for F_2 is: $F_2(g) \rightarrow F(g) + F(g)$ $\Delta H°$ = 156.9 kJ

The energy for the process $F_2(g) \rightarrow F^+(g) + F^-(g)$ can be found by Hess's law. Thus,

Reaction	$\Delta H°$	Source
$F_2(g) \rightarrow F(g) + F(g)$	$\Delta H°$ = 156.9 kJ	(From Table 9.4)
$F(g) \rightarrow F^+(g) + e^-$	$\Delta H°$ = 1680 kJ	(From Table 8.4)
$F(g) + e^- \rightarrow F^-(g)$	$\Delta H°$ = -333 kJ	(From Table 8.5)
$F_2(g) \rightarrow F^+(g) + F^-(g)$	$\Delta H°$ = 1504 kJ	

It is much easier to dissociate F_2 into two neutral F atoms than it is to dissociate it into a fluorine cation and anion.

9.105 The compound exists as two resonance structures.

Solution: The Lewis structures are:

$$H-\underset{H}{\overset{H}{C}}-\ddot{N}=C=\ddot{O} \leftrightarrow H-\underset{H}{\overset{H}{C}}-\overset{+}{N}\equiv C-\ddot{\underset{..}{O}}:^{-}$$

9.106 This problem is long and somewhat tedious, but it is exactly analogous to other problems such as 9.102 above.

Solution: (a) Calculating ΔH_f^o for MgCl, $MgCl_2$, and $MgCl_3$:

For MgCl

(1) $Mg(s) \rightarrow Mg(g)$	$\Delta H_1^o = 150.2$ kJ
(2) $\frac{1}{2}Cl_2(g) \rightarrow Cl(g)$	$\Delta H_2^o = 121.4$ kJ
(3) $Mg(g) \rightarrow Mg^+(g) + e^-$	$\Delta H_3^o = 738.1$ kJ
(4) $Cl(g) + e^- \rightarrow Cl^-(g)$	$\Delta H_4^o = -348$ kJ
(5) $Mg^+(g) + Cl^-(g) \rightarrow MgCl(s)$	$\Delta H_5^o = -753$ kJ
$Mg(s) + \frac{1}{2}Cl_2(g) \rightarrow MgCl(s)$	$\Delta H_f^o = ?$

$$\Delta H_f^o = \Delta H_1^o + \Delta H_2^o + \Delta H_3^o + \Delta H_4^o + \Delta H_5^o$$

$$\Delta H_f^o = (150.2) + (121.4) + (738.1) + (-348) + (-753) = -91.3 \text{ kJ/mol}$$

For $MgCl_2$

(1) $Mg(s) \rightarrow Mg(g)$	$\Delta H_1^o = 150.2$ kJ
(2) $Cl_2(g) \rightarrow 2Cl(g)$	$\Delta H_2^o = 121.4 \times 2$ kJ
(3) $Mg(g) \rightarrow Mg^+(g) + e^-$	$\Delta H_3^o = 738.1$ kJ
(4) $Mg^+(g) \rightarrow Mg^{2+}(g) + e^-$	$\Delta H_4^o = 1450$ kJ
(5) $2Cl(g) + 2e^- \rightarrow 2Cl^-(g)$	$\Delta H_5^o = -348 \times 2$ kJ
(6) $Mg^{2+}(g) + 2Cl^-(g) \rightarrow MgCl_2(s)$	$\Delta H_6^o = -2527$ kJ
$Mg(s) + Cl_2(g) \rightarrow MgCl_2(s)$	$\Delta H_f^o = ?$

$$\Delta H_f^o = \Delta H_1^o + \Delta H_2^o + \Delta H_3^o + \Delta H_4^o + \Delta H_5^o + \Delta H_6^0$$

$$\Delta H_f^o = (150.2) + (242.8) + (738.1) + (1450) + (-696) + (-2527) = -641.9 \text{ kJ/mol}$$

For $MgCl_3$

(1) $Mg(s) \rightarrow Mg(g)$	$\Delta H_1^o = 150.2$ kJ
(2) $\frac{3}{2}Cl_2(g) \rightarrow 3Cl(g)$	$\Delta H_2^o = 121.4 \times 3$ kJ
(3) $Mg(g) \rightarrow Mg^+(g) + e^-$	$\Delta H_3^o = 738.1$ kJ

(4) $Mg^+(g) \rightarrow Mg^{2+}(g) + e^-$ $\qquad \Delta H_4^\circ = 1450$ kJ

(5) $Mg^{2+}(g) \rightarrow Mg^{3+}(g) + e^-$ $\qquad \Delta H_5^\circ = 7730$ kJ

(6) $3Cl(g) + 3e^- \rightarrow 3Cl^-(g)$ $\qquad \Delta H_6^\circ = -348 \times 3$ kJ

(7) $Mg^{3+}(g) + 3Cl^-(g) \rightarrow MgCl_3(s)$ $\qquad \Delta H_7^\circ = -5440$ kJ

$Mg(s) + \frac{3}{2}Cl_2(g) \rightarrow MgCl_3(s)$ $\qquad \Delta H_f^\circ = ?$

$$\Delta H_f^\circ = \Delta H_1^\circ + \Delta H_2^\circ + \Delta H_3^\circ + \Delta H_4^\circ + \Delta H_5^\circ + \Delta H_6^\circ + \Delta H_7^\circ$$

$$\Delta H_f^\circ = (150.2) + (364.2) + (738.1) + (1450) + (7730) + (-1044) + (-5440) = 3949 \text{ kJ/mol}$$

Therefore $MgCl_2$, as expected, is the most stable of the three chlorides with respect to the elements. Note that $MgCl_3$ is endothermic and therefore unstable.

(b) Calculating the standard enthalpy of reaction: $2MgCl(s) \rightarrow MgCl_2(s) + Mg(s)$

$$\Delta H^\circ = \Delta H_f^\circ[MgCl_2(s)] + \Delta H_f^\circ[Mg(s)] - 2\Delta H_f^\circ[MgCl(s)]$$

$$\Delta H^\circ = (1 \text{ mol})(-641.9 \text{ kJ/mol}) + 0 - (2 \text{ mol})(-91.3 \text{ kJ/mol}) = -459.3 \text{ kJ}$$

Thus, even though MgCl is exothermic with respect to the elements, it is unstable relative to disproportionation to $MgCl_2$ and Mg.

9.107 Assume that the skeletal structure is $ClONO_2$.

Solution: A reasonable Lewis structure is:

```
                :O:
                 ||        ..  -
 : Cl—O—N—O :
   ..    ..   +     ..
```

9.109 These Lewis structures are all similar to methane CH_4.

Solution: The nonbonding electron pairs around Cl and F are omitted for simplicity.

```
     Cl              Cl              F              F   H
     |               |               |              |   |
 F — C — Cl      F — C — F       H — C — F      F — C — C — F
     |               |               |              |   |
     Cl              Cl              Cl             F   F
```

9.110 As you will find in Chapter 25 organic chemistry, these compounds are all called *alkenes*.

Solution: The structures are (the nonbonding electron pairs on fluorine have beem omitted for simplicity):

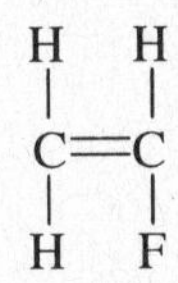

```
H   H   H         H   H   H   H
|   |   |         |   |   |   |
C═══C───C───H     C═══C───C───C───H
|       |         |       |   |
H       H         H       H   H
```

9.111 With a little practice, drawing these organic Lewis structures is quite easy.

Solution: Note that the nonbonding electron pairs have been deleted from oxygen, nitrogen, sulfur, and chlorine for simplicity.

(a)

```
    H
    |
H───C───O───H
    |
    H
```

(b)

```
    H   H
    |   |
H───C───C───O───H
    |   |
    H   H
```

(c)

```
          C2H5
           |
H5C2 ─── Pb ─── C2H5
           |
          C2H5
```

(d)

```
    H   H
    |   |
H───C───N───H
    |
    H
```

(e) (f) (g)

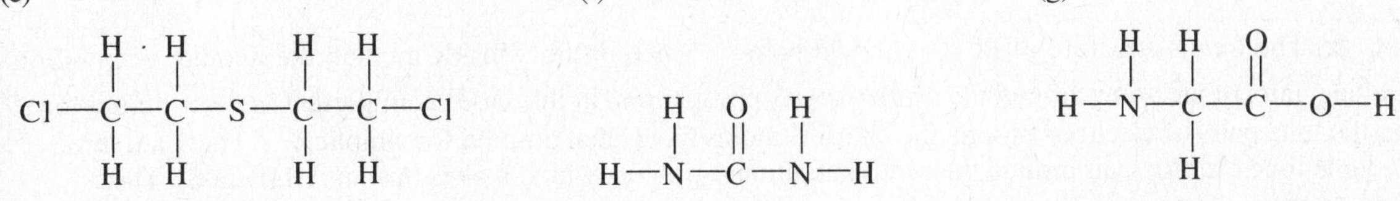

(h)

```
    H   H   O
    |   |   ‖
H───C───C───C───O───H
    |   |
    H   O───H
```

Note: in part (c) above, ethyl = C_2H_5 =

```
    H   H
    |   |
H───C───C───
    |   |
    H   H
```

Note: If you have not already done so, please read the "Introduction for the Student" at the beginning of this Student Solutions Manual.

Chapter 10

CHEMICAL BONDING II: MOLECULAR GEOMETRY AND MOLECULAR ORBITALS

10.8 The first step in applying the VSEPR method is always to work out the Lewis structure of the molecule. This gives you the number of bonding and unshared electron pairs without which the VSEPR method cannot be applied. If you are not sure of what to do, you must go back and master the material in Section 9.6 and especially Examples 9.3-9.5. Work out the Lewis structures of the molecules in this problem and then classify each by structure type using either Table 10.1 or 10.2. The VSEPR method is described in detail in Section 10.1. Study this material carefully if you haven't already done so. If you have difficulties, study the application of these principles in Example 10.1.

Solution: (a) The Lewis structure of PCl_3 is shown below. Since in the VSEPR method the number of bonding pairs and lone pairs of electrons around the *central atom* (phosphorus, in this case) is important in determining the structure, the lone pairs of electrons around the chlorine atoms have been omitted for simplicity. There are three bonds and one lone electron pair around the central atom, phosphorus, which makes this an AB_3E case. The information in Table 10.2 shows that the structure is a trigonal pyramid like ammonia.

```
       ..
  Cl — P — Cl
       |
       Cl
```

What would be the structure of the molecule if there were no lone pairs and three bonds?

(b) The Lewis structure of $CHCl_3$ is shown below. There are four bonds and no lone pairs around carbon which makes this an AB_4 case. The molecule should be tetrahedral like methane (Table 10.1).

```
       H
       |
  Cl — C — Cl
       |
       Cl
```

(c) The Lewis structure of SiH_4 is shown below. Like part (b), it is a tetrahedral AB_4 molecule.

```
       H
       |
   H— Si— H
       |
       H
```

What would be the structure of this molecule if there were a lone pair in addition to the four bonds? Can Si form molecules with more than four electron pairs (octet)?

(d) The Lewis structure of $TeCl_4$ is shown below. There are four bonds and one lone pair which make this an AB_4E case. Consulting Table 10.2 shows that the structure should be that of a distorted tetrahedron like SF_4.

```
        ..
   Cl— Te— Cl
       /  \
     Cl    Cl
```

Are $TeCl_4$ and SF_4 isoelectronic? Should isoelectronic molecules have similar VSEPR structures?

10.10 If you need help with VSEPR, refer to Problem 10.8 above and the Example 10.1.

Solution: The lone pairs of electrons on the bromine atoms have been omitted for simplicity.

Br — Hg — Br	:N≡N—Ö:	:S̈—C≡N:⁻
linear	linear	linear

10.12 What combination of bonds and lone pairs gives molecules with tetrahedral (not distorted tetrahedral) geometry? Which of these molecules possess that combination?

Solution: Only molecules with four bonds and no lone pairs (AB_4, Table 10.1) are tetrahedral.

```
       Cl              I              ┌    Cl    ┐ 2-
       |               |              │    |     │
  Cl— Si — Cl      I — C — I          │Cl—Cd— Cl │
       |               |              │    |     │
       Cl              I              └    Cl    ┘
```

What are the Lewis structures and shapes for XeF_4 and SeF_4?

10.13 Review Problems 10.10 and 10.11.

Solution: The Lewis structure is

```
       H   :O:
       |   ||    ..
   H — C — C — O — H
       |         ..
       H
```

H–C(–H)(–H)– AB_4 tetrahedral

–C(=O:)– AB_3 trigonal planar

–Ö–H AB_2E_2 bent

Why isn't the geometry around the oxygen atom in the OH group linear?

10.18 What are the geometries of these two molecules? Review Problems 10.8-10.13 and Example 10.1 if you don't know. Will CS_2 have a dipole moment (see Review Question 10.17)? Are the C-S and C-O bond moments exactly equal?

Solution: Both molecules are linear (AB_2). Carbon disulfide will have no dipole moment because the C-S bond moments exactly cancel. Even though OCS is linear, the C-O and C-S bond moments are not exactly equal (which is larger?), and there will be a small net dipole moment. OCS has a larger dipole moment than CS_2 (zero).

10.19 This is like Problem 10.18. Do all the molecules have the same structure? Are they isoelectronic? The dipole moment of each should be proportional to the electronegativity difference between hydrogen and the other element.

Solution: All four molecules have two bonds and two lone pairs (AB_2E_2) and therefore the bond angles are not linear. Since electronegativity decreases going down a column (group) in the periodic table, the electronegativity differences between hydrogen and the other Group VI element will increase in the order Te<Se<S<O. The dipole moments will increase in the same order. Would this conclusion be as easy if the elements were in different groups?

10.20 This is similar to Problem 10.19

Solution: $CO_2 = CBr_4$ ($\mu = 0$ for both) $< H_2S < NH_3 < H_2O < HF$

10.22 First work out the molecular geometries using the VSEPR method (Problems 10.8-10.13), then apply the principles discussed in Section 10.2. Study Example 10.2 before beginning this problem.

Solution:

Dot structure	*Label*	*Shape*	*Bond dipole*	*Resultant dipole moment*
H–Ö–H (bent)	AB_2E_2	bent		↑ $\mu > 0$
:P(Cl)(Cl)(Cl)	AB_3E	trigonal pyramidal		↓ $\mu > 0$

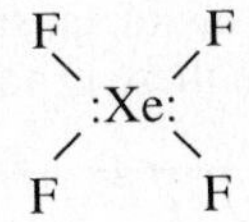 F_4:Xe:	AB_4E_2	square planar		$\mu = 0$
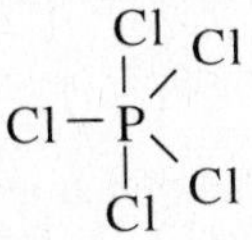 $Cl-PCl_4$	AB_5	trigonal bipyramid	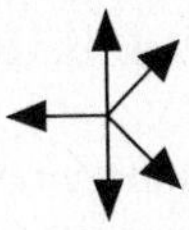	$\mu = 0$

Why do the bond dipoles add to zero in PCl_5?

10.24 Look at the arrangement of the bond moments in each molecule. Be sure the individual moments are pointing the right way (check the electronegativities of the elements).

Solution: The molecules shown in (b) and (d) are nonpolar because the bond moments in each molecule cancel one another. The molecules shown in (a) and (c) are polar. The dipole moment of the molecule in (c) is smaller than that in (a), therefore (b) = (d) < (c) < (a).

10.26 The Lewis theory tells us that a covalent bond is the result of the sharing of a pair of electrons between two atoms. The quantum theory tells us that in atoms electrons occupy allowed states called orbitals. The valence bond theory permits us to picture the sharing of these pairs of electrons in terms of overlapping orbitals belonging to the bonded atoms.

Solution: In the Lewis theory we represent the formation of the chlorine molecule in terms of the Lewis structures below:

$$Cl\cdot + \cdot Cl \rightarrow Cl\text{–}Cl$$

The quantum theory of atomic structure tells us that the unpaired electron on each chlorine atom occupies a 3p orbital (review orbital shapes in Section 7.8 and Figures 7.24-7.26). The valence bond theory depicts the chlorine-chlorine shared electron pair bond in terms of the overlapping of the two 3p chlorine orbitals. The two shared electrons spend most of their time in the region between the two atoms where the orbital overlap is greatest.

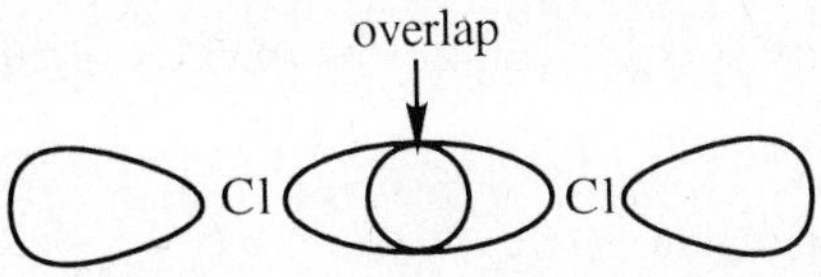

overlap

Cl Cl

3p orbitals

The quantum theory tells us that the unpaired electrons in hydrogen and chlorine are found in 1s and 3p orbitals, respectively.

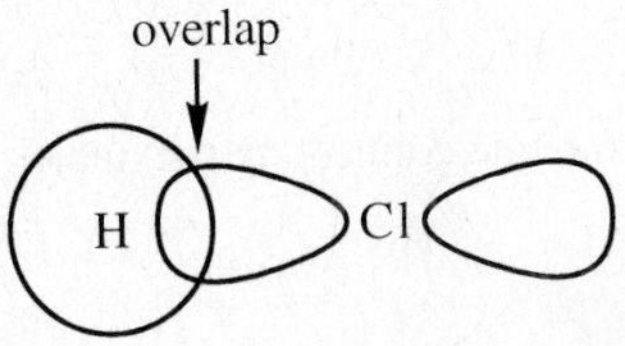

overlap

H Cl

10.31 What types of molecular geometry call for the description of covalent bonds in terms of sp, sp^2, or sp^3 hybrid orbitals? Study the examples in Section 10.5. What are the expected bond angles in these geometries?

Solution: (a) Two sp hybrid orbitals. In the valence bond theory when a molecule has a linear geometry (AB_2 type, Table 10.1), the central atom uses sp hybrid orbitals to form covalent bonds. Since the geometry is linear, the angle between the two sp hybrid orbitals is 180°.

(b) Two sp^2 hybrid orbitals. In the valence bond theory when a molecule has a trigonal planar arrangement of electron pairs around the central atom (AB_3 and AB_2E type molecules) the atom uses sp^2 hybrid orbitals to form covalent bonds and contain lone pairs. A trigonal planar arrangement of electron pairs requires an angle of 120° between the hybrid orbitals.

(c) Two sp^3 hybrid orbitals. In the valence bond theory a tetrahedral arrangement of electron pairs around a central atom (AB_4, AB_3E, or AB_2E_2 type molecules) requires the use of sp^3 hybrid orbitals to form covalent bonds and contain lone pairs. A tetrahedral arrangement of electron pairs requires an angle of 109.5° between the hybrid orbitals.

10.33 The key to recognizing sigma and pi bond types is in the way the two orbitals overlap with each other. Study the first two paragraphs of Section 10.6 and focus on the distinction in terms of the orbital overlaps or where the greatest electron densities are found. Where is the region of maximum overlap in a sigma bond?

Solution: To form a sigma bond, two atomic orbitals must overlap in such a way that the region of maximum electron density, namely the overlap region, is found along the line connecting the two nuclei (atom centers). In a pi bond the overlap region is above and below this line. If neither type of overlap is possible, no bond can form (except in certain special cases outside the scope of a first-year chemistry course).

(a) Sigma bond. The overlap is directly between the atoms.

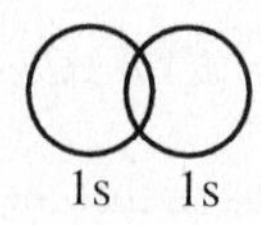

(b) Sigma bond. The overlap is on the line connecting the nuclei.

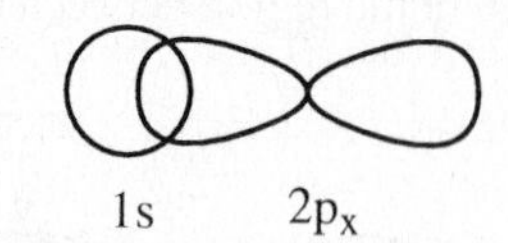

(c) No bond. There is no common overlap region.

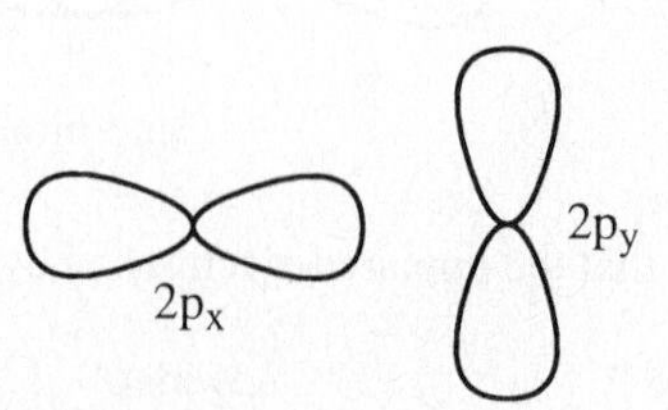

(d) Pi bond. The overlap is above and below the line connecting the nuclei.

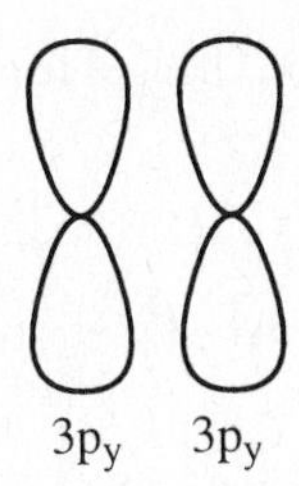

(e) Sigma bond. If the line connecting the nuclei is the x axis, the bond will be a sigma type. It is just like the first part of Problem 10.26.

(f) Sigma bond. This is just like (a) above, except that the two orbitals are not the same size. Draw a picture.

10.34 Study and compare Tables 10.1 and 10.4. Notice that there is one and only one type of hybrid orbital assigned to each of the electron pair arrangements in Table 10.1. For instance, a trigonal planar electron pair arrangement is always associated with sp^2 hybrid orbitals (row 2, Table 10.4). This same sort of correspondence applies in all the other cases in Table 10.1 as well. Begin this problem by working out the Lewis structure for AsH_3. How many bond pairs are there? How many lone pairs are there? What type of electron pair arrangement does this represent? What type of hybrid orbital set corresponds to this electron pair arrangement? See Example 10.3 in the text.

Solution: AsH_3 has the Lewis structure shown below. There are three bond pairs and one lone pair. The four electron pairs have a tetrahedral arrangement (Table 10.1), and the molecular geometry is trigonal pyramidal (AB_3E) like ammonia (see Table 10.2). The As (arsenic) atom is in an sp^3 hybridization state.

```
   ..
H— As— H
    |
    H
```

Three of the sp^3 hybrid orbitals form bonds to the hydrogen atoms by overlapping with the hydrogen 1s orbitals (sigma or pi bonds?). The fourth sp^3 hybrid orbital holds the lone pair.

10.35 Please read and study Problem 10.34 above. Draw Lewis structures of the two compounds, count bond and lone pairs around each Si (silicon) atom, determine the electron pair arrangement about each Si (Table 10.1), and assign the hybrid orbital type (Table 10.4). Example 10.3 in the textbook is a similar problem.

Solution: The Lewis structures are shown below. Each silicon atom has four bonds and zero lone pairs, giving a tetrahedral electron pair arrangement in each case. This corresponds to sp^3 hybridization.

```
     H                H    H
     |                |    |
H— Si— H         H— Si— Si— H
     |                |    |
     H                H    H
```

10.36 Analyze the Lewis structures of the two aluminum compounds as in Problems 10.34 and 10.35 above.

Solution: The Lewis structures of $AlCl_3$ and $AlCl_4^-$ are shown below. By the reasoning of the two problems above, the hybridization changes from sp^2 to sp^3.

```
Cl—Al—Cl            ┌    Cl    ┐ −
    |               │    |     │
    Cl              │Cl—Al—Cl  │
                    │    |     │
                    └    Cl    ┘
```

What are the geometries of these molecules?

10.38 You must begin the problem by drawing the Lewis structure.

Solution: (a) NH_3 is an AB_3E type molecule just as AsH_3 in Problem 10.34. Referring the Table 10.4, the nitrogen is sp^3 hybridized.

(b) N_2H_4 has two equivalent nitrogen atoms. Centering attention on just one nitrogen atom shows that it is an AB_3E molecule, so the nitrogen atoms are sp^3 hybridized. From structural considerations, how can N_2H_4 be considered to be a derivative of NH_3?

(c) The nitrate anion NO_3^- is isoelectronic and isostructural with the carbonate anion CO_3^{2-} that is discussed in Section 9.8. There are three resonances structures, and the ion is of type AB_3; thus the nitrogen hydridization is sp^2.

10.40 Analyze the Lewis structures of these molecules as in Problems 10.34 and 10.35 above. The problem is similar to Example 10.5

Solution: (a) Each carbon has four bond pairs and no lone pairs and therefore has a tetrahedral electron pair arrangement. This implies sp^3 hybrid orbitals.

```
   H  H
   |  |
H— C— C— H
   |  |
   H  H
```

(b) The leftmost carbon is tetrahedral and therefore has sp^3 hybrid orbitals. The two carbon atoms connected by the double bond are trigonal planar with sp^2 hybrid orbitals.

```
   H  H  H
   |  |  |
H— C— C= C— H
   |
   H
```

(c) This is just like (a) above. Both have sp^3 hybrid orbitals. What are the hybrid orbitals on the oxygen?

```
   H  H
   |  |
H— C— C— O— H
   |  |
   H  H
```

(d) The leftmost carbon is tetrahedral (sp^3 hybrid orbitals). The carbon connected to oxygen is trigonal planar (why?)

and has sp^2 hybrid orbitals.

```
    H  H
    |  |
H — C — C = O
    |
    H
```

(e) The leftmost carbon is tetrahedral (sp^3 hybrid orbitals). The other carbon is trigonal planar sp^2 hybridized orbitals.

```
    H  O
    |  ||
H — C — C — O — H
    |
    H
```

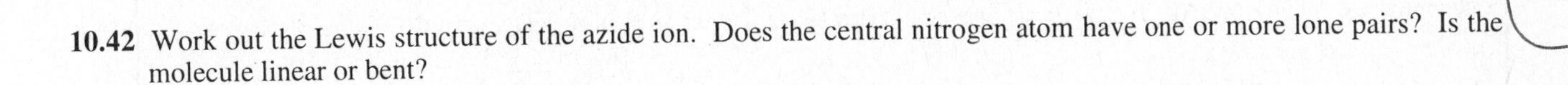

10.42 Work out the Lewis structure of the azide ion. Does the central nitrogen atom have one or more lone pairs? Is the molecule linear or bent?

Solution: The Lewis structure is shown below (several resonance forms).

```
     +  2-             -   +   -            2-   +
 : N≡N—N :    ↔    : N=N=N :     ↔     : N—N≡N :
```

Since there are no lone pairs on the central nitrogen, the molecule must be linear and sp hybrid orbitals (Table 10.4) must be used.

10.43 Analyze the electron pair arrangement around each carbon atom as in Problem 10.40.

Solution: The Lewis structure is shown below. The two end carbons are trigonal planar and therefore use sp^2 hybrid orbitals. The central carbon is linear and must use sp hybrid orbitals. Study Figures 10.16, 10.17, and 10.19 to see the hybrid orbital overlaps.

```
    H       H
    |       |
H — C = C = C — H
```

A Lewis drawing does not necessarily show actual molecular geometry. Notice that the two CH_2 groups at the ends of the molecule must be perpendicular. This is because the two double bonds must use different 2p orbitals on the middle carbon, and these two 2p orbitals are perpendicular. Is the allene molecule polar?

10.44 This is like Example 10.4.

Solution: The PF_5 molecule has a trigonal bipyramidal geometry (AB_5). Thus the phosphorus atom is sp^3d hybridized (see Example 10.4 and PCl_5 in Problem 10.22).

10.46 Study Section 10.5. In a double bond, how many sigma bonds are there? How many pi bonds are in a triple bond? How many pi bonds are in a single bond?

Solution: It is almost always true that a single bond is a sigma bond, that a double bond is a sigma bond and a pi bond, and that a triple bond is *always* a sigma bond and two pi bonds.

(a) sigma bonds: 4; pi bonds: 0 (b) sigma bonds: 5; pi bonds: 1; (c) sigma bonds: 10; pi bonds: 3

10.51 Study Section 10.7 on molecular orbital electron configurations. What is the bond order in each of the three hydrogen molecules? How does bond order relate to internuclear distance (bond length)?

Solution: The molecular orbital electron configuration and bond order of each species is shown below.

	H_2	H_2^+	H_2^{2+}
σ^*_{1s}	☐	☐	☐
σ_{1s}	↑↓	↑	☐
	bond order 1	bond order $\frac{1}{2}$	bond order 0

The internuclear distance in the +1 ion should be greater than that in the neutral hydrogen molecule. The distance in the +2 ion will be arbitrarily large because there is no bond (bond order zero).

10.52 Hint: How does the Pauli exclusion principle affect the occupancy pattern of electrons in atomic or molecular orbitals?

Solution: In order for the two hydrogen atoms to combine to form a H_2 molecule, the electrons must have opposite spins. Furthermore, the combined energy of the two atoms must not be too great. Otherwise, the H_2 molecule will possess too much energy and will break apart into two hydrogen atoms.

10.53 Use a molecular orbital energy level diagram like Figure 10.23.

Solution: The energy level diagrams are shown below.

	He_2	HHe	He_2^+
σ^*_{1s}	↑↓	↑	↑
σ_{1s}	↑↓	↑↓	↑↓

He_2 has a bond order of zero; the other two have bond orders of 1/2. Based on bond orders alone, He_2 has no stability, while the other two have roughly equal stabilities.

10.54 This problem is similar to Example 10.6.

Solution: The electron configurations are listed. Refer to Table 10.5 for the molecular orbital diagram.

Li_2: $(\sigma_{1s})^2(\sigma^*_{1s})^2(\sigma_{2s})^2$ bond order = 1

Li_2^+: $(\sigma_{1s})^2(\sigma^*_{1s})^2(\sigma_{2s})^1$ bond order = $\frac{1}{2}$

Li_2^-: $(\sigma_{1s})^2(\sigma^*_{1s})^2(\sigma_{2s})^2(\sigma^*_{2s})^1$ bond order = $\frac{1}{2}$

Order of increasing stability: $Li_2^- \approx Li_2^+ < Li_2$

In reality, Li_2^+ is more stable than Li_2^- because there is less electrostatic repulsion in Li_2^+.

10.56 Refer to Table 10.5 that shows the orbital diagram for B_2. Does B_2^+ have one more electron or one less than B_2?

Solution: Removing an electron from B_2 gives B_2^+ which has a bond order of $\frac{1}{2}$. Therefore, B_2^+ has a weaker and longer bond than B_2.

10.57 Use molecular orbital energy level diagrams like Figures 10.26 and 10.27 to solve this problem.

Solution: The energy level diagrams are shown below.

Orbital	C_2^{2-}	C_2
$\sigma^*_{2p_x}$	☐	☐
$\pi^*_{2p_y}$, $\pi^*_{2p_z}$	☐ ☐	☐ ☐
σ_{2p_x}	↑↓	☐
π_{2p_y}, π_{2p_z}	↑↓ ↑↓	↑↓ ↑↓
σ^*_{2s}	↑↓	↑↓
σ_{2s}	↑↓	↑↓
σ^*_{1s}	↑↓	↑↓
σ_{1s}	↑↓	↑↓

The bond order of the carbide ion is 3 and that of C_2 is only 2. With what homonuclear diatomic molecule is the carbide ion isoelectronic? With which heteronuclear diatomics is it isoelectronic?

10.58 Work out the Lewis structure and the molecular orbital energy level diagram for the oxygen molecule.

Solution: In both the Lewis structure and the molecular orbital energy level diagram (Table 10.5) the oxygen molecule has a double bond (bond order = 2). The principal difference is that the molecular orbital treatment predicts that the molecule will have two unpaired electrons (paramagnetic). Experimentally this is found to be true.

10.60 Check out Examples 10.6 and 10.7 for similar problems. Hint: Work out molecular orbital energy level diagrams like Table 10.5.

Solution: We refer to Table 10.5.

O_2 has a bond order of 2 and is paramagnetic (two unpaired electrons).

O_2^+ has a bond order of 2.5 and is paramagnetic (one unpaired electron).

O_2^- has a bond order of 1.5 and is paramagnetic (one unpaired electron).

O_2^{2-} has a bond order of 1 and is diamagnetic.

Based on molecular orbital theory, the stability of these molecules increases as follows:

$$O_2^{2-} < O_2^- < O_2 < O_2^+$$

10.62 You could use the general molecular orbital scheme to construct the molecular orbital diagrams for B_2 and C_2 from Figure 10.27. However, the diagrams have been worked out for you in Table 10.5.

Solution: As discussed in the text (see Table 10.5), the single bond in B_2 is a pi bond (the electrons are in a pi bonding molecular orbital) and the double bond in C_2 is made up of two pi bonds (the electrons are in the pi bonding molecular orbitals).

10.66 You should read Section 10.8 before trying to work this problem.

Solution: Benzene is stabilized by delocalized molecular orbitals. The C-C bonds are equivalent, rather than alternating single and double bonds. The additional stabilization makes the bonds in benzene much less reactive chemically than isolated double bonds such as those in ethylene.

10.67 Read Problem 10.66 above before you try this problem.

Solution: The symbol on the left shows the pi bond delocalized over the entire molecule. The symbol on the right shows only one of the two resonance structures of benzene; it is an incomplete representation.

10.68 If the planes of the two rings in biphenyl can be rotated so that they are perpendicular, can the 2p orbitals on the two carbons connecting the rings overlap to form a pi bond?

Solution: If the two rings happen to be perpendicular in biphenyl, the pi molecular orbitals are less delocalized. In naphthalene the pi molecular orbital is always delocalized over the entire molecule. What do you think is the most stable structure for biphenyl: both rings in the same plane or both rings perpendicular?

10.69 See Problem 10.34 for suggestions on assigning hybrid orbital types. Study the discussion of the carbonate ion in Section 10.8.

Solution: (a) Two Lewis resonance forms (with lone electron pairs on fluorine omitted) are shown below:

```
       F                          F
       |                          |
  ..      .. -               - ..      ..
  O==N+—O:        <-->        :O—N+==O
  ..      ..                    ..
```

(b) There are no lone pairs on the nitrogen atom; it should have a trigonal planar electron pair arrangement and therefore use sp^2 hybrid orbitals.

(c) The bonding consists of sigma bonds joining the nitrogen atom to the fluorine and oxygen atoms. In addition there is a pi molecular orbital delocalized over the entire molecule (like the nitrate ion). Is nitryl fluoride isoelectronic with the carbonate ion?

10.70 This is like Problem 10.69 above.

Solution: The ion contains 24 valence electrons. Of these, six are involved in three sigma bonds between the nitrogen and oxygen atoms. The hybridization of the nitrogen atoms is sp^2. Four valence electrons on each oxygen (12 total) are non-bonding. The remaining six electrons are in delocalized pi molecular orbitals which result from the overlap of the p_z orbital of nitrogen and the p_z orbitals of the three oxygen atoms. The molecular orbitals are similar to those of the carbonate ion (Section 10.8).

10.72 This problem is similar to Problems 10.8 and 10.12.

Solution: Only (c) will not be tetrahedral. All the others have AB_4-type Lewis structures and will therefore be tetrahedral. For SF_4 the Lewis structure below is of the AB_4E type which gives rise to a distorted tetrahedral geometry (Section 10.2).

10.73 See Table 10.1 for the Lewis structure and geometry of mercury(II) chloride. What measurable physical property discussed in this chapter depends upon molecular geometry?

Solution: The molecule is linear (see Table 10.1). If mercury(II) bromide were bent, it would have a measurable dipole moment (Section 10.2). It has no dipole moment and therefore must be linear.

10.74 This problem is similar to Problem 10.46.

> ***Solution:*** A single bond is usually a sigma bond, a double bond is usually a sigma bond and a pi bond, and a triple bond is always a sigma bond and two pi bonds. Therefore, there are nine pi bonds and nine sigma bonds in the molecule.

10.75 Because silicon atoms are larger than carbon atoms by approximately 40%, the internuclear Si-Si distance is significantly greater than the corresponding C-C distance. Can you suggest a reason why a Si-Si pi bond might be much weaker than a C-C pi bond?

> ***Solution:*** According to valence bond theory, a pi bond is formed through the side-to-side overlap of a pair of p orbitals (see Section 10.5 and Figure 10.17). If two orbitals overlap poorly, i.e. they share very little space in common, then the resulting bond will be very weak. This situation applies in the case of pi bonds between silicon atoms as well as between any other elements not found in the second period. It is usually far more energetically favorable for silicon, or any other heavy element, to form two single (sigma) bonds to two other atoms than to form a double bond (sigma + pi) to only one other atom.

10.76 This problem is similar to Problem 9.8.

> ***Solution:*** The Lewis structure is shown below.
>
> $$:\underset{..}{\ddot{\text{Cl}}}-\underset{..}{\ddot{\text{S}}}-\underset{..}{\ddot{\text{Cl}}}:$$
>
> The central sulfur atom is of the AB_2E_2 type which requires a bent geometry. The sulfur atom should have sp^3 hybrid orbitals.

10.77 Note that the two oxygen atoms are connected to the carbon and not to each other. This problem is similar in part to Problem 10.69.

> ***Solution:*** (a) The Lewis resonance structures are shown below.
>
> $$\text{H}-\text{C}\begin{matrix} // :\ddot{\text{O}}: \\ \backslash :\underset{..}{\ddot{\text{O}}}:^{-} \end{matrix} \quad \leftrightarrow \quad \text{H}-\text{C}\begin{matrix} / :\ddot{\text{O}}:^{-} \\ \backslash\backslash :\ddot{\text{O}}: \end{matrix}$$
>
> (b) In the molecular orbital description there is a delocalized pi orbital connecting the O-C-O part of the molecule. The orbital picture would be very similar to Figure 10.30 except that there is no pi bond between carbon and hydrogen (why?).
>
> Note that resonance structures and delocalized pi molecular orbitals are really just two different ways of trying to describe the same phenomenon: the sharing of electrons between between more than two atoms.

10.78 Work out the Lewis structures of these species and then apply VSEPR principles. This problem is like Problems 10.8-10.12.

Solution: The Lewis structures and VSEPR geometries of these species are shown below. The three nonbonding pairs of electrons on each fluorine atom have been omitted for simplicity.

[F—Xe—F with F below]$^+$ [Xe with five F]$^+$ [Sb with six F]$^-$

AB_3E_2	AB_5E	AB_6
T-shaped	Square Pyramid	Octahedral

10.80 This problem is similar to Problems 10.8 and 10.18-10.19.

Solution: (a) The Lewis structure is

F— B— F
     |
     F

The shape will be trigonal planar (AB_3)

(b) The Lewis structure is

[O— Cl—O]$^-$ (with =O below Cl)

The molecule will be a trigonal pyramid (nonplanar).

(c) The Lewis structure and the dipole moment for H_2O is presented in Problem 10.22. The dipole moment is directed from the positive hydrogen end to the more negative oxygen.

d) The Lewis structure is

O (bonded to F and F)

The molecule is bent and therefore polar.

(e) The Lewis structure is

N (bonded to O and O)

The nitrogen atom is of the AB_2E type, but there is only one unshared electron rather than the usual pair. As a result, the repulsion will not be as great and the O-N-O angle will be greater than the 120° expected for AB_2E geometry. Experiment shows the angle to be around 135°.

Which of the species in this problem has resonance structures?

10.82 What is the location of phosphorus and arsenic in the periodic table?

Solution: Since arsenic and phosphorus are both in the same group, then this problem is exactly like Problem 10.44. AsF_5 is an AB_5 type molecule, so the geometry is trigonal bipyramidal.

10.84 This is similar to Problems 10.8-10.12. As in all of this type of problem, you much first determine the Lewis structure in order to determine the geometry.

Solution: Only ICl_2^- and $CdBr_2$ will be linear. The rest are bent. What are the geometries of the other species?

10.86 The simplist Lewis structures are given to you. Do the nitrogen atoms have an octet of electrons?

Solution: (a) The geometry around each nitrogen is identical. From the viewpoint of each nitrogen atom, the molecule is an AB_2E type (one F– group, one FN– group, and one nonbonding pair of electrons). The hydridization of each nitrogen is therefore sp^2.

(b) The structure on the right has a dipole moment because the two N–F bond dipoles do not cancel each other out as do the N–F bond dipoles in the left-hand structure.

10.87 What is the bond angle for a carbon atom with sp^3 hydridization?

Solution: (a) The structures for cyclpropane and cubane are

Cyclopropane

Cubane

(b) The C–C–C bond in cyclopropane is 60° and in cubane is 90°. Both are smaller than the 109.5° expected for sp^3 hybridized carbon. Consequently, there is considerable strain the molecules.

(c) They would be more difficult to make than sp^3 hybridized carbon in an unconstrained (that is, not in a small ring) system in which the carbons can adopt bond angles closer to 109.5°.

10.89 The problem is a direct application of Hess's law (Section 6.3).

Solution:

$$F_2^-(g) \rightarrow F_2(g) + e^- \qquad \Delta H^\circ_{rxn} = 290 \text{ kJ}$$

$$F_2(g) \rightarrow 2F(g) \qquad \Delta H^\circ_{rxn} = 156.9 \text{ kJ}$$

$$F(g) + e^- \rightarrow F^-(g) \qquad \Delta H^\circ_{rxn} = -333 \text{ kJ}$$

$$F_2^-(g) \rightarrow F(g) + F^-(g) \qquad \Delta H^\circ_{rxn} = 114 \text{ kJ}$$

Thus, the bond energy of the F_2^- ion is 114 kJ.

(b) We see that the bond energy of F_2^- is smaller than the bond dissociation energy of F_2 (156.9 kJ). As can be seen from the molecular orbital scheme below, the bond order of F_2 is 1.0, while the bond order of F_2^- is 0.5 because the extra electron enters the antibonding ($\sigma^*_{2p_x}$) orbital. Based on the molecular orbital description, we expect F_2^- to have a weaker bond than F_2.

Electron configuration of F_2 $\quad (\sigma_{1s})^2(\sigma^*_{1s})^2(\sigma_{2s})^2(\sigma^*_{2s})^2(\sigma_{2p_x})^2(\pi_{2p_y})^2(\pi_{2p_z})^2(\pi^*_{2p_y})^2(\pi^*_{2p_z})^2$

Bond order = 1

Electron configuration of F_2^-

$(\sigma_{1s})^2(\sigma^*_{1s})^2(\sigma_{2s})^2(\sigma^*_{2s})^2(\sigma_{2p_x})^2(\pi_{2p_y})^2(\pi_{2p_z})^2(\pi^*_{2p_y})^2(\pi^*_{2p_z})^2(\sigma^*_{2p_x})^1$ $\quad$ Bond order $= \frac{1}{2}$

10.90

10.91 This problem is similar to 10.90(b) above. You can use Figure 10.6 to compare the neutral homonuclear diatomics to their ions. Other similar problems are Problems 10.54 and 10.56.

Solution: (a) Be_2^+ with a bond order of 0.5 would have a higher dissociation energy than Be_2 with a bond order of zero.

(b) B_2^- with a bond order of 1.5 would have a higher dissociation energy than B_2 with a bond order of 1.0.

(c) C_2 with a bond order of 2.0 would have a higher dissociation energy than C_2^+ with a bond order of 1.5.

10.92 Follow the hint.

Solution: The analysis in Problem 10.43 shows the molecule to be nonplanar (why?). The geometric planes containing the CHCl groups at each end of the molecule are mutually perpendicular. This means that the two chlorine atoms can be considered to be on one side of the molecule and the two hydrogen atoms on the other. The molecule has a dipole moment. Draw the end-on view if you aren't convinced.

10.93 Visualizing molecules in three dimensions is difficult and only becomes easy with a lot of experience. In these examples focus on the differences between the molecules, especially the positions of the two Y substituents relative to each other.

> ***Solution:*** For an octahedral AX_4Y_2 molecule only two different structures are possible: one with the two Y's next to each other like (b) and (d), and one with the two Y's on opposite sides of the molecule like (a) and (c). The different looking drawings simply depict the same molecule seen from a different angle or side.
>
> It would help to develop your power of spatial visualization to make some simple models and convince yourself of the validity of these answers. How many different structures are possible for octahedral AX_5Y or AX_3Y_3 molecules? Would an octahedral AX_2Y_4 molecule have a different number of structures from AX_4Y_2? Ask your instructor if you aren't sure.

10.95 What is the size of carbon relative to silicon?

> ***Solution:*** Because of the relatively large size of silicon compared to carbon, orbital overlap with oxygen is less favorable to form pi bonds. It is energetically favorable for SiO_2 to form a three-dimensional network consisting of all Si-O single bonds, as compared to the double bonds in CO_2.

10.97 Fluorine has a number of unique properties, partially because of its small size, that are often difficult to predict.

> ***Solution:*** The smaller size of F compared to Cl results in a shorter F-F bond than a Cl-Cl bond. The closer proximity of the lone pairs of electrons on the F atoms results in greater electron-electron repulsions that weaken the bond.

10.98 Refer to Problem 10.42 above.

> ***Solution:*** Since nitrogen is a second row element, it cannot exceed an octet. Since there are no lone pairs on the central nitrogen, the molecule must be linear and sp hybrid orbitals (Table 10.4) must be used.
>
> $$:\!N\equiv\overset{+}{N}-\overset{2-}{\underset{\cdot\cdot}{\overset{\cdot\cdot}{N}}}\!: \;\leftrightarrow\; :\!\overset{-}{\overset{\cdot\cdot}{N}}=\overset{+}{N}=\overset{-}{\overset{\cdot\cdot}{N}}\!: \;\leftrightarrow\; :\!\overset{2-}{\underset{\cdot\cdot}{\overset{\cdot\cdot}{N}}}-\overset{+}{N}\equiv N\!:$$
>
> The $2p_y$ orbital on the central nitrogen atom overlaps with the $2p_y$ orbitals on the terminal nitrogen atoms, and the $2p_z$ orbital on the central nitrogen atom overlaps with the $2p_z$ orbitals on the terminal nitrogen atoms to form delocalized molecular orbitals.

10.99 The resonance structure that has the smallest formal charges and that has negative charges on the most electronegative elements is the most likely. Review Section 9.7 if necessary.

Solution: (a)

$$:\mathrm{N{\equiv}C{-}\ddot{\underset{..}{O}}:}^{-} \leftrightarrow {}^{-}\mathrm{\ddot{\underset{..}{N}}{=}C{=}\ddot{\underset{..}{O}}} \leftrightarrow {}^{2-}:\mathrm{\ddot{\underset{..}{N}}{-}C{\equiv}O}^{+}:$$

(b)

$$^{-}:\mathrm{C{\equiv}\overset{+}{N}{-}\ddot{\underset{..}{O}}:}^{-} \leftrightarrow {}^{2-}\mathrm{\ddot{\underset{..}{C}}{=}\overset{+}{N}{=}\ddot{\underset{..}{O}}} \leftrightarrow {}^{3-}:\mathrm{\ddot{\underset{..}{C}}{-}\overset{+}{N}{\equiv}\overset{+}{O}:}$$

In both cases, the most likely structure is on the left and the least likely structure is on the right

10.100 Be sure to consider the lone pairs of electrons in determining the hydridization of the nitrogen atoms.

Solution: The Lewis structure is:

```
                                 O        H
                                  \\ *   /
         HO—CH2      H             C—N
            |   O    |            /  ♦ \  *
            C       C————————N♦        C=O
        H /  \     /          \       /
        H—C——C—H               *C=C*
          |   |               /     \
        * N   H              H       CH3
          ||
          N
          ||
        * N
```

The carbon atoms and nitrogen atoms marked with an asterisk (C* and N*) are sp^2 hybridized; unmarked carbon atoms and nitrogen atoms marked with a diamond (C and N♦) are sp^3 hybridized; and the unmarked nitrogen atom is sp hybridized.

Note: If you have not already done so, please read the "Introduction for the Student" at the beginning of this Student Solutions Manual.

Chapter 11

INTERMOLECULAR FORCES AND LIQUIDS AND SOLIDS

11.7 Which molecule would be polar? Does the presence of a dipole moment increase the intermolecular forces in a substance? Does this affect the melting and boiling points (see Section 11.2)?

> ***Solution:*** ICl has a dipole moment and Br_2 does not. The dipole moment increases the intermolecular attractions between ICl molecules and causes that substance to have a higher melting point than bromine.

11.8 Which compound should have the highest boiling point? Which the lowest? Why? Look up the boiling points to be sure.

> ***Solution:*** On a very cold day propane and butane would be liquids (boiling points -44.5°C and -0.5°C, respectively); only methane would still be a gas (boiling point -161.6°C).
>
> Other factors being equal, the molecule with the greater number of electrons will exert greater intermolecular attractions (which type?). Just by looking at the molecular formulas you can predict that the order of increasing boiling points will be $CH_4 < C_3H_8 < C_4H_{10}$.

11.9 The molecular structures of these compounds should be the same (why?). Use VSEPR principles to deduce the common structure. Are the molecules polar? Can any form hydrogen bonds? What intermolecular forces can exist between these molecules?

> ***Solution:*** All are tetrahedral (AB_4 type) and are nonpolar. Therefore, the only intermolecular forces possible are dispersion forces (Section 11.2). Without worrying about what causes dispersion forces, you only need to know that the strength of the dispersion force increases with the number of electrons in the molecule (all other things being equal). As a consequence, the magnitude of the intermolecular attractions and of the boiling points should increase with increasing molar mass.

11.10 You must decide whether the listed compounds are ionic or molecular, and, if molecular, whether polar or nonpolar. Example 11.1 in the text is a similar problem.

Solution: (a) Benzene (C_6H_6) is a nonpolar molecular compound. Only dispersion forces will be present.

(b) Chloroform (CH_3Cl) is a polar (why?) molecular compound. Dispersion and dipole-dipole forces will be present.

(c) Phosphorus trifluoride (PF_3) will be like (b).

(d) Sodium chloride (NaCl) is an ionic compound. Ion-ion (and dispersion) forces will be present.

(e) Carbon disulfide (CS_2) will be like (a).

11.11 Use Figure 11.6 as a model for your drawing.

Solution: The center ammonia molecule is hydrogen-bonded to two other ammonia molecules.

```
      H          H          H
      |          |          |
  H— N····H— N····H— N
      |          |          |
      H          H          H
```

11.12 Study the definition and description of the hydrogen bond in Section 11.3. What atoms must be present in a molecule in order to have hydrogen bonds? To what atoms must hydrogen be connected in order to form hydrogen bonds? Example 11.2 is similar to this problem.

Solution: In this problem you must identify the species capable of hydrogen bonding among themselves, not with water as in Example 11.2. In order for a molecule to be capable of hydrogen bonding with itself, it must have at least one hydrogen atom bonded to N, O, or F. Of the choices only (e) CH_3COOH shows this structural feature. The others cannot form hydrogen bonds among themselves.

Which can form hydrogen bonds with water?

11.13 Classify each substance by the types of intermolecular forces present. The compound with the strongest forces will have the highest boiling point.

Solution: CO_2 is a nonpolar molecular compound. The only intermolecular force present is a relatively weak dispersion force (small molar mass). CO_2 will have the lowest boiling point.

CH_3Br is a polar molecule. Dispersion forces (present in all matter) and dipole-dipole forces will be present. This compound has the next highest boiling point.

CH_3OH is polar and can form hydrogen bonds, which are especially strong. Dispersion forces and hydrogen bonding (a very strong dipole-dipole force) are present to give this substance the next highest boiling point.

RbF is an ionic compound (why?). Ion-ion attractions are much stronger than any intermolecular force. RbF has the highest boiling point.

11.15 What happens to the volume of a sample of water as it freezes? See Section 11.3.

> ***Solution:*** Water expands when it freezes. The pool walls and plumbing could easily break when this expansion occurs.

11.17 What types of intermolecular forces are present in these compounds? Which forces are strongest?

> ***Solution:*** (a) CH_4 has a lower boiling point because NH_3 is polar and can form hydrogen bonds; CH_4 is nonpolar and can only form weak attractions through dispersion forces.
>
> (b) KCl is an ionic compound. Ion-ion forces are much stronger than any intermolecular forces. I_2 is a nonpolar molecular substance; only weak dispersion forces are possible.

11.18 If you are unsure of the meanings of any of these terms, study their definitions and explanations in Section 11.2.

> ***Solution:*** (a) False. Permanent dipoles are usually much stronger than temporary dipoles.
>
> (b) False. The hydrogen atom must be bonded to N, O, or F.
>
> (c) True.
>
> (d) False. The magnitude of the attraction depends on both the ion charge and the polarizability of the neutral atom or molecule.

11.19 This is similar to Problems 11.12 and 11.13.

> ***Solution:*** (a) Hydrogen bonding, dipole-dipole attractions, and dispersion forces.
>
> (b) Dispersion forces only (nonpolar molecules).
>
> (c) Dispersion forces only (nonpolar molecules).
>
> (d) The attractive forces of the covalent bond.

11.21 Given the hint, which molecular shape could be packed together more easily?

> ***Solution:*** The linear structure (n-butane) has a higher boiling point ($-0.5^\circ C$) than the branched structure (2-methylpropane, boiling point $-11.7^\circ C$) because the linear form can be stacked together more easily.

11.33 Study Section 11.3 to find what molecular properties affect the magnitude of the surface tension of a liquid. What intermolecular forces are possible in ethanol and dimethyl ether?

Solution: Ethanol molecules can attract each other with strong hydrogen bonds; dimethyl ether molecules cannot (why?). The surface tension of enthanol is greater than that of dimethyl ether because of stronger intermolecular forces (the hydrogen bonds). Note that ethanol and dimethyl ether have identical molar masses and molecular formulas so attractions resulting from dispersion forces will be equal.

11.35 This problem is similar to Problem 11.33 above.

Solution: Water molecules can attract each other with strong hydrogen bonds; diethyl ether molecules cannot (why?). The surface tension of water is greater than that of diethyl ether because of stronger intermolecular forces (the hydrogen bonds).

11.47 Study the meaning of coordination number in Section 11.4 and carefully examine the different structures in Figures 11.16-11.20. In each case try to focus on one sphere and count how many other spheres it is touching. It may help to look at a three-dimensional model if you can.

Solution: (a) In a simple cubic structure each sphere touches six others on the $\pm$x, $\pm$y and $\pm$ z axes.

(b) In a body-centered cubic lattice each sphere touches eight others. Visualize the body-center sphere touching the eight corner spheres.

(c) In a face-centered cubic lattice each sphere touches twelve others.

11.48 In calculating the number of spheres in a specific type of unit cell, it is necessary to remember that the spheres are often shared between several adjacent cells. Among how many cells is a corner sphere shared? Among how many cells is a sphere at the center of a face shared? What about a sphere located at the center of a unit cell (body-center)? Study Figure 11.24 to see these relationships. Packing efficiencies are discussed in Section 11.4.

Solution: A corner sphere is shared equally among eight unit cells, so only one-eighth of each corner sphere "belongs" to any one unit cell. A face-centered sphere is divided equally between the two unit cells sharing the face. A body-centered sphere belongs entirely to its own unit cell.

In a simple cubic cell there are eight corner spheres. One-eighth of each belongs to the individual cell giving a total of one whole sphere per cell. In a body-centered cubic cell there are eight corner spheres and one body-center sphere giving a total of two spheres per unit cell (one from the corners and one from the body-center). In a face-centered cell there are eight corner spheres and six face-centered spheres (six faces). The total number of spheres would be four: one from the corners and three from the faces.

Figure 11.24 shows the mathematical relationships between sphere radii and cube cell edges for the three types of cell in this problem. The packing efficiencies are calculated below:

(a) Simple cubic cell cell edge = a sphere radius $= \frac{a}{2}$

$$\text{Packing efficiency} = \frac{\left(\frac{4\pi r^3}{3}\right) \times 100\%}{(2r)^3} = \frac{4\pi r^3 \times 100\%}{24r^3} = \frac{\pi}{6} \times 100\% = 52.4\%$$

(b) Body-centered cubic cell $\qquad$ cell edge $= \dfrac{4r}{\sqrt{3}}$ $\qquad$ sphere radius = r

$$\text{Packing efficiency} = \frac{2 \times \left(\frac{4\pi r^3}{3}\right) \times 100\%}{\left(\frac{4r}{\sqrt{3}}\right)^3} = \frac{2 \times \left(\frac{4\pi r^3}{3}\right) \times 100\%}{\left(\frac{64r^3}{3\sqrt{3}}\right)} = \frac{2\pi\sqrt{3}}{16} \times 100\% = 68.0\%$$

(c) Face-centered cubic cell $\qquad$ cell edge $= \sqrt{8}r$ $\qquad$ sphere radius = r

$$\text{Packing efficiency} = \frac{4 \times \left(\frac{4\pi r^3}{3}\right) \times 100\%}{\left(\sqrt{8}\,r\right)^3} = \frac{\left(\frac{16\pi r^3}{3}\right) \times 100\%}{8r^3\sqrt{8}} = \frac{2\pi}{3\sqrt{8}} \times 100\% = 74.0\%$$

11.49 Given the density, what is the mass of the iron in a cube 287 pm on edge? How many iron atoms does it take to make up this mass? Example 11.3 in the text is a similar problem.

Solution: The mass of one cube of edge 287 pm can be found easily from the mass of one cube of edge 1.00 cm (7.87 g):

$$\left(\frac{7.87 \text{ g}}{1 \text{ cm}^3}\right)\left(\frac{100 \text{ cm}}{1 \text{ m}}\right)^3\left(\frac{1 \text{ m}}{10^{12} \text{ pm}}\right)^3(287 \text{ pm})^3 = 1.86 \times 10^{-22} \text{ g}$$

The mass of one iron atom can be found by dividing the molar mass of iron (55.85 g) by Avogadro's number:

$$\left(\frac{55.85 \text{ g}}{1 \text{ mol}}\right)\left(\frac{1 \text{ mol}}{6.022 \times 10^{23}}\right) = 9.27 \times 10^{-23} \text{ g}$$

The number of iron atoms in one cell is the quotient of these numbers:

$$\frac{1.86 \times 10^{-22} \text{ g}}{9.28 \times 10^{-23} \text{ g}} = 2.00$$

What type of cubic unit cell is this?

11.50 Hints on the hint: How many atoms are contained in one body-centered cubic unit cell, and what is the packing efficiency in the cell (see Problem 11.48) ?

Solution: The packing efficiency in a body-centered cubic cell is 68.0%, and there are two atoms per cell. The volume occupied by one mole of barium atoms is:

$$\left(\frac{68.0\%}{100\%}\right)\left(\frac{137.3 \text{ g Ba}}{1 \text{ mol Ba}}\right)\left(\frac{1 \text{ cm}^3}{3.50 \text{ g}}\right) = 26.7 \text{ cm}^3/\text{mol}$$

The volume occupied by one barium atom in one unit cell is:

$$\frac{(502\text{ pm})^3}{1\text{ cell}}\left(\frac{1\text{ m}}{10^{12}\text{ pm}}\right)^3\left(\frac{100\text{ cm}}{1\text{ m}}\right)^3\left(\frac{1\text{ cell}}{2\text{ atoms}}\right)\left(\frac{68.0\%}{100\%}\right) = 4.30\times10^{-23}\text{ cm}^3\text{/atom}$$

The number of barium atoms in one mole is:

$$\left(\frac{26.7\text{ cm}^3}{1\text{ mol}}\right)\left(\frac{1\text{ atom}}{4.30\times10^{-23}\text{ cm}^3}\right) = 6.21\times10^{23}\text{ atom/mol}$$

11.52 This problem is similar to Examples 11.3, 11.4, and 11.7 in the text. Rather than being given the cell edge and solving for density, you are given the unit cell edge and being asked to solve for density.

Solution: The mass of the unit cell is the mass in grams of two europium atoms

$$m = \left(\frac{2\text{ Eu atoms}}{1\text{ unit cell}}\right)\left(\frac{152.0\text{ g Eu}}{1\text{ mol Eu}}\right)\left(\frac{1\text{ mol Eu}}{6.022\times10^{23}\text{ Eu atoms}}\right) = 5.048\times10^{-22}\text{ g Eu/unit cell}$$

$$V = \left(\frac{5.048\times10^{-22}\text{ g}}{1\text{ unit cell}}\right)\left(\frac{1\text{ cm}^3}{5.26\text{ g}}\right) = 9.60\times10^{-23}\text{ cm}^3\text{/unit cell}$$

The edge length a is $a = V^{1/3} = (9.60\times10^{-23}\text{ cm}^3)^{1/3} = 4.58\times10^{-8}\text{ cm} = 458\text{ pm}$

11.53 This problem is very similar to Problem 11.49 above.

Solution: The volume of the unit cell is

$$V = a^3 = (543\text{ pm})^3\left(\frac{1\text{ m}}{1\times10^{12}\text{ pm}}\right)^3\left(\frac{100\text{ cm}}{1\text{ m}}\right)^3 = 1.60\times10^{-22}\text{ cm}^3$$

$$m = dV = (2.33\text{ g/cm}^3)(1.60\times10^{-22}\text{ cm}^3) = 3.73\times10^{-22}\text{ g}$$

The mass of one silicon atom is $\left(\frac{28.09\text{ g Si}}{1\text{ mol Si}}\right)\left(\frac{1\text{ mol Si}}{6.022\times10^{23}\text{ atoms Si}}\right) = 4.665\times10^{-23}\text{ g}$

The number of silicon atoms in one unit cell is

$$\frac{\text{mass of one unit cell}}{\text{mass of one Si atom}} = \frac{3.73\times10^{-22}\text{ g}}{4.665\times10^{-23}\text{ g}} = 8$$

11.55 Study Section 11.6 and the contents of Table 11.5.

Solution: (a) Carbon dioxide forms molecular crystals; it is a molecular compound and can only exert weak dispersion type intermolecular attractions because of its lack of polarity.

(b) Boron is a nonmetal with an extremely high melting point. It forms covalent crystals like carbon (diamond).

(c) Sulfur forms molecular crystals; it is a molecular substance (S_8) and can only exert weak dispersion type intermolecular attractions because of its lack of polarity.

(d) KBr forms ionic crystals because it is an ionic compound.

(e) Mg is a metal; it forms metallic crystals.

(f) SiO_2 (quartz) is a hard, high melting nonmetallic compound; it forms covalent crystals like boron and C (diamond).

(g) LiCl is an ionic compound; it forms ionic crystals.

(h) Cr (chromium) is a metal and forms metallic crystals.

11.56 Study the comparison of the crystal structures of diamond and graphite in Section 11.6.

Solution: In diamond each carbon atom is covalently bonded to four other carbon atoms. Because these bonds are strong and uniform, diamond is a very hard substance. In graphite the carbon atoms in each layer are linked by strong bonds, but the layers are bound by weak dispersion forces. As a result, graphite may be cleaved easily between layers and is not hard.

11.58 Is the energy needed to separate two opposite charges proportional to the interionic distance or the inverse of the interionic distance (see Section 9.2 and Equation (9.2))?

Solution: The two graphs are shown below:

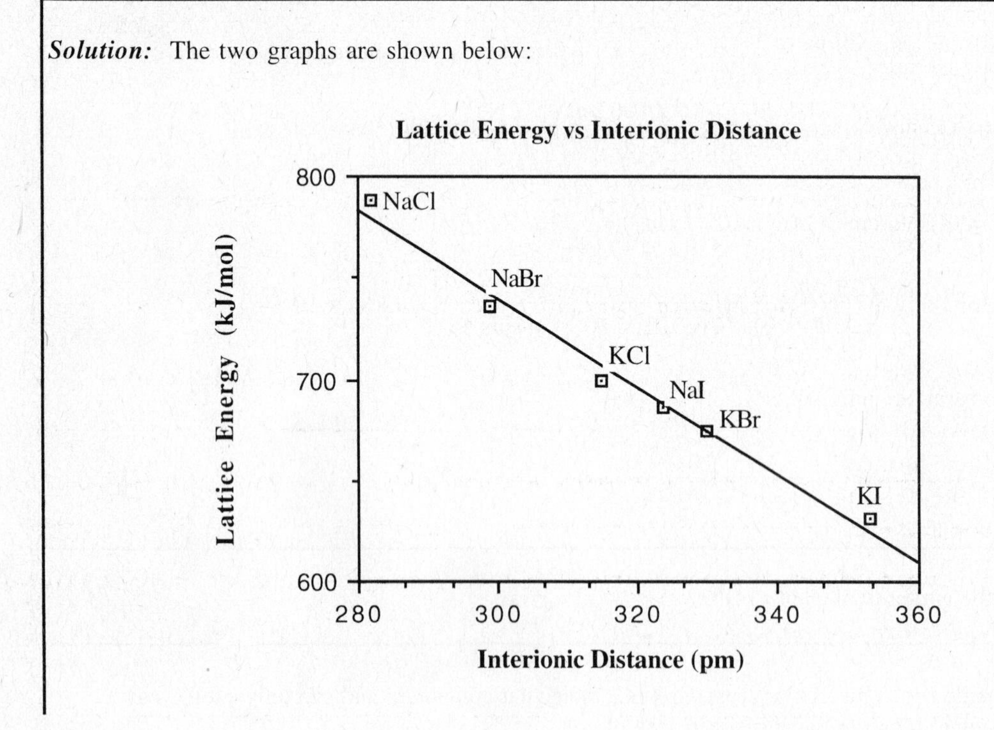

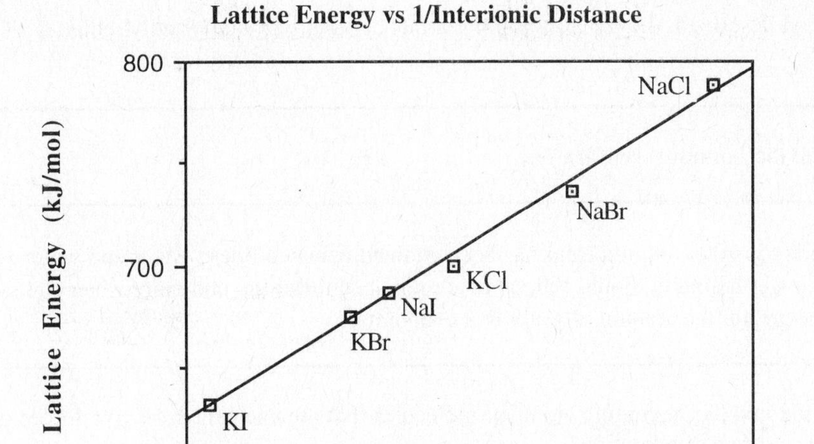

The second graph is more linear (look closely). The energy required to separate two opposite charges is given by Equation (9.2):

$$E = k\frac{Q_+Q_-}{r}$$

As the separation increases, less work is needed to pull the ions apart, therefore the lattice energies become smaller as the interionic distances become larger.

From these data what can you conclude about the relationship between lattice energy and the size of the negative ion? What about lattice energy versus positive ion size (compare KCl with NaCl, KBr with NaBr, etc.)?

11.61 This is like Example 11.5. Study that example and the discussion of X-ray diffraction in Section 11.5.

Solution: From Equation 11.1 we can write

$$d = \frac{n\lambda}{2\sin\theta} = \frac{\lambda}{2\sin\theta} = \frac{0.090\text{ nm}\left(\frac{1000\text{ pm}}{1\text{ nm}}\right)}{2\sin(15.2°)} = 172\text{ pm}$$

11.62 This is like Problem 11.61. Rearrange the Bragg equation to find the wavelength.

Solution: Rearranging the Bragg equation, we have

$$\lambda = \frac{2d\sin\theta}{n} = \frac{2(282\text{ pm})(\sin 23.0°)}{1} = 220\text{ pm} = 0.220\text{ nm}$$

1.83 The boiling points at 0.5 atm can be estimated by drawing a horizontal line across the graph at 0.5 atm and reading the temperatures below the points of intersection with the vapor pressure curves.

Solution: Using the method described, the 0.5 atm boiling points are approximately: ethyl ether, 17°C; water, 75°C; mercury, 310°C.

11.84 Will water evaporate when the humidity is 100%?

Solution: When the humidity is 100%, equilibrium has been attained between liquid water and water vapor, and no evaporation occurs. At lower humidity, liquid and vapor are not at equilibrium, and more water can evaporate. The lower the humidity, the greater the amount of water that evaporates.

11.85 Both fusion and vaporization involve separating atoms or molecules that are exerting attractive forces on each other. Which process involves greater separation?

Solution: The melting of most solids is accompanied by volume expansions of no more than a few percent. On the other hand vaporization always involves a large volume increase. In other words in the vaporization process molecules are separated to a much larger extent. Since molecules attract each other, vaporization always requires more work (energy) than fusion.

11.86 Which is commonly used to heat buildings: steam or liquid water?

Solution: The heat or enthalpy of vaporization of water (Table 11.7) is much larger than the corresponding heat or enthalpy of fusion of water (Table 11.9). Condensing steam releases much more heat than freezing water.

11.87 Does the boiling point of water depend on the number of Bunsen burners used to supply heat?

Solution: Adding another burner would result in the water reaching the boiling point sooner, but would not change the temperature at which the water boils. Study Section 11.8 if you are unsure of the exact definition of boiling point. What is the normal boiling point of a liquid? Does the normal boiling point depend upon the rate of heating?

11.88 You are being asked to find the amount of heat or thermal energy required to convert 74.6 g of liquid water at 100°C to steam at 100°C.

Solution: The molar heat of vaporization of water is 40.79 kJ/mol (Table 11.7). One must find the number of moles of water in the sample:

$$\text{Moles } H_2O = 74.6 \text{ g } H_2O\left(\frac{1 \text{ mol } H_2O}{18.02 \text{ g } H_2O}\right) = 4.14 \text{ mol } H_2O$$

We can then calculate the amount of heat

$$q = 4.14 \text{ mol } H_2O\left(\frac{40.79 \text{ kJ}}{1 \text{ mol } H_2O}\right) = 169 \text{ kJ}$$

11.89 This is like Example 11.9. The heats of vaporization and of fusion for water can be found in Tables 11.6 and 11.9, respectively. The specific heat of liquid water is given in Example 11.9.

Solution: Step 1: Warming ice to the melting point

$$q_1 = ms\Delta t = 866 \text{ g } H_2O \times 2.03 \text{ J/g°C} \times [0°C - (-10°C)] = 17.6 \text{ kJ}$$

Step 2: Converting ice at the melting point to liquid water at 0°C.

$$q_2 = 866 \text{ g } H_2O \left(\frac{1 \text{ mol}}{18.02 \text{ g } H_2O}\right)\left(\frac{6.01 \text{ kJ}}{1 \text{ mol}}\right) = 289 \text{ kJ}$$

Step 3: Heating water from 0°C to 100°C

$$q_3 = ms\Delta t = 866 \text{ g } H_2O \times 4.184 \text{ J/g°C} \times (100°C - 0°C) = 362 \text{ kJ}$$

Step 4: Converting water at 100°C to steam at 100°C

$$q_4 = 866 \text{ g } H_2O \left(\frac{1 \text{ mol}}{18.02 \text{ g } H_2O}\right)\left(\frac{40.79 \text{ kJ}}{1 \text{ mol}}\right) = 1960 \text{ kJ}$$

Step 5: Heating steam from 100°C to 126°C

$$q_5 = ms\Delta t = 866 \text{ g } H_2O \times 1.99 \text{ J/g°C} \times (126°C - 100°C) = 44.8 \text{ kJ}$$

$$q_{total} = q_1 + q_2 + q_3 + q_4 + q_5 = 2670 \text{ kJ (3 significant figures)}$$

How would you set up and work this problem if you were computing the heat lost in cooling steam from 126°C to ice at -10°C?

11.91 After studying Section 11.8, try to visualize the evaporation of a sample of water (a) on a cold day and on a very warm day, and (b) in a very narrow glass and in a wide bowl. A substance with very weak intermolecular forces is liquified air. Does this substance evaporate slowly or rapidly?

Solution: (a) Other factors being equal, liquids evaporate faster at higher temperatures.

(b) The greater the surface area, the greater the rate of evaporation.

(c) Weak intermolecular forces imply a high vapor pressure and rapid evaporation.

11.93 Study Section 11.8, especially the "Chemistry in Action" segment on freeze-drying.

Solution: Two phase changes occur in this process. First, the liquid is turned to solid (freezing), then the solid ice is turned to gas (sublimation).

11.95 This problem is similar to Problem 11.86 above.

Solution: When steam condenses to liquid water at 100°C, it releases a large amount of heat equal to the enthalpy of vaporization. Thus steam at 100°C exposes one to more heat than an equal amount of water at 100°C.

11.97 See the "Chemistry in Action" segment in Section 11.9.

Solution: The pressure exerted by the blades on the ice lowers the melting point of the ice. A film of liquid water between the blades and the solid ice provides lubrication for the motion of the skater.

11.99 This problem involves the graphical application of the Clausius-Clapyron relationship (Equation (11.2)). The molar heat of vaporization can be found from the slope of a graph of ln P versus 1/T.

Solution: The graph is shown below:

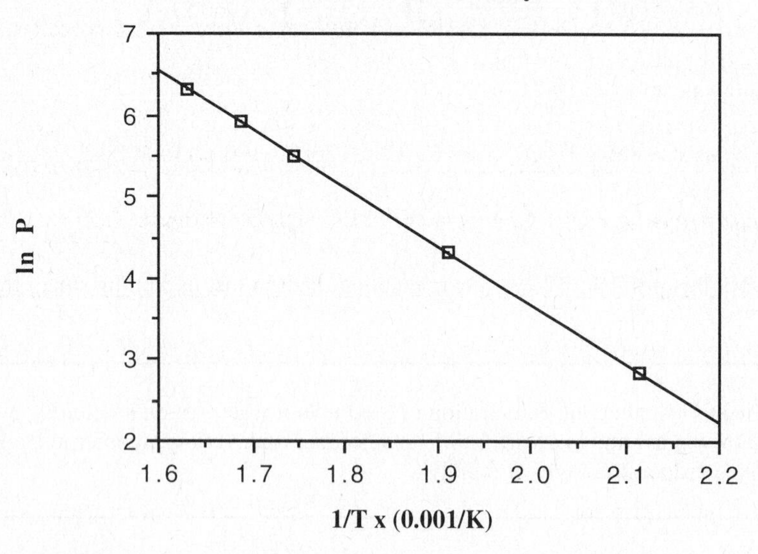

Using the first and last points to determine the slope, we have

$$\text{Slope} = \frac{\ln P_i - \ln P_f}{\frac{1}{T_i} - \frac{1}{T_f}} = \frac{6.32 - 2.85}{1.63 \times 10^{-3} - 2.11 \times 10^{-3}} = -7230 = \frac{-\Delta H_{vap}}{R} = \frac{-\Delta H_{vap}}{8.314 \text{ J/K}\cdot\text{mol}}$$

$$\Delta H_{vap} = 60.1 \text{ kJ/mol}$$

11.103 On a phase diagram, changing the temperature at constant pressure means moving along a straight line parallel to the temperature axis. Changing the pressure at constant temperature means moving along a straight line parallel to the pressure axis.

Solution: Region labels: The region containing point A is the solid region. The region containing point B is the liquid region. The region containing point C is the gas region.

(a) Raising the temperature at constant pressure beginning at A implies starting with solid ice and warming until melting occurs. If the warming continued, the liquid water would eventually boil and change to steam. Further warming would increase the temperature of the steam.

(b) At point C water is in the gas phase. Cooling without changing the pressure would eventually result in the formation of solid ice. Liquid water would never form.

(c) At B the water is in the liquid phase. Lowering the pressure without changing the temperature would eventually result in boiling and conversion to water in the gas phase.

11.104 Compare solids, liquids, and gases in terms the arrangements and motions of their particles. For which are the molecules or atoms the furthest apart on average? For which are the particles usually arranged in 3-dimensional repeating lattices or structures?

Solution: (a) The average separation between particles decreases from gases to liquids to solids, so the ease of compressibility decreases in the same order.

(b) In solids the molecules or atoms are usually locked in a rigid 3-dimensional structure which determines the shape of the crystal. In liquids and gases the particles are free to move relative to each other.

(c) The trend in volume is due to the same effect as part (a).

11.105 You need to identify the types of intermolecular forces present in the listed substances. Refer to Problems 11.10-11.20.

Solution: (a) Boiling liquid ammonia requires breaking hydrogen bonds between molecules. Dipole-dipole and dispersion forces must also be overcome.

(b) P_4 is a nonpolar molecule, so the only intermolecular forces are of the dispersion type.

(c) CsI is an ionic solid. To dissolve in any solvent ion-ion interparticle forces must be overcome.

(d) Metallic bonds must be broken. See Table 11.5.

11.107 This is like Problem 11.12.

Solution: The HF molecules are held together by strong intermolecular hydrogen bonds. Therefore, liquid HF has a lower vapor pressure than liquid HI. (The HI molecules do not form hydrogen bonds with each other.)

11.108 How does each of these properties depend upon whether intermolecular attractions are strong or weak? If you aren't sure, reread the appropriate section in this chapter.

Solution: (a) A low surface tension means the attraction between molecules making up the surface is weak. Water has a high surface tension; water bugs could not "walk" on the surface of a liquid with a low surface tension.

(b) A low critical temperature means a gas is very difficult to liquify by cooling. This is the result of weak interparticle attractions. Helium has the lowest known critical temperature (5.3 K).

(c) A low boiling point means weak interparticle attractions. It takes little energy to separate the particles. All ionic compounds have extremely high boiling points.

(d) A low vapor pressure means it is difficult to remove molecules from the liquid phase because of high intermolecular attractions. Substances with low vapor pressures have high boiling points (why?).

11.109 Behaving ideally means obeying the ideal gas equation, PV = nRT. Refer to Problem 5.104. Which molecule would exert the weakest intermolecular attraction? Which will have the lowest value of "a" in the van der Waals equation?

Solution: CH_4 is a tetrahedral, nonpolar molecule that can only exert weak dispersion type attractive forces. SO_2 is bent (why?) and possesses a dipole moment which gives rise to stronger dipole-dipole attractions. Sulfur dioxide will have a larger value of "a" in the van der Waals equation (a is a measure of the strength of the interparticle attraction) and will behave less like an ideal gas than methane.

11.110 See Section 11.8 and Problem 11.108.

Solution: The kinetic energy of molecules increases with temperature. However, the intermolecular forces between molecules do not depend upon temperature. Thus, for each substance there exists a temperature above which its gas molecules will possess enough kinetic energy to break away from intermolecular forces. This temperature is the critical temperature.

11.111 See Section 11.3 and Problem 11.108.

Solution: Oil is made up of nonpolar molecules and therefore does not mix with water. To minimize contact, the oil drop assumes a spherical shape. (For a given volume, the sphere has the smallest surface area.)

11.112 This problem is much simpler than it may appear. Under standard thermodynamic conditions molecular bromine is a liquid. Write a chemical equation for the formation of gaseous molecular bromine under standard conditions. From what does gaseous molecular bromine form? What is the standard enthalpy change for this reaction?

Solution: The standard enthalpy change for the formation of gaseous bromine from liquid bromine is simply the difference between the standard enthalpies of formation of the products and the reactants in the equation:

$$Br_2(l) \rightarrow Br_2(g)$$

$$\Delta H^\circ_{vap} = \Delta H^\circ_f(Br_2(g)) - \Delta H^\circ_f(Br_2(l)) = 30.7 \text{ kJ/mol} - 0 \text{ kJ/mol} = 30.7 \text{ kJ/mol}$$

11.113 The formula is not X_8YZ_6! Study Problems 11.48 and 11.54.

Solution: Each of the eight corner atoms in a unit cell is shared among eight adjoining unit cells. Each body-centered atom is completely within its own unit cell. Each of the six face-centered atoms (six faces on any unit cell) is shared between two adjoining unit cells. We can write

X: 8 corners/cell $\times \frac{1}{8}$ atom/corner = 1 X atom/cell

Y: 1 body center/cell × atom/body center = 1 Y atom/cell

Z: 6 faces/cell $\times \frac{1}{2}$ atom/face = 3 Z atoms/cell

The formula is XYZ_3.

11.115 Review Section 11.9.

Solution: Reading directly from the graph: (a) solid; (b) vapor.

11.117 See Section 11.2 if you need review on intermolecular forces. What kind of information does the melting point provide?

Solution: LiF, ionic bonding and dispersion forces; BeF_2, ionic bonding and dispersion forces; BF_3, dispersion forces; CF_4, dispersion forces; NF_3, dipole-dipole interaction and dispersion forces, OF_2, dipole-dipole interaction and dispersion forces, F_2, dispersion forces.

11.119 What is the definition of heat of hydration (Section 6.6)?

Solution: Smaller ions have more concentrated charges (charge densities) and are more effective in ion-dipole interaction. The greater the ion-dipole interaction, the larger is the heat of hydration.

11.121 This is a chemical detective problem.

Solution: W must be a reasonably nonreactive metal. It conducts electricity and is malleable, but doesn't react the nitric acid. Of the choices, it must be gold.

X is nonconducting (and therefore isn't a metal), is brittle, high melting, and reacts with nitric acid. Of the choices, it must be lead sulfide.

Y doesn't conduct and is soft (and therefore is a nonmetal). It melts at a low temperature with sublimation. Of the choices, it must be iodine.

Z doesn't conduct, is chemically inert, and is high melting (network solid). Of the choices, it must be mica (SiO_2).

Would the colors of the species have been any help in determing their identity?

11.123 While this problem is qualitative, Example 11.8 is a quantitative application of how vapor pressure varies with temperature and the molar heat of vaporization.

> ***Solution:*** Application of the Clausius-Clapeyron Equation (11.5) predicts that the less the vapor pressure rises over a temperature range, the greater the heat of vaporization will be. Thus the molar heat of vaporization of X < Y.

11.124 Review VSEPR theory and dipole moments in Chapter 10 and hydrogen bonding in this chapter if you need help.

> ***Solution:*** (a) If water were linear, the two O-H bond dipoles would cancel each other as in CO_2. Thus a linear water molecule would not be polar.
>
> (b) Hydrogen bonding would still occur between water molecules even if were linear.

11.125 The process occuring in part (a) is vaporization; in part (b) is bond dissociation. Which of these processes would you guess would require more energy?

> ***Solution:*** (a) For the process: $Br_2(l) \rightarrow Br_2(g)$
>
> $$\Delta H° = \Delta H_f^\circ[Br_2(g)] - \Delta H_f^\circ[Br_2(l)]$$
>
> $$\Delta H° = (1\ \text{mol})(30.7\ \text{kJ/mol}) - 0 = 30.7\ \text{kJ}$$
>
> (b) For the process: $Br_2(g) \rightarrow 2Br(g)$ $\Delta H° = 192.5$ kJ (from Table 9.4)
>
> As expected, the bond dissociation energy represented in part (b) is much greater than the energy of vaporization represented in part (a). It requires more energy to break the bond than to vaporize the molecule.

11.128 This problem is an application of Equation (11.5) and is similar to Example 11.8.

> ***Solution:*** Using Equation (11.5):
>
> $$\ln\left(\frac{P_1}{P_2}\right) = \left(\frac{\Delta H_{vap}}{R}\right)\left(\frac{T_1 - T_2}{T_1T_2}\right)$$
>
> $$\ln\left(\frac{1}{2}\right) = \left(\frac{\Delta H_{vap}}{8.314\ \text{J/K·mol}}\right)\left(\frac{358\text{K} - 368\text{K}}{358\text{K} \times 368\text{K}}\right) = \Delta H_{vap}\left(\frac{-7.59 \times 10^{-5}}{8.314\ \text{J/mol}}\right)$$
>
> $$\Delta H_{vap} = 7.59 \times 10^4\ \text{J} = 75.9\ \text{kJ}$$

Note: If you have not already done so, please read the "Introduction for the Student" at the beginning of this Student Solutions Manual.

Chapter 12

PHYSICAL PROPERTIES OF SOLUTIONS

12.8 How do solid naphthalene and CsF differ in terms of intermolecular attractions?

Solution: CsF is an ionic solid; the ion-ion attractions are too strong to be overcome in the dissolving process in benzene. The ion-induced dipole interaction is too weak to stabilize the ion. Nonpolar naphthalene molecules form a molecular solid in which the only interparticle forces are of the weak dispersion type. The same forces operate in liquid benzene causing naphthalene to dissolve with relative ease. Like dissolves like.

12.9 Are the intermolecular forces in ethanol different from those in cyclohexane?

Solution: Strong hydrogen bonding is the principal intermolecular attraction in liquid ethanol, but in liquid cyclohexane the intermolecular forces are of the weak dispersion type (why?). Cyclohexane cannot form hydrogen bonds with ethanol (why?), and therefore cannot attract ethanol molecules strongly enough to form a solution.

12.10 What intermolecular forces are present in each substance? What intermolecular forces are present in water? This problem is like Example 12.1.

Solution: The order of increasing solubility is: $O_2 < Br_2 < LiCl < CH_3OH$. Methanol is miscible with water (see Section 12.3 for definition) because of strong hydrogen bonding. LiCl is an ionic solid and is very soluble because of the high polarity of the water molecules. Both oxygen and bromine are nonpolar and exert only weak dispersion forces. Bromine is the larger molecule and is therefore more polarizable and susceptible to dipole-induced dipole attractions (Section 11.2).

12.11 Think about the structures of the alcohols. Each has an -OH group capable of hydrogen bond formation. The rest of the molecule is a hydrocarbon chain capable only of weak dispersion attractions. How does the length of this chain affect the water solubility?

Solution: The longer the chain gets, the more the molecule "looks like" a hydrocarbon and the less important the -OH group becomes. The -OH group can form strong hydrogen bonds with water molecules, but the rest of the molecule cannot.

12.12 Consider the types of intermolecular forces present in the alcohols and in the solutes. Example 12.1 is a similar problem.

Solution: As the chain becomes longer, the alcohols become more like hydrocarbons (nonpolar) in their properties. The alcohol with five carbons (n-pentanol) would be the best solvent for iodine (a) and n-pentane (c) (why?). Methanol (CH_3OH) is the most water-like and is the best solvent for an ionic solid like KBr.

12.14 Study the discussion of percent by mass in Section 12.5. This problem is similar to Example 12.2.

Solution: Percent mass equals the mass of solute divided by the mass of the solution (that is, solute plus solvent) times 100 (to convert to percentage).

(a) $$\frac{5.50 \text{ g NaBr}}{78.2 \text{ g soln}} \times 100\% = 7.03\%$$

(b) $$\frac{31.0 \text{ g KCl}}{(31.0 + 152) \text{ g soln}} \times 100\% = 16.9\%$$

(c) $$\frac{4.5 \text{ g toluene}}{(4.5 + 29) \text{ g soln}} \times 100\% = 13\%$$

12.16 This is like Example 12.3. Use the given densities to calculate the masses of benzene and toluene.

Solution: $$\text{moles benzene} = 62.6 \text{ mL} \left(\frac{0.879 \text{ g}}{1 \text{ mL}}\right)\left(\frac{1 \text{ mol}}{78.11 \text{ g}}\right) = 0.704 \text{ mol}$$

$$\text{moles toluene} = 80.3 \text{ mL} \left(\frac{0.867 \text{ g}}{1 \text{ mL}}\right)\left(\frac{1 \text{ mol}}{92.13 \text{ g}}\right) = 0.756 \text{ mol}$$

$$X_{\text{benzene}} = \frac{\text{moles benzene}}{\text{moles benzene + moles toluene}} = \frac{0.704 \text{ mol}}{0.704 \text{ mol} + 0.756 \text{ mol}} = 0.482$$

$$X_{\text{toluene}} = \frac{\text{moles toluene}}{\text{moles benzene + moles toluene}} = \frac{0.756 \text{ mol}}{0.704 \text{ mol} + 0.756 \text{ mol}} = 0.518$$

12.17 Study the definition of molality in Section 12.5. This problem is similar to Example 12.4.

Solution: (a) The molality is the number of moles of sucrose (molar mass 342.3 g/mol) divided by the mass of the solvent (water) in kg.

$$\text{Molality} = 14.3 \text{ g sucrose} \left(\frac{1 \text{ mol}}{342.3 \text{ g sucrose}}\right)\left(\frac{1}{0.676 \text{ kg } H_2O}\right) = 0.0618 \text{ m}$$

(b) $$\text{Molality} = \frac{7.20 \text{ mol ethylene glycol}}{3.546 \text{ kg } H_2O} = 2.03 \text{ m}$$

12.18 In this problem assume that the density of the dilute solution is the same as that of water, namely 1.0 g/mL. What is the level of accuracy (number of significant figures) in this problem? Find the molality of the 0.010 M solution. What is the mass of 1.0 L of the solution? What is the mass of the solvent?

Solution: This problem should be worked at a two significant figure level of accuracy (why?). The mass of 1.0 L of the solution is

$$\text{Mass of 1.0 L} = 1.0\ \text{L}\left(\frac{1000\ \text{mL}}{1\ \text{L}}\right)\left(\frac{1.0\ \text{g}}{1\ \text{mL}}\right) = 1.0 \times 10^3\ \text{g}$$

The mass of 0.010 mol of urea is

$$0.010\ \text{mol urea}\left(\frac{60.06\ \text{g urea}}{1\ \text{mol urea}}\right) = 0.60\ \text{g urea}$$

The mass of the solvent is

$$(\text{solution mass}) - (\text{solute mass}) = (1.0 \times 10^3\ \text{g}) - (0.60\ \text{g}) = 1.0 \times 10^3\ \text{g} = 1.0\ \text{kg}$$

$$\text{Molality} = \frac{\text{moles solute}}{\text{mass solvent}} = \frac{0.010\ \text{mol}}{1.0\ \text{kg}} = 0.010\ m$$

12.20 This is like Problem 12.18 above and like Example 12.6.

Solution: In each case we consider one liter of solution. mass of solution = volume × density

(a) $$\text{mass of sugar} = 1.22\ \text{mol sugar}\left(\frac{342.3\ \text{g sugar}}{1\ \text{mol sugar}}\right) = 418\ \text{g sugar}$$

$$\text{mass of soln} = 1000\ \text{mL}\left(\frac{1.12\ \text{g}}{1\ \text{mL}}\right) = 1120\ \text{g}$$

$$\text{molality} = \frac{1.22\ \text{mol sugar}}{(1120 - 418)\ \text{g}\ H_2O}\left(\frac{1000\ \text{g}\ H_2O}{1\ \text{kg}\ H_2O}\right) = 1.74\ m$$

(b) $$\text{mass of NaOH} = 0.87\ \text{mol NaOH}\left(\frac{40.00\ \text{g NaOH}}{1\ \text{mol NaOH}}\right) = 35\ \text{g NaOH}$$

$$\text{molality} = \frac{0.87\ \text{mol NaOH}}{(1040 - 35)\ \text{g}\ H_2O}\left(\frac{1000\ \text{g}\ H_2O}{1\ \text{kg}\ H_2O}\right) = 0.87\ m$$

(c) $$\text{mass of}\ NaHCO_3 = 5.24\ \text{mol}\ NaHCO_3\left(\frac{84.01\ \text{g}\ NaHCO_3}{1\ \text{mol}\ NaHCO_3}\right) = 440\ \text{g}$$

$$\text{molality} = \frac{5.24\ \text{mol}\ NaHCO_3}{(1190 - 440)\ \text{g}\ H_2O}\left(\frac{1000\ \text{g}\ H_2O}{1\ \text{kg}\ H_2O}\right) = 6.99\ m$$

12.21 Assume that the solution volume is just the sum of the volumes of the water and the ethanol.

Solution: We find the volume of ethanol in 1.00 L of 75 proof gin. Note that 75 proof means $\left(\frac{75}{2}\right)\%$.

$$\text{Volume} = 1.00\text{ L}\left(\frac{75}{2}\right)\% = 0.38\text{ L} = 3.8 \times 10^2\text{ mL}$$

$$\text{Ethanol mass} = 3.8 \times 10^2\text{ mL}\left(\frac{0.798\text{ g}}{1\text{ mL}}\right) = 3.0 \times 10^2\text{ g}$$

12.22 This is similar to Examples 12.6 and 12.7.

Solution: A 100 g sample of 98.0% sulfuric acid contains 98.0 g H_2SO_4 and 2.0 g H_2O. We assume water is the solvent.

$$\text{molality} = 98.0\text{ g } H_2SO_4\left(\frac{1\text{ mol } H_2SO_4}{98.09\text{ g } H_2SO_4}\right)\left(\frac{1000\text{ g } H_2O}{1\text{ kg } H_2O}\right)\left(\frac{1}{2.0\text{ g } H_2O}\right) = 5.0 \times 10^2\ m$$

The volume of 100 g of this solution is

$$V = 100\text{ g}\left(\frac{1\text{ mL}}{1.83\text{ g}}\right)\left(\frac{1\text{ L}}{1000\text{ mL}}\right) = 0.0546\text{ L}$$

$$\text{molarity} = 98.0\text{ g } H_2SO_4\left(\frac{1\text{ mol } H_2SO_4}{98.09\text{ g } H_2SO_4}\right)\left(\frac{1}{0.0546\text{ L}}\right) = 18.3\text{ M}$$

12.24 In parts (a), (b), and (d) assume you have 100.0 g of solution. The problem is similar to Example 12.7.

Solution: (a) The mass of ethanol in the solution is 10.0% × 100.0 g = 10.0 g. The mass of the water is 100.0 g - 10.0 g = 90.0 g = 0.0900 kg. The amount of ethanol in moles is

$$10.0\text{ g ethanol}\left(\frac{1\text{ mol}}{46.07\text{ g}}\right) = 0.217\text{ mol ethanol}$$

$$\text{Molality} = \frac{\text{moles solute}}{\text{mass solvent}} = \frac{0.217\text{ mol}}{0.0900\text{ kg}} = 2.41\ m$$

(b) The volume of the solution is

$$100.0\text{ g}\left(\frac{1\text{ mL}}{0.984\text{ g}}\right) = 102\text{ mL} = 0.102\text{ L}$$

The amount of ethanol in moles is 0.217 mol (part (a)).

$$\text{Molarity} = \frac{\text{moles solute}}{\text{volume solution}} = \frac{0.217\text{ mol}}{0.102\text{ L}} = 2.13\text{ M}$$

(c) Solution volume $= 0.125\text{ mol}\left(\frac{1\text{ L}}{2.13\text{ mol}}\right) = 0.0587\text{ L}$

(d) From part (a) the masses of ethanol and water are 10.0 g and 90.0 g, respectively. The amount of ethanol is 0.217 mol. The amount of water is

$$90.0\ \text{g}\left(\frac{1\ \text{mol}}{18.02\ \text{g}}\right) = 4.99\ \text{mol}\ H_2O$$

$$\text{Mole fraction } H_2O = \frac{(\text{moles } H_2O)}{(\text{moles } H_2O) + (\text{moles ethanol})} = \frac{4.99\ \text{mol}}{4.99\ \text{mol} + 0.217\ \text{mol}} = 0.958$$

What percent of the molecules in the solution are water molecules?

12.32 Read the problem carefully. Note that the given data refer to the amounts of KNO_3 that 100 g of water will dissolve, while the problem centers on 100 g of solution. Calculate the mass of KNO_3 that will crystallize from a solution made from 100 g of water. Once this is found, how can you find the corresponding mass crystallizing from 100 g of solution?

Solution: At 75°C, 155 g of KNO_3 dissolve in 100 g of water to form 255 g of solution. When cooled to 25°C, only 38.0 g of KNO_3 remain dissolved so that (155 - 38.0) g = 117 g of KNO_3 will crystallize. The amount of KNO_3 formed by 100 g of saturated solution can be found by simple proportion.

$$117\ \text{g}\ KNO_3\left(\frac{100\ \text{g}}{255\ \text{g}}\right) = 45.9\ \text{g}\ KNO_3$$

12.33 How many grams of KCl are present in the impure sample? How much water does it take to dissolve this quantity of KCl? How much $KClO_3$ will dissolve in this amount of water?

Solution: The mass of KCl is 10% of the mass of the whole sample, namely 5.0 g. The $KClO_3$ mass is 45 g. If 100 g of water will dissolve 25.5 g of KCl, then the amount of water to dissolve 5.0 g of KCl is

$$5.0\ \text{g KCl}\left(\frac{100\ \text{g}\ H_2O}{25.5\ \text{g KCl}}\right) = 20\ \text{g}\ H_2O$$

The 20 g of H_2O will dissolve: $20\ \text{g}\ H_2O\left(\frac{7.1\ \text{g}\ KClO_3}{100\ \text{g}\ H_2O}\right) = 1.4\ \text{g}\ KClO_3$

The $KClO_3$ remaining undissolved will be: 45 g $KClO_3$ - 1.4 g $KClO_3$ = 44 g $KClO_3$

12.38 Study the subsection on Gas Solubility and Temperature in Section 12.6.

Solution: The bubbles in the two beakers are not of the same origin. The bubbles in the 30°C beaker are air bubbles. (The solubility of air decreases with increasing temperature.) The bubbles in the beaker at 100°C are the bubbles formed at the boiling point. At this temperature, most or all of the dissolved air has been driven off.

12.40 How does the solubility of a gas in water depend upon the pressure of that gas? Study Henry's law in Section 12.7 and Example 12.8.

Solution: We first find the value of k for Henry's law [Equation (12.1)].

$$k = \frac{c}{P} = \frac{0.034 \text{ mol/L}}{1 \text{ atm}} = 0.034 \text{ mol/L·atm}$$

For atmospheric conditions we write

$$c = kP = (0.034 \text{ mol/L·atm})(0.00030 \text{ atm}) = 1.0 \times 10^{-5} \text{ mol/L}$$

12.41 Study the kinetic molecular theory analysis of Henry's law in Section 12.7. When helium is bubbling through the solution, is the dissolved air still in dynamic equilibrium with the surrounding atmosphere?

Solution: When a dissolved gas is in dynamic equilibrium with its surroundings, the number of gas molecules entering the solution (dissolving) is equal to the number of dissolved gas molecules leaving and entering the gas phase. When the surrounding air is replaced by helium, the number of air molecules leaving the solution is greater than the number dissolving. As time passes the concentration of dissolved air becomes very small or zero, and the concentration of dissolved helium increases to a maximum.

12.42 Study the reasons for deviations from Henry's law in Section 12.7.

Solution: Both HCl and NH_3 react with water as follows.

$$HCl(g) + H_2O(l) \rightarrow H^+(aq) + Cl^-(aq)$$

$$NH_3(g) + H_2O(l) \rightleftharpoons NH_4^+(aq) + OH^-(aq)$$

Because of these reactions, neither compound obeys Henry's law. (Keep in mind that in aqueous solution HCl is a strong electrolyte and NH_3 is a weak electrolyte.)

12.43 Does atmospheric pressure increase or decrease as altitude decreases? How does this connect with Henry's law?

Solution: The soft drink tastes flat at the bottom of the mine because the carbon dioxide pressure is greater and the dissolved gas is not released from the solution. As the miner goes up in the elevator, the atmospheric carbon dioxide pressure decreases and dissolved gas is released from his stomach.

12.44 Use Henry's law to find the amount of nitrogen dissolved in 5.0 L of blood under both sets of conditions.

Solution: First calculate the Henry's law constant k:

$$c = kP$$

$$(5.6 \times 10^{-4} \text{ mol/L}) = k(0.80 \text{ atm})$$

$$k = 7.0 \times 10^{-4} \text{ mol/L·atm}$$

The concentration of N_2 in blood at 4.0 atm is

$$c = (7.0 \times 10^{-4}\ \text{mol/L·atm})(4.0\ \text{atm}) = 2.8 \times 10^{-3}\ \text{mol/L}$$

The number of moles of N_2 in 5.0 L of blood at 0.80 atm is

$$(5.6 \times 10^{-4}\ \text{mol/L})\ 5.0\ \text{L} = 2.8 \times 10^{-3}\ \text{mol}$$

The number of moles of N_2 in 5.0 L of blood at 4.0 atm is

$$(2.8 \times 10^{-3}\ \text{mol/L})\ 5.0\ \text{L} = 1.4 \times 10^{-2}\ \text{mol}$$

The amount of N_2 released in moles when the diver returns to the surface (total pressure 1.0 atm) is

$$1.4 \times 10^{-2}\ \text{mol} - 2.8 \times 10^{-3}\ \text{mol} = 1.1 \times 10^{-2}\ \text{mol}$$

The volume is $V = \dfrac{nRT}{P} = \dfrac{(1.1 \times 10^{-2}\ \text{mol})(0.0821\ \text{L·atm/K·mol})(310\ \text{K})}{(1.0\ \text{atm})} = 0.28\ \text{L}$

12.60 Study the subsection on Vapor Pressure Lowering in Section 12.8. This problem is a direct application of Raoult's law, Equation (12.2) and is similar to Example 12.9.

Solution: Since Equation (12.2) uses mole fraction, the first step is to find the number of moles of sucrose and of water.

$$\text{Moles sucrose} = 396\ \text{g}\left(\frac{1\ \text{mol}}{342.3\ \text{g}}\right) = 1.16\ \text{mol}$$

$$\text{Moles water} = 624\ \text{g}\left(\frac{1\ \text{mol}}{18.02\ \text{g}}\right) = 34.6\ \text{mol}$$

The mole fraction of water is

$$X_{H_2O} = \frac{34.6\ \text{mol}}{34.6\ \text{mol} + 1.16\ \text{mol}} = 0.968$$

The vapor pressure of the solution is found from Equation (12.2)

$$P_{\text{solution}} = X_{H_2O} \times P^{\circ}_{H_2O} = 0.968 \times 31.8\ \text{mmHg} = 30.8\ \text{mmHg}$$

12.62 This is like Problem 12.60 above.

Solution: Let us call benzene component 1 and camphor component 2.

$$P_1 = X_1 P^{\circ}_1 = \left(\frac{n_1}{n_1 + n_2}\right) P^{\circ}_1$$

$$n_1 = 98.5 \text{ g benzene} \left(\frac{1 \text{ mol}}{78.11 \text{ g}}\right) = 1.26 \text{ mol benzene}$$

$$n_2 = 24.6 \text{ g camphor} \left(\frac{1 \text{ mol}}{152.2 \text{ g}}\right) = 0.162 \text{ mol camphor}$$

$$P_1 = \frac{1.26 \text{ mol}}{(1.26 + 0.162) \text{ mol}} \times 100.0 \text{ mmHg} = 88.6 \text{ mmHg}$$

12.63 Study Raoult's law in Section 12.8. What is the mole fraction of the 1-propanol?

Solution: For any solution the sum of the mole fractions of the components is always 1.00, so the mole fraction of the 1-propanol is 0.700. The partial pressures are:

$$P_{ethanol} = X_{ethanol} \times P^{\circ}_{ethanol} = 0.300 \times 100 \text{ mmHg} = 30.0 \text{ mmHg}$$

$$P_{1\text{-propanol}} = X_{1\text{-propanol}} \times P^{\circ}_{1\text{-propanol}} = 0.700 \times 37.6 \text{ mmHg} = 26.3 \text{ mmHg}$$

Is the vapor phase richer in one of the components than the solution? Which component? Should this always be true for ideal solutions?

12.64 Part (a) is similar to Problem 12.63 above.

Solution: (a) First find the mole fractions of the solution components.

$$\text{Moles methanol} = 30.0 \text{ g} \left(\frac{1 \text{ mol}}{32.04 \text{ g}}\right) = 0.936 \text{ mol } CH_3OH$$

$$\text{Moles ethanol} = 45.0 \text{ g} \left(\frac{1 \text{ mol}}{46.07 \text{ g}}\right) = 0.977 \text{ mol } C_2H_5OH$$

$$X_{methanol} = \frac{0.936 \text{ mol}}{0.936 \text{ mol} + 0.977 \text{ mol}} = 0.489$$

$$X_{ethanol} = 1 - X_{methanol} = 0.511$$

The vapor pressures of the methanol and ethanol are:

$$P_{methanol} = 0.489 \times 94 \text{ mmHg} = 46 \text{ mmHg}$$

$$P_{ethanol} = 0.511 \times 44 \text{ mmHg} = 22 \text{ mmHg}$$

(b) Since n = PV/RT and V and T are the same for both vapors, the number of moles of each substance is proportional to the partial pressure. We can then write for the mole fractions:

$$X_{methanol} = \frac{P_{methanol}}{P_{methanol} + P_{ethanol}} = \frac{46 \text{ mmHg}}{46 \text{ mmHg} + 22 \text{ mmHg}} = 0.68$$

$X_{ethanol} = 1 - X_{methanol} = 0.32$

(c) As shown above, the vapor phase above a solution is always richer in the substance with the higher vapor pressure (the more volatile component). Substances of differing volatilities can be separated by a process of repeatedly vaporizing part of the mixture and then condensing the vapor. The more this process is repeated, the richer is the condensate in the more volatile substance. This process is called fractional distillation (Section 12.8).

12.65 Think about the two liquids (they are in a common closed vessel) in terms of the kinetic molecular theory model of equilibrium vapor pressure.

Solution: The glucose solution has a lower vapor pressure than pure water. This means that fewer water molecules escape from the glucose solution per second than from the pure water. Because of the differing rates of evaporation, with the passage of time there will be a gradual transfer of water molecules from the pure water container to the glucose solution container. Eventually the pure water will evaporate completely.

12.66 This is like Problem 12.60 above.

Solution: $\Delta P = X_2 P^{\circ}_1 = 2.50 \text{ mmHg} = X_2(31.8 \text{ mmHg})$

$X_2 = 0.0786$

The number of moles of water is

$$n_1 = 450 \text{ g } H_2O \left(\frac{1 \text{ mol } H_2O}{18.02 \text{ g } H_2O}\right) = 25.0 \text{ mol } H_2O$$

$$X_2 = \frac{n_2}{n_1 + n_2} = 0.0786 = \frac{n_2}{25.0 + n_2}$$

$n_2 = 2.13 \text{ mol}$

$$\text{mass of urea} = 2.13 \text{ mol urea} \left(\frac{60.06 \text{ g urea}}{1 \text{ mol urea}}\right) = 128 \text{ g urea}$$

12.68 This problem is a direct application of Equation (12.5). See Example 12.10 for the use of this equation.

Solution: $m = \dfrac{\Delta T_f}{K_f} = \dfrac{1.1°C}{1.86°C/m} = 0.59\ m$

Does it make any difference in this problem what the solute is?

12.69 The first part of this problem involves finding an empirical formula. This part is similar to Problem 2.88. The balance of the problem is like Example 12.12.

Solution: The empirical formula can be found from the percent by mass data assuming a 100.0 g sample:

$$\text{Moles C} = 80.78\text{ g}\left(\frac{1\text{ mol}}{12.01\text{ g}}\right) = 6.726\text{ mol C}$$

$$\text{Moles H} = 13.56\text{ g}\left(\frac{1\text{ mol}}{1.008\text{ g}}\right) = 13.45\text{ mol H}$$

$$\text{Moles O} = 5.66\text{ g}\left(\frac{1\text{ mol}}{16.00\text{ g}}\right) = 0.354\text{ mol O}$$

The empirical formula is $C_{6.726}H_{13.45}O_{0.354}$. Dividing through by the smallest subscript (0.354) gives the formula $C_{19}H_{38}O$.

The freezing point depression is $\Delta T_f = 5.5°C - 3.37°C = 2.1°C$. This implies a solution molality of

$$m = \frac{\Delta T_f}{K_f} = \frac{2.1°C}{5.12°C/m} = 0.41\ m$$

Since the solvent mass is 8.50 g or 0.00850 kg, the amount of solute is

$$\frac{0.41\text{ mol}}{1\text{ kg benzene}} \times 0.00850\text{ kg benzene} = 3.5 \times 10^{-3}\text{ mol}$$

Since 1.00 g of the sample represents 3.5×10^{-3} mol, the molar mass is

$$\frac{1.00\text{ g}}{3.5 \times 10^{-3}\text{ mol}} = 286\text{ g/mol.}$$

The mass of the empirical formula is 282 g/mol, so this is also the molecular formula.

12.70 This problem is similar to Problem 12.68 and Example 12.10.

Solution: We want a freezing point depression of 20°C. Using Equation (12.5) we write

$$m = \frac{\Delta T_f}{K_f} = \frac{20°C}{1.86°C/m} = 10.8\ m$$

The mass of ethylene glycol (EG) in 6.5 L or 6.5 kG of water is

$$\text{mass EG} = 6.50\text{ kg }H_2O\left(\frac{10.8\text{ mol EG}}{1\text{ kg }H_2O}\right)\left(\frac{62.07\text{ g EG}}{1\text{ mol EG}}\right) = 4.36 \times 10^3\text{ g}$$

The volume of EG needed is

$$V = 4.36 \times 10^3\text{ g EG}\left(\frac{1\text{ mL EG}}{1.11\text{ g EG}}\right)\left(\frac{1\text{ L}}{1000\text{ mL}}\right) = 3.93\text{ L}$$

Finally, we calculate the boiling point

$$\Delta T_b = mK_b = \left(\frac{10.8 \text{ mol EG}}{1 \text{ kg}}\right)\left(\frac{0.52°\text{C kg}}{1 \text{ mol}}\right) = 5.6°\text{C}$$

The boiling point of the solution will be 100.0°C + 5.6°C = 105.6°C

12.71 This is similar to Problem 12.69 above.

Solution: First we find the freezing point depression.

$$\Delta T_f = (5.5 - 4.3)°\text{C} = 1.2°\text{C}$$

$$m = \frac{\Delta T_f}{K_f} = \frac{1.2°\text{C}}{5.12°\text{C}/m} = 0.23\ m$$

Let x be the molar mass of the compound. We write for the molality

$$m = 0.23 \text{ mol/kg} = \frac{\text{mass of compound/molar mass}}{\text{solvent mass (kg)}} = \frac{2.50 \text{ g}/x}{0.025 \text{ kg}}$$

$$x = 4.3 \times 10^2 \text{ g/mol}$$

The molar mass of C_6H_5P is 108.1 g/mol. Therefore, the molecular formula is $4(C_6H_5P)$ or $C_{24}H_{20}P_4$.

12.73 This problem is similar to Problem 12.69 above.

Solution: First we find the empirical formula. We assume 100 g of compound.

$$\text{C:} \quad 40.0 \text{ g C}\left(\frac{1 \text{ mol C}}{12.01 \text{ g C}}\right) = 3.33 \text{ mol C}$$

$$\text{H:} \quad 6.7 \text{ g H}\left(\frac{1 \text{ mol H}}{1.008 \text{ g H}}\right) = 6.6 \text{ mol H}$$

$$\text{O:} \quad 53.3 \text{ g O}\left(\frac{1 \text{ mol O}}{16.00 \text{ g O}}\right) = 3.33 \text{ mol O}$$

The empirical formula is $C_{3.33}H_{6.6}O_{3.33}$ or CH_2O. To find the molecular formula we must use the freezing point data to determine the molar mass.

$$\text{molality} = \frac{\Delta T_f}{K_f} = \frac{1.56°\text{C}}{8.00°\text{C}/m} = 0.195\ m$$

Let x be the molar mass of the compound. We write

$$\text{molality} = 0.195 \text{ mol/kg} = \frac{\text{mass of compound/molar mass}}{\text{solvent mass (kg)}} = \frac{0.650 \text{ g}/x}{0.0278 \text{ kg}}$$

$$x = 120 \text{ g/mol}$$

The molar mass of the empirical formula is 30.0 g. The molar mass is four times this amount, or $(CH_2O)_4$ or $C_4H_8O_4$.

12.75 Use Example 12.13 as a model to find the molar mass of the compound. Finding the empirical formula from the percent by mass data is like Problem 12.69.

Solution: As in Example 12.13, first find the concentration of the solution, then work out the molar mass. The concentration is found from

$$\text{Molarity} = \frac{\pi}{RT} = \frac{1.43\text{ atm}}{(0.0821\text{ L·atm/K·mol})(300\text{ K})} = 0.0581\text{ mol/L}$$

The solution volume is 0.3000 L so the number of moles of solute is

$$\frac{0.0581\text{ mol}}{1\text{ L}} \times 0.3000\text{ L} = 0.0174\text{ mol}$$

The molar mass is then

$$\frac{7.480\text{ g}}{0.0174\text{ mol}} = 430\text{ g/mol}$$

The empirical formula can be found most easily by assuming a 100.0 g sample of the substance:

$$\text{Moles C} = 41.8\text{ g} \times \frac{1\text{ mol}}{12.01\text{ g}} = 3.48\text{ mol C}$$

$$\text{Moles H} = 4.7\text{ g} \times \frac{1\text{ mol}}{1.008\text{ g}} = 4.7\text{ mol H}$$

$$\text{Moles O} = 37.3\text{ g} \times \frac{1\text{ mol}}{16.00\text{ g}} = 2.33\text{ mol O}$$

$$\text{Moles N} = 16.3\text{ g} \times \frac{1\text{ mol}}{14.01\text{ g}} = 1.16\text{ mol N}$$

The empirical formula is $C_{3.48}H_{4.7}O_{2.33}N_{1.16}$. Dividing through by the smallest subscript (1.16) gives $C_3H_4O_2N$, which has a mass of 86.0 g per formula unit. The molar mass is five times this amount (430 ÷ 86.0 = 5.0), so the molecular formula is $(C_3H_4O_2N)_5$ or $C_{15}H_{20}O_{10}N_5$.

12.77 This is similar to Problem 12.75 above and Example 12.13 in the text.

Solution: We first find the molarity of the solution.

$$M = \frac{\pi}{RT} = \frac{(5.20/760)\text{ atm}}{(0.0821\text{ L·atm/K·mol})(25 + 273)\text{ K}} = 2.80 \times 10^{-4}\text{ mol/L}$$

Let x be the molar mass of the protein

$$M = 2.80 \times 10^{-4}\text{ mol/L} = \frac{\text{mass of protein/molar mass}}{\text{solution volume (L)}} = \frac{(0.8330\text{ g}/x)}{0.1700\text{ L}}$$

$x = 1.75 \times 10^4$ g/mol

12.84 Carefully study Section 12.9. How do solutions of electrolytes differ from nonelectrolyte solutions in their effect on properties like vapor pressure, etc.? Which of the two compounds is an electrolyte? Assume it to be a strong electrolyte. What is the total ion concentration?

Solution: $CaCl_2$ is an ionic compound (why?) and is therefore an electrolyte in water. Assuming it is a strong electrolyte (no ion pairs, van't Hoff factor i = 3), the total ion concentration will be $3 \times 0.35 = 1.05$ m, which is larger than the urea (nonelectrolyte) concentration of 0.90 m.

(a) The $CaCl_2$ solution will show a larger boiling point elevation.

(b) The $CaCl_2$ solution will show a larger freezing point depression. The freezing point of the urea solution will be higher (read the question!).

(c) The $CaCl_2$ solution will have a larger vapor pressure lowering.

12.85 What properties depend on the dissolved particle concentration?

Solution: Boiling point elevation, vapor pressure, and osmotic pressure all depend on particle concentration.

12.86 This problem is like Problem 12.84 above. Assume complete ionization for all the electrolytes and calculate the concentration of dissolved particles for each solution. CH_3COOH is a weak electrolyte.

Solution: Assume that all the salts are completely dissociated or ionized. Calculate the molality of the ions in the solutions.

(a) 0.10 m Na_3PO_4: $0.10 \text{ m} \times 4$ ions/unit = 0.40 m

(b) 0.35 m NaCl: $0.35 \text{ m} \times 2$ ions/unit = 0.70 m

(c) 0.20 m $MgCl_2$: $0.20 \text{ m} \times 3$ ions/unit = 0.60 m

(d) 0.15 m $C_6H_{12}O_6$: nonelectrolyte, 0.15 m

(e) 0.15 m CH_3COOH: weak electrolyte, slightly greater than 0.15 m

The solution with the lowest molality will have the highest freezing point (smallest freezing point depression): (d) > (e) > (a) > (c) > (b)

12.87 What effect would these substances have on the freezing point of water?

Solution: Both NaCl and $CaCl_2$ are strong electrolytes. Urea and sucrose are nonelectrolytes. The NaCl or $CaCl_2$ will yield more particle per mole of solid dissolved, resulting in greater freezing point depression. Sucrose and urea would also make a mess when the ice melts.

12.89 Which substance is an electrolyte? How does this affect boiling and freezing point computations? When electrolyte properties are taken into account, this problem is like Example 12.9 and Problem 12.60. The molecular formula of urea is CH_4N_2O.

Solution: (a) NaCl is a strong electrolyte. The concentration of particles (ions) is double the concentration of NaCl. Note that 135 mL of water has a mass of 135 g (why?).

The number of moles of NaCl is

$$21.2 \text{ g NaCl}\left(\frac{1 \text{ mol}}{58.44 \text{ g}}\right) = 0.363 \text{ mol NaCl}$$

We use Equations (12.7) and (12.8), respectively, to find the changes in boiling and freezing points ($i = 2$)

$$\Delta T_b = iK_b m = 2(0.52°\text{C}/m)\left(\frac{0.363 \text{ mol}}{0.135 \text{ kg}}\right) = 2.8°\text{C}$$

$$\Delta T_f = iK_f m = 2(1.86°\text{C}/m)\left(\frac{0.363 \text{ mol}}{0.135 \text{ kg}}\right) = 10.0°\text{C}$$

The boiling point is 102.8°C; the freezing point is -10.0°C.

(b) Urea is a nonelectrolyte. The particle concentration is just equal to the urea concentration.

$$\Delta T_b = iK_b m = 1(0.52°\text{C}/m)(15.4 \text{ g urea})\left(\frac{1 \text{ mol}}{60.06 \text{ g}}\right)\left(\frac{1}{0.0667 \text{ kg}}\right) = 2.0°\text{C}$$

$$\Delta T_f = iK_f m = 1(1.86°\text{C}/m)(15.4 \text{ g urea})\left(\frac{1 \text{ mol}}{60.06 \text{ g}}\right)\left(\frac{1}{0.0667 \text{ kg}}\right) = 7.15°\text{C}$$

The boiling point is 102.0°C; the freezing point is -7.15°C.

12.90 Study Section 12.8 and Example 12.9. The only difference here is that the solute is a strong electrolyte. Note that you are asked for the concentration of NaCl, not the total concentration of all dissolved particles. Assume the solution is made using 1000 g of water.

Solution: Application of Equation (12.3) allows us to find the mole fraction of the NaCl. We use 1 for H_2O and 2 for NaCl.

$$P_1^o - P_1 = 23.76 \text{ mmHg} - 22.98 \text{ mmHg} = X_2 \times P_1^o = X_2 \times 23.76 \text{ mmHg}$$

$$X_2 = \frac{23.76 \text{ mmHg} - 22.98 \text{ mmHg}}{23.76 \text{ mmHg}} = 0.03283$$

Assuming we have 1000 g of water as the solvent, we can find the number of moles of particles dissolved in the water.

$$X_2 = \frac{n_2}{n_1 + n_2}$$

$$n_1 = 1000 \text{ g } H_2O\left(\frac{1 \text{ mol } H_2O}{18.02 \text{ g } H_2O}\right) = 55.49 \text{ mol } H_2O$$

$$\frac{n_2}{55.49 + n_2} = 0.03283$$

$$n_2 = 1.884 \text{ mol}$$

Since NaCl ionizes to form two particles (ions), the number of moles of NaCl is half of the above result.

$$\text{Moles NaCl} = \frac{1.884 \text{ mol}}{2} = 0.9420 \text{ mol}$$

The molality of the solution is

$$\frac{0.9420 \text{ mol}}{1.000 \text{ kg}} = 0.9420\ m$$

12.91 Urea is a nonelectrolyte ($i = 1$). Both solutions have the same concentration in terms of moles of solute per liter of solution. If $CaCl_2$ were completely ionized with no ion pairs, what would be the osmotic pressure of its solution relative to the urea solution? This problem is similar to Example 12.14.

Solution: The temperature and molarity of the two solutions are the same. If we divide Equation (12.9) for one solution by the same equation for the other, we can find the ratio of the van't Hoff factors in terms of the ratio of the osmotic pressures:

$$\frac{\pi_{CaCl_2}}{\pi_{urea}} = \frac{iMRT}{MRT} = i = \frac{0.605 \text{ atm}}{0.245 \text{ atm}} = 2.47$$

12.92 Find the molarity of 1000 mL of this solution.

Solution: The mass of 1000 mL of the solution is 1005 g. The mass of NaCl in the solution is

$$1005 \text{ g} \times \frac{0.86\%}{100\%} = 8.6 \text{ g NaCl}$$

The molarity of the solution is then

$$\left(\frac{8.6 \text{ g NaCl}}{1.000 \text{ L}}\right)\left(\frac{1 \text{ mol NaCl}}{58.44 \text{ g NaCl}}\right) = 0.15 \text{ M}$$

Since NaCl is a strong electrolyte, the van't Hoff factor is 2. We use Equation (12.9) to find the osmotic pressure

$$\pi = iMRT = 2(0.15 \text{ mol/L})(0.0821 \text{ L·atm/K·mol})(310 \text{ K}) = 7.6 \text{ atm}$$

12.93 Study the discussion of osmosis in Section 12.8. Will there be an osmotic pressure if the membrane allows everything to pass through? If one ion can pass through the membrane but not the other, will a separation of electric charge occur?

Solution: (a) If the membrane is permiable to all the ions and to the water, the result will be the same as just removing the membrane. You will have two solutions of equal NaCl concentration.

(b) This part is tricky. The movement of one ion but not the other would result in one side of the apparatus acquiring a positive electric charge and the other side becoming equally negative. This has never been known to happen, so we must conclude that migrating ions always drag other ions of the opposite charge with them. In this hypothetical situation only water would move through the membrane from the dilute to the more concentrated side.

(c) This is the classic osmosis situation. Water would move through the membrane from the dilute to the concentrated side.

12.94 Study the discussion of Raoult's law and ideal solutions in Section 12.8. The conditions involving the interparticle attractive forces in the first row of the table is a Case 1 situation.

Solution: The completed table is shown below:

Attractive Forces	Deviation from Raoult's Law	$\Delta H_{solution}$
A↔A, B↔B > A↔B	Positive	Positive (endothermic)
A↔A, B↔B < A↔B	Negative	Negative (exothermic)
A↔A, B↔B = A↔B	Zero	Zero

The first row represents a Case 1 situation in which A's attract A's and B's attract B's more strongly than A's attract B's. As described in Section 12.8, this results in positive deviation from Raoult's law (higher vapor pressure than calculated) and positive heat of solution (endothermic).

In the second row a negative deviation from Raoult's law (lower than calculated vapor pressure) means A's attract B's better than A's attract A's and B's attract B's. This causes a negative (exothermic) heat of solution.

In the third row a zero heat of solution means that A-A, B-B, and A-B interparticle attractions are all the same. This corresponds to an ideal solution which obeys Raoult's law exactly.

What sorts of substances form ideal solutions with each other?

12.95 Appropriate study material for each part: (a) Section 12.8 and Example 12.10, (b) Henry's law in Section 12.7 and the adjacent "Chemistry in Action" segment, (c) Problem 12.18, (d) Section 12.8 (how does the molality of a solution depend on temperature?), (e) Section 12.8 and Example 12.10.

Solution: (a) Seawater has a larger number of ionic compounds dissolved in it; thus the boiling point is elevated.

(b) Carbon dioxide escapes from an opened soft drink bottle because gases are less soluble in liquids at lower pressure (Henry's law). Refer to the Chemistry in Action in Section 12.7.

(c) As you proved in Problem 12.18, at dilute concentrations molality and molarity are almost the same because the density of the solution is almost equal to that of the pure solvent.

(d) For colligative properties we are concerned with the number of solute particles in solution relative to the number of solvent particles. Since in colligative particle measurements we frequently are dealing with changes in temperature (and since density varies with temperature), we need a concentration unit that is temperature invarient. We use units of moles per kilogram of mass (molality) rather than moles per liter of solution (molarity).

(e) Methanol is very water soluble (why?) and effectively lowers the freezing point of water. However in the summer, the temperatures are sufficiently high so that most of the methanol would be lost to vaporization.

12.97 What is the molality of the solution calculated from the given masses of aluminum chloride and water? What is the molality of the solution calculated from the freezing point data?

Solution: The molality of the solution assuming $AlCl_3$ to be a nonelectrolyte is

$$1.00 \text{ g } AlCl_3 \left(\frac{1 \text{ mol } AlCl_3}{133.3 \text{ g } AlCl_3}\right)\frac{1}{0.0500 \text{ kg}} = 0.150\ m$$

The molality calculated with Equation (12.5) from experimental freezing point data is

$$m = \frac{\Delta T_f}{K_f} = \frac{1.11°C}{1.86°C/m} = 0.597\ m$$

The ratio $\frac{0.597\ m}{0.150\ m}$ is 4. Thus each $AlCl_3$ dissociates as follows

$$AlCl_3(s) \rightarrow Al^{3+}(aq) + 3Cl^-(aq)$$

12.98 Study the discussion of osmotic pressure in Section 12.8. What is crenation?

Solution: Water migrates through the semipermiable cell walls of the cucumber into the concentrated salt solution.

When we go swimming in the ocean, why don't we shrivel up like the cucumber? When we swim in a fresh water pool, why don't we swell up and burst?

12.99 Look at Equation (12.6). How does the osmotic pressure depend upon the solution molarity when the temperature remains constant? Assume the volumes of the solutions are additive.

Solution: At constant temperature the osmotic pressure of a solution is proportional to the molarity. When equal volumes of the two solutions are mixed, the molarity will just be the mean of the molarities of the two solutions (assuming additive volumes). Since the osmotic pressure is proportional to the molarity, the osmotic pressure of the solution will be the mean of the osmotic pressure of the two solutions.

$$\pi = \frac{2.4 \text{ atm} + 4.6 \text{ atm}}{2} = 3.5 \text{ atm}$$

12.101 How does reverse osmosis work (see the "Chemistry in Action" segment in Section 12.9)? Calculate the osmotic pressure of seawater. See Equations (12.6) and (12.9). Which should be applied here?

Solution: To reverse the osmotic migration of water across a semipermiable membrane, an external pressure exceeding the osmotic pressure must be applied. To find the osmotic pressure of 0.70 M NaCl solution, we must use Equation (12.9) because NaCl is a strong electrolyte and the total ion concentration becomes 2(0.70 M) = 1.4 M. The osmotic pressure of sea water is

$$\pi = i\text{MRT} = 2(0.70 \text{ mol/L})(0.0821 \text{ L·atm/K·mol})(298 \text{ K}) = 34 \text{ atm}$$

To cause reverse osmosis a pressure in excess of 34 atm must be applied.

12.102 For part (a) find the molar mass using Equation (12.6). For part (b) find the molar mass using Equation (12.9). Assume complete ionization.

Solution: (a) Using Equation (12.6), we find the molarity of the solution

$$\text{M} = \frac{\pi}{\text{RT}} = \frac{0.257 \text{ atm}}{(0.0821 \text{ L·atm/K·mol})(298 \text{ K})} = 0.0105 \text{ mol/L}$$

This is the combined concentrations of all the ions. The amount dissolved in 10.0 mL (0.01000 L) is

$$\text{moles} = \left(\frac{0.0105 \text{ mol}}{1 \text{ L}}\right) 0.0100 \text{ L} = 1.05 \times 10^{-4} \text{ mol}$$

Since the mass of this amount of protein is 0.225 g, the apparent molar mass is

$$\frac{0.225 \text{ g}}{1.05 \times 10^{-4} \text{ mol}} = 2.14 \times 10^{3} \text{ g/mol}$$

(b) We use Equation (12.9) to take into account the fact that the protein is a strong electrolyte. The van't Hoff factor will be $i = 21$ (why?).

$$\text{M} = \frac{\pi}{i\text{RT}} = \frac{0.257 \text{ atm}}{(21)(0.0821 \text{ L·atm/K·mol})(298 \text{ K})} = 5.00 \times 10^{-4} \text{ mol/L}$$

This is the actual concentration of the protein. The amount in 10.0 mL (0.0100 L) is

$$\frac{5.00 \times 10^{-4} \text{ mol}}{1 \text{ L}} \times 0.0100 \text{ L} = 5.00 \times 10^{-6} \text{ mol}$$

Therefore the actual molar mass is

$$\frac{0.225\text{ g}}{5.00 \times 10^{-6}\text{ mol}} = 4.50 \times 10^4 \text{ g/mol}$$

12.103 Although this is a somewhat tedious problem, it is not difficult. It is a direct application of the definition of molarity.

Solution: First, we tabulate the concentration of all of the ions. Notice that the chloride concentration comes from more than one source.

$MgCl_2$:	If $[MgCl_2] = 0.054$ M,	$[Mg^{2+}] = 0.054$ M	$[Cl^-] = 2 \times 0.054$ M
Na_2SO_4:	if $[Na_2SO_4] = 0.051$ M,	$[Na^+] = 2 \times 0.051$ M	$[SO_4^{2-}] = 0.051$ M
$CaCl_2$:	if $[CaCl_2] = 0.010$ M,	$[Ca^{2+}] = 0.010$ M	$[Cl^-] = 2 \times 0.010$ M
KCl:	if $[KCl] = 0.0090$ M,	$[K^+] = 0.0090$ M	$[Cl^-] = 0.0090$ M
$NaHCO_3$:	if $[NaHCO_3] = 0.0020$ M	$[Na^+] = 0.0020$ M	$[HCO_3^-] = 0.0020$ M

The subtotal of chloride ion concentration is

$$[Cl^-] = (2 \times 0.0540) + (2 \times 0.010) + (0.0090) = 0.137 \text{ M}$$

Since the required $[Cl^-]$ is 2.60 M, the difference (2.6 - 0.137 = 2.46 M) must come from NaCl.

The subtotal of sodium ion concentration is

$$[Na^+] = (2 \times 0.051) + (0.0020) = 0.104 \text{ M}$$

Since the required $[Na^+]$ is 2.56 M, the difference (2.56 - 0.104 = 2.46 M) must come from NaCl.

Now, calculating the mass of the compounds required:

NaCl: $$2.46 \text{ mol} \times \frac{58.44 \text{ g NaCl}}{1 \text{ mol NaCl}} = 143.8 \text{ g}$$

$MgCl_2$: $$0.054 \text{ mol} \times \frac{95.21 \text{ g } MgCl_2}{1 \text{ mol } MgCl_2} = 5.14 \text{ g}$$

Na_2SO_4: $$0.051 \text{ mol} \times \frac{142.1 \text{ g } Na_2SO_4}{1 \text{ mol } Na_2SO_4} = 7.25 \text{ g}$$

$CaCl_2$: $$0.010 \text{ mol} \times \frac{111.0 \text{ g } CaCl_2}{1 \text{ mol } CaCl_2} = 1.11 \text{ g}$$

KCl: $$0.0090 \text{ mol} \times \frac{74.55 \text{ g KCl}}{1 \text{ mol KCl}} = 0.67 \text{ g}$$

$NaHCO_3$: $$0.0020 \text{ mol} \times \frac{84.01 \text{ g } NaHCO_3}{1 \text{ mol } NaHCO_3} = 0.17 \text{ g}$$

12.105 This problem is a direct application of Raoult's law presented in Equations (12.2 and 12.3).

Solution: (a) Since there is one mole of both compound A and compound B

$$X_A = 0.0500 \text{ and } X_B = 0.0500$$

$$P_A = X_A P^{\circ}_A = 0.500 \text{ x } 76 \text{ mmHg} = 38 \text{ mmHg}$$

$$P_B = X_B P^{\circ}_B = 0.500 \text{ x } 132 \text{ mmHg} = 66 \text{ mmHg}$$

(b) Since there are two moles of A and 7 moles of B

$$X_A = \frac{2}{7} \text{ and } X_B = \frac{5}{7}$$

$$P_A = X_A P^{\circ}_A = \frac{2}{7} \text{ x } 76 \text{ mmHg} = 22 \text{ mmHg}$$

$$P_B = X_B P^{\circ}_B = \frac{5}{7} \text{ x } 132 \text{ mmHg} = 94 \text{ mmHg}$$

12.107 This problem is similar to Example 12.13.

Solution: Letting x equal the molar mass, then the molarilty

$$M = \left(\frac{1.22 \text{ g}}{x \text{ g/mol}}\right)\left(\frac{1000\text{mL/L}}{262 \text{ mL}}\right)$$

Using Equation (12.6): $\pi = MRT$

$$\frac{30.3 \text{ mmHg}}{760 \text{ mmHg}} = \left(\frac{1.22 \text{ g}}{x \text{ g/mol}}\right)\left(\frac{1000\text{mL/L}}{262 \text{ mL}}\right) \times (0.0821 \text{ L·atm/mol·K}) \times (308 \text{ K}) = 2.95 \times 10^3 \text{ g/mol}$$

$$x = 2.95 \times 10^3 \text{ g/mol.}$$

12.109 This problem is similar to Problem 12.67.

Solution: The freezing point constant for benzene can be found in Table 12.2.

$\Delta T_f = K_f m$ $\quad (5.5 - 3.9)°C = (5.12°C/m) \times m$ $\quad m = 0.313$ moles solute/kg benzene

Solving for the molar mass M : $\quad 0.313 \text{ moles solute/kg benzene} = \dfrac{0.50 \text{ g solute}/M \text{ g/mol}}{0.008 \text{ kg benzene}}$

$$M = 200 \text{ g/mol}$$

The molar mass for cocaine $C_{17}H_{21}NO_4$ = 303 g/mol, so the compound is not cocaine. We assume in our analysis that the compound is a pure, monomeric, nonelectrolyte.

12.110 If the two liquids were totally immiscible, would the volumes add to 1000 mL?

Solution: Since the total volume is less than the sum of the two volumes, the ethanol and water must have an intermolecular attraction that results in an overall smaller volume.

12.111 As in any stoichiometry problem, we must begin with a balanced equation.

Solution:

$$2H_2O_2 \rightarrow 2H_2O + O_2$$

$$10\text{ mL}\left(\frac{3.0\text{ g }H_2O_2}{100\text{ mL}}\right)\left(\frac{\text{mol }H_2O_2}{34.02\text{ g }H_2O_2}\right)\left(\frac{\text{mol }O_2}{2\text{ mol }H_2O_2}\right) = 4.4 \times 10^{-3}\text{ mol }O_2$$

(a) Using the ideal gas law:

$$V = \frac{nRT}{P} = \frac{(4.4 \times 10^{-3}\text{ mol }O_2)(0.0821\text{ L·atm/mol·K})(273\text{K})}{1.0\text{ atm}} = 99\text{ mL}$$

(b) The ratio of the volumes:

$$\frac{99\text{ mL}}{10\text{ mL}} = 9.9$$

Could we have made the calculation in part (a) simpler if we used the fact that 1 mole of all ideal gases at STP occupies a volume of 22.4 L?

12.112 This problem is an interesting application of Rauolt's law. See Sections 12.8 and 12.9.

Solution: One manometer has pure water over the mercury, one manometer has a 1.0 M solution of NaCl and the other manometer has a 1.0 M solution of urea. The pure water will have the highest vapor pressure and will thus force the mercury column down the most; column X. Both the salt and the urea will lower the overall pressure of the water. However, the salt dissolves into sodium and chloride ions (van't Hoff factor $i = 2$), whereas urea is a molecular compound with a van't Hoff factor of 1. Therefore the urea solution will lower the pressure only half as much as the salt solution. Y is the NaCl solution and Z is the urea solution.

Assuming that you knew the temperature, could you actually calculate the distance from the top of the solution to the top of the manometer?

Note: If you have not already done so, please read the "Introduction for the Student" at the beginning of this Student Solutions Manual.

Chapter 13

CHEMICAL KINETICS

13.6 This problem is directly analogous to Example 13.1. If you need review, study the subsection on Reaction Rate and Stoichiometry in Section 13.1. What is the relationship between the coefficients in the balanced equation and the rate expression?

Solution: In general for a reaction $aA + bB \rightarrow cC + dD$

$$\text{rate} = -\frac{1}{a}\frac{\Delta[A]}{\Delta t} = -\frac{1}{b}\frac{\Delta[B]}{\Delta t} = \frac{1}{c}\frac{\Delta[C]}{\Delta t} = \frac{1}{d}\frac{\Delta[D]}{\Delta t}$$

(a) $$\text{rate} = -\frac{\Delta[H_2]}{\Delta t} = -\frac{\Delta[I_2]}{\Delta t} = \frac{1}{2}\frac{\Delta[HI]}{\Delta t}$$

(b) $$\text{rate} = -\frac{1}{2}\frac{\Delta[H_2]}{\Delta t} = -\frac{\Delta[O_2]}{\Delta t} = \frac{1}{2}\frac{\Delta[H_2O]}{\Delta t}$$

(c) $$\text{rate} = -\frac{1}{5}\frac{\Delta[Br^-]}{\Delta t} = -\frac{\Delta[BrO_3^-]}{\Delta t} = -\frac{1}{6}\frac{\Delta[H^+]}{\Delta t} = \frac{1}{3}\frac{\Delta[Br_2]}{\Delta t}$$

Note that because the reaction is carried out in the aqueous phase, we do not monitor the concentration of water.

13.7 In Section 13.1 study the subsection on Reaction Rate and Stoichiometry and work Example 13.1. How are the rates at which the concentrations of different reactants and products related through the balanced equation for the reaction? If you know the rate at which the concentration of one substance is changing, can you calculate the rates at which the others are changing at that moment? Note that since H_2 is disappearing, the rate of change is -0.074 M/s.

Solution: From the given balanced equation the rates at which the molarities of nitrogen, hydrogen, and ammonia are changing are connected by the relationship

$$\text{rate} = -\frac{\Delta[N_2]}{\Delta t} = -\frac{1}{3}\frac{\Delta[H_2]}{\Delta t} = \frac{1}{2}\frac{\Delta[NH_3]}{\Delta t}$$

(a) If hydrogen is reacting at the rate of -0.074 M/s, the rate at which ammonia is being formed is

$$\frac{\Delta[NH_3]}{\Delta t} = -\frac{2}{3}\frac{\Delta[H_2]}{\Delta t} = -\frac{2}{3}(-0.074\text{ M/s}) = 0.049\text{ M/s}$$

(b) The rate at which nitrogen is reacting must be

$$\frac{\Delta[N_2]}{\Delta t} = \frac{1}{3}\frac{\Delta[H_2]}{\Delta t} = \frac{1}{3}(0.074 \text{ M/s}) = 0.025 \text{ M/s}$$

Will the rate at which ammonia forms always be twice the rate of reaction of nitrogen, or is this true only at the instant described in this problem?

13.8 In the equation for the reaction which substances are and which are not electrolytes? How well might the solution conduct electricity before the reaction starts? How well might it conduct after the reaction is over?

Solution: Neither of the reactants is an electrolyte, so initially the solution will be a poor conductor. As the reaction progresses, hydrogen ions and iodide ions appear (HI is a strong electrolyte), and the conductivity increases. One could relate the change in conductivity to the fact that two moles of ions form per mole of C_2H_5I consumed, and thus follow the progress of the reaction.

13.18 This problem is similar to Example 13.2.

Solution: By comparing the first and second sets of data, we see that changing [B] does not affect the rate of the reaction. Therefore, the reaction is zero order in B. By comparing the first and third sets of data, we see that doubling [A] doubles the rate of the reaction. This shows that the reaction is first order in A.

$$\text{rate} = k[A]$$

From the first set of data: 3.20×10^{-1} M/s = k(1.50 M); $k = 0.213 \text{ s}^{-1}$

What would be the value of k if you had used the second or third set of data? Should k be constant?

13.20 Study the discussion of first-order reactions in Section 13.2. This problem is a straightforward application of Equation (13.3). When the reaction is 35.5% complete, how much A remains? What will be the units of this rate constant? Refer to Example 13.3 in the text for a similar problem.

Solution: In terms of the initial concentration of A, the concentration of A remaining after 4.90 min is 0.645 $[A]_o$. Equation (13.3) becomes

$$\ln\frac{[A]_o}{[A]} = \ln\frac{[A]_o}{0.645[A]_o} = 0.439 = kt$$

The rate constant k is

$$k = \frac{0.439}{t} = \frac{0.439}{4.90 \text{ min}} = 0.0896 \text{ min}^{-1}$$

What are the units of k in s^{-1}? Are the values of k different when calculated with base 10 and base e logarithms?

13.22 Study the subsection of Section 13.2 on the Determination of Reaction Order and work Example 13.2. This problem also requires the use of the method of initial rates. The rate in part (c) can be found once the rate law is known.

Solution: (a) As in Example 13.3 assume the rate law has the form

$$\text{rate} = k[F_2]^x[ClO_2]^y$$

If we take the ratio of the rates in the first two experiments we find that the rate quadruples when the concentration of ClO_2 quadruples. The order in this reactant must be one.

$$\frac{\text{rate 2}}{\text{rate 1}} = \frac{4.8 \times 10^{-3}}{1.2 \times 10^{-3}} = 4.0 = \frac{k[0.10]^x[0.040]^y}{k[0.10]^x[0.010]^y} = [4]^y$$

$$4^y = 4$$

$$y = 1$$

Comparing the third and first experiments in the same way shows that the rate doubles when the concentration of F_2 is doubled. The order in that reactant must also be one. The rate law is

$$\text{rate} = k[F_2][ClO_2]$$

(b) The value of k can be found using the data from any of the experiments. If we take the numbers from the second experiment we have

$$k = \frac{\text{rate}}{[F_2][ClO_2]} = \frac{4.8 \times 10^{-3}\ \text{M/s}}{(0.10\ \text{M})(0.040\ \text{M})} = 1.2\ /\text{M}\cdot\text{s}$$

Verify that the same value of k can be obtained from the other sets of data.

(c) Since we now know the rate law and the value of the rate constant, we can calculate the rate at any concentration of reactants

$$\text{rate} = k[F_2][ClO_2] = (1.2\ /\text{M}\cdot\text{s})(0.010\ \text{M})(0.020\ \text{M}) = 2.4 \times 10^{-4}\ \text{M/s}$$

13.23 Butadiene is the only reactant in the equation, and most probably the rate law is either first or second-order. Study the discussion of first- and second-order reactions in Section 13.2. Example 13.5 is a model for the graphical proof of the first-order character of a reaction. In this case what must be plotted against what? What will the graph look like if the reaction is first-order? What will the graph look like if the reaction is not first order? For second-order reactions what must be plotted against what?

Solution: Let us test for a first-order reaction. If the rate law for the conversion of butadiene to cyclobutane is first-order, a plot of ln[butadiene] versus time will be a straight line of slope -k (Equation (13.4)). The graph is shown below:

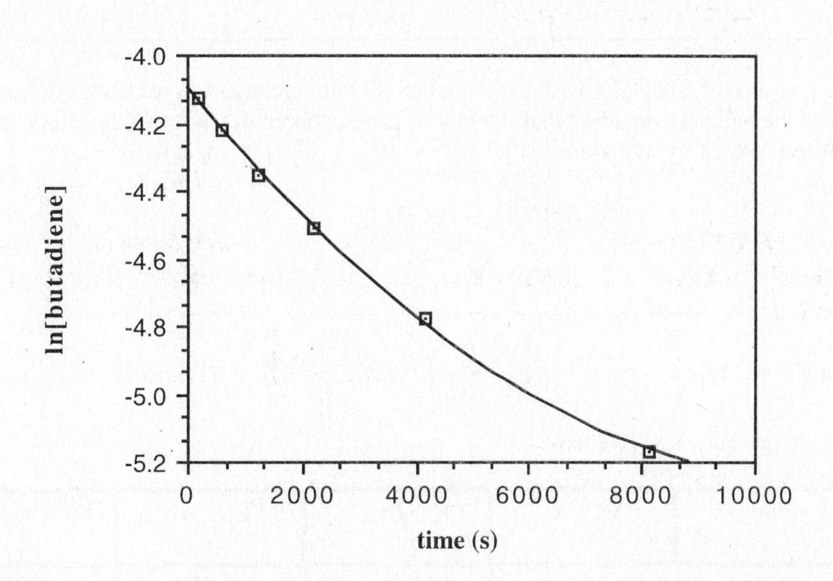

The data do not graph to a straight line, so the rate law cannot be first-order. If the rate law is second-order, a graph of 1/[butadiene] versus time will give a straight line with a slope of +k (Equation (13.6)). The graph is shown below:

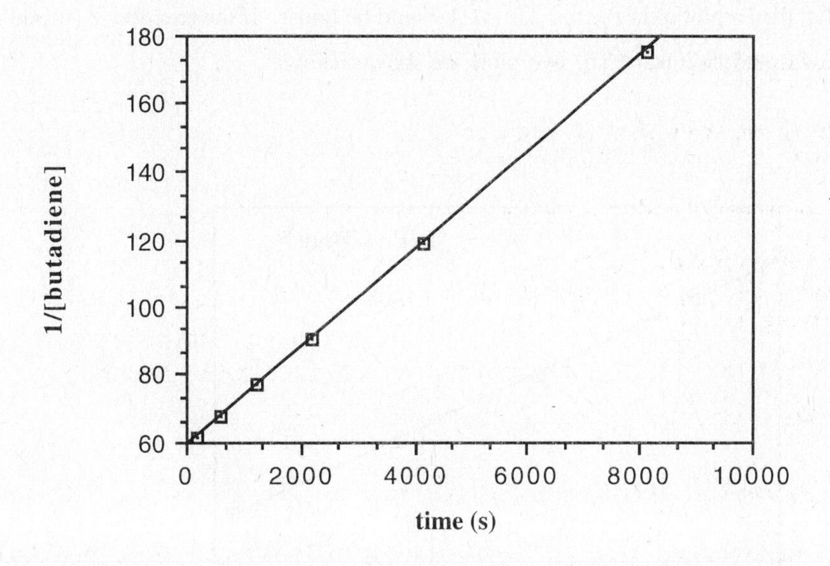

The data fit a straight-line plot, and the slope and rate constant are (taking the first and last points):

$$\text{slope} = \frac{(175 - 61.7)\text{M}^{-1}}{(8135 - 195)\text{s}^{-1}} = 0.0143\ /\text{M}\cdot\text{s} = k$$

13.24 This can be solved like Problem 13.23 above. Exammple 13.5 in the text is similar.

Solution: Let P_o be the pressure of $ClCO_2CCl_3$ at t = 0, and let x be the decrease in pressure after time t. Note that from the coefficients in the balanced equation that the loss of 1 atmosphere of $ClCO_2CCl_3$ results in the formation of two atmospheres of $COCl_2$. We write

$$ClCO_2CCl_3 \rightarrow 2COCl_2$$

Time	$[ClCO_2CCl_3]$	$[COCl_2]$
t = 0	P_o	0
t = t	P_o - x	2x

Thus the change (increase) in pressure is x. We have

t (s)	P (mmHg)	ΔP = x	$P_{ClCO_2CCl_3}$ = P - 2x	ln $P_{ClCO_2CCl_3}$	$1/P_{ClCO_2CCl_3}$
0	15.76	0.00	15.76	2.757	0.0635
181	18.88	3.12	12.64	2.537	0.0791
513	22.79	7.03	8.73	2.167	0.115
1164	27.08	11.32	4.44	1.491	0.225

If the reaction is first order, then a plot of ln $P_{ClCO_2CCl_3}$ vs. t would be linear. If the reaction is second order, a plot of $1/P_{ClCO_2CCl_3}$ vs., t would be linear. The two plots are shown below

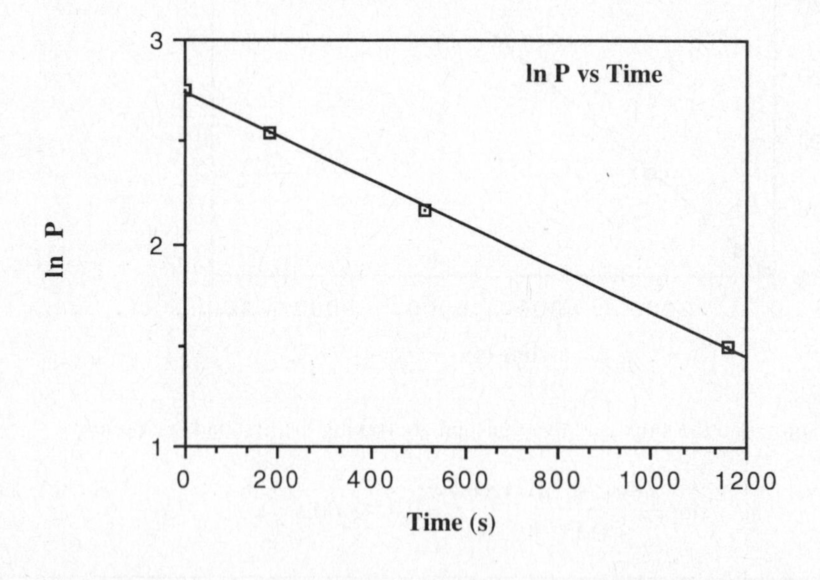

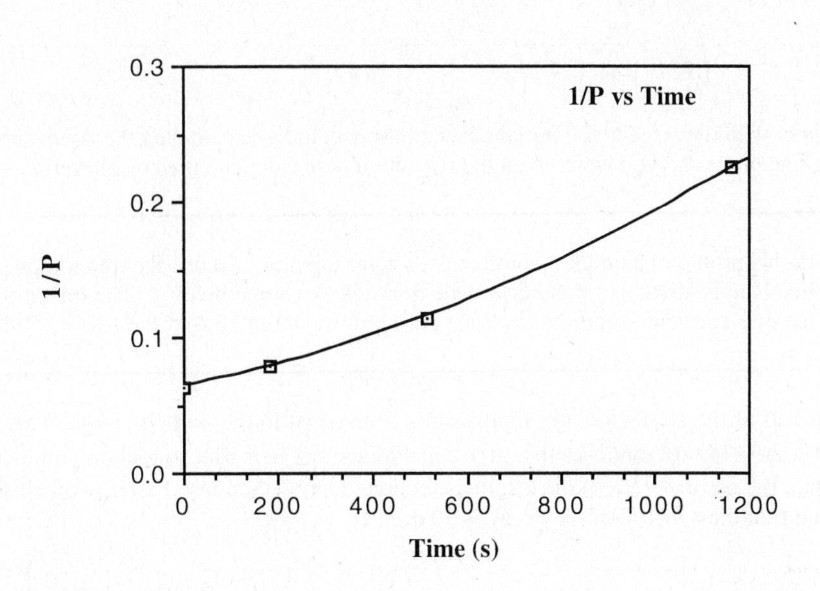

From the graphs we see that the reaction must be first order.

13.25 This is similar to Problem 13.23 above.

Solution: (a) Compare the second and fifth set of data. Doubling [X] at constant [Y] increases the rate by a factor of 0.508/0.127 or 4. Therefore, the reaction is second-order in X. Compare the second and fourth set of data. Doubling [Y] at constant [X] increases the rate by a factor of 1.016/0.127 or 8 times. Therefore, the reaction is third order in Y. The rate law is

$$\text{rate} = k[X]^2[Y]^3$$

(b)Using the first set of data to calculate k

$$k = \frac{\text{rate}}{[X]^2[Y]^3} = \frac{0.147\ M/s}{(0.10)^2(0.50)^3} = 1.2 \times 10^2\ /M^4 s$$

Finally,

$$\text{rate} = (1.2 \times 10^2\ /M^4 s)(0.20\ M)^2(0.30\ M)^3 = 0.13\ M/s$$

13.26 Study the discussion of second-order reactions in Section 13.2. This problem involves the application of Equation (13.6).

Solution: Since the reaction is known to be second-order, the relationship between reactant concentration and time is given by Equation (13.6). The problem supplies the rate constant and the initial (time = 0) concentration of NOBr. The concentration after 22 s can be found easily:

$$\frac{1}{[NOBr]} = \frac{1}{[NOBr]_o} + kt$$

$$= \frac{1}{0.086\ M} + (0.80\ /M\cdot s)(22\ s)$$

$$= 29\ M^{-1}$$

$$[NOBr] = 0.034\ M$$

If the reaction were first order with the same k and initial concentration, could you calculate the concentration after 22 s? If the reaction were first order and you were given the $t_{1/2}$, could you calculate the concentration after 22 s?

13.30 There are two ways to work this problem; a quick method and a longer method. To use the quick method, you must realize that the time involved is an integer number of half-lives (as explained below). The longer method involves first solving for the rate constant k and then solving for $t_{1/2}$. Example 13.4 in the text is similar.

Solution: We know that half of the substance decomposes in a time equal to the half-life $t_{1/2}$.. This leaves half of the compound. Half of what is left decomposes in a time equal to another half-life, so that only one quarter of the original compound remains. We see that 75% of the original compound has decomposed after two half-lives. Thus two half-lives equal one hour, or the half-life of the decay is 30 min.

$$100\%\text{ starting compound} \xrightarrow{t_{1/2}} 50\%\text{ starting compound} \xrightarrow{t_{1/2}} 25\%\text{ starting compound}$$

Using the longer method you could, using first order kinetics, solve for k using Equation (13.3), with $[A]_o = 100$ and $[A] = 25$,

$$\ln \frac{[A]_o}{[A]} = kt$$

$$\ln \frac{100}{25} = k(60\ \text{min})$$

$$k = \frac{\ln(4)}{60\ \text{min}} = 0.023\ \text{min}^{-1}$$

Then, substituting k into Equation (13.5), you arrive at the same answer for $t_{1/2}$.

$$t_{1/2} = \frac{0.693}{k} = \frac{0.693}{0.023\ \text{min}^{-1}} = 30\ \text{min}$$

13.31 This problem is similar to Example 13.4 and Problem 13.30 above.

Solution: (a) For any first order reaction the rate constant can be found from the half-life with Equation (13.5)

$$k = \frac{0.693}{t_{1/2}} = \frac{0.693}{35.0\ \text{s}} = 0.0198\ \text{s}^{-1}$$

(b) The elapsed time can be found with Equation (13.3). The value of [A] after 95% of the reactant has decomposed is $0.0500\ [A]_o$.

$$t = \frac{1}{k} \ln \frac{[A]_o}{[A]} = \frac{1}{0.0198\ \text{s}^{-1}} \ln \frac{[A]_o}{0.0500[A]_o} = \frac{\ln 20}{0.0198\ \text{s}^{-1}} = 151\ \text{s}$$

Could you calculate the time if this were a second order reaction with the same k? Would you need to know anything else? What?

13.32 What is the concentration of A after one half-life? Study the discussion of second order reactions in Section 13.2. This problem is similar in part to Example 13.6.

Solution: Equation (13.7) gives the relationship between the initial reactant concentration, rate constant, and half-life for a second order reaction.

$$t_{1/2} = \frac{1}{k[A]_o} = \frac{1}{(1.46\ M/s)(0.86\ M)} = 0.80\ s$$

Could this problem be solved if the initial concentration of A were unknown? Could it be solved without knowing the initial concentration of A if this were a first order reaction?

13.33 This problem is similar to Example 13.5.

Solution: We know from Equation (13.4) for first order reactions that the slope of the line (plot of ln[A] vs time) equals -k. Therefore

$$k = 6.18 \times 10^{-4}\ min^{-1}$$

The half-life can be found using Equation (13.5).

$$t_{1/2} = \frac{0.693}{k} = \frac{0.693}{6.18 \times 10^{-4}\ min^{-1}} = 1.12 \times 10^{3}\ min$$

13.41 Study Section 13.3 on the Arrhenius equation. Example 13.7 is similar to this problem. Try to work through the example and use it as a model to solve this problem.

Solution: Graphing Equation (13.9) requires plotting ln k versus 1/T. The graph is shown below.

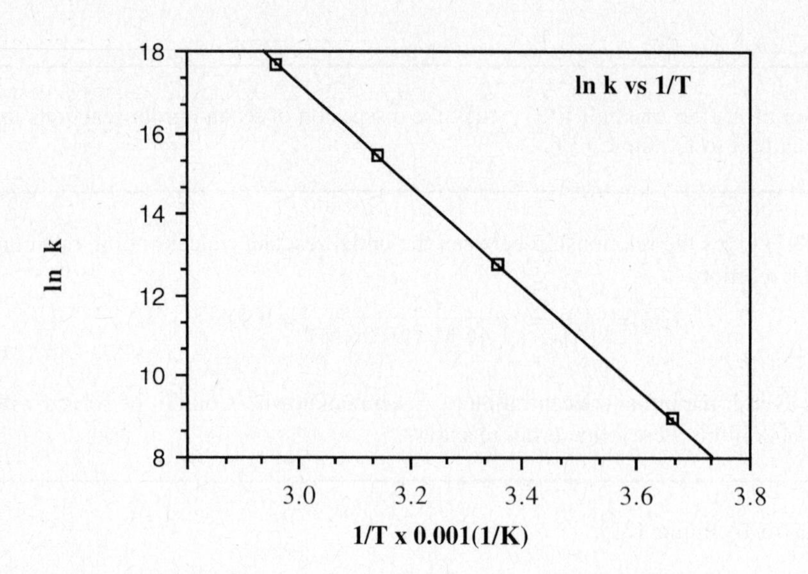

The slope of the line is -1.25×10^4 K, which is $-E_a/R$. The activation energy is

$$-E_a = \text{slope} \times R = (-1.25 \times 10^4 \text{ K}) \times (8.314 \text{ J/K·mol})$$

$$E_a = 104 \text{ kJ/mol}$$

Do you need to know the order of the reaction to find the activation energy? Is it possible to have a negative activation energy? What would a potential energy versus reaction coordinate diagram look like in such a case?

13.42 Another way of stating this problem is to say that the rate constant for the reaction at 250°C is 1.50×10^3 times larger than the rate constant for the same reaction at 150°C. In this problem assume that the activation energy is independent of temperature. You can solve the problem by using Equation (13.10).

Solution: In deriving Equation (13.10) the frequency factor A drops out if it is constant. The ratio of rate constants is just 1.50×10^3. Let T_1 = 250°C = 523 K and T_2 = 150°C = 423 K. We can then solve Equation (13.10) for E_a

$$\ln\frac{k_1}{k_2} = \ln(1.50 \times 10^3) = \frac{E_a}{R}\left(\frac{T_1 - T_2}{T_1T_2}\right)$$

$$\ln(1.50 \times 10^3) = \frac{E_a}{8.314 \text{ J/mol·K}}\left(\frac{523 \text{ K} - 423 \text{ K}}{(523 \text{ K})(423 \text{ K})}\right)$$

$$7.313 = \frac{E_a}{8.314 \text{ J/mol·K}}(4.52 \times 10^{-4} \text{K}^{-1})$$

$$E_a = 135{,}000 \text{ J/mol} = 135 \text{ kJ/mol}$$

13.43 This is a "plug-in" calculation using Equation (13.8). Watch out for units!

Solution: The appropriate value of R is 8.314 J/K mol, not 0.08206 L atm/mol K. You must also use the activation energy value of 63000 J/mol (why?). Once the temperature has been converted to K, the rate constant is

$$k = Ae^{-E_a/RT} = (8.7 \times 10^{12}\ s^{-1})\ e^{-63000/(8.314)(348)} = (8.7 \times 10^{12}\ s^{-1})(3.5 \times 10^{-10}) = 3.0 \times 10^{3}\ s^{-1}$$

Can you tell from the units of k what the order of the reaction is?

13.44 This problem involves the Arrhenius equation (Section 13.3). Find the value of A from the given data and then use that number in the Arrhenius equation to solve for T. Study the previous problem for help with units. Example 13.8 in the text is a similar problem.

Solution: First use Equation (13.8) to find the value of the frequency factor A with the given data

$$A = \frac{k}{e^{-E_a/RT}} = \frac{4.60 \times 10^{-4}\ s^{-1}}{e^{-104000/(8.314 \times 623)}} = 2.41 \times 10^{5}\ s^{-1}$$

The Arrhenius equation can be solved for T as shown

$$T = \frac{-E_a}{R \cdot \ln(k/A)}$$

At the given value of k, the temperature must be

$$T = \frac{-104000\ \text{J/mol}}{(8.314\ \text{J/mol·K}) \times \ln\left(\frac{8.80 \times 10^{-4}}{2.41 \times 10^{5}}\right)} = 644\ \text{K or } 371°\text{C}$$

13.45 This problem is similar to Problem 13.42 above.

Solution: Let k_1 be the rate constant at 295 K and $2k_1$ the rate constant at 305 K. We write

$$\ln\frac{k_1}{2k_1} = \frac{E_a}{R}\left(\frac{T_1 - T_2}{T_1T_2}\right)$$

$$-0.693 = \frac{E_a}{8.314\ \text{J/K·mol}}\left(\frac{295\ \text{K} - 305\ \text{K}}{(295\ \text{K})(305\ \text{K})}\right)$$

$$E_a = 51.9\ \text{kJ/mol}$$

13.55 Study Section 13.4 and try working Example 13.9. What is the relationship between the slow or rate-determining step in a mechanism and the appearance of the rate law?

Solution: (a) The order of the reaction is simply the sum of the exponents in the rate law (Section 13.2). The order of this reaction is 2.

(b) The rate law reveals the identity of the substances participating in the slow or rate-determining step of a reaction mechanism. This rate law implies that the slow step involves the reaction of a molecule of NO with a molecule of Cl_2. If this is the case, then the first reaction shown must be the rate-determining (slow) step, and the second reaction must be much faster.

13.56 The task of finding the rate law from the information given is like Problem 13.22 above. What is the order of the reaction with respect to Z? What does the rate law reveal about the rate-determining step (see Problem 13.56 above)?

Solution: (a) Using reasoning similar to that in Problem 13.22, we can see that the reaction is first order in X_2 and first order in Y (second order overall). The rate law is

$$\text{rate} = k[X_2][Y]$$

(b) If a change in the concentration of Z has no effect on the rate, the concentration of Z is not a term in the rate law. This implies that Z does not participate in the rate-determining step of the reaction mechanism.

(c) The rate law shows that the slow step involves reaction of a molecule of X_2 with a molecule of Y. Since Z is not present in the rate law, it does not take part in the slow step and must appear in a fast step at a later time. (If the fast step involving Z happened before the rate-determining step, the rate law would involve Z in a more complex way.) A mechanism that is consistent with the rate law could be

$$X_2 + Y \rightarrow XY + X \quad \text{(slow)}$$

$$X + Z \rightarrow XZ \quad \text{(fast)}$$

The rate law only tells us about the slow step. Other mechanisms with different subsequent fast steps are possible. Try to invent one. Ask your instructor for help.

13.57 Remember that the process shown is an elementary step, not the overall reaction. Assume it is also the rate-determining step. What does a rate law reveal about the rate-determining step? What is the molecularity of this step? Does the molecularity affect the order of the reaction? Study Section 13.4 to answer these questions before attempting to answer this problem. Part (b) is a simple "plug-in" computation with the rate law from part (a).

Solution: (a) In this rate-determining elementary step three molecules must collide simultaneously (one X and two Y's). This makes the reaction termolecular (see Section 13.4), and consequently the rate law must be third order: first order in X and second order in Y. The rate law is rate = $k[X][Y]^2$.

(b) The value of the rate constant can be found by solving algebraically for k

$$k = \frac{\text{rate}}{[X][Y]^2} = \frac{3.8 \times 10^{-3}\ \text{M/s}}{(0.26\ \text{M})(0.88\ \text{M})^2} = 1.9 \times 10^{-2}\ /\text{M}^2\cdot\text{s}$$

If the reaction shown in the problem were not the slow elementary step, could you write the rate law? Could you write the rate law if the reaction shown were the overall balanced equation and not an elementary step?

13.58 Study the discussion of the hydrogen-iodine reaction mechanism in Section 13.4. What is the rate-determining elementary step in the one-step mechanism? What is the rate-determining elementary step in the two-step mechanism? The ultraviolet light is sufficiently energetic to split iodine molecules into atoms. How would this affect the rate of the reaction (consider the effect of increasing the number of free iodine atoms on each rate law)? Be sure to read the last part of Section 13.4 on Experimental Proof of Reaction Mechanisms.

Solution: (a) In the two-step mechanism the rate-determining step is the collision of a hydrogen molecule with two iodine atoms. If visible light increases the concentration of iodine atoms, then the rate must increase. If the true rate-determining step were the collision of a hydrogen molecule with an iodine molecule (the one-step mechanism), then the visible light would have no effect (it might even slow the reaction by depleting the number of available iodine molecules).

(b) To split hydrogen molecules into atoms one needs ultraviolet light of much higher energy.

13.59 This is a difficult problem. The basic mechanism is similar to the two-step scheme in the hydrogen-iodine reaction (Section 13.4). Study this mechanism and the derivation of the rate law. Write the rate laws for all the steps and substitute as in the hydrogen-iodine mechanism.

Solution: As in the hydrogen-iodine case, a two-step mechanism is proposed. The first step involves forward and reverse reactions that are much faster than the second step. The rates of the reactions in the first step are given by

$$\text{forward rate} = k_1[O_3]$$

$$\text{reverse rate} = k_{-1}[O][O_2]$$

It is assumed that these two processes rapidly reach a state of dynamic equilibrium in which the rates of the forward and reverse reactions are equal:

$$k_1[O_3] = k_{-1}[O][O_2]$$

If we solve this equality for [O] we have

$$[O] = \frac{k_1[O_3]}{k_{-1}[O_2]}$$

The equation for the rate of the second step is

$$\text{rate} = k_2[O][O_3]$$

If we substitute the expression for [O] derived from the first step, we have the experimentally verified rate law.

$$\text{overall rate} = \frac{k_1 k_2}{k_{-1}}\frac{[O_3]^2}{[O_2]} = k\frac{[O_3]^2}{[O_2]}$$

The above rate law predicts that higher concentrations of O_2 will decrease the rate. This is because of the reverse reaction in the first step of the mechanism. Notice that if more O_2 molecules are present, they will serve to scavenge free O atoms and thus slow the disappearance of O_3.

13.60 To work this problem you must study the relationship between reaction mechanisms and rate laws in Section 13.4. Write the rate law implied by each of the three proposed mechanisms. You will need to study the discussion of the hydrogen-iodine reaction in order to write the rate law for Mechanism III.

Solution: The experimentally determined rate law is first order in H_2 and second order in NO. In Mechanism I the slow step is bimolecular and the rate law would be

$$\text{rate} = k[H_2][NO]$$

Mechanism I can be discarded.

The rate-determining step in Mechanism II involves the simultaneous collision of two NO molecules with one H_2 molecule. The rate law would be

$$\text{rate} = k[H_2][NO]^2$$

Mechanism II is a possibility.

In Mechanism III we assume the forward and reverse reactions in the first fast step are in dynamic equilibrium, so their rates are equal:

$$k_f[NO]^2 = k_r[N_2O_2]$$

The slow step is bimolecular and involves collision of a hydrogen molecule with a molecule of N_2O_2. The rate would be

$$\text{rate} = k_2[H_2][N_2O_2]$$

If we solve the dynamic equilibrium equation of the first step for $[N_2O_2]$ and substitute into the above equation, we have the rate law

$$\text{rate} = \frac{k_2 k_f}{k_r}[H_2][NO]^2 = k[H_2][NO]^2$$

Mechanism III is also a possibility. Can you suggest an experiment that might help to decide between the two mechanisms?

13.67 What might happen to the structure of an enzyme at high temperatures?

Solution: Higher temperatures may disrupt the intricate three dimensional structure of the enzyme, thereby reducing or totally destroying its catalytic activity.

13.69 All the reactions shown involve gases. Does any measurable property of the system as a whole change as the reaction proceeds?

Solution: In each case the gas pressure will either increase or decrease. The pressure can be related to the progress of the reaction through the balanced equation.

13.71 Compare the given statement with Problem 13.44.

> ***Solution:*** Strictly, the temperature must be given whenever the rate or rate constant of a reaction is quoted.

13.73 Study the discussion of the Experimental Proof of Reaction Mechanisms in Section 13.4. Draw the structures of all the reactants and products in this problem. Where does the oxygen come from in each product?

> ***Solution:*** Since the methanol contains no oxygen-18, the oxygen atom must come from the phosphate group and not the water. The mechanism must involve a bond-breaking process like
>
> ```
>              O
>            ≷ ||
> CH3 — O ≷ P — O — H
>            ≷ |
>              O — H
> ```

13.75 What properties of metals are important relative to their behavior as catalysts?

> ***Solution:*** Most transition metals have several stable oxidation states. This allows the metal atoms to act as either a source or a receptor of electrons in a broad range of reactions.

13.77 This is a "plug-in" type problem using the rate law.

> ***Solution:*** Since the reaction is first order in both A and B, then we can write the rate law expression
>
> $$\text{rate} = k[A][B]$$
>
> Substituting in the values for the rate and [A] and [B]
>
> $$4.1 \times 10^{-4}\ M/s = k(1.6 \times 10^{-2})(2.4 \times 10^{-3})$$
>
> $$k = 10.7\ M^{-1}s^{-1}$$
>
> Knowing that the overall reaction was second order, could you have predicted the units for k?

13.78 To work this problem, you must begin with a balanced equation.

> ***Solution:*** The balanced equation is: $2N_2O \rightarrow 2N_2 + O_2$
>
> For every one mole of N_2O that is lost, one mole of N_2 and 0.5 moles of O_2 will be formed. If we had 2 moles of N_2O at $t = 0$, after one half-life there will be one mole of N_2O remaining and one mole of N_2 and 0.5 moles of O_2. The total pressure will be
>
> $$\left(\frac{1 + 1 + 0.5}{2}\right) \times 2.10\ \text{atm} = 2.63\ \text{atm}$$

13.79 Review Problem 13.75 above. Why might Fe(III) behave as a catalyst?

Solution: Fe^{3+} undergoes a redox cycle. $Fe^{3+} \rightarrow Fe^{2+} \rightarrow Fe^{3+}$

$$2Fe^{3+} + 2I^- \rightarrow 2Fe^{2+} + I_2$$

$$2Fe^{2+} + S_2O_8^{2-} \rightarrow 2Fe^{3+} + 2SO_4^{2-}$$

$$2I^- + S_2O_8^{2-} \rightarrow I_2 + 2SO_4^{2-}$$

The uncatalyzed reaction is slow because both I^- and $S_2O_8^{2-}$ are negatively charged which makes their mutual approach unfavorable.

13.80 Use the rate constant k from the Chemistry in Action in Section 13.3.

Solution: For a first order reacion:
$$t = \frac{1}{k}\ln\frac{\text{decay rate at } t = 0}{\text{decay rate at } t = t}$$

$$t = \frac{1}{1.21 \times 10^{-4}\ \text{yr}}\ln\frac{0.260}{0.186} = 2.77 \times 10^3\ \text{yr}$$

13.81 Is it necessary that we know the concentration of the potassium permanganate in order to work this problem?

Solution: Since the volume of gas produced is directly proportional to the change in concention of hydrogen peroxide, we plot ln (volume) versus time.

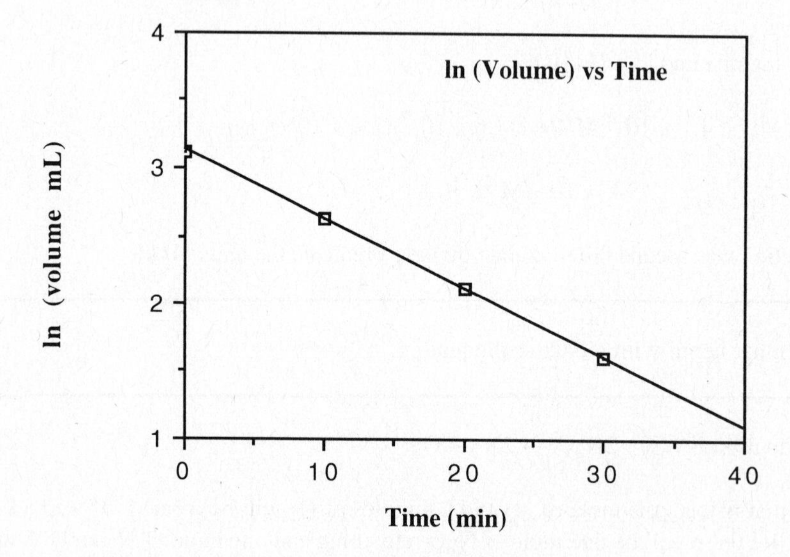

Since the plot is linear, the reaction is first order. For a first order reaction, the slope of the line equals -k. Taking any two points on the line

$$\text{slope} = -0.0504\ \text{min}^{-1} \quad \text{and} \quad k = 0.0504\ \text{min}^{-1}$$

Can you prove that it is not necessary to know the concentration of potassium permanganate? (Hint: assume some arbitrary concentration of potassium permanganate and repeat the problem, plotting $\ln[KMnO_4]$ versus time.)

13.83 What does *zero order* mean with respect to the exponents in the rate law?

Solution: For a rate law, *zero order* means that the exponent is zero (see the last part of Section 13.2). In other words, the reaction rate is just equal to a constant; it doesn't change as time passes. (a) The rate law would be

$$\text{rate} = k[A]^0 = k$$

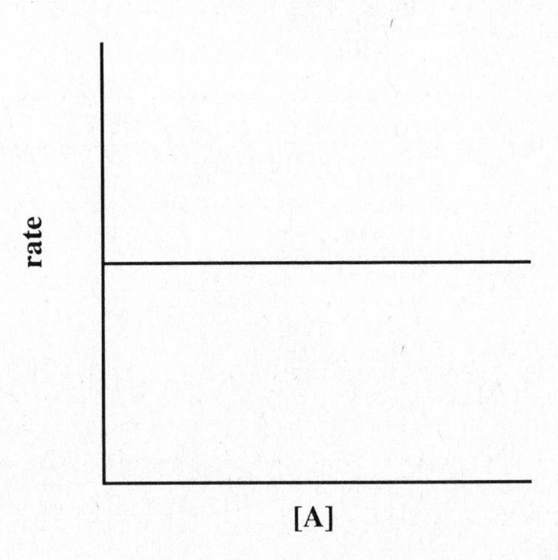

(b) The change in concentration of A per unit time is equal to the rate constant k.

$$-\frac{\Delta[A]}{\Delta t} = k \quad \Delta[A] = -\Delta tk$$

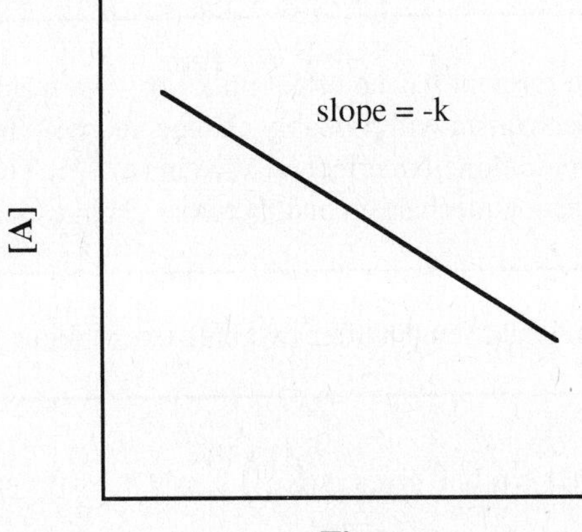

That is, a plot of [A] versus time t should be a straight line with a slope equal to -k.

13.85 This problem is a rather indirect application of the Arrhenius equation.

Solution: Writing the Arrhenius equation for both cases

$$k_1 = A_1 e^{-E_{a1}/RT}$$

$$k_2 = A_2 e^{-E_{a2}/RT}$$

Rearranging:

$$\frac{k_1 A_2}{k_2 A_1} = e^{(E_{a2} - E_{a1})/RT}$$

$$\ln \frac{k_1 A_2}{k_2 A_1} = \frac{E_{a2} - E_{a1}}{RT}$$

Since $k_1 = k_2$ at 300 K, we have:

$$\ln \frac{A_2}{A_1} = \frac{(6.13 \times 10^4) - (4.86 \times 10^4)}{(8.314)(300)} = 5.09$$

At a different T

$$\ln \frac{k_1 A_2}{k_2 A_1} = \frac{1.528 \times 10^3}{T}$$

If $k_1 = 2k_2$:

$$\ln \frac{2A_2}{A_1} = \frac{1.528 \times 10^3}{T}$$

$$\ln 2 + \ln \frac{A_2}{A_1} = \ln 2 + 5.09 = \frac{1.528 \times 10^3}{T}$$

$$T = \frac{1.528 \times 10^3}{\ln 2 + 5.09} = 265 \text{ K}$$

13.87 What can change the value of the rate constant k?

Solution: (a) Changing the concentration of a reactant has no effect on k. (b) If a reaction is run in a solvent other than in the gas phase, then the reaction mechanism will probably change and will thus change k. (c) Doubling the pressure simply changes the concentration. No effect on k, as in (a). (d) The rate constant k changes with temperature. (e) A catalyst changes the reaction mechanism and therefore changes k as in (b).

13.89 What is the definition of half-life? How much would remain after two half-lives? Four half-lives?

Solution: Mathematically, the amount left after ten half-lives is $\left(\frac{1}{2}\right)^{10} = 9.8 \times 10^{-4} = 0.098\%$

13.91 In order to answer this question you need to decide how the changes effect the concentration of reactants.

Solution: The net ionic equation is: $Zn(s) + 2H^+(aq) \rightarrow Zn^{2+}(aq) + H_2(g)$

(a) Changing from the same mass of granulated zinc to powdered zinc will increase the rate because the surface area of the zinc (and thus its concentration) has increased.

(b) Changing the mass of zinc (in the same granulated form) will decrease the rate because the concentration of zinc has decreased.

(c) The concentration of protons has decreased in changing from the strong acid (hydrochloric) to the weak acid (acetic); the rate will decrease.

(d) An increase in temperature will increase the rate of reaction.

13.92 This problem is similar to Example 13.2 in the text.

Solution: The overall rate law is of the general form: $\text{rate} = k[H_2]^x[NO]^y$

(a) Comparing Experiment #1 and Experiment #2, we see that the concentration of NO is constant and the concentration of H_2 has decreased by one-half. The initial rate has also decreased by one-half. Therefore the initial rate is directly proportional to the concentration of H_2; $x = 1$.

Comparing Experiment #1 and Experiment #3, we see that the concentration of H_2 is constant and the concentration of NO has decreased by one-half. The initial rate has decreased by one-fourth. Therefore the initial rate is proportional to the squared concentration of NO; $y = 2$.

The overall rate law is: $\text{rate} = k[H_2][NO]^2$

(b) Using Experiment #1 to calculate the rate constant, $\text{rate} = k[H_2][NO]^2$ or $k = \dfrac{\text{rate}}{[H_2][NO]^2}$

$$k = \frac{2.4 \times 10^{-6}\ M/s}{(0.010\ M)(0.025\ M)^2} = 0.38\ /M^2s$$

(c) Consulting the rate law, we assume that the slow step in the reaction mechanism will probably involve one H_2 molecule and two NO molecules. Additionally the hint tells us that O atoms are an intermediate.

$$H_2 + 2NO \rightarrow N_2 + H_2O + O \qquad \text{slow step}$$
$$O + H_2 \rightarrow H_2O \qquad \text{fast step}$$

$$2H_2 + 2NO \rightarrow N_2 + 2H_2O$$

13.93 N_2O_5 is the only reactant in the equation, and most probably the rate law is either first or second-order. Study the discussion of first- and second-order reactions in Section 13.2. In this case we must plot $[N_2O_5]$ versus initial rate. What will the graph look like if the reaction is first-order? What will the graph look like if the reaction is not first order?

Solution: First we plot the data for the reaction: $2N_2O_5 \rightarrow 4NO_2 + O_2$

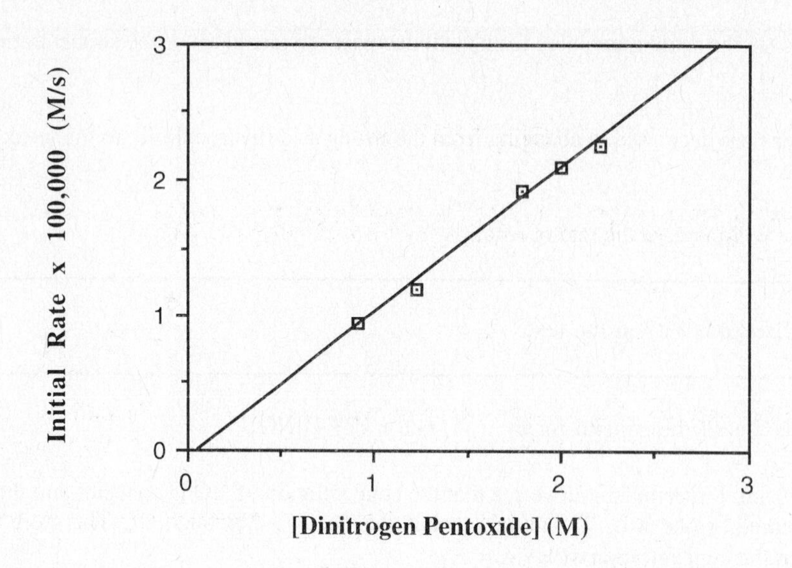

The data in linear, which means that the initial rate is directly proportional to the concentration of N_2O_5. Thus the rate law is:

$$\text{Rate} = k[N_2O_5]$$

The rate constant k can be determined from the slope of the graph $\left(\frac{\Delta(\text{Initial Rate})}{\Delta[N_2O_5]}\right)$ or by using any set of data.

$$k = 1.0 \times 10^{-5}\ s^{-1}$$

Note that the rate law is ***not*** Rate = $k[N_2O_5]^2$, as we might expect from the balanced equation. In general, the order of a reaction must be determined by experiment; it cannot be deduced from the coefficients in the balanced equation.

Note: If you have not already done so, please read the "Introduction for the Student" at the beginning of this Student Solutions Manual.

Chapter 14

CHEMICAL EQUILIBRIUM

14.7 This problem deals with writing equilibrium constant expressions. Study the material in Section 14.3 and Examples 14.1 and 14.5. Note that the problem involves both homogeneous and heterogeneous systems. (What is the difference?)

Solution: (a) $K_c = [H_2O]$ $\qquad K_p = P_{H_2O}$

(b) $K_c = \dfrac{[H_2][CO_2]}{[H_2O][CO]}$ $\qquad K_p = \dfrac{P_{H_2}P_{CO_2}}{P_{H_2O}P_{CO}}$

(c) $K_c = \dfrac{1}{[O_2]}$ $\qquad K_p = \dfrac{1}{P_{O_2}}$

(d) $K_c = \dfrac{[PCl_3][Cl_2]}{[PCl_5]}$ $\qquad K_p = \dfrac{P_{PCl_3}P_{Cl_2}}{P_{PCl_5}}$

If you missed any of these, you must go back and thoroughly learn the procedures for writing K_p's and K_c's for both homogeneous and heterogeneous systems.

14.12 What is the relationship between the equilibrium constant written for a reaction and the equilibrium constant for the reverse of that reaction. Review Section 14.3 if you can't answer the question.

Solution: When the equation for a reversible reaction is written in the opposite direction, the equilibrium constant becomes the reciprocal of the original equilibrium constant.

$$K' = \frac{1}{K} = \frac{1}{4.17 \times 10^{-34}} = 2.40 \times 10^{33}$$

14.13 This is similar to Example 14.2. Study Section 14.3 and this example. You must convert the given data to molarities.

Solution: The molarities can be found by simply dividing by the volume of the flask.

$$[H_2] = \frac{2.50 \text{ mol}}{12.0 \text{ L}} = 0.208 \text{ M}$$

$$[S_2] = \frac{1.35 \times 10^{-5} \text{ mol}}{12.0 \text{ L}} = 1.12 \times 10^{-6} \text{ M}$$

$$[H_2S] = \frac{8.70 \text{ mol}}{12.0 \text{ L}} = 0.725 \text{ M}$$

Once the molarities are known, K_c can be found by simple substitution into the equilibrium constant expression.

$$K_c = \frac{[H_2S]^2}{[H_2]^2[S_2]} = \frac{(0.725)^2}{(0.208)^2(1.12 \times 10^{-6})} = 1.08 \times 10^7$$

If you forget to convert moles to moles/liter, will you get a different answer? Under what circumstances will the two answers be the same?

14.14 This problem is a direct conversion from K_p to K_c like Example 14.4.

Solution: Using Equation (14.5): $K_p = K_c(0.0821\ T)^{\Delta n}$ where $\Delta n = 2 - 3 = -1$ and $T = (1273 + 273)$ K $= 1546$ K

$$K_p = (2.24 \times 10^{22})(0.0821 \times 1546)^{-1} = 1.76 \times 10^{20}$$

14.15 Write out the equilibrium constant expression for each equation and substitute the given concentrations. See Example 14.7 if you are unsure of the appropriate expressions.

Solution: The equilibrium constant expressions are

(a) $$K_c = \frac{[NH_3]^2}{[N_2][H_2]^3}$$

(b) $$K_c = \frac{[NH_3]}{[N_2]^{1/2}[H_2]^{3/2}}$$

Substituting the given equilibrium concentrations gives

(a) $$K_c = \frac{(0.25)^2}{(0.11)(1.91)^3} = 0.082$$

(b) $$K_c = \frac{(0.25)}{(0.11)^{1/2}(1.91)^{3/2}} = 0.29$$

According to the material at the last of Section 14.3, the equilibrium constant that you caluclated for part (a) above 0.082) should equal the square of the equilibrium constant for part (b). Does it (taking into account round-off errors)? Refer to part (d) in Example 14.7 in the text if you need additional help.

14.16 This is similar to Problem 14.13 and to Example 14.3.

Solution: We can write the equilibrium constant expression from the balanced equation and substitute in the pressures.

$$K_p = \frac{P_{NO}^2}{P_{N_2} P_{O_2}} = \frac{(0.050)^2}{(0.15)(0.33)} = 0.051$$

Do we need to know the temperature?

14.17 This problem has two parts. The first involves finding the relationship between the two K_c expressions. Write the equilibrium constant expression for the two equations. How are they related? For help on this type of question, study Example 14.7. The second part of this problem is like Example 14.4.

Solution: The equilibrium constant expressions for the two forms of the equation are

$$K_c = \frac{[I]^2}{[I_2]} \quad \text{and} \quad K_c' = \frac{[I_2]}{[I]^2}$$

The relationship between the two equilibrium constants is

$$K_c' = \frac{1}{K_c} = \frac{1}{3.8 \times 10^{-5}} = 2.6 \times 10^4$$

K_p can be found as shown below

$$K_p = K_c'(0.0821T)^{\Delta n} = (2.6 \times 10^4)(0.0821 \times 1000)^{-1} = 3.2 \times 10^2$$

14.18 Review the subsection Heterogeneous Equilibria in Section 14.3. This problem is exactly analogous to Example 14.6.

Solution: Using Equation (14.8): $K_p = P_{CO_2} = 0.105$

From Equation (14.5), we know: $K_c = \dfrac{K_p}{(0.0821\ T)^{\Delta n}}$ where $\Delta n = 1$, and $T = (350 + 273)$ K = 623 K

$$K_c = \frac{0.105}{(0.0821 \times 623)} = 2.05 \times 10^{-3}$$

14.19 To solve this problem, you must find the concentrations of the reactants and products at equilibrium. You are given the amount (not the concentration) of NOCl before the reaction starts. The initial amounts of the products are zero. How does the balanced equation relate changes in reactant concentration to changes in product concentration? Before attempting this problem, study Section 14.5 and Examples 14.9 and 14.10.

Solution: First we find the initial concentration of NOCl

$$[NOCl]_o = \frac{2.50 \text{ mol}}{1.50 \text{ L}} = 1.67 \text{ M}$$

If 28.0% of the NOCl had dissociated at equilibrium, the concentrations of all species at equilibrium must be

$$[NOCl] = (1 - 0.280)(1.67 \text{ M}) = 1.20 \text{ M}$$

$$[NO] = (0.280)(1.67 \text{ M}) = 0.468 \text{ M}$$

$$[Cl_2] = \frac{1}{2}(0.280)(1.67 \text{ M}) = 0.234 \text{ M}$$

The equilibrium constant K_c is found by simple substitution

$$K_c = \frac{[NO]^2[Cl_2]}{[NOCl]^2} = \frac{(0.468)^2(0.234)}{(1.20)^2} = 0.0356 \text{ or } 3.56 \times 10^{-2}$$

14.20 This is similar to Example 14.8. Study that example and then evaluate the reaction quotient Q_p for the species in this problem. Which way will the reaction shift?

Solution: We substitute the given pressures into the reaction quotient expression

$$Q_p = \frac{P_{PCl_3} P_{Cl_2}}{P_{PCl_5}} = \frac{(0.223)(0.111)}{(0.177)} = 0.140$$

The calculated value of Q_p is less than K_p for this system. The system will change in a way to increase Q_p until it is equal to K_p. To achieve this the pressures of PCl_3 and Cl_2 must increase, and the pressure of PCl_5 must decrease.

Could you actually determine the final pressure of each gas?

14.22 This similar to Problems 14.15 and 14.17 and Example 14.7.

Solution: Given the following:

$$K_c = \frac{[NH_3]^2}{[N_2][H_2]^3} = 0.65$$

(a) From Equation (14.5):

$$K_p = K_c(0.0821 \text{ T})^{\Delta n}$$

where $\Delta n = 2 - 4 = -2$
and $T = (200 + 273) \text{ K} = 473 \text{ K}$

$$K_p = (0.65)(0.0821 \times 473)^{-2} = 4.3 \times 10^{-4}$$

(b)Recalling that:

$$K_{forward} = \frac{1}{K_{reverse}}$$

Therefore:

$$K_c' = \frac{1}{0.65} = 1.5$$

(c) Since the equation $\frac{1}{2}N_2(g) + \frac{3}{2}H_2(g) \rightleftharpoons NH_3(g)$ is equal to $\frac{1}{2}[N_2(g) + 3H_2(g) \rightleftharpoons 2NH_3(g)]$,

then, K_c' for the reaction: $\frac{1}{2}N_2(g) + \frac{3}{2}H_2(g) \rightleftharpoons NH_3(g)$

equals $K_c^{1/2}$ for the reaction: $N_2(g) + 3H_2(g) \rightleftharpoons 2NH_3(g)$

Thus $K_c' = K_c^{1/2}$

$$K_c' = \sqrt{0.65} = 0.81$$

(d) For K_p in part (b): $K_p = (1.5)(0.0821 \times 473)^{+2} = 2.3 \times 10^3$

and for K_p in part (c): $K_p = (0.81)(0.0821 \times 473)^{-1} = 0.021$

14.24 This similar to Problem 14.19 above and 14.25 below.

Solution: Of the original 1.05 moles of Br_2, 1.20% has dissociated and 98.8% is undissociated at equilibrium. Calculating the molar concentrations in the 0.980 L flask at equilibrium

$$[Br_2] = (0.988 \times 1.05/0.980) = 1.06\ M$$

$$[Br] = (0.012 \times 2 \times 1.05/0.980) = 0.0257\ M$$

$$K_c = \frac{[Br]^2}{[Br_2]} = \frac{0.0257^2}{1.06} = 6.2 \times 10^{-4}$$

14.25 This problem is similar in part to Problem 14.19. What are the partial pressures of the reactants and products at equilibrium? You must express all of them in terms of a common unknown, the initial pressure of NOBr.

Solution: Let x be the initial pressure of NOBr. Using the balanced equation, we can write expressions for the partial pressures at equilibrium

$$P_{NOBr} = (1 - 0.34)x = 0.66x$$

$$P_{NO} = 0.34x$$

$$P_{Br_2} = 0.17x$$

The sum of these is the total pressure

$$0.66x + 0.34x + 0.17x = 1.17x = 0.25\ atm\ ; \qquad x = 0.21\ atm$$

The equilibrium pressures are then

$$P_{NOBr} = 0.66(0.21) = 0.14\ atm$$

$$P_{NO} = 0.34(0.21) = 0.071 \text{ atm}$$

$$P_{Br_2} = 0.17(0.21) = 0.036 \text{ atm}$$

We find K_p by substitution

$$K_p = \frac{(P_{NO})^2 P_{Br_2}}{(P_{NOBr})^2} = \frac{(0.071)^2(0.036)}{(0.14)^2} = 9.3 \times 10^{-3}$$

The relationship between K_p and K_c is given by

$$K_p = K_c(RT)^{\Delta n}$$

We find K_c (for this system $\Delta n = +1$)

$$K_c = \frac{K_p}{(RT)^{\Delta n}} = \frac{K_p}{RT} = \frac{9.3 \times 10^{-3}}{0.0821 \times 298} = 3.8 \times 10^{-4}$$

14.26 This is similar in part to Problem 14.25 above. What is the initial pressure of phosgene? What is the phosgene pressure at equilibrium? What is the chlorine pressure at equilibrium?

Solution: If the CO pressure at equilibrium is 0.497 atm, the balanced equation requires the chlorine pressure to have the same value. The initial pressure of phosgene gas can be found from the ideal gas equation

$$P = \frac{nRT}{V} = \frac{(3.00 \times 10^{-2})(0.0821)(800)}{(1.50)} = 1.31 \text{ atm}$$

The phosgene pressure at equilibrium must be

$$P_{COCl_2} = (1.31 - 0.497) \text{ atm} = 0.81 \text{ atm}$$

The value of K_p is then found by substitution

$$K_p = \frac{P_{COCl_2}}{P_{CO}P_{Cl_2}} = \frac{0.81}{(0.497)^2} = 3.3$$

14.28 Study the subsection on Multiple Equilibria in Section 14.3. How is the last equation in this problem related to the first two? How is the equilibrium constant for the last equation related to the K_c's for the first two?

Solution: The target equation is the sum of the first two:

$$H_2S \rightleftharpoons H^+ + HS^-$$

$$HS^- \rightleftharpoons H^+ + S^{2-}$$

$$H_2S \rightleftharpoons 2H^+ + S^{2-}$$

Since this is the case, the equilibrium constant for the combined reaction is the product of the constants for the component reactions (Section 14.3). The equilibrium constant is therefore

$$K_c = K_c'K_c'' = 9.5 \times 10^{-27}$$

What happens in the special case when the two component reactions are the same? Can you generalize this relationship to adding more than two reactions? What happens if one takes the difference between two reactions?

14.29 This problem is similar in part to Problem 14.28 above. How is the final reaction related to the two given reactions? How is the equilibrium constant of a reaction that is the sum of two reactions related to the equilibrium constants of those reactions? What happens to the equilibrium constant of a reaction when the balanced equation is multiplied through by a constant?

Solution: The target equation is the sum of the two given equations each multiplied through by 1/2.

$$\frac{1}{2}(2SO_2 + O_2 \rightleftharpoons 2SO_3) \qquad K_1$$

$$\frac{1}{2}(2NO_2 \rightleftharpoons 2NO + O_2) \qquad K_2$$

$$SO_2 + NO_2 \rightleftharpoons SO_3 + NO \qquad K_c$$

where K_1 and K_2 are the equilibrium constants for the reactions in parentheses. According to the rules for expressing equilibrium constants in multiple equilibria situations, the equilibrium constant must be

$$K_c = (K_1)^{1/2}(K_2)^{1/2}$$

14.30 This is like Problem 14.29 above.

Solution: Given:

$$K_p' = \frac{P_{CO}^2}{P_{CO_2}} = 1.3 \times 10^{14} \qquad K_p'' = \frac{P_{COCl_2}}{P_{CO}P_{Cl_2}} = 6.0 \times 10^{-3}$$

For the overall reaction

$$K_p = \frac{P_{COCl_2}^2}{P_{CO_2}P_{Cl_2}^2} = K_p'(K_p'')^2 = (1.3 \times 10^{14})(6.0 \times 10^{-3})^2 = 4.7 \times 10^9$$

14.33 Study Section 14.4. What is the relationship between the rate constants k_f and k_r for a reaction and the equilibrium constant K_c? What is the equilibrium constant expression for the reaction in this problem?

Solution: (a) Assuming the self-ionization of water occurs by a single elementary-step mechanism (Section 13.4), the equilibrium constant is just the ratio of the forward and reverse rate constants

$$K = \frac{k_f}{k_r} = \frac{k_1}{k_{-1}} = \frac{2.4 \times 10^{-5}}{1.3 \times 10^{11}} = 1.8 \times 10^{-16}$$

(b) The product can be written

$$[H^+][OH^-] = K[H_2O]$$

What is $[H_2O]$? It is the concentration of pure water. One liter of water has a mass of 1000 g (density = 1.00 g/mL). The number of moles of H_2O is

$$1000 \text{ g} \left(\frac{1 \text{ mol}}{18.0 \text{ g}}\right) = 55.5 \text{ mol}$$

The concentration of water is 55.5 mol/1.00 L or 55.5 M. The product is

$$[H^+][OH^-] = (1.8 \times 10^{-16})(55.5) = 1.0 \times 10^{-14}$$

We assume the concentrations of hydrogen ion and hydroxide ion are equal

$$[H^+] = [OH^-] = (1.0 \times 10^{-14})^{1/2} = 1.0 \times 10^{-7} \text{ M}$$

14.34 Study Section 14.4 to find the relationship between the equilibrium constant of a reaction and the rate constants of the forward and reverse reactions. What is the order of the forward reaction? What are the units of the rate constant for this type of reaction?

Solution: At equilibrium the value of K_c is equal to the ratio of the forward rate constant to the rate constant for the reverse reaction.

$$K_c = \frac{k_f}{k_r} = \frac{k_f}{5.1 \times 10^{-2}} = 12.6$$

$$k_f = (12.6)(5.1 \times 10^{-2}) = 0.64$$

The forward reaction is third order, so the units of k_f must be

$$\text{rate} = k_f(A)^2(B)$$

$$k_f = \frac{\text{rate}}{(\text{concentration})^3} = \frac{M/s}{M^3} = 1/M^2 \cdot s$$

$$k_f = 0.64/M^2 \cdot s$$

14.37 This is similar to Example 14.8. If the reactants were completely converted to sulfur trioxide, would the pressure increase or decrease?

Solution: Given:

$$K_p = \frac{P^2_{SO_3}}{P^2_{SO_2}P_{O_2}} = 5.60 \times 10^4$$

Initially, the total pressure is (0.350 + 0.762) atm or 1.112 atm. As the reaction progresses from left to right toward equilibrium there will be a decrease in the number of moles of molecules present. (Note that 2 moles pf SO_2 react with 1 mole of O_2 to produce 2 moles of SO_3, or, at constant pressure, three atmolspheres of reactants forms two atmospheres of products.) Since pressure is directly proportional to the number of molecules present, at equilibrium the total pressure will be less than 1.112 atm.

14.38 This problem is like Example 14.8.

Solution: The reaction is

$$N_2(g) + 3H_2(g) \rightleftharpoons 2NH_3(g)$$

$$K_c = \frac{[NH_3]^2}{[N_2][H_2]^3} = 0.65$$

$$Q_c = \frac{(0.48)^2}{(0.60)(0.76)^3} = 0.87$$

To reach equilibrium, the net reaction must proceed from right to left. Therefore, $[NH_3]$ will decrease and $[N_2]$ and $[H_2]$ will increase at equilibrium.

Is it possible to calculate the equilibrium concentrations from the given information?

14.39 Use Example 14.9 as a model to work out the solution to this problem. Note that you must first calculate the initial concentrations.

Solution: The balanced equation shows that one mole of carbon monoxide will combine with one mole of water to form hydrogen and carbon dioxide. Following the method of Example 14.9, let x be the depletion in the concentration of either CO or H_2O at equilibrium (why can x serve to represent either quantity?). The equilibrium concentration of hydrogen must then be equal to x (why not 2x like the example?). The changes are summarized as shown in the table

	H_2 +	CO_2	$\rightleftharpoons$	H_2O +	CO
Initial (M):	0	0		0.0300	0.0300
Change (M):	+x	+x		-x	-x
Equilibrium (M):	x	x		(0.0300 - x)	(0.0300 - x)

The equilibrium constant is

$$K_c = \frac{[H_2O][CO]}{[H_2][CO_2]} = 0.534$$

Substituting

$$\frac{(0.0300 - x)^2}{x^2} = 0.534$$

Taking the square root of both sides, we obtain

$$\frac{(0.0300 - x)}{x} = \sqrt{0.534} = 0.731$$

$$x = 0.0173 \text{ M}$$

The number of moles of H_2 formed is 0.0173 mol/L × 10.0 L = 0.173 mol.

14.40 In this problem you are to assume that the NO_2 is the only source of the oxygen gas product. If the partial pressure of oxygen at equilibrium is 0.25 atm, what is the corresponding partial pressure of NO?

Solution: If the equilibrium partial pressure of oxygen is 0.25 atm, the pressure of NO must be 0.50 atm (why?). We can find the equilibrium NO_2 pressure from the equilibrium constant expression

$$P_{NO_2} = \sqrt{\frac{(P_{NO})^2 P_{O_2}}{K_p}} = \sqrt{\frac{(0.50)^2(0.25)}{158}} = 0.020 \text{ atm}$$

14.41 Use Example 14.9 as a model to set up the solution to this problem. Note that you must first calculate the starting concentration of HBr.

Solution: Notice that the balanced equation requires that for every two moles of HBr consumed, one mole of H_2 and one mole of Br_2 must be formed. Let 2x be the depletion in the concentration of HBr at equilibrium. The equilibrium concentrations of H_2 and Br_2 must therefore each be x. (Why?) The changes are shown in the table

	H_2	+ Br_2	$\rightleftharpoons$ 2HBr
Initial (M):	0	0	0.267
Change (M):	+x	+x	-2x
Equilibrium (M):	x	x	(0.267 - 2x)

The equilibrium constant relationship is given by

$$K_c = \frac{[HBr]^2}{[H_2][Br_2\}}$$

Substitution of the equilibrium concentration expressions gives

$$K_c = \frac{(0.267 - 2x)^2}{x^2} = 2.18 \times 10^6$$

Taking the square root of both sides we obtain

$$\frac{0.267 - 2x}{x} = 1.48 \times 10^3$$

$$x = 1.80 \times 10^{-4}$$

The equilibrium concentrations are

$$[H_2] = [Br_2] = 1.80 \times 10^{-4} \text{ M}$$
$$[HBr] = 0.267 - 2(1.80 \times 10^{-4}) = 0.267 \text{ M}$$

If the depletion in the concentration of HBr at equilibrium were defined as x, rather than 2x, what would be the appropriate expressions for the equilibrium concentrations of H_2 and Br_2? Should the final answers be different in this case?

14.42 This like Problem 14.41 above.

Solution: The initial concentration of I_2 is 0.0456 mol/2.30 L or 0.0198 mol/L. Let x be the amount (in mol/L) of I_2 dissociated. We construct the table:

	$I_2(g)$	$\rightleftharpoons$	$2I(g)$
Initial (M):	0.0198		0.000
Change (M):	-x		+2x
Equilibrium (M):	(0.0198 - x)		2x

$$K_c = \frac{[I]^2}{[I_2]} = \frac{(2x)^2}{(0.0198 - x)} = 3.8 \times 10^{-5}$$

$$4x^2 + (3.8 \times 10^{-5})x - (7.52 \times 10^{-7}) = 0 \qquad x = 4.29 \times 10^{-4}\ M$$

At equilibrium we have

$$[I] = 2 \times 4.29 \times 10^{-4}\ M = 8.58 \times 10^{-4}\ M$$

$$[I_2] = (0.0198 - 4.29 \times 10^{-4})\ M = 0.0194\ M$$

14.44 Part (a) of this problem is like Problem 14.15 above. In part (b) you are to assume that more CO_2 has been added to the system, and that before anything changes, the CO_2 concentration becomes 0.50 mol/L. What will happen to the concentrations of the other reactants and products when the system moves to a new equilibrium? Study Problem 14.37 above and Examples 14.8 and 14.11.

Solution: (a) The equilibrium constant can be found by simple substitution

$$K_c = \frac{[H_2O][CO]}{[CO_2][H_2]} = \frac{(0.040)(0.050)}{(0.086)(0.045)} = 0.52$$

(b) The magnitude of the reaction quotient Q_c for the system after the concentration of CO_2 becomes 0.50 mol/L is

$$Q_c = \frac{(0.040)(0.050)}{(0.50)(0.045)} = 0.089$$

The value of Q_c is smaller than K_c, therefore the concentrations of CO and H_2O will increase and the concentrations of CO_2 and H_2 will decrease (why?). Let x be the depletion in the concentration of CO_2 at equilibrium. The stoichiometry of the balanced equation then requires that the decrease in the concentration of H_2 must also be x, and that the concentration increases of CO and H_2O be equal to x as well. The changes in the original concentrations are shown in the table.

	CO_2	+	H_2	$\rightleftharpoons$	CO	+	H_2O
Initial (M):	0.50		0.045		0.050		0.040
Change (M):	-x		-x		+x		+x
Equilibrium (M):	(0.50 - x)		(0.045 - x)		(0.050 + x)		(0.040 + x)

The equilibrium constant expression is:

$$K_c = \frac{[H_2O][CO]}{[CO_2][H_2]} = \frac{(0.040 + x)(0.050 + x)}{(0.50 - x)(0.045 - x)} = 0.52$$

As in Example 14.10, the expression must be reduced to a quadratic equation.

$$0.48x^2 + 0.373x - 9.7 \times 10^{-3} = 0$$

The positive root of the equation is x = 0.025. The equilibrium concentrations are

$$[CO_2] = (0.50 - 0.025)\ M = 0.48\ M$$
$$[H_2] = (0.045 - 0.025)\ M = 0.020\ M$$
$$[CO] = (0.050 + 0.025)\ M = 0.075\ M$$
$$[H_2O] = (0.040 + 0.025)\ M = 0.065\ M$$

14.45 Write the equilibrium constant expression for this system, then express the total pressure in terms of the partial pressures of CO and CO_2. Can you make an appropriate substitution and solve the problem?

Solution: The equilibrium constant expression for the system is

$$K_p = \frac{(P_{CO})^2}{P_{CO_2}}$$

The total pressure can be expressed as

$$P_{total} = P_{CO_2} + P_{CO}$$

If we let the partial pressure of CO be x, then the partial pressure of CO_2 is

$$P_{CO_2} = P_{total} - x = (4.50 - x)\ \text{atm}$$

Substitution gives the equation

$$K_p = \frac{(P_{CO})^2}{P_{CO_2}} = \frac{x^2}{(4.50 - x)} = 1.52$$

This can be rearranged to the quadratic:

$$x^2 + 1.52x - 6.84 = 0.$$

The solutions are x = 1.96 and x = -3.48; only the positive result has physical significance (why?). The equilibrium pressures are

$$P_{CO} = x = 1.96\ \text{atm}$$
$$P_{CO_2} = (4.50 - 1.96) = 2.54\ \text{atm}$$

14.47 This is similar to Example 14.10 and Problem 14.41. Evaluate the reaction quotient Q_c to see which way the system will shift to attain equilibrium. Once you know the direction of the reaction, define the changes in reactant and product concentrations in terms of the common factor x. Note that you must compute the concentrations.

Solution: We first must find the initial concentrations of all the species in the system.

$$[H_2]_o = \frac{0.714 \text{ mol}}{2.40 \text{ L}} = 0.298 \text{ M}$$

$$[I_2]_o = \frac{0.984 \text{ mol}}{2.40 \text{ L}} = 0.410 \text{ M}$$

$$[HI]_o = \frac{0.886 \text{ mol}}{2.40 \text{ L}} = 0.369 \text{ M}$$

The reaction quotient is then found by substitution

$$Q_c = \frac{[HI]_o^2}{[H_2]_o[I_2]_o} = \frac{(0.369)^2}{(0.298)(0.410)} = 1.11$$

We find that Q_c is less than K_c, which implies that the concentrations of H_2 and I_2 will decrease and that the concentration of HI will increase.

Let 2x (why times 2?) be the increase in the concentration of HI. We set up the usual table

	H_2 +	I_2 ⇌	2HI
Initial (M):	0.298	0.410	0.369
Change (M):	-x	-x	+2x
Equilibrium (M):	(0.298 - x)	(0.410 - x)	(0.369 + 2x)

The equilibrium constant expression is:

$$K_c = \frac{[HI]^2}{[H_2][I_2]} = \frac{(0.369 + 2x)^2}{(0.298 - x)(0.410 - x)} = 54.3$$

This becomes the quadratic equation $50.3x^2 - 39.9x + 6.48 = 0$

The smaller root is x = 0.228 M (The larger root is physically impossible). At equilibrium

$$[H_2] = (0.298 - 0.228) \text{ M} = 0.070 \text{ M}$$

$$[I_2] = (0.410 - 0.228) \text{ M} = 0.182 \text{ M}$$

$$[HI] = 0.369 + 2(0.228) \text{ M} = 0.825 \text{ M}$$

14.53 Make sure that you have read and understand Section 14.6 and understand Le Chatelier's Principle.

Solution: (a) Addition of more $Cl_2(g)$ (a reactant) would shift the position of equilibrium to the right.

(b) Removal of $SO_2Cl_2(g)$ (a product) would shift the position of equilibrium to the right.

(c) Removal of $SO_2(g)$ (a reactant) would shift the position of equilibrium to the left.

14.55 Study Section 14.6 to find the relationship between the sign of the enthalpy change for a reaction and the change in magnitude of the equilibrium constant in response to a change in temperature.

Solution: (a) This reaction is endothermic. (Why?) According to Section 14.6, an increase in temperature favors an endothermic reaction, so the equilibrium constant should become larger.

(b) This reaction is exothermic. Such reactions are favored by decreases in temperature. The magnitude of K_c should decrease.

(c) In this system heat is neither absorbed nor released. A change in temperature should have no effect on the magnitude of the equilibrium constant.

14.56 Study Section 14.6 to learn the effects of pressure changes on the equilibrium concentrations of reactants and products. Under what circumstances do changes in pressure have no effect on equilibrium concentrations? This problem is similar in part to Example 14.12.

Solution: (a) All reactants and products are in condensed phases. Pressure change should have no effect on this system.

(b) Same situation as (a).

(c) Only the product is in the gas phase. According to Section 14.6, an increase in pressure should favor the reaction (forward reaction or reverse reaction) that decreases the total number of moles of gas. The equilibrium should shift to the left, i.e., the amount of B should decrease and that of A should increase. This situation can be found in the reaction in Problem 14.18.

(d) In this system there are equal amounts of gaseous reactants and products. A shift in either direction will have no effect on the total number of moles of gas present. There will be no change. This is like the reaction in Problem 14.46.

(e) A shift in the direction of the reverse reaction (left) will have the result of decreasing the total number of moles of gas present. This situation can be seen in the reaction in Problem 14.40.

14.57 Part (a) is like Problem 14.56 above. In part (b) will the reaction quotient Q_c increase or decrease when more I_2 is added? What will happen then? Study Example 14.13 for help. Part (c) is similar to Problem 14.55 above.

Solution: (a) A pressure increase will favor the reaction (forward or reverse?) that decreases the total number of moles of gas. The equilibrium should shift to the right, i.e., more I_2 will be produced at the expense of the I.

(b) If the concentration of I_2 is suddenly altered, the system is no longer at equilibrium. Evaluating the magnitude of the reaction quotient Q_c allows us to predict the direction of the resulting equilibrium shift. The reaction quotient for this system is

$$Q_c = \frac{[I_2]_o}{[I]_o^2}$$

Increasing the concentration of I_2 will increase Q_c. The equilibrium will be reestablished in such a way that Q_c is again equal to the equilibrium constant. More I will form.

(c) The forward reaction is exothermic. A decrease in temperature will shift the equilibrium from left to right.

14.59 This is like Problem 14.57 above and 14.61 below.

Solution: (a) Increasing the temperature favors the endothermic reaction so that the concentrations of SO_2 and O_2 will increase while that of SO_3 will decrease.

(b) Increasing the pressure favors the reaction that decreases the number of moles of gas. The concentration of SO_3 will increase.

(c) Increasing the concentration of SO_2 will lead to an increase in the concentration of SO_3 and a decrease in the concentration of O_2.

(d) A catalyst has no effect on the position of equilibrium.

(e) Adding an inert gas at constant volume has no effect on the position of equilibrium.

14.61 Study Section 14.6 carefully to master the ways that changes in volume or pressure can affect the position of equilibrium. In part (a) how can more gas be added to the system without changing the total pressure or the temperature? Will the partial pressures of the reactants and products change in this process? In part (b) will the partial pressures of the reactants and products change? Does the pressure of helium appear in K_p for this reaction? Refer to Example 14.13 for help.

Solution: (a) If helium gas is added to the system without changing the pressure or the temperature, the volume of the container must necessarily be increased. This will decrease the partial pressures of all the reactants and products. A pressure decrease will favor the reaction that increases the number of moles of gas. The position of equilibrium will shift to the left.

(b) If the volume remains unchanged, the partial pressures of all the reactants and products will remain the same. The reaction quotient Q_c will still equal the equilibrium constant, and there will be no change in the position of equilibrium.

14.62 Write out the expression for K_p for this system. Remember that K_p is a constant as long as the temperature does not change. What happens when carbon dioxide comes into contact with a solution of NaOH? What about a solution of hydrochloric acid? Is the reaction in question exo- or endothermic?

Solution: For this system K_p is $\qquad K_p = [CO_2]$

This means that to remain at equilibrium, the pressure of carbon dioxide must stay at a fixed value as long as the temperature remains the same.

(a) If the volume is increased, the pressure of CO_2 will drop (Boyle's law, Section 5.3 and some $CaCO_3$ will break down to form more CO_2 and CaO.

(b) Assuming that the amount of added solid CaO is not so large that the volume of the system is altered significantly, there should be no change at all. If a huge amount of CaO were added, this would have the effect of reducing the volume of the container. What would happen then?

(c) Assuming that the amount of $CaCO_3$ removed doesn't alter the container volume significantly, there should be no change. Removing a huge amount of $CaCO_3$ will have the effect of increasing the container volume. The result in that case will be the same as in part (a).

(d) The pressure of CO_2 will be greater and will exceed the value of K_p. Some CO_2 will combine with CaO to form more $CaCO_3$.

(e) Carbon dioxide combines with aqueous NaOH according to the equation

$$CO_2(g) + NaOH(aq) \rightarrow NaHCO_3(aq)$$

This will have the effect of reducing the CO_2 pressure and causing more $CaCO_3$ to break down to CO_2 and CaO.

(f) Carbon dioxide does not react with hydrochloric acid, but $CaCO_3$ does

$$CaCO_3(s) + 2HCl(aq) \rightarrow CaCl_2(aq) + CO_2(g) + H_2O(l)$$

The CO_2 produced by the action of the acid will combine with CaO as discussed in (d) above.

(g) This is a decomposition reaction. Decomposition reactions are endothermic (Section 9.10). Increasing the temperature will favor this reaction and produce more CO_2 and CaO.

14.63 Is the reaction exo- or endothermic? Will a temperature increase shift the position of equilibrium to the right or to the left? How will an increase in pressure affect the position of equilibrium in this system? What combination of high or low temperature and high or low pressure will maximize the formation of NOCl?

Solution: Using data from Appendix 3 we calculate the enthalpy change for the reaction

$$\Delta H° = 2\Delta H_f^o(NOCl) - 2\Delta H_f^o(NO) - \Delta H_f^o(Cl_2) = 2(51.7) - 2(90.4) - (0) = -77.4 \text{ kJ}$$

The enthalpy change is negative, so the reaction is exothermic. The formation of NOCl will be favored by low temperature.

A pressure increase favors the reaction forming fewer moles of gas. The formation of NOCl will be favored by high pressure.

14.64 Compare the given information with Problems 14.46 and 14.47.

Solution: (i) The temperature of the system is not given. (ii) It is not stated whether the equilibrium constant is K_p or K_c (would they be different for this reaction?). (iii) A balanced equation is not given. (iv) The phases of the reactants and products are not given.

14.65 What is the sum of the partial pressures of NO and Cl_2? Part (b) is like Problem 14.16.

Solution: (a) Since the total pressure is 1.00 atm, the sum of the partial pressures of NO and Cl_2 is (1.00 - 0.64) atm = 0.36 atm. The stoichiometry of the reaction requires that the partial pressure of NO be 0.24 atm and that the partial pressure of Cl_2 be 0.12 atm.

(b) The equilibrium constant K_p is found by substitution

$$K_p = \frac{(P_{NO})^2 P_{Cl_2}}{(P_{NOCl})^2} = \frac{(0.24)^2(0.12)}{(0.64)^2} = 0.017$$

Is this reaction exo- or endothermic?

14.66 Is the formation of NO favored at high temperature? Study the discussion of the effects of temperature change in Section 14.6.

Solution: Since the equilibrium constant is larger at the higher temperature, the formation of NO (the product) is favored by a temperature increase. The reaction is endothermic.

14.67 Write out the expression for K_p for this system. How will the amount of CO_2 produced be affected if the container volume is fixed? Study the analysis of Problem 14.62.

Solution: The equilibrium expression for this system is given by

$$K_p = P_{CO_2} P_{H_2O}$$

(a) In a closed vessel the decomposition will stop when the product of the partial pressures of CO_2 and H_2O equals K_p. Adding more sodium bicarbonate will have no effect.

(b) In an open vessel the partial pressures of CO_2 and H_2O will never become large enough for their product to equal K_p. Adding more sodium bicarbonate will result in more products.

14.69 Part (a) is like Problems 14.16 and 14.17.

Solution: (a) Using Equation (14.5), we write

$$K_p = K_c(0.0821T)^{\Delta n}, \text{ where } \Delta n = +1$$

$$K_c = \frac{K_p}{(0.0821T)} = \frac{2 \times 10^{-42}}{(0.0821 \times 298)} = 8 \times 10^{-44}$$

(b) Because of a very large activation energy, the reaction of hydrogen with oxygen is infinitely slow without a catalyst or an initiator. The action of a single spark on a mixture of these gases results in the explosive formation of water.

14.70 Part (a) is like Problem 14.16. In part (b) you are to assume that K_p has the same value. Using the given information, can you express P_B in terms of P_A and solve for P_A?

Solution: (a) The value of K_p is found by substitution.

$$K_p = \frac{P_B}{P_A^2} = \frac{(0.60)}{(0.60)^2} = 1.7$$

(b) Under the new conditions we can write:

$$P_A + P_B = 1.5 \text{ atm}$$

$$P_B = 1.5 - P_A$$

Substituting into the expression for K_p:

$$K_p = \frac{(1.5 - P_A)}{P_A^2} = 1.7$$

$$1.7P_A^2 + P_A - 1.5 = 0$$

Solving the quadratic, we obtain: $P_A = 0.69$ atm (the other root is negative)

By difference: $P_B = 0.81$ atm

14.71 What are the partial pressures of ammonia and hydrogen sulfide at equilibrium in the vessel? How many moles of these gases are present at equilibrium? If the volume of the vessel were doubled at constant temperature, what would be the total pressure inside the vessel when equilibrium was reestablished?

Solution: (a) The balanced equation shows that equal amounts of ammonia and hydrogen sulfide are formed in this decomposition. The partial pressures of these gases must just be half the total pressure, i.e., 0.355 atm. The value of K_p is

$$K_p = P_{NH_3}P_{H_2S} = (0.355)^2 = 0.126$$

(b) We find the number of moles of ammonia (or hydrogen sulfide) and ammonium hydrogen sulfide

$$n_{NH_3} = \frac{PV}{RT} = \frac{(0.355 \text{ atm})(4.000 \text{ L})}{(0.0821 \text{ L·atm/K·mol})(297 \text{ K})} = 0.0582 \text{ mol}$$

$$n_{NH_4HS} = 6.1589 \text{ g} \left(\frac{1 \text{ mol}}{51.12 \text{ g}}\right) = 0.1205 \text{ mol (before decomposition)}$$

From the balanced equation the percent decomposed is

$$\frac{0.0582 \text{ mol}}{0.1205 \text{ mol}} \times 100\% = 48.3\%$$

(c) If the temperature does not change, K_p has the same value. The total pressure will still be 0.709 atm at equilibrium. In other words the amounts of ammonia and hydrogen sulfide will be twice as great, and the amount of solid ammonium hydrogen sulfide will be

$$0.1205 - 2(0.0582) \text{ mol} = 0.0041 \text{ mol}$$

14.73 Is this a heterogeneous or homogeneous reaction?

Solution: Since the reactant is a solid, we can write

$$K_p = (P_{NH_3})^2 P_{CO_2}$$

The total pressure is the sum of the ammonia and carbon dioxide pressures.

$$P_{total} = P_{NH_3} + P_{CO_2}$$

From the stoichiometry

$$P_{NH_3} = 2P_{CO_2}$$

Therefore

$$P_{total} = 2P_{CO_2} + P_{CO_2} = 3P_{CO_2} = 0.318 \text{ atm}$$

$$P_{CO_2} = 0.106 \text{ atm}$$

$$P_{NH_3} = 0.212 \text{ atm}$$

Substituting into the equilibrium expression

$$K_p = (0.212)^2(0.106) = 4.76 \times 10^{-3}$$

14.75 This problem is not as difficult as it appears. Simply set up you equilibrium concentrations, substitute into the equilibrium expression, and solve for α.

Solution: (a) From the balanced equation

	N_2O_4	$\rightleftharpoons$	$2NO_2$
initial (mol)	1		0
Change (mol):	$-\alpha$		$+2\alpha$
Equilibrium (mol):	$(1 - \alpha)$		2α

The total moles in the system = (moles N_2O_4 + moles NO_2) = $[(1 - \alpha) + 2\alpha] = 1 + \alpha$. If the total pressure in the system is P, then

$$P_{N_2O_4} = \frac{1 - \alpha}{1 + \alpha} P \quad \text{and} \quad P_{NO_2} = \frac{2\alpha}{1 + \alpha} P$$

$$K_p = \frac{P_{NO_2}^{2}}{P_{N_2O_4}} = \frac{\left(\frac{2\alpha}{1+\alpha}\right)^2 P^2}{\left(\frac{1-\alpha}{1+\alpha}\right) P}$$

$$K_p = \frac{\left(\frac{4\alpha^2}{1+\alpha}\right) P}{1-\alpha} = \frac{4\alpha^2}{1-\alpha^2} P$$

(b) Rearranging the K_p expression

$$4\alpha^2 P = K_p - \alpha^2 K_p$$

$$\alpha^2(4P + K_p) = K_p$$

$$\alpha^2 = \frac{K_p}{4P + K_p}$$

$$\alpha = \sqrt{\frac{K_p}{4P + K_p}}$$

K_p is a constant (at constant temperature). Thus, as P increases, α must decrease. This is just what one would predict based on Le Chatelier's principle.

14.77 This is a "plug-in" equilibrium problem.

Solution: For the balanced equation:

$$K_c = \frac{[H_2]^2[S_2]}{[H_2S]^2}$$

$$[S_2] = \frac{[H_2S]^2}{[H_2]^2} K_c = \left(\frac{4.84 \times 10^{-3}}{1.50 \times 10^{-3}}\right)^2 \times (2.25 \times 10^{-4}) = 2.34 \times 10^{-3}\ \text{M}$$

14.79 See the subsection entitled The Manufacture of Sulfuric Acid in Section 13.6 for more information.

Solution: For a 100% yield, 2.00 moles of SO_3 would be formed (why?). An 80% yield means 2.00 moles x (0.80) = 1.60 moles SO_3.

Th amount of SO_2 remaining at equilibrium= (2.00 - 1.60) mol = 0.40 mol

The amount of O_2 reacted = $\frac{1}{2}$ x (amount of SO_2 reacted) = ($\frac{1}{2}$ x 1.60) mol = 0.80 mol

The amount of O_2 remaining at equilibrium = (2.00 - 0.80) mol = 1.20 mol

Total moles at equilibrium = moles SO_2 + moles O_2 + moles SO_3 =(0.40 + 1.20 + 1.60) mol = 3.20 moles

$$P_{SO_2} = \frac{0.40}{3.20} P_{total} = 0.125\ P_{total}$$

$$P_{O_2} = \frac{1.20}{3.20} P_{total} = 0.375\ P_{total}$$

$$P_{SO_3} = \frac{1.60}{3.20} P_{total} = 0.500\ P_{total}$$

$$K_p = \frac{P_{SO_3}^{\ 2}}{P_{SO_2}^{\ 2}\ P_{O_2}} = \frac{(0.500\ P_{total})^2}{(0.125\ P_{total})^2(0.375\ P_{total})} = 0.13$$

$$P_{total} = 328\ \text{atm}$$

14.82 This is not a trivial problem. It is similar to but more difficult than Problem 14.73 above.

Solution: According to the ideal gas law, we can use pressure as the coefficients in the balanced equation when the reaction is at constant volume and temperature. The reaction is

	N_2	+	$3H_2$	$\rightleftharpoons$	$2NH_3$
Initial (atm):	0.862		0.373		0
Change (atm):	$-x$		$-3x$		$2x$
Equilibrium (atm):	$(0.862 - x)$		$(0.373 - 3x)$		$2x$

$$K_p = \frac{P_{NH_3}^2}{P_{H_2}^3 P_{N_2}} \qquad 4.31 \times 10^{-4} = \frac{(2x)^2}{(0.373 - 3x)^3(0.862 - x)} \approx \frac{(2x)^2}{(0.373)^3(0.862)}$$

Solving for x gives $\qquad x = 2.20 \times 10^{-3}$ atm

The equilibrium pressures are:

$P_{N_2} = (0.862 - 2.20 \times 10^{-3})$ atm $= 0.860$ atm

$P_{H_2} = (0.373 - 3 \times 2.20 \times 10^{-3})$ atm $= 0.366$ atm

$P_{NH_3} = 2(2.20 \times 10^{-3})$ atm $= 4.40 \times 10^{-3}$ atm

Notice that in solving the equilibrium equation above that we made an assumption about the magnitude of x relative to the initial pressures of hydrogen and nitrogen. Was the assumption valid? Why did we make the assumption?

14.83 What is the total of the mole fractions of carbon dioxide and carbon monoxide?

Solution: (a) The sum of the mole fractions must equal one.

$$X_{CO} + X_{CO_2} = 1 \qquad \text{and} \qquad X_{CO_2} = 1 - X_{CO}$$

According to the hint, the average molar mass is the sum of the products of the mole fraction of each gas and its molar mass.

$$(X_{CO} \times 28.01 \text{ g}) + (1 - X_{CO}) \times 44.01 \text{ g} = 35 \text{ g}$$

Solving, $X_{CO} = 0.56$ and $X_{CO_2} = 0.44$

(b) Solving for the pressures

$$P_{total} = P_{CO} + P_{CO_2} = 11 \text{ atm}$$

$$P_{CO} = X_{CO}P_{total} = 0.56 \times 11 \text{ atm} = 6.2 \text{ atm}$$

$$P_{CO_2} = X_{CO_2}P_{total} = 0.44 \times 11 \text{ atm} = 4.8 \text{ atm}$$

$$K_P = \frac{P_{CO}^2}{P_{CO_2}} = \frac{(6.2)^2}{4.8} = 8.0$$

14.85 How would you determine if an iodine sample were radioactive?

Solution: If you started with radioactive iodine in the solid phase, then you should find radioactive iodine in the vapor phase. Conversely, if you started with radioactive iodine in the vapor phase, you should find radioactive iodine in the solid phase at equilibrium. Both of these observations indicate a dynamic equilibrium between solid and vapor phase.

14.87 How much CO_2 reacts? How much NO?

Solution: If there were 0.88 mole of CO_2 initially and at equilibrium there were 0.11 moles, then (0.88 - 0.11) moles = 0.77 moles reacted.

	NO	+	CO_2	$\rightleftharpoons$	NO_2	+	CO
Initial (mol):	3.9		0.88		0		0
Change (mol):	-0.77		-0.77		+0.77		+0.77
Equilibrium (mol):	(3.9 - 0.77)		0.11		0.77		0.77

Solving for the equilibrium constant:

$$K_C = \frac{(0.77)(0.77)}{(3.9 - 0.77)(0.11)} = 1.7$$

Note that since the problem did not state the volume, we can solve for the equilibrium constant in terms of moles.

14.88 This problem is similar to 14.87 above. What can we deduce about the pressures of gases A and B?

Solution: Since we started with pure A, then any A that is lost forms equal amounts of B and C. Since the total pressure is P, the pressure of B + C = P - 0.14 P = 0.86 P. The pressure of B = C = 0.43 P.

$$K_P = \frac{P_B P_C}{P_A} = \frac{(0.43 \text{ P})(0.43 \text{ P})}{0.14 \text{ P}} = 1.3 \text{ P}$$

14.90 This problem requires both considerable thought and understanding.

Solution: Since the catalyst is exposed to the reacting system, it would catalyze the 2A → B reaction. This shift would result in a decrease in the number of gas molecules, so the gas pressure decreases. The piston would be pushed down by the atmospheric pressure. When the cover is over the box, the catalyst is no longer able to favor the forward reaction. To reestablish equilibrium, the B → 2A step would dominate. This would increase the gas pressure so the piston rises and so on.

Conclusion: Such a catalyst would result in a perpetual motion machine (the piston would move up and down forever) which can be used to do work without input of energy or net consumption of chemicals. Such a machine cannot exist.

Note: If you have not already done so, please read the "Introduction for the Student" at the beginning of this Student Solutions Manual.

Chapter 15

ACIDS AND BASES: GENERAL PROPERTIES

15.7 What are the definitions of Brønsted acids and bases? How do you recognize each? Can a molecule be both an acid and a base in the Brønsted system? Study Section 15.1 if you can't answer any of these questions.

Solution: Table 15.2 contains a list of important Brønsted acids and bases. (a) both (why?), (b) base, (c) acid, (d) base, (e) acid, (f) base, (g) base, (h) base, (i) acid, (j) acid.

15.8 Refer to Table 15.2.

Solution: (a) nitrite ion: NO_2^-

(b) hydrogen sulfate ion (also called bisulfate ion): HSO_4^-

(c) hydrogen sulfide ion (also called bisulfide ion): HS^-

(d) cyanide ion: CN^-

(e) formate ion: $HCOO^-$

15.9 What is a conjugate acid-base pair? How can you tell which is which? If you don't know, study Section 15.1 and Example 15.1.

Solution: In general the components of the conjugate acid-base pair are on opposite sides of the reaction arrow. The base always has one less proton than the acid.

(a) The conjugate acid-base pairs are (1) HCN (acid) and CN^- (base) and (2) CH_3COO^- (base) and CH_3COOH (acid).

(b) (1) HCO_3^- (acid) and CO_3^{2-} (base) and (2) HCO_3^- (base) and H_2CO_3 (acid).

(c) (1) $H_2PO_4^-$ (acid) and HPO_4^{2-} (base) and (2) NH_3 (base) and NH_4^+ (acid).

(d) (1) HClO (acid) and ClO^- (base) and (2) CH_3NH_2 (base) and $CH_3NH_3^+$ (acid).

(e) (1) H_2O (acid) and OH^- (base) and (2) CO_3^{2-} (base) and HCO_3^- (acid).

(f) (1) H_2O (acid) and OH^- (base) and (2) CH_3COO^- (base) and CH_3COOH (acid).

15.10 What is the relationship between a base and its conjugate acid? Study Section 15.1 if you are uncertain.

Solution: The conjugate acid of any base is just the base with a proton added.

(a) H_2S (c) HCO_3^- (e) $H_2PO_4^-$ (g) H_2SO_4 (i) H_2SO_3

(b) H_2CO_3 (d) H_3PO_4 (f) HPO_4^{2-} (h) HSO_4^- (j) HSO_3^-

15.11 What is the relationship between an acid and its conjugate base? See Section 15.1 if you are not sure.

Solution: The conjugate base of any acid is simply the acid minus one proton.

(a) CH_2ClCOO^- (d) HPO_4^{2-} (g) SO_4^{2-} (j) NH_3 (m) OCl^-

(b) IO_4^- (e) PO_4^{3-} (h) HSO_3^- (k) HS^-

(c) $H_2PO_4^-$ (f) HSO_4^- (i) SO_3^{2-} (l) S^{2-}

.12 The carbon-oxygen framework of oxalic acid does not change when the protons are lost. Review Section 9.6 if you don't remember how to work out Lewis structures. Part (b) is like Problem 15.7.

Solution: (a) The Lewis structures are

$$^-O-\overset{\overset{O}{||}}{C}-\overset{\overset{O}{||}}{C}-O-H \quad \text{and} \quad ^-O-\overset{\overset{O}{||}}{C}-\overset{\overset{O}{||}}{C}-O^-$$

(b) H^+ and $C_2H_2O_4$ can act only as acids, $C_2HO_4^-$ can act as both an acid and a base, and $C_2O_4^{2-}$ can act only as a base.

15.22 In parts (a) and (b), which substance is an acid and which a base? For parts (c) and (d), what is the concentration of hydrogen ion in a neutral solution at 25°C (see Sections 15.2 and 15.3)? Example 15.2 is similar

Solution: (a) Basic.

(b) Acidic.

(c) Basic. The product of the hydrogen ion and hydroxide ion concentrations equals K_w. For the given conditions the hydroxide ion concentration must exceed the hydrogen ion concentration. What is the value of $[OH^-]$?

(d) Acidic. The reasoning is the same as in part (c). What is the value of $[H^+]$?

15.24 Refer to Examples 15.3-15.5 for similar problems.

Solution: For (a) and (b) we use the same equation representing the definition of pH.

(a) $pH = -\log [H^+]$; $[H^+] = 10^{-5.20} = 6.3 \times 10^{-6}$ M

(b) $pH = -\log [H^+]$; $[H^+] = 10^{-16.00} = 1.0 \times 10^{-16}$ M

(c) We can first find pOH, then convert to pH and solve as in (a) and (b)

$pOH = -\log[OH^-] = -\log(3.7 \times 10^{-9}) = 8.43$; $pH = 14.00 - pOH = 5.57$

$[H^+] = 10^{-5.57} = 2.7 \times 10^{-6}$ M

An alternate, but equivalent, method of solving (c) would be to use the relationship $[H^+][OH^-] = K_w = 1.0 \times 10^{-14}$.

15.25 Study Section 15.3 and Examples 15.3-15.6.

Solution: (a) HCl is a strong acid, so the concentration of hydrogen ion is also 0.0010 M. (What is the concentration of chloride ion?) We use the definition of pH

$$pH = -\log[H^+] = -\log(0.0010) = 3.00$$

(b) KOH is an ionic compound (why?) and is fully ionized. We first find the concentration of hydrogen ion as in Example 15.2.

$$[H^+] = \frac{K_w}{[OH^-]} = \frac{1.0 \times 10^{-14}}{0.76} = 1.3 \times 10^{-14} \text{ M}$$

The pH is then found from its defining equation

$$pH = -\log[H^+] = -\log[1.3 \times 10^{-14}] = 13.89$$

(c) $Ba(OH)_2$ is ionic and fully ionized in water. The concentration of hydroxide ion is 5.6×10^{-4} M (Why? What is the concentration of Ba^{2+}?) We find the hydrogen ion concentration as in Example 15.2.

$$[H^+] = \frac{K_w}{[OH^-]} = \frac{1.0 \times 10^{-14}}{5.6 \times 10^{-4}} = 1.8 \times 10^{-11} \text{ M}$$

The pH is then

$$pH = -\log[H^+] = -\log(1.8 \times 10^{-11}) = 10.74$$

(d) Nitric acid is a strong acid, so the concentration of hydrogen ion is also 5.2×10^{-4} M. The pH is found using its defining equation

$$pH = -\log[H^+] = -\log(5.2 \times 10^{-4}) = 3.28$$

15.26 What are the concentrations of hydrogen and hydroxide ion at this temperature? Study the subsection on the Ion Product of Water in Section 15.2.

Solution: As shown in Section 15.2, the concentration of hydrogen ion in pure water at 40°C is 1.9×10^{-7} M. The pH is then

$$pH = -\log[H^+] = -\log(1.9 \times 10^{-7}) = 6.72$$

Is this water sample neutral? (What is the hydroxide ion concentration?)

15.28 Study the equations defining pOH in Section 15.3. How can you find pH if you know pOH? How do you convert pH to hydrogen ion concentration?

Solution: The pH can be found using Equation (15.5)

$$pH = 14.00 - pOH = 14.00 - 9.40 = 4.60$$

The hydrogen ion concentration can be found as in Example 15.4.

$$4.60 = -\log[H^+]$$

Taking the antilog of both sides: $[H^+] = 2.5 \times 10^{-5}$ M

15.37 This problem can be solved by reference to Table 15.2.

Solution: (a) strong acid, (b) weak acid, (c) strong acid (first stage of ionization), (d) weak acid, (e) waek acid, (f) weak acid, (g) strong acid, (h) weak acid, (i) weak acid.

15.39 In this problem you are being asked to compare the pH, the electrical conductance, and the hydrogen gas evolution rates of the two acid solutions. Assume that the concentrations of HA and HB are the same.

Solution: (a) The pH of the solution of HA would be lower. (Why?)

(b) The electrical conductance of the HA solution would be greater. (Why?)

(c) The rate of hydrogen evolution from the HA solution would be greater. Presumably, the rate of the reaction between the metal and hydrogen ion would depend on the hydrogen ion concentration (i.e., this would be part of the rate law). The hydrogen ion concentration will be greater in the HA solution.

15.40 What is the difference between a weak and a strong acid? What is the maximum possible hydrogen ion concentration for a 0.10 M solution of any acid HA? What is the pH corresponding to this hydrogen ion concentration? Study Sections 15.3 and 15.6 if you are not certain on any of these questions.

Solution: The maximum possible concentration of hydrogen ion in a 0.10 M solution of HA is 0.10 M. This is the case if HA is a strong acid. If HA is a weak acid, the hydrogen ion concentration is less than 0.10 M. The pH corresponding to 0.10 M $[H^+]$ is 1.00. (Why three digits?) For a smaller $[H^+]$ the pH is larger than 1.00 (why?).

(a) false, the pH is greater than 1.00; (b) false, they are equal; (c) true, (d) false.

15.41 The principles needed to understand this problem are the same as for the previous problem.

Solution: (a) false, they are equal; (b) true, find the value of log(1.00) on your calculator; (c) true; (d) false, if the acid is strong, [HA] = 0.00 M.

15.44 This is similar to Example 15.8. Study the discussion of ternary acids in Section 15.5. Into which category do the listed oxoacids fall?

Solution: All the listed pairs are oxoacids that contain different central atoms whose elements are in the same group of the periodic table and have the same oxidation number. In this situation the acid with the most electronegative central atom will be the strongest.

(a) $H_2SO_4 > H_2SeO_4$, (b) $H_2SO_3 > H_2SeO_3$, (c) $H_3PO_4 > H_3AsO_4$, (d) $HBrO_4 > HIO_4$.

15.45 Study the discussion of binary acids in Section 15.5. What factor determines the relative strengths of binary acids? Example 15.8 in the text is a similar problem.

Solution: The strength of the H-X bond is the dominant factor in determining the strengths of binary acids. As with the hydrogen halides (see Section 15.5), the H-X bond strength decreases going down the column in Group 6A. The compound with the weakest H-X bond will be the strongest binary acid: $H_2Se > H_2S > H_2O$.

15.46 This problem is similar to Example 15.8.

Solution: The $CH_2ClCOOH$ is a stronger acid than CH_3COOH. Chlorine is more electronegative than hydrogen and attracts electrons toward itself, making the O-H bond more polar. The hydrogen atom in the COOH group is more easily ionized.

15.48 Study the different categories of acid-base reactions in Section 15.6. What are differences in the ways the net ionic equations are written?

Solution: (a) NaOH is a strong base (100% ionized) and ammonium ion is a weak acid (see Table 15.2). The net ionic equation is

$$OH^-(aq) + NH_4^+(aq) \rightarrow H_2O(l) + NH_3(aq)$$

(b) HCl is a strong acid and acetic acid is a weak acid (Table 15.2). The net ionic equation is

$$H^+(aq) + CH_3COO^-(aq) \rightarrow CH_3COOH(aq)$$

15.52 Study the discussion and example reactions offered in Section 15.7. To be a Lewis base, an atom or molecule must have one or more pairs of electrons to donate. To be a Lewis acid an atom or molecule must participate in reactions in which an electron pair is accepted from some Lewis base to form a covalent bond. Study the examples of $B(OH)_3$, CO_2, Ag^+, etc. in Section 15.7. Study Example 15.9.

Solution: (a) Lewis acid; see the reaction with water shown in Section 15.7.

(b) Lewis base; water combines with H^+ to form H_3O^+.

(c) Lewis base; see example with Cd^{2+} in Section 15.7.

(d) Lewis acid; SO_2 reacts with water to form H_2SO_3. Compare to CO_2 above. Actually, SO_2 can also act as a Lewis base under some circumstances.

(e) Lewis base; see the reaction with H^+ to form ammonium ion.

(f) Lewis base; see the reaction with H^+ to form water.

(g) Lewis acid; does H^+ have any electron pairs to donate?

(h) Lewis acid; compare to the example of NH_3 reacting with BF_3.

15.54 A Lewis acid is defined as an electron pair acceptor, i.e. it is an atom or molecule that is attracted to lone pairs of electrons on other atoms or molecules (Lewis acids are sometimes called electrophiles). Consider the two pairs of Lewis acids in this problem. Which one of each pair might be more strongly attracted to unshared electrons in other atoms or molecules?

Solution: (a) Both molecules have the same acceptor atom (boron) and both have exactly the same structure (trigonal planar). Fluorine is more electronegative than chlorine so we would predict based on electronegativity arguments that boron trifluoride would have a greater affinity for unshared electron pairs than boron trichloride.

(b) Since it has the larger positive charge, iron(III) should be a stronger Lewis acid than iron(II).

15.56 What chemical entity must be possessed by a Brønsted acid? Find some Lewis acids among the choices in Problem 15.52 that do not possess this entity.

Solution: By definition Brønsted acids are proton donors, therefore such compounds must contain at least one hydrogen atom. In Problem 15.52 Lewis acids that do not contain hydrogen are CO_2, SO_2, and BCl_3. Can you name others?

15.58 What is dynamic equilibrium? Review Section 14.1 if you don't remember.

Solution: The ionization of HCN is

$$HCN(aq) \rightleftharpoons H^+(aq) + CN^-(aq)$$

At every instant a number of HCN molecules break up to form H^+ and CN^- ions, and at the same time, a number of H^+ and CN^- ions recombine to form HCN units. However, at equilibrium the forward rate is exactly equal to the reverse rate so that there is no net change in the concentration of any of the three species.

15.60 Review the material in Chapter 5 if you need help on the gas laws.

Solution: We first find the number of moles of CO_2 produced in the reaction:

$$0.350 \text{ g NaHCO}_3 \left(\frac{1 \text{ mol NaHCO}_3}{84.01 \text{ g NaHCO}_3}\right)\left(\frac{1 \text{ mol CO}_2}{1 \text{ mol NaHCO}_3}\right) = 4.17 \times 10^{-3} \text{ mol CO}_2$$

$$V = \frac{nRT}{P} = \frac{(4.17 \times 10^{-3} \text{ mol})(0.0821 \text{ L·atm/K·mol})(37.0 + 273) \text{ K}}{(1.00 \text{ atm})} = 0.106 \text{ L}$$

15.62 Study the subsection on Acidic, Basic, and Amphoteric Oxides in Section 15.6. What determines the base strength of a metallic oxide?

Solution: The most basic oxides occur with metal ions having the lowest positive charges (or lowest oxidation numbers). (a) $Al_2O_3 < BaO < K_2O$; (b) $CrO_3 < Cr_2O_3 < CrO$.

15.64 Study the last part of Section 15.6.

Solution: The $Al(OH)_3$ is an amphoteric hydroxide. The reaction is

$$Al(OH)_3(s) + OH^-(aq) \rightarrow Al(OH)_4^-(aq)$$

This is a Lewis acid-base reaction. Can you identify the acid and base?

15.65 What determines whether a nonmetallic oxide is acidic or neutral? Are nonmetallic oxides ever basic? What determines whether a metallic oxide is basic, amphoteric, or acidic? Study the last parts of Section 15.6 if you can't answer any of these questions.

Solution: (a) acidic, (b) basic, (c) basic, (d) acidic, (e) neutral, (f) neutral, (g) amphoteric, (h) acidic, (i) amphoteric, (j) basic.

15.66 Consider the hint. How would the electronegativities of F and H affect the abilities of the two compounds to donate a pair of electrons?

Solution: The fact that fluorine attracts electrons in a molecule more strongly than hydrogen should cause NF_3 to be a poor electron pair donor and a poor base. NH_3 is the stronger base.

15.67 Consider the hint. Which base will bind a proton more strongly?

> ***Solution:*** Because the P-H bond is weaker, there is a greater tendency for PH_4^+ to ionize. Therefore, PH_3 is a weaker base than NH_3.

15.68 Review Section 9.6 on drawing Lewis structures if you can't answer the question.

> ***Solution:*** The Lewis structure of H_3O^+ is
>
> $$\left[\begin{array}{c} \;\;\;\mathrm{H} \\ \;\;\;| \\ \mathrm{H}-\ddot{\mathrm{O}}: \\ \;\;\;| \\ \;\;\;\mathrm{H} \end{array}\right]^+$$
>
> Note that this structure is very similar to the Lewis structure of HN_3. The geometry is pyramidal.

15.69 Review the material in Section 15.2. The definitions are exactly analogous to "normal" water. What is D?

> ***Solution:*** The autoionization for deuterium-substituted water is: $D_2O \rightleftharpoons D^+ + OD^-$
>
> $$[D^+][OD^-] = 1.35 \times 10^{-15} \qquad (1)$$
>
> (a) The definition of pH is $\quad pD = -\log[D^+] = -\log(1.35 \times 10^{-15})^{1/2} = 7.43$
>
> (b) To be acidic, the pD must be < 7.43.
>
> (c) Taking - log of both sides of equation (1) above
>
> $$-\log[D^+] + -\log[OD^-] = -\log(1.35 \times 10^{-15})$$
>
> $$pD + pOD = 14.87$$

15.70 Remember that pH calculations involve molar concentrations.

> ***Solution:*** First we must calculate the molarity of the trifluoromethane sulfonic acid. (Molar mass = 150.1 g)
>
> $$\text{Molarity} = \frac{0.616\ \text{g}}{150.1\ \text{g/mol}} \times \frac{1000\ \text{mL}}{250\ \text{mL}} \times \frac{1}{1\ \text{L}} = 0.0164\ \text{M}$$
>
> Since trifluoromethane sulfonic acid is a strong acid and is 100% ionized, the $[H^+]$ is 0.0164 M (why?).
>
> $$pH = -\log(0.0164) = 1.79$$

15.72 What is the definition of a strong acid?

Solution: If we assume that the unknown monoprotic acid is a strong acid that is 100% ionized, then the $[H^+]$ concentration will be 0.0642 M.

$$pH = -\log(0.0642) = 1.19$$

Since the actual pH is higher, the acid must be a weak acid.

15.74 Hint: Write the equilibria involved and apply Le Chatelier's principle.

Solution: The reactions are

$$HF \rightleftharpoons H^+ + F^- \quad (1)$$

$$F^- + HF \rightleftharpoons HF_2^- \quad (2)$$

Note that for equation (2), the equilibrium constant is relatively large with a value of 5.2. This means that the equilibrium lies to the right. Applying Le Chatelier's principle, as HF ionizes in the first step, the F^- that is produced is partially removed in the second step. More HF must ionize to compensate for the removal of the F^-, at the same time producing more H^+.

15.76 This is a simple question, but a very important one. If you cannot quickly understand this problem, you should review Section 15.7 on the reaction of acids and bases.

Solution: The reaction of a weak acid with a strong base is driven to completion by the formation of water. Irrespective of whether the strong base is reacting with a strong monoprotic acid or a weak monoprotic acid, the same number of moles of acid is required to react with a constant number of moles of base. Therefore the volume of base required to react with the same concentration of acid solutions (either both weak, both strong, or one strong and one weak) will be the same.

15.77 Is ammonia volatile? Is an ammonium salt volatile? What is necessary for a compound to be smelled?

Solution: In Chapter 11, we found that salts with their formal electrostatic intermolecular attractions had low vapor pressures and thus high boiling points. Ammonia and its derivatives (amines) are molecules with dipole-dipole attractions; as long as the nitrogen has one direct N-H bond, the molecule will have hydrogen bonding. Even so, these molecules will have much higher vapor pressures than ionic species. Thus if we could convert the neutral ammonia-type molecules into salts, their vapor pressures, and thus associated odors, would decrease. Lemon juice contains acids which can react with neutral ammonia-type (amine) molecules to form ammonium salts.

$$NH_3 + H^+ \rightarrow NH_4^+ \qquad\qquad RNH_2 + H^+ \rightarrow RNH_3^+$$

15.79 This problem is an interesting application of Le Chatelier's principle.

Solution: Given the equation:

$$HbH^+ + O_2 \rightleftharpoons HbO_2 + H^+$$

(a) From the equilibrium equation, high oxygen concentration puts stress on the left side of the equilibrium and thus shifts the concentrations to the right to compensate. HbO_2 is favored.

(b) High acid H^+ concentration places stress on the right side of the equation forcing concentrations on the left side to increase, thus releasing oxygen and increasing the concentration of HbH^+.

(c) Removal of CO_2 decreases H^+ (in the form of carbonic acid), thus shifting the reaction to the right. More HbO_2 will form. Breathing into a paper bag increases the concentration of CO_2 (re-breathing the exhaled CO_2), thus causing more O_2 to be released as explained above.

15.80 The question states that the formula of the metal carbonate is MCO_3. What is the charge on the metal cation?

Solution: (a)The equations are:

$$MCO_3(aq) + 2HCl(aq) \rightarrow MCl_2(aq) + CO_2(g) + H_2O(l)$$

$$HCl(aq) + NaOH(aq) \rightarrow NaCl(aq) + H_2O(l)$$

(b) Total number of moles of HCl added: $\text{mol} = M \times \text{vol (L)} = 0.0800\ M \times 0.0200\ L = 1.60 \times 10^{-3}$ mol HCl

Excess HCl titrated with NaOH: $\text{mol} = M \times \text{vol (L)} = 0.1000\ M \times 0.00564\ L = 5.64 \times 10^{-4}$ mol HCl

Moles of HCl reacted with MCO_3: (total HCl - excess HCl) $= (1.60 - 0.564) \times 10^{-3} = 1.04 \times 10^{-3}$ mol

From the balanced equation: $1\ MCO_3 \equiv 2\ HCl$; moles of $MCO_3 = \dfrac{1.04 \times 10^{-3}\ \text{mol}}{2} = 5.20 \times 10^{-4}$ mol

$$\text{Molar mass} = \frac{\text{g } MCO_3}{\text{mol } MCO_3} = \frac{0.1022\ \text{g } MCO_3}{5.20 \times 10^{-4}\ \text{mol } MCO_3} = 197\ \text{g/mol}$$

15.81 This problem is a simple stoichiometry problem that has been complicated by details. We must use pH to determine the initial concentration of HCl. We calculate how much of the HCl solution must be used to react with the Mg, and we thus know how much HCl will be left. We thus can determine the pH of the remaining solution.

Solution: The balanced equation is: $Mg + 2HCl \rightarrow MgCl_2 + H_2$

$$\text{mol of Mg} = \frac{\text{g Mg}}{\text{molar mass Mg}} = \frac{1.87\ \text{g Mg}}{24.31\ \text{g/mol}} = 0.0769\ \text{mol}$$

From the balanced equation

mol of HCl required for reaction $= 2 \times$ mol Mg $= 2 \times 0.0769$ mol $= 0.154$ mol HCl

The concentration of HCl:

pH = -0.544, thus $[H^+] = 3.50$ M
mol HCl $= M \times \text{vol (L)} = 3.50\ M \times 0.0800\ L = 0.280$ mol HCl

Moles of HCl left after reaction:

total mol HCl - mol HCl reacted = 0.280 mol - 0.154 mol = 0.126 mol HCl

Molarity of HCl left after reaction:

mol/L = 0.126 mol/0.080 L = 1.58 M   pH = -0.20

Note: If you have not already done so, please read the "Introduction for the Student" at the beginning of this Student Solutions Manual.

Chapter 16

ACID - BASE EQUILIBRIA

16.6 This problem is a classic example of a weak acid equilibrium calculation. It is similar to Examples 16.1 and 16.2.

Solution: We set up a table for the dissociation

	$C_6H_5COOH(aq)$ ⇌	$H^+(aq)$ +	$C_6H_5COO^-(aq)$
Initial (M):	0.10	0.00	0.00
Change (M):	-x	+x	+x
Equilibrium (M):	(0.10 - x)	x	x

$$K_a = \frac{[H^+][C_6H_5COO^-]}{[C_6H_5COOH]} = \frac{x^2}{(0.10 - x)} = 6.5 \times 10^{-5}$$

$$x^2 + (6.5 \times 10^{-5})x - (6.5 \times 10^{-6}) = 0$$

Solving the quadratic equation: $x = 2.5 \times 10^{-3}$ M

$[H^+] = [C_6H_5COO^-] = 2.5 \times 10^{-3}$ M $\qquad [C_6H_5COOH] = (0.10 - 2.5 \times 10^{-3})$ M $= 0.10$ M

$$[OH^-] = \frac{1.0 \times 10^{-14}}{2.5 \times 10^{-3}} = 4.0 \times 10^{-12}\ M$$

This problem could be solved more easily if we could assume that $(0.10 - x) \cong 0.10$. If the assumption is mathmatically valid, then it would not be necessary to solve a quadratic equation, as we did above. Re-solve the problem above, making the assumption. Was the assumption valid? What is our criterion for deciding?

16.7 Find the value of the equilibrium constant K_a for HF in Table 16.1. Solve the equilibrium constant expression for the [conjugate base]/[acid] ratio and substitute the hydrogen ion concentration computed from the pH.

Solution: First we find the hydrogen ion concentration

$$[H^+] = 10^{-pH} = 10^{-6.20} = 6.3 \times 10^{-7}\ M$$

We then rearrange the acid ionization constant expression

$$\frac{[F^-]}{[HF]} = \frac{K_a}{[H^+]} = \frac{7.1 \times 10^{-4}}{6.3 \times 10^{-7}} = 1.1 \times 10^3$$

16.8 This is a classic weak acid equilibrium problem. First compute the molarity of the acetic acid solution, then proceed as in Example 16.1.

Solution: The concentration of acetic acid (before ionization) is 0.0187 M. It is a weak monoprotic acid. Using Example 16.1 as a model, let x be the equilibrium concentration of H^+ and CH_3COO^- ions. (Why are they equal?) The equilibrium concentration of un-ionized acid must then be (0.0187 - x) M. (Why?) Summarize the changes in a table to make the situation clearer.

	$CH_3COOH(aq)$	⇌ $H^+(aq)$	+ $CH_3COO^-(aq)$
Initial (M):	0.0187	0.0000	0.0000
Change (M):	-x	+x	+x
Equilibrium (M)	(0.0187 - x)	x	x

Substitute the equilibrium expressions from this table into the formula for K_a

$$K_a = \frac{[H^+][CH_3COO^-]}{[CH_3COOH]} = \frac{x^2}{(0.0187 - x)} = 1.8 \times 10^{-5}$$

Assuming $(0.0187 - x) \cong 0.0187$ (see Example 16.1) we have

$$\frac{x^2}{0.0187} = 1.8 \times 10^{-5}$$

$$x = 5.8 \times 10^{-4} \text{ M}$$

Note that x is less than 5 percent of 0.0187 so the approximation is valid. At equilibrium

$$[H^+] = [CH_3COO^-] = 5.8 \times 10^{-4} \text{ M}$$

$$[CH_3COOH] = (0.0187 - 0.00058) \text{ M} = 0.0181 \text{ M}$$

16.9 Find the value of K_a for formic acid in Table 16.1. Write the equilibrium constant expression for the ionization of this acid and find the hydrogen ion concentration from the given pH. What must be the concentration of formate ion? Study the solution to Examples 16.1-16.3 and set up a table showing the expressions for the concentrations of all species for this problem. What is the unknown here?

Solution: A pH of 3.26 corresponds to a $[H^+]$ of 5.5×10^{-4} M. Let the original concentration of formic acid be x so that

$$\frac{[H^+][HCOO^-]}{[HCOOH]} = 1.7 \times 10^{-4}$$

$$\frac{(5.5 \times 10^{-4})^2}{(x - 5.5 \times 10^{-4})} = 1.7 \times 10^{-4} \qquad x = 2.3 \times 10^{-3}\ M$$

16.10 This is similar to Example 16.4. Do you think the concentration of the acid has any effect on the percent ionization?

Solution: (a) Set up a table showing initial and equilibrium concentrations.

	$C_6H_5COOH(aq)$	$\rightleftharpoons$	$H^+(aq)$	+	$C_6H_5COO^-(aq)$
Initial (M):	0.20		0.00		0.00
Change (M):	-x		+x		+x
Equilibrium (M):	(0.20 - x)		x		x

Using the value of K_a from Table 16.1

$$K_a = \frac{[H^+][C_6H_5COO^-]}{[C_6H_5COOH]} = \frac{x^2}{(0.20 - x)} = 6.5 \times 10^{-5}$$

We assume that x is small so $(0.20 - x) \cong 0.20$

$$\frac{x^2}{0.20} = 6.5 \times 10^{-5}$$

$$x = 3.6 \times 10^{-3}\ M = [H^+] = [C_6H_5COO^-]$$

$$\text{Percent ionization} = \frac{3.6 \times 10^{-3}\ M}{0.20\ M} \times 100\% = 1.8\%$$

(b) Set up a table as above.

	$C_6H_5COOH(aq)$	$\rightleftharpoons$	$H^+(aq)$	+	$C_6H_5COO^-(aq)$
Initial (M):	0.00020		0.00000		0.00000
Change (M):	-x		+x		+x
Equilibrium (M):	(0.00020 - x)		x		x

Using the value of K_a from Table 16.1

$$K_a = \frac{[H^+][C_6H_5COO^-]}{[C_6H_5COOH]} = \frac{x^2}{(0.00020 - x)} = 6.5 \times 10^{-5}$$

In this case we cannot apply the approximation that $(0.00020 - x) \cong 0.00020$ (see the discussion in Example 16.1). We obtain the quadratic equation

$$x^2 + (6.5 \times 10^{-5})x - (1.3 \times 10^{-8}) = 0$$

The positive root of the equation is $x = 8.6 \times 10^{-5}$ M. (Is this less than 5% of the original concentration, 0.00020

M? That is, is the acid more than 5% ionized?) The percent ionization is then:

$$\text{Percent ionization} = \frac{8.6 \times 10^{-5}\ \text{M}}{0.00020\ \text{M}} \times 100\% = 43\%$$

16.11 This is like Problem 16.10 above and like Example 16.5. Each part is done the same way. Does the weak acid concentration have any influence on the percent ionization?

Solution: (a) We set up the usual table for the ionization.

	$HF(aq)$	$\rightleftharpoons$	$H^+(aq)$	+	$F^-(aq)$
Initial (M):	1.00		0.000		0.000
Change (M):	-x		+x		+x
Equilibrium (M):	(1.00 - x)		x		x

$$K_a = \frac{[H^+][F^-]}{[HF]} = \frac{x^2}{(1.00 - x)} = 7.1 \times 10^{-4}$$

Assuming (1.00 - x) ≅ 1.00, we have $x = \sqrt{7.1 \times 10^{-4}} = 2.7 \times 10^{-2}\ \text{M} = [H^+]$

$$\text{Percent ionization} = \frac{2.7 \times 10^{-2}\ \text{M}}{1.00\ \text{M}} \times 100\% = 2.7\%$$

(b) This is the same as Example 16.4(a).

Percent ionization = 3.5%

(c)

$$K_a = \frac{[H^+][F^-]}{[HF]} = \frac{x^2}{(0.080 - x)} = 7.1 \times 10^{-4}$$

$$x^2 + (7.1 \times 10^{-4})x - (5.7 \times 10^{-5}) = 0 \qquad x = 7.2 \times 10^{-3}\ \text{M}$$

$$\text{Percent ionization} = \frac{7.2 \times 10^{-3}\ \text{M}}{0.080\ \text{M}} \times 100\% = 9.0\%$$

(d)

$$K_a = \frac{[H^+][F^-]}{[HF]} = \frac{x^2}{(0.0046 - x)} = 7.1 \times 10^{-4}$$

$$x^2 + (7.1 \times 10^{-4})x - (3.3 \times 10^{-6}) = 0 \qquad x = 1.5 \times 10^{-3}\ \text{M}$$

$$\text{Percent ionization} = \frac{1.5 \times 10^{-3}\ \text{M}}{0.0046\ \text{M}} \times 100\% = 33\%$$

(e)

$$K_a = \frac{[H^+][F^-]}{[HF]} = \frac{x^2}{(0.00028 - x)} = 7.1 \times 10^{-4}$$

$$x^2 + (7.1 \times 10^{-4})x - (2.0 \times 10^{-7}) = 0 \qquad x = 2.2 \times 10^{-4}\ \text{M}$$

$$\text{Percent ionization} = \frac{2.2 \times 10^{-4}\ \text{M}}{0.00028\ \text{M}} \times 100\% = 79\%$$

As the solution becomes more dilute, the percent ionization increases.

16.12 This is similar to Example 16.4. If you know the percent ionization is 14%, what are the concentrations of hydrogen ion, conjugate base, and un-ionized acid?

Solution: Given 14% ionization, the concentrations must be

$$[H^+] = [A^-] = 0.14 \times 0.040\ \text{M} = 0.0056\ \text{M}$$

$$[HA] = (0.040 - 0.0056)\ \text{M} = 0.034\ \text{M}$$

The value of K_a can be found by substitution

$$K_a = \frac{[H^+][A^-]}{[HA]} = \frac{(0.0056)^2}{0.034} = 9.2 \times 10^{-4}$$

16.13 Part (a) is like Example 16.4. In part (b) assume that the hydrogen ion concentration is just that of gastric juice. Under those conditions what is the ratio of ionized to unionized acid? The neutral, unionized acid molecules are more readily absorbed by the tissues making up the lining of the stomach.

Solution: (a) Using Problem 16.6 as a model:

$$K_a = 3.0 \times 10^{-4} = \frac{[H^+][C_9H_7O_4^-]}{[C_9H_8O_4]} = \frac{x^2}{(0.20 - x)}$$

Assuming $(0.20 - x) \cong 0.20$, $x = [H^+] = 7.7 \times 10^{-3}\ \text{M}$

$$\text{Percent ionization} = \frac{[C_9H_7O_4^-]}{[C_9H_8O_4]} \times 100\% = \frac{7.7 \times 10^{-3}\ \text{M}}{0.20\ \text{M}} \times 100\% = 3.9\ \%$$

(b) At pH 1.00 the concentration of hydrogen ion is 0.10 M. (Why only two significant figures?) This will tend to suppress to ionization of the weak acid (Le Chatelier's principle, Section 14.5). The extra hydrogen ion shifts the position of equilibrium in the direction of the un-ionized acid, and to two-significant-figure accuracy, we can safely ignore the contribution of the weak acid to the total hydrogen ion concentration. The percent ionization of the acid is then

$$\text{Percent ionization} = \frac{[C_9H_7O_4^-]}{[C_9H_8O_4]} \times 100\% = \frac{K_a}{[H^+]} \times 100\%$$

$$= \frac{3.0 \times 10^{-4}}{0.10} \times 100\% = 0.30\%$$

The high acidity of the gastric juices appears to enhance the rate of absorption of unionized aspirin molecules through the stomach lining. In some cases this can irritate these tissues and cause bleeding.

16.17 This is like Example 16.6.

Solution: (a) We construct the usual table

	$NH_3(aq)$ + $H_2O(l)$	⇌ $NH_4^+(aq)$	+ $OH^-(aq)$
Initial (M):	0.10	0.00	0.00
Change (M)	-x	+x	+x
Equilibrium (M):	(0.10 - x)	x	x

$$K_b = \frac{[NH_4^+][OH^-]}{[NH_3]} = \frac{x^2}{(0.10 - x)} = 1.8 \times 10^{-5}$$

Assuming (0.10 - x) ≅ 0.10, we have $\frac{x^2}{0.10} = 1.8 \times 10^{-5}$ $x = 1.3 \times 10^{-3}$ M

$$pOH = -\log(1.3 \times 10^{-3}) = 2.89 \qquad pH = 14.00 - 2.89 = 11.11$$

By following the identical procedure, we can show: (b) pH = 8.96.

16.18 Write the equilibrium expression for the reaction of a weak base with water. Use the equation for ammonia in water in Section 16.2 as a model. What is the hydroxide ion concentration? What is the concentration of the conjugate acid of the weak base? What is the concentration of the unprotonated base? Example 16.6 in the text is a similar problem.

Solution: If the pH is 10.66, the pOH is 14.00 - 10.66 = 3.34. The hydroxide ion concentration is then

$$[OH^-] = 10^{-3.34} = 4.6 \times 10^{-4} \text{ M}$$

The concentration of the conjugate acid of the weak base is the same. (Why?) At two-significant-figure accuracy the concentration of the unprotonated base is still 0.30 M. The value of K_b can be found by simple substitution

$$K_b = \frac{[BH^+][OH^-]}{[B]} = \frac{(4.6 \times 10^{-4})^2}{0.30} = 7.1 \times 10^{-7}$$

16.19 This is like Problem 16.9 for the case of a weak acid. Find K_b for ammonia in Table 16.2. What are the equilibrium concentrations of hydroxide ion, ammonium ion, and ammonia?

Solution: A pH of 11.22 corresponds to a $[H^+]$ of 6.03×10^{-12} M and a $[OH^-]$ of 1.66×10^{-3} M (why?)

The equilibrium is: $NH_3(aq) + H_2O(l) \rightleftharpoons NH_4^+(aq) + OH^-(aq)$

The concentration of $[OH^-] = [NH_4^+]$ (why?) If we let x equal the original concentration of ammonia

$$K_b = 1.8 \times 10^{-5} = \frac{(1.66 \times 10^{-3})(1.66 \times 10^{-3})}{x - 1.66 \times 10^{-3}}$$

Assuming the 1.66×10^{-3} is small relative to x, then

$$x = 0.15\ M$$

16.23 $KHSO_4$ is an ionic compound and is fully dissociated to K^+ and HSO_4^- ions. The concentration of each is 0.20 M. This problem is therefore like Examples 16.1 and 16.7; the acid is HSO_4^-.

Solution: There is no H_2SO_4 in the solution because HSO_4^- has no tendency to accept a proton to produce H_2SO_4. (Why?) We are only concerned with the ionization

$$HSO_4^-(aq) \rightleftharpoons H^+(aq) + SO_4^{2-}(aq)$$

	$HSO_4^-(aq)$	$H^+(aq)$	$SO_4^{2-}(aq)$
Initial (M):	0.20	0.00	0.00
Change (M):	-x	+x	+x
Equilibrium (M)	(0.20 - x)	+x	+x

$$K_a = 1.3 \times 10^{-2} = \frac{[H^+][SO_4^{2-}]}{[HSO_4^-]} = \frac{(x)(x)}{(0.20 - x)}$$

Solving the quadradic equation $\quad x = [H^+] = [SO_4^{2-}] = 0.045\ M$

$$[HSO_4^-] = (0.20 - 0.045)\ M = 0.16\ M$$

16.24 This is like Example 16.8.

Solution: We follow the same procedure as outlined in Example 16.8. For the first stage of ionization

$$H_2CO_3(aq) \rightleftharpoons H^+(aq) + HCO_3^-(aq)$$

	$H_2CO_3(aq)$	$H^+(aq)$	$HCO_3^-(aq)$
Initial (M):	0.025	0.00	0.00
Change (M):	-x	+x	+x
Equilibrium (M):	(0.025 - x)	x	x

$$K_{a_1} = \frac{[H^+][HCO_3^-]}{[H_2CO_3]} = \frac{x^2}{(0.025 - x)} = 4.2 \times 10^{-7}$$

Assuming $(0.025 - x) \cong 0.025$, we get $x = 1.0 \times 10^{-4}$ M

For the second ionization $\quad HCO_3^-(aq) \rightleftharpoons H^+(aq) + CO_3^{2-}(aq)$

$$K_{a_2} = \frac{[H^+][CO_3^{2-}]}{[HCO_3^-]} = 4.8 \times 10^{-11}$$

Since HCO_3^- is a very weak acid, there is little ionization at this stage. Therefore we have:

$[H^+] = [HCO_3^-] = 1.0 \times 10^{-4}$ M and $[CO_3^{2-}] = 4.8 \times 10^{-11}$ M

16.25 Don't panic! This problem isn't as bad as it might appear. Study Example 16.8 and the discussion following in Section 16.4. One normally works through this sort of problem by treating each successive ionization step as a new exercise; the concentrations calculated in the first step carry over as starting conditions in the next step. Solve the first ionization like Example 16.1. The answers for the second ionization can be written down using the generalization following Example 16.8. In the third ionization the concentration of phosphate ion can be found using the monohydrogen phosphate concentration from the second step and the hydrogen ion concentration from the first.

Solution: The solution for the first step is standard:

$$H_3PO_4(aq) \rightleftharpoons H^+(aq) + H_2PO_4^-(aq)$$

	H_3PO_4	H^+	$H_2PO_4^-$
Initial (M):	0.100	0.000	0.000
Change (M):	-x	+x	+x
Equilibrium (M):	(0.100 - x)	x	x

$$K_{a_1} = \frac{[H^+][H_2PO_4^-]}{[H_3PO_4]} = \frac{x^2}{(0.100 - x)} = 7.5 \times 10^{-3}$$

In this case we probably cannot say that $(0.100 - x) \cong 0.100$ (see Example 16.1). We obtain the quadratic equation

$$x^2 + (7.5 \times 10^{-3})x - (7.5 \times 10^{-4}) = 0$$

The positive root is x = 0.0239 M. We have

$$[H^+] = [H_2PO_4^-] = 0.0239 \text{ M}$$

$$[H_3PO_4] = (0.100 - 0.0239) \text{ M} = 0.076 \text{ M}$$

Second stage of ionization

$$H_2PO_4^-(aq) \rightleftharpoons H^+(aq) + HPO_4^{2-}(aq)$$

	$H_2PO_4^-$	H^+	HPO_4^{2-}
Initial (M):	0.0239	0.0239	0.000
Change (M):	-y	+y	+y
Equilibrium (M):	(0.0239 - y)	(0.0239 + y)	y

$$K_{a_2} = \frac{[H^+][HPO_4^{2-}]}{[H_2PO_4^-]} = \frac{(0.0239 + y)(y)}{(0.0239 - y)} = 6.2 \times 10^{-8}$$

Since K_{a_2} is small, we can assume $(0.0239 + y) \cong (0.0239 - y) \cong 0.0239$, so that $y = 6.2 \times 10^{-8}$ M. Thus

$$[H^+] = [H_2PO_4^-] = 0.0239 \text{ M}$$

$$[HPO_4^{2-}] = K_{a_2} = 6.2 \times 10^{-8} \text{ M}$$

We set up the problem for the third ionization in the same manner

$$HPO_4^{2-}(aq) \rightleftharpoons H^+(aq) + PO_4^{3-}(aq)$$

	$HPO_4^{2-}(aq)$	$H^+(aq)$	$PO_4^{3-}(aq)$
Initial (M):	6.2×10^{-8}	0.0239	0
Change (M):	-z	+z	+z
Equilibrium (M):	$[(6.2 \times 10^{-8}) - z]$	$(0.0239 + z)$	z

$$K_{a_3} = \frac{[H^+][PO_4^{3-}]}{[HPO_4^{2-}]} = \frac{(0.0239 + z)(z)}{[(6.2 \times 10^{-8}) - z]} = 4.8 \times 10^{-13}$$

We can assume $(0.0239 + z) \cong 0.0239$ and $[(6.2 \times 10^{-8}) - z] \cong 6.2 \times 10^{-8}$.

Therefore $$\frac{(0.0239)(z)}{(6.2 \times 10^{-8})} = 4.8 \times 10^{-13}$$

$$z = 1.2 \times 10^{-18}\ M$$

The equilibrium concentrations are: $[H^+] = [H_2PO_4^-] = 0.0239\ M$

$[H_3PO_4] = 0.076\ M$

$[HPO_4^{2-}] = 6.2 \times 10^{-8}\ M$

$[PO_4^{3-}] = 1.2 \times 10^{-18}\ M$

See also Problem 16.76.

16.32 Study Section 16.5 and Review Question 16.29.

Solution: There are two possibilities: (i) MX is the salt of a strong acid and a strong base so that neither the cation nor the anion react with water to alter the pH and (ii) MX is the salt of a weak acid and a weak base with K_a for the acid equal to K_b for the base. The hydrolysis of one would be exactly offset by the hydrolysis of the other.

16.33 Is KOH a strong or a weak base? Is the conjugate base of a weak acid a good proton acceptor? As an acid becomes weaker, what happens to the strength of its conjugate base? Consult Section 16.5 if you are not certain.

Solution: Find and study Table 16.1. Notice that as acid strength decreases, the proton accepting power of the conjugate base increases. In general the weaker the acid, the stronger the conjugate base. The order of increasing acid strength is HZ < HY < HX.

16.34 Find K_b for acetate ion in Table 16.1 in the text. The rest is similar to Example 16.9.

Solution: The salt sodium acetate completely ionizes upon dissolution, producing 0.36 M $[Na^+]$ and 0.36 M $[CH_3COO^-]$ ions. The $[CH_3COO^-]$ ions will undergo hydrolysis because they are a weak base.

	$CH_3COO^-(aq)$ + $H_2O(l)$	⇌ $CH_3COOH(aq)$	+ $OH^-(aq)$
Initial (M):	0.36	0.00	0.00
Change (M):	-x	+x	+x
Equilibrium (M):	(0.36 - x)	+x	+x

$$K_b = \frac{[CH_3COOH][OH^-]}{[CH_3COO^-]} = \frac{x^2}{0.36 - x} = 5.6 \times 10^{-10}$$

Assuming $(0.36 - x) \cong 0.36$, then $x = [OH^-] = 1.4 \times 10^{-5}$

$$pOH = -\log(1.4 \times 10^{-5}) = 4.85$$

$$pH = 14.00 - 4.85 = 9.15.$$

16.35 Find K_a for the ammonium ion in Table 16.2 in the text. The problem is similar to Example 16.10 and Problem 16.34 above.

Solution: The salt ammonium chloride completely ionizes upon dissolution, producing 0.42 M $[NH_4^+]$ and 0.42 M $[Cl^-]$ ions. The $[NH_4^+]$ ions will undergo hydrolysis because they are a weak acid.

	$NH_4^+(aq)$ + $H_2O(l)$	⇌ $NH_3(aq)$	+ $H_3O^+(aq)$
Initial (M):	0.42	0.00	0.00
Change (M):	-x	+x	+x
Equilibrium (M):	(0.42 - x)	+x	+x

$$K_a = \frac{[NH_3][H_3O^+]}{[NH_4^+]} = \frac{x^2}{0.42 - x} = 5.6 \times 10^{-10}$$

Assuming $(0.42 - x) \cong 0.42$, then $x = [H^+] = 1.5 \times 10^{-5}$ M

$$pH = -\log(1.5 \times 10^{-5}) = 4.82$$

Since NH_4Cl is the salt of a weak base (aqueous ammonia) and a strong acid (HCl), we expect that the solution to be slightly acidic, which is confirmed by the calculation.

16.36 In general, this sort of problem is difficult, but this particular exercise is simple. Find K_a for ammonium ion in Table 16.2. Find K_b for acetate ion in Table 16.1. Which is stronger: the acid or the base?

Solution: In this specific case the K_a of ammonium ion is the same as the K_b of acetate ion. The two are of exactly (to two significant figures) equal strength. The solution will have pH 7.00.

What would the pH be if the concentration were 0.1 M in ammonium acetate? 0.4 M?

16.37 This problem involves a solution of a salt of a weak acid and a strong base. Find K_a for hydrocyanic acid and K_b for cyanide ion in Table 16.1. (Which K should be used in this problem?) This is a variation on Example 16.9;

you know the pH and must find the concentration (and convert this to a mass of NaCN). Set up a table like the one in Example 16.9. What is the hydroxide concentration at pH = 10.00? What is the concentration of HCN?

Solution: When the pH is 10.00, the pOH is 4.00 and the concentration of hydroxide ion is 1.0×10^{-4} M. The concentration of HCN must be the same. (Why?) If the concentration of NaCN is x, the table looks like

	$CN^-(aq)$ + $H_2O(l)$	⇌	$HCN(aq)$	+	$OH^-(aq)$
Initial (M):	x		0		0
Change (M):	-1.0×10^{-4}		$+1.0 \times 10^{-4}$		$+1.0 \times 10^{-4}$
Equilibrium (M):	$(x - 1.0 \times 10^{-4})$		(1.0×10^{-4})		(1.0×10^{-4})

$$K_b = \frac{[HCN][OH^-]}{[CN^-]} = \frac{(1.0 \times 10^{-4})^2}{(x - 1.0 \times 10^{-4})} = 2.0 \times 10^{-5}$$

$$x = 6.0 \times 10^{-4} \text{ M} = [CN^-]_o$$

$$\text{Amount of NaCN} = \frac{6.0 \times 10^{-4} \text{ mol}}{1 \text{ L}} \times 0.250 \text{ L} \times \frac{49.01 \text{ g}}{1 \text{ mol}} = 7.4 \times 10^{-3} \text{ g}$$

16.38 The monohydrogen phosphate ion can act as either a proton donor (acid) or a proton acceptor (base). Write out the balanced equations for both processes and find the appropriate K's in Table 16.3.

Solution: The acid and base reactions are

$$\text{acid: } HPO_4^{2-}(aq) \rightleftharpoons H^+(aq) + PO_4^{3-}(aq)$$

$$\text{base: } HPO_4^{2-}(aq) + H_2O(l) \rightleftharpoons H_2PO_4^-(aq) + OH^-(aq)$$

Table 16.3 shows that the value of K_a for HPO_4^{2-} is 4.8×10^{-13}. Note that HPO_4^{2-} is the conjugate base of $H_2PO_4^-$, so K_b is 1.6×10^{-7} (see Table 16.3). Comparing the two K's we conclude that the monohydrogen phosphate ion is a much stronger proton acceptor (base) than a proton donor (acid). The solution will be basic.

16.41 The first part of this problem just involves the ionization of a weak acid which is covered in Section 16.1. Part (b) involves the common ion acetate (where does the acetate ion come from) is like Example 16.12.

Solution: (a) This is a simple weak acid problem which is solved like Problem 16.6 and Example 16.1

	$CH_3COOH(aq)$	⇌	$H^+(aq)$	+	$CH_3COO^-(aq)$
Initial (M):	0.40		0.00		0.00
Change (M):	-x		+x		+x
Equilibrium (M):	(0.40 - x)		+x		+x

$$K_a = 1.8 \times 10^{-5} = \frac{[H^+][CH_3COO^-]}{[CH_3COOH]} = \frac{x^2}{(0.40 - x)} \cong \frac{x^2}{0.40}$$

$x = 2.7 \times 10^{-3}$ M pH = 2.57

(b) In addition to the acetate ion formed from the ionization of acetic acid, we also have acetate ion formed from the sodium acetate dissolving.

$$CH_3COONa(aq) \rightarrow CH_3COO^-(aq) + Na^+(aq)$$

Dissolving 0.20 M sodium acetate initially produces 0.20 M CH_3COO^- and 0.20 M Na^+. The sodium ions are not involved in any further equilibrium (why?), but the acetate ions must be added to the equilibrium in part (a).

	$CH_3COOH(aq)$	$\rightleftharpoons$ $H^+(aq)$	+ $CH_3COO^-(aq)$
Initial (M):	0.40	0.00	0.20
Change (M):	-x	+x	+x
Equilibrium (M):	(0.40 - x)	+x	(0.20 + x)

$$K_a = 1.8 \times 10^{-5} = \frac{[H^+][CH_3COO^-]}{[CH_3COOH]} = \frac{(x)(0.20 + x)}{(0.40 - x)} \cong \frac{x(0.20)}{0.40}$$

$x = 3.6 \times 10^{-5}$ M pH = 4.44.

Could you have predicted whether the pH should have increased or decreased after the additon of the sodium acetate to the pure 0.40 M acetic acid in part (a)?

An alternate way, which is somewhat quicker, to work part (b) of this problem is to use the Henderson-Hasselbalch equation.

$$pH = pK_a + \log\frac{[\text{conjugate base}]}{[\text{acid}]} = -\log(1.8 \times 10^{-5}) + \log\frac{0.20}{0.40} = 4.74 - 0.30 = 4.44$$

16.42 Part (a) is like Example 16.6. Part (b) is like Example 16.12 except that the solution contains a weak base and its conjugate acid. Use Equation (16.5), but remember that this relationship calls for K_a, not K_b. If you need more detail, Problem 16.41 is similar and is presented in more detail.

Solution: (a) This part of the problem is exactly like Example 16.6. It is a simple weak base calculation. The problem is also analogous to 16.44 (a) above (except that problem involves a weak acid). Solving for the pH of this solution as in Example 16.6

pH = 11.28.

(b) Table 16.2 gives the value of K_a for the ammonium ion. Using this and the given concentrations with the Henderson-Hasselbalch equation gives

$$pH = pK_a + \log\frac{[\text{conjugate base}]}{[\text{acid}]} = -\log(5.6 \times 10^{-10}) + \log\frac{(0.20)}{(0.30)} = 9.25 - 0.18 = 9.07$$

Is there any difference in the Henderson-Hasselbalch equation in the cases of a weak acid and its conjugate base and a weak base and its conjugate acid? Can the same equation be used for any such solution?

16.43 Part (a) is like Example 16.1. In part (b), HCl is a strong acid and can be assumed to be 100% ionized. The problem can be set up like Example 16.7. Will the percent ionization of acetic acid be larger in part (a) or (b)?

Solution: (a) This part of the problem is exactly like the first part of Problem 16.41.

$$CH_3COOH(aq) \rightleftharpoons H^+(aq) + CH_3COO^-(aq)$$

	CH_3COOH	H^+	CH_3COO^-
Initial (M):	0.100	0.000	0.000
Change (M):	-x	+x	+x
Equilibrium (M):	(0.100 - x)	x	x

$$K_a = 1.8 \times 10^{-5} = \frac{[H^+][CH_3COO^-]}{[CH_3COOH]} = \frac{x^2}{(0.100 - x)} \cong \frac{x^2}{0.100}$$

$$x = 1.3 \times 10^{-3}\ M = [H^+] = [CH_3COO^-]$$

(b) Let x be the concentration of hydrogen ion and acetate ion produced by the ionization of the acetic acid. We summarize

$$CH_3COOH(aq) \rightleftharpoons H^+(aq) + CH_3COO^-(aq)$$

	CH_3COOH	H^+	CH_3COO^-
Initial (M):	0.100	0.100	0.000
Change (M):	-x	+x	+x
Equilibrium (M):	(0.100 - x)	(0.100 +x)	x

$$K_a = \frac{[H^+][CH_3COO^-]}{[CH_3COOH]} = \frac{(0.100 + x)(x)}{(0.100 - x)} = 1.8 \times 10^{-5}$$

Because of the common ion effect (Le Chatelier's principle), x should be much less than 5% of the initial concentration (see Example 16.1) so we can write $(0.100 - x)\ M \cong (0.100 + x)\ M \approx 0.100\ M$ The equation then becomes

$$\frac{(0.100)(x)}{(0.100)} = x = 1.8 \times 10^{-5}\ M$$

The concentration of hydrogen ion will be 0.100 M and the concentration of acetate ion will be 1.8×10^{-5} M.

16.50 Review the definition of a buffer solution at the beginning of Section 16.7. This is like Example 16.13.

Solution: All choices except (a) and (e) are buffer systems. (Why?)

16.51 This like Problem 16.42 above.

Solution: $NH_4^+(aq) \rightleftharpoons NH_3(aq) + H^+(aq)$ $K_a = 5.6 \times 10^{-10}$; $pK_a = 9.25$

$$pH = pK_a + \log \frac{[NH_3]}{[NH_4^+]} = 9.25 + \log \frac{0.15}{0.35} = 8.88$$

16.53 Write out the Henderson-Hasselbalch equation for this system and solve algebraically for the desired ratio. At pH = 7.40 which component of the buffer system is present in higher concentration?

Solution: At pH = 7.40, the Henderson-Hasselbalch equation for this system is

$$pH = 7.40 = -\log(4.2 \times 10^{-7}) + \log\frac{[HCO_3^-]}{[H_2CO_3]} = 6.38 + \log\frac{[HCO_3^-]}{[H_2CO_3]}$$

The [conjugate base]/[acid] ratio is

$$\log \frac{[HCO_3^-]}{[H_2CO_3]} = 7.40 - 6.38 = 1.02$$

$$\frac{[HCO_3^-]}{[H_2CO_3]} = 10^{1.02} = 10$$

The buffer should be more effective against an added acid because more base is present. Note that a pH of 7.40 is only a two significant figure number (Why?); the final result should only have two significant figures.

16.55 In this problem which compound is the weak acid and which is its salt? Which K_a from Table 16.3 should you choose for the Henderson-Hasselbalch equation?

Solution: The acid (proton donor) will always be the substance with more protons. Choose the value of K_a for the dihydrogen phosphate ion from Table 16.3. The pH of this solution can be found by substitution into the Henderson-Hasselbalch equation

$$pH = -\log(6.2 \times 10^{-8}) + \log \frac{[HPO_4^{2-}]}{[H_2PO_4^-]} = 7.21 + \log \frac{(0.10)}{(0.15)} = 7.21 - 0.18 = 7.03$$

16.57 This problem is like Example 16.14.

Solution: Before the addition of an acid or a base, using the Henderson-Hasselbalch equation,

$$pH = pK_a + \log \frac{[\text{conjugate base}]}{[\text{acid}]} = 4.74 + \log \frac{1.00}{1.00} = 4.74$$

(a) Here the neutralization reaction is

$CH_3COOH(aq)$ +	$OH^-(aq)$	→	$CH_3COO^-(aq) + H_2O(l)$
0.080 mol	0.080 mol		0.080 mol

$$[CH_3COOH] = (\text{initial - amount reacted}) = (1.00 - 0.080)\ M = 0.92\ M$$

$[CH_3COO^-]$ = (initial + amount formed) = (1.00 + 0.080) M = 1.08 M

$$pH = 4.74 + \log \frac{1.08}{0.92} = 4.81$$

(b) The neutralization is:

$$CH_3COO^-(aq) + H^+(aq) \rightarrow CH_3COOH(aq)$$

$$0.12 \text{ mol} \quad 0.12 \text{ mol} \quad 0.12 \text{ mol}$$

$[CH_3COOH]$ = (initial + amount formed) = (1.00 + 0.12) M = 1.12 M

$[CH_3COO^-]$ = (initial - amount reacted) = (1.00 - 0.12) M = 0.88 M

$$pH = 4.74 + \log \frac{0.88}{1.12} = 4.64$$

16.58 The first part is like Review Question 16.49 (use the correct K!). The second part is similar to Example 16.14. Notice, however, that you must recalculate all concentrations because of the volume change.

Solution: For the first part we use K_a for ammonium ion. (Why?) The Henderson-Hasselbalch equation is

$$pH = -\log(5.6 \times 10^{-10}) + \log\frac{(0.20)}{(0.20)} = 9.25$$

For the second part, the acid-base reaction is

$$NH_3(g) + H^+(aq) \rightarrow NH_4^+(aq)$$

We find the number of moles of HCl added

$$10.0 \text{ mL} \left(\frac{1 \text{ L}}{1000 \text{ mL}}\right)\left(\frac{0.10 \text{ mol HCl}}{1 \text{ L}}\right) = 0.0010 \text{ mol HCl}$$

The number of moles of NH_3 and NH_4^+ originally present are

$$65.0 \text{ mL} \left(\frac{1 \text{ L}}{1000 \text{ mL}}\right)\left(\frac{0.20 \text{ mol}}{1 \text{ L}}\right) = 0.013 \text{ mol}$$

Using the acid-base reaction, we find the number of moles of NH_3 and NH_4^+ after addition of the HCl

$$\text{moles } NH_3 = (0.013 - 0.0010) \text{ mol} = 0.012 \text{ mol } NH_3$$

$$\text{moles } NH_4^+ = (0.013 + 0.0010) \text{ mol} = 0.014 \text{ mol } NH_4^+$$

We find the new pH: $pH = 9.25 + \log\frac{(0.012)}{(0.014)} = 9.18$

16.59 This problem is similar to Example 16.15. Study the discussion of buffer range in Section 16.7.

Solution: We write

$K_{a_1} = 1.1 \times 10^{-3}$; $pK_{a_1} = 2.96$ $K_{a_2} = 2.5 \times 10^{-6}$; $pK_{a_2} = 5.60$

In order for the buffer solution to behave effectively, the pK_a of the acid component must be close to the desired pH. Therefore, the proper buffer system is Na_2A/NaHA.

16.61 Does this problem involve any weak acid of weak base equilibria?

Solution: The neutralization reaction is

$$H_2SO_4(aq) + 2NaOH(aq) \rightarrow Na_2SO_4(aq) + 2H_2O(l)$$

Since one mole of sulfuric acid combines with two moles of sodium hydroxide, we write

$$12.5 \text{ mL } H_2SO_4 \left(\frac{1 \text{ L soln}}{1000 \text{ mL soln}}\right)\left(\frac{0.500 \text{ mol } H_2SO_4}{1 \text{ L soln}}\right)\left(\frac{2 \text{ mol NaOH}}{1 \text{ mol } H_2SO_4}\right)\left(\frac{1000 \text{ mL/L}}{50.0 \text{ mL}}\right) = 0.25 \text{ M}$$

16.62 How many moles of acid are represented by a 0.2688 g sample?

Solution: Since the acid is monoprotic, the number of moles of KOH is equal to the number of moles of acid.

$$\text{Moles acid} = 16.4 \text{ mL} \left(\frac{1 \text{ L}}{1000 \text{ mL}}\right)\left(\frac{0.08133 \text{ mol}}{1 \text{ L}}\right) = 0.00133 \text{ mol}$$

$$\text{Molar mass} = \frac{0.2688 \text{ g}}{0.00133 \text{ mol}} = 202 \text{ g/mol}$$

16.63 This problem is similar to Problem 16.62 above.

Solution: The neutralization reaction is: $2KOH(aq) + H_2A(aq) \rightarrow K_2A(aq) + 2H_2O(l)$

The number of moles of H_2A reacted is

$$11.1 \text{ mL KOH} \left(\frac{1.00 \text{ mol KOH}}{1000 \text{ mL}}\right)\left(\frac{1 \text{ mol } H_2A}{2 \text{ mol KOH}}\right) = 5.55 \times 10^{-3} \text{ mol } H_2A$$

The molar mass of H_2A is given by: $\dfrac{5.00 \text{ g}}{10 \times 5.6 \times 10^{-3} \text{ mol}} = 89 \text{ g/mol}$

16.65 Part (a) is like Problem 16.62 above. In part (b) you are dealing with a buffer solution. Find the [conjugate base]/[acid] ratio from the given data and use the Henderson-Hasselbalch equation to calculate pK_a.

Solution: (a) The molar mass of the acid is found as in Problem 16.62.

$$\text{Moles acid} = 18.4 \text{ mL} \left(\frac{1 \text{ L}}{1000 \text{ mL}}\right)\left(\frac{0.0633 \text{ mol}}{1 \text{ L}}\right) = 0.00116 \text{ mol}$$

$$\text{Molar mass} = \frac{0.1276 \text{ g}}{0.00116 \text{ mol}} = 110 \text{ g/mol}$$

(b) The amount of NaOH in 10.0 mL of solution is

$$10.0 \text{ mL} \left(\frac{1 \text{ L}}{1000 \text{ mL}}\right)\left(\frac{0.0633 \text{ mol}}{1 \text{ L}}\right) = 0.000633 \text{ mol}$$

After addition of the NaOH, the amounts of acid and its conjugate base are

$$\text{Moles acid} = (0.00116 - 0.000633) \text{ mol} = 0.00053 \text{ mol}$$

$$\text{Moles conjugate base} = 0.000633 \text{ mol (Why?)}$$

Using Equation (16.5) we write

$$\text{pH} = 5.87 = \text{pK}_a + \log \frac{0.000633 \text{ mol}}{0.00053 \text{ mol}} = \text{pK}_a + 0.077$$

$$K_a = 1.6 \times 10^{-6}$$

16.66 Is the resulting solution a buffer system? Which part of Example 16.16 best represents this situation? What is the final volume of the combined solutions?

Solution: The resulting solution is not a buffer system. There is excess NaOH and the neutralization is well past the equivalence point. We have a solution of sodium acetate and sodium hydroxide like part (c) of Example 16.16.

$$\text{Moles NaOH} = 0.500 \text{ L} \left(\frac{0.167 \text{ mol}}{1 \text{ L}}\right) = 0.0835 \text{ mol}$$

$$\text{Moles } CH_3COOH = 0.500 \left(\frac{0.100 \text{ mol}}{1 \text{ L}}\right) = 0.0500 \text{ mol}$$

After neutralization, the amount of NaOH remaining is 0.0835 - 0.0500 = 0.0335 mol. The volume of the resulting solution is 1.00 L.

$$[OH^-] = \frac{0.0335 \text{ mol}}{1.00 \text{ L}} = 0.0335 \text{ M} \qquad [Na^+] = \frac{0.0835 \text{ mol}}{1.00 \text{ L}} = 0.0835 \text{ M}$$

$$[H^+] = \frac{1.0 \times 10^{-14}}{0.0335} = 3.0 \times 10^{-13} \text{ M} \qquad [CH_3COO^-] = \frac{0.0500 \text{ mol}}{1.00 \text{ L}} = 0.0500 \text{ M}$$

$$[CH_3COOH] = \frac{[H^+][CH_3COO^-]}{K_a} = 8.3 \times 10^{-10} \text{ M}$$

16.67 You need to determine whether the pH at the equivalence point is less than 7, equal to 7, or greater than 7 before you can work this problem.

Solution: (a) This is a strong acid-strong base titration. The equivalence point is pH = 7. See Figure 16.8.

(b) This is a strong acid-weak base titration. The equivalence point is pH < 7. See Figure 16.9(b).

(c) This is a weak acid-strong base tiration. The equivalence point is pH > 7. See Figure 16.9(a).

16.73 In this problem HIn is a weak acid. Find the $[HIn]/[In^-]$ ratio.

Solution: From the given information the pK_a of the indicator is 6.00. We write

$$\frac{[HIn]}{[In^-]} = \frac{[H^+]}{K_a} = \frac{1.0 \times 10^{-4}}{1.0 \times 10^{-6}} = 100$$

The color will be that of the nonionized form (red).

16.75 How do these two acids compare in terms of strength? In what way are they different?

Solution: The pH of a 0.040 M HCl solution (strong acid) is 1.40 (why?). Following the procedure in Example 16.7, we find that the pH of a 0.040 M H_2SO_4 solution is 1.31.

Without doing any calculations, could you have known that the pH of the sulfuric acid would be lower (more acidic) than that of the hydrochloric acid?

16.76 Study the solution of Example 16.8. In the analysis of the second stage of ionization, why can we say $0.054 \pm y \cong 0.054$?

Solution: If $K_{a_1} >> K_{a_2}$, we can assume that the equilibrium concentration of hydrogen ion results only from the first stage of ionization. In the second stage this always leads to an expression of the type

$$\frac{(c + y)(y)}{(c - y)} = K_{a_2}$$

where c represents the equilibrium hydrogen ion concentration found in the first stage. If $c >> K_{a_2}$, we can assume $(c \pm y) \cong c$, and consequently $y = K_{a_2}$.

Is this conclusion also true for the second stage ionization of a triprotic acid like H_3PO_4? See Problem 16.25.

16.77 Write out the equation for the ionization of formic acid. To what class of reactions does it belong (i.e., combination, metathesis, etc.)? Is this type of reaction endo- or exothermic? How does the equilibrium constant for that type of process depend upon temperature? Will your conclusion be true for most acids?

Solution: The ionization of any acid is an endothermic process. The higher the temperature, the greater the K_a value. Formic acid will be a stronger acid at 40°C than at 25°C.

16.78 How can you calculate the equilibrium constant for a reaction that is the sum of two other reactions (assuming you know the equilibrium constants for these reactions)? If you don't remember this, review Section 14.2.

Solution: From Table 16.1 we have the equilibria below

$$CH_3COOH(aq) \rightleftharpoons H^+(aq) + CH_3COO^-(aq) \qquad K_a = 1.8 \times 10^{-5}$$

$$H^+(aq) + NO_2^-(aq) \rightleftharpoons HNO_2(aq) \qquad K_a' = \frac{1}{K_a} = 2.2 \times 10^3$$

$$CH_3COOH(aq) + NO_2^-(aq) \rightleftharpoons CH_3COO^-(aq) + HNO_2(aq)$$

$$K = (1.8 \times 10^{-5})(2.2 \times 10^3) = 4.0 \times 10^{-2}$$

The equation in this problem is the sum of these two reactions. The equilibrium constant for this sum is the product of the equilibrium constants of the component reactions (Section 14.2).

16.79 Study the discussion of distribution curves in Section 16.7. Examine Figure 16.5. What do the curves represent? Why are they different in the two figures? How can you predict the points at which the curves intersect? How many curves will there be for oxalic acid, and where will the intersections occur?

Solution: In Figure 16.5 each curve represents the fraction of a particular solution species present at a given pH. The curves intersect when the fractions are equal, i.e. 0.50. In terms of the equilibrium constant expression this means that at this point pH = pK_a.

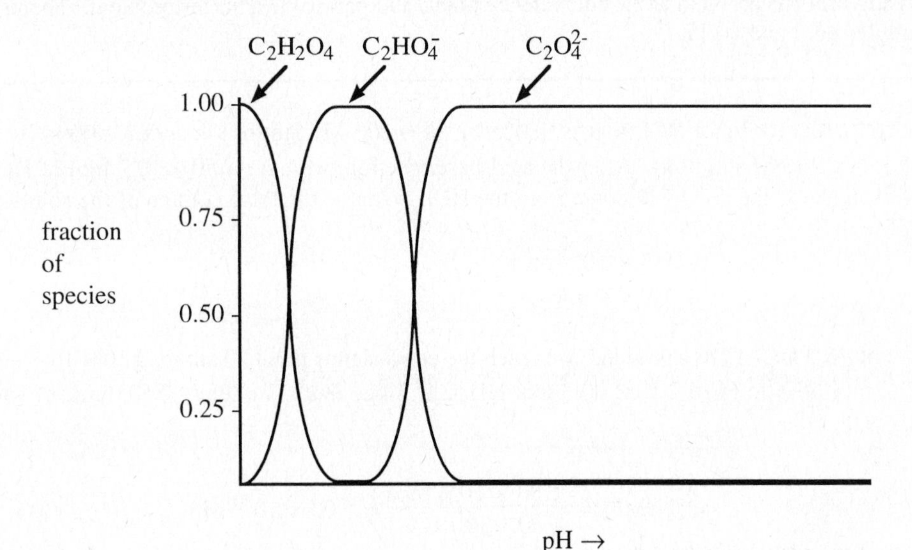

$$pH = pK_a + \log\frac{[A^-]}{[HA]} = pK_a + 0 = pK_a$$

For oxalic acid the distribution curves track the fractions of three species: $C_2H_2O_4$, $C_2HO_4^-$, and $C_2O_4^{2-}$. The curves intersect at the points where the pH equals pK_{a_1} (1.19) and pK_{a_2} (4.21).

16.81 What happens when sodium carbonate reacts with hydrochloric acid? Write a balanced equation. Does sodium carbonate react with KOH? Does the KOH neutralize all the acid?

Solution: The amount of HCl initially present is

$$25.0 \text{ mL} \left(\frac{1 \text{ L}}{1000 \text{ mL}}\right)\left(\frac{1.00 \text{ mol HCl}}{1 \text{ L}}\right) = 0.0250 \text{ mol HCl}$$

The amount of KOH added is

$$0.560 \text{ g KOH} \left(\frac{1 \text{ mol}}{56.11 \text{ g}}\right) = 0.00998 \text{ mol KOH}$$

The amount of HCl remaining is

$$(0.0250 - 0.00998) \text{ mol} = 0.0150 \text{ mol}$$

The balanced equation for the reaction is

$$Na_2CO_3(aq) + 2HCl(aq) \rightarrow CO_2(g) + 2NaCl(aq) + H_2O(l)$$

The mass of CO_2 is

$$0.0150 \text{ mol HCl} \left(\frac{1 \text{ mol } CO_2}{2 \text{ mol HCl}}\right)\left(\frac{44.01 \text{ g } CO_2}{1 \text{ mol } CO_2}\right) = 0.330 \text{ g } CO_2$$

16.83 We are titrating a strong acid HCl with a weak base NH_3. Part (a) involves the calculation of the pH of a strong acid; part (b) is a hydrolysis problem at the equivalence point; and part (c) is a buffer problem. The problem is similar to Examples 16.6 and 16.17.

Solution: (a) To 2.50×10^{-3} mol HCl (that is, 0.0250 L of 0.100 M solution) is added 1.00×10^{-3} mol NH_3 (that is, 0.0100 L of 0.100 M solution). After the acid-base reaction, we have 1.50×10^{-3} mol of HCl remaining. Since HCl is a strong acid, the $[H^+]$ will come from the HCl. What is the total volume of the solution?

$$[H^+] = \frac{1.50 \times 10^{-3} \text{ mol}}{0.0350 \text{ L}} = 0.0429 \text{ M} \qquad \text{pH} = 1.37$$

(b) When a total of 25.0 mL of NH_3 is added, we reach the equivalence point. That is, 2.50×10^{-3} mol HCl reacts with 2.50×10^{-3} mol NH_3 to form 2.50×10^{-3} mol NH_4Cl. Since there is a total of 50.0 mL of solution, the concentration of NH_4^+ is

$$[NH_4^+] = \frac{2.50 \times 10^{-3} \text{ mol}}{0.0500 \text{ L}} = 5.00 \times 10^{-2} \text{ M}$$

This is a problem involving the hydrolysis of the weak acid NH_4^+ (see Example 16.10).

$$K_a = 5.6 \times 10^{-10} = \frac{[NH_3][H^+]}{[NH_4^+]} = \frac{x^2}{(5.00 \times 10^{-2} - x)} \cong \frac{x^2}{5.00 \times 10^{-2}}$$

$2.80 \times 10^{-11} = x^2$ $\quad x = 5.2 \times 10^{-6}$ M $\quad$ pH = 5.28

(c) When a total of 35.0 mL of 0.100 M NH_3 (3.50×10^{-3} mol) is added to the 25 mL of 0.100 mol HCl (2.50×10^{-3} mol), the acid-base reaction produces 2.50×10^{-3} mol NH_4Cl with 1.00×10^{-3} mol of NH_3 in excess. Using the Henderson-Hasselbalch equation

$$pH = pK_a + \log \frac{[\text{conjugate base}]}{[\text{acid}]}$$

$$pH = -\log(5.6 \times 10^{-10}) + \log \frac{(1.00 \times 10^{-3})}{(2.50 \times 10^{-3})} = 8.85$$

16.85 The K_a for formic acid is in Table 16.1.

Solution: Converting the pH value of 2.53 to $[H^+] = 2.95 \times 10^{-3}$ M. Using the balanced equation

$$\begin{array}{ccccc} HCOOH(aq) & \rightleftharpoons & H^+(aq) & + & HCOO^-(aq) \\ y - (2.95 \times 10^{-3}) & & (2.95 \times 10^{-3}) & & (2.95 \times 10^{-3}) \end{array}$$

$$K_a = 1.7 \times 10^{-4} = \frac{[H^+][HCOO^-]}{[HCOOH]} = \frac{(2.95 \times 10^{-3})^2}{(y - 2.95 \times 10^{-3})}$$

$$y = 0.054 \text{ M}$$

The molar mass of HCOOH is 46.03 g. Calculating the number of grams in 100 mL is

Molarity = moles/L = [g/(molar mass)]/L $\quad$ g = Molarity × L × (molar mass)

g = (0.054 mol/L) x (0.100L) x (46.03 g/mol) = 0.25 g

16.87 Write an equilibrium expression for the reaction, and see if you can recognize K_w and K_a in the expression.

Solution: Writing K for the reaction

$$K = \frac{[CH_3COO^-]}{[CH_3COOH][OH^-]}$$

If we multiply both top and bottom by $[H^+]$

$$K = \frac{[CH_3COO^-][H^+]}{[CH_3COOH][OH^-][H^+]}$$

This expression is the same as

$$K = \frac{K_a}{K_w} = \frac{1.8 \times 10^{-5}}{1.0 \times 10^{-14}} = 1.8 \times 10^9$$

The very large equilibrium constant means that the reaction goes almost to completion, which is exactly what is expected from and acid-base neutralization reaction.

16.89 The first part of the problem is a Henderson-Hasselbalch buffer problem. The second part is a weak acid calculation. The K_a for phenolphthalein is 9.10.

Solution: (a)

$$pH = pK_a + \log \frac{[\text{conjugate base}]}{[\text{acid}]}$$

$$8.00 = 9.10 + \log \frac{[\text{ionized}]}{[\text{un-ionized}]}$$

$$\frac{[\text{un-ionized}]}{[\text{ionized}]} = 12.6 \qquad (1)$$

(b) The total concentration of the indicator is

$$(2 \text{ drops}) \times \left(\frac{0.050 \text{ mL}}{1 \text{ drop}}\right) \times (0.060 \text{ M}) \times \left(\frac{1}{50 \text{ mL}}\right) = 1.2 \times 10^{-4} \text{ M}$$

Using equation (1) above and letting y = [ionized]

$$\frac{1.0 \times 10^{-4} - y}{y} = 12.6$$

$$y = 8.8 \times 10^{-6} \text{ M}$$

16.91 Remember that when you mix equal volumes of two solutions, since the volume is doubled, the concentrations will be halved.

Solution: (a) Mix 500 mL of 0.40 M CH_3COOH with 500 mL of 0.40 M CH_3COONa. Since the final volume is 1.00 L, then the concentrations of the two solutions that were mixed must be one-half of their initial concentrations.

(b) Mix 500 mL of 0.80 M CH_3COOH with 500 mL of 0.40 M NaOH. (Note: half of the acid reacts with all of the base to make a solution identical to that in part (a) above.

$$CH_3COOH + NaOH \rightarrow CH_3COONa + H_2O$$

(c) Mix 500 mL of 0.80 M CH_3COONa with 500 mL of 0.40 M HCl. (Note: half of the salt reacts with all of the acid to make a solution identical to that in part (a) above.

$$CH_3COO^- + H^+ \rightarrow CH_3COOH$$

16.93 This problem is a very direct application of the Henderson-Hasselbalch equation.

Solution: $K_b = 8.91 \times 10^{-6}$ $\quad pK_b = 5.05$ $\quad pK_a = 14.00 - pK_b = 8.95$

Using equation 16.5 $\quad pH = pK_a + \log \frac{[\text{conjugate base}]}{[\text{acid}]}$ $\quad 7.40 = 8.95 + \log \frac{[\text{conjugate base}]}{[\text{acid}]}$

$$\frac{[\text{conjugate base}]}{[\text{acid}]} = 0.0282$$

16.94 This problem is quite similar to 16.93 above in that it is a straightforward application of Equation 16.5.

Solution: $$pH = pK_a + \log \frac{[\text{conjugate base}]}{[\text{acid}]}$$

Solving for the pH with 90% of the indicator in the HIn form:

$$pH = 3.46 + \log \frac{[10]}{[90]} = 3.46 - 0.95 = 2.51$$

Solving for the pH with 90% of the indicator in the In^- form:

$$pH = 3.46 + \log \frac{[90]}{[10]} = 3.46 + 0.95 = 4.41$$

Thus the pH range varies from 2.51 to 4.41 as the [HIn] varies from 90% to 10%.

16.96 What type of salt is $NaNO_2$? In order to work the problem you must realize that $NaNO_2$ is a salt of a weak acid (HNO_2) and a strong base (NaOH).

Solution: The important equation is the hydrolysis of NO_2^-: $NO_2^- + H_2O \rightleftharpoons HNO_2 + OH^-$

(a) Addition of HCl will result in the reaction of the H^+ from the HCl with the OH^- that was present in the solution. The OH^- will effectively be removed and the equilibrium will shift to the right to compensate (more hydrolysis).

(b) Addition of NaOH is effectively addition of more OH^- which places stress on the right hand side of the equilibrium. The equilibrium will shift to the left (less hydroysis) to compensate for the addition of OH^-.

(c) Addition of NaCl will have no effect.

(d) Recall that the percent ionization of a weak acid increases with dilution (see Figure 15.1). The same is true for weak bases. Thus dilution will cause more hydrolysis, shifting the equilibrium to the right.

16.97 This is not a trivial problem. In effect, it is a titration problem that is turned around. You first find that the amount of NaOH added is less than the equivalence point, so you can consider the problem to be a buffer problem and use the Henderson-Hasselbalch equation.

Solution: The original pH of the 2.00 M weak acid HNO_2 solution:

$$K_a = \frac{[H^+][NO_2^-]}{[HNO_2]} = \frac{x^2}{2.00 - x} = 4.5 \times 10^{-4} \qquad x = 3.0 \times 10^{-2} = [H^+] \quad pH = 1.52$$

Since the pH after the addition is 1.5 pH units greater, the new pH = 1.52 + 1.50 = 3.02. $[H^+] = 9.55 \times 10^{-4}$ M

When the NaOH is added, we dilute our original 2.00 M HNO_2 solution to: $\frac{400 \text{ mL}}{600 \text{ mL}} \times 2.00 \text{ M} = 1.33 \text{ M}$

Since we have not reached the equivalence point, we have a buffer solution. The decease in $[HNO_2]$ is the same as the [NaOH], as is the increase in $[NO_2^-]$.

$$\begin{array}{cccc} HNO_2 & \rightleftharpoons & H^+ & + & NO_2^- \\ 1.33 - x & & 9.55 \times 10^{-4} & & x \end{array}$$

Using Equation (16.5) $\quad pH = pK_a + \log \frac{[\text{conjugate base}]}{[\text{acid}]} \qquad 3.02 = 3.35 + \log \frac{x}{1.33 - x}$

Solving, $\quad x = 0.424 \text{ M} \quad$ Correcting for the dilution from 200 mL to 600 mL, $\frac{600 \text{ mL}}{200 \text{ mL}} \times 0.424 \text{ M} = 1.27 \text{ M}$

Note: If you have not already done so, please read the "Introduction for the Student" at the beginning of this Student Solutions Manual.

Chapter 17

SOLUBILITY EQUILIBRIA

17.7 Find the K_{sp} values in Table 17.1 and write out the solubility product expressions.

Solution: (a) The solubility equilibrium is given by the equation

$$AgI(s) \rightleftharpoons Ag^{+}(aq) + I^{-}(aq)$$

The expression for K_{sp} is given by

$$K_{sp} = [Ag^{+}][I^{-}]$$

The value of K_{sp} can be found in Table 17.1. If the equilibrium concentration of silver ion is the value given, the concentration of iodide ion must be

$$[I^{-}] = \frac{K_{sp}}{[Ag^{+}]} = \frac{8.3 \times 10^{-17}}{9.1 \times 10^{-9}} = 9.1 \times 10^{-9}\ M$$

(b) The value of K_{sp} for aluminum hydroxide can be found in Table 17.1. The equilibrium expressions are

$$Al(OH)_3(s) \rightleftharpoons Al^{3+}(aq) + 3OH^{-}(aq)$$

$$K_{sp} = [Al^{3+}][OH^{-}]^3$$

Using the given value of the hydroxide ion concentration, the equilibrium concentration of aluminum ion is

$$[Al^{3+}] = \frac{K_{sp}}{[OH^{-}]^3} = \frac{1.8 \times 10^{-33}}{(2.9 \times 10^{-9})^3} = 7.4 \times 10^{-8}\ M$$

What is the pH of this solution? Will the aluminum concentration change if the pH is altered?

17.9 What ions are produced when $MnCO_3$ dissolves? If s moles of $MnCO_3$ dissolve, how many moles of each of the ions will be produced. If you have difficulty with this problem, study Example 17.1 and then try this problem again.

Solution: For $MnCO_3$ dissolving, we write

$$MnCO_3(s) \rightleftharpoons Mn^{2+}(aq) + CO_3^{2-}(aq)$$

For every mole of $MnCO_3$ that dissolves, one mole of Mn^{2+} will be produced and one mole of CO_3^{2-} will be produced. If the molar solubility of $MnCO_3$ is s mol/L, then the concentration of

$$[Mn^{2+}] = [CO_3^{2-}] = s = 4.2 \times 10^{-6}\ M$$

$$K_{sp} = [Mn^{2+}][CO_3^{2-}] = s^2 = (4.2 \times 10^{-6})^2 = 1.8 \times 10^{-11}$$

17.11 What are the charges of the M and X ions? Find the concentrations of the M and X ions in 1.0 L of saturated solution. Example 17.2 in the text is a similar problem.

Solution: The charges of the M and X ions are +3 and -2, respectively (are other values possible?). We first calculate the number of moles of M_2X_3 that dissolve in 1.0 L of water

$$\text{Moles } M_2X_3 = 3.6 \times 10^{-17}\ g \left(\frac{1\ mol}{288\ g}\right) = 1.3 \times 10^{-19}\ mol$$

The molar solubility s of the compound is therefore 1.3×10^{-19} M. At equilibrium the concentration of M^{3+} must be 2s and that of X^{2-} must be 3s. (See Example 17.4 and Table 17.2.)

$$K_{sp} = [M^{3+}]^2[X^{2-}]^3 = [2s]^2[3s]^3 = 108s^5$$

Since these are equilibrium concentrations, the value of K_{sp} can be found by simple substitution

$$K_{sp} = 108s^5 = 108(1.3 \times 10^{-19})^5 = 4.0 \times 10^{-93}$$

17.13 You must use the data in Table 17.1 to solve this problem, which is similar to Example 17.4.

Solution: First we find the molar solubility s and then convert moles to grams. The solubility equilibrium for silver carbonate is

$$Ag_2CO_3(s) \rightleftharpoons 2Ag^+(aq) + CO_3^{2-}(aq)$$

$$K_{sp} = [Ag^+]^2[CO_3^{2-}] = (2s)^2(s) = 4s^3 = 8.1 \times 10^{-12}$$

$$s = \left(\frac{8.1 \times 10^{-12}}{4}\right)^{1/3} = 1.3 \times 10^{-4}\ M$$

Convering from mol/L to g/L: $\left(\frac{1.3 \times 10^{-4}\ mol}{1\ L\ soln}\right)\left(\frac{275.8\ g}{1\ mol}\right) = 0.036\ g/L$

17.14 This problem is similar to Example 17.5 in the text.

Solution: According to the solubility rules, the only precipitate that might form is $BaCO_3$.

$$Ba^{2+}(aq) + CO_3^{2-}(aq) \rightarrow BaCO_3(s)$$

The number of moles of Ba^{2+} present in the original 20.0 mL of $Ba(NO_3)_2$ solution is

$$20.0 \text{ mL} \times \frac{0.10 \text{ mol Ba}^{2+}}{1 \text{ L soln}} \times \frac{1 \text{ L}}{1000 \text{ mL}} = 2.0 \times 10^{-3} \text{ mol Ba}^{2+}$$

The total volume after combining the two solutions is 70.0 mL. The concentration of Ba^{2+}in the 70 mL volume is

$$[Ba^{2+}] = \frac{2.0 \times 10^{-3} \text{ mol Ba}^{2+}}{70.0 \text{ mL}} \times \frac{1000\text{mL}}{1\text{L soln}} = 2.9 \times 10^{-2} \text{ M}$$

The number of moles of CO_3^{2-} present in the original 50.0 mL Na_2CO_3 solution is

$$50.0 \text{ mL} \times \frac{0.10 \text{ mol CO}_3^{2-}}{1 \text{ L soln}} \times \frac{1 \text{ L}}{1000 \text{ mL}} = 5.0 \times 10^{-3} \text{ mol CO}_3^{2-}$$

The concentraion of CO_3^{2-} in the 70.0 mL of combined solution is

$$[CO_3^{2-}] = \frac{5.0 \times 10^{-3} \text{ mol CO}_3^{2-}}{70.0 \text{ mL}} \times \frac{1000\text{mL}}{1\text{L soln}} = 7.1 \times 10^{-2} \text{ M}$$

Now we must compare Q and K_{sp}. From Table 17.1, the K_{sp} for $BaCO_3$ is 8.1×10^{-9}. As for Q,

$$Q = [Ba^{2+}][CO_3^{2-}] = (2.9 \times 10^{-2})(7.1 \times 10^{-2}) = 2.1 \times 10^{-3}$$

Since $(2.1 \times 10^{-3}) > (8.1 \times 10^{-9})$, then $Q > K_{sp}$. Therefore $BaCO_3$ will precipitate.

17.15 Find the hydroxide ion concentration for the saturated solution and convert this to pH.

Solution: Let s be the molar solubility of $Zn(OH)_2$. The equilibrium concentrations of the ions are then

$$[Zn^{2+}] = s \text{ and } [OH^-] = 2s$$

$$K_{sp} = [Zn^{2+}][OH^-]^2 = (s)(2s)^2 = 4s^3 = 1.8 \times 10^{-14}$$

$$s = \left(\frac{1.8 \times 10^{-14}}{4}\right)^{1/3} = 1.7 \times 10^{-5}$$

$$[OH^-] = 2s = 3.4 \times 10^{-5} \text{ M and pOH} = 4.47$$

$$pH = 14.00 - 4.47 = 9.53$$

If the K_{sp} of $Zn(OH)_2$ were smaller by many more powers of ten, would 2s still be the hydroxide ion concentration in the solution?

17.16 What is the hydroxide ion concentration in this solution? What about the concentration of M^+?

Solution: First we need to calculate the OH^- concentration from the pH.

$$pH = 9.68 \qquad [H^+] = 2.1 \times 10^{-10}\ M$$

$$[OH^-] = \frac{K_w}{[H^+]} = \frac{1.0 \times 10^{-14}}{2.1 \times 10^{-10}} = 4.8 \times 10^{-5}\ M$$

From the balanced equation we know that $[M^+] = [OH^-]$, so we can write

$$K_{sp} = [M^+][OH^-] = (4.8 \times 10^{-5})^2 = 2.3 \times 10^{-9}$$

17.17 In this problem a precipitate of SrF_2 will form; Na^+ and NO_3^- will not precipitate (why?). Which ion is the limiting reagent, Sr^{2+} or F^-? Find the concentrations of all the ions after mixing, but before precipitation starts (notice that the new volume is 100 mL).

Solution: We first find the limiting reagent in the precipitation reaction.

$$\text{Moles NaF} = 75\ \text{mL}\left(\frac{1\ \text{L}}{1000\ \text{mL}}\right)\left(\frac{0.060\ \text{mol}}{1\ \text{L}}\right) = 0.0045\ \text{mol}$$

$$\text{Moles Sr(NO}_3)_2 = 25\ \text{mL}\left(\frac{1\ \text{L}}{1000\ \text{mL}}\right)\left(\frac{0.15\ \text{mol}}{1\ \text{L}}\right) = 0.0038\ \text{mol}$$

To form 0.0038 mol of SrF_2 we require 0.0076 mol of NaF. The NaF is present in short supply and is therefore the limiting reagent. The pre-equilibrium concentrations of the ions are:

$$[F^-]_o = [Na^+]_o = \frac{0.0045\ \text{mol}}{0.10\ \text{L}} = 0.045\ M; \quad [Sr^{2+}]_o = \frac{0.0038\ \text{mol}}{0.10\ \text{L}} = 0.038\ M; \quad [NO_3^-]_o = \frac{0.0076\ \text{mol}}{0.10\ \text{L}} = 0.076\ M$$

The sodium ion and nitrate ion do not precipitate; their concentrations are the values calculated above.

$$[Na^+] = 0.045\ M \text{ and } [NO_3^-] = 0.076\ M$$

The concentration of Sr^{2+} will be that remaining after precipitation. The 0.0045 mol of F^- will combine with 0.0023 mol of Sr^{2+} leaving (0.0038 - 0.0023) mol = 0.0015 mol. The concentration will be

$$[Sr^{2+}] = \frac{0.0015\ \text{mol}}{0.100\ \text{L}} = 0.015\ M = 1.5 \times 10^{-2}\ M$$

We use the method of Example 17.7 to find $[F^-]$.

$$K_{sp} = [Sr^{2+}][F^-]^2 = [0.015][F^-]^2 = 2.0 \times 10^{-10}$$

$$[F^-] = \left(\frac{2.0 \times 10^{-10}}{0.015}\right)^{1/2} = 1.2 \times 10^{-4}\ M$$

17.18 How can you tell which compound will precipitate first, CuI or AgI? Study Section 17.2 and Example 17.6.

Solution: (a) The solubility product expressions for both substances have exactly the same mathematical form and are therefore directly comparable. The substance having the smaller K_{sp} (AgI) will precipitate first. (Why?)

(b) When CuI just begins to precipitate the solubility product expression will just equal K_{sp} (saturated solution). The concentration of Cu^+ at this point is 0.010 M (given in the problem), so the concentration of iodide ion must be

$$K_{sp} = [Cu^+][I^-] = (0.010)[I^-] = 5.1 \times 10^{-12}$$

$$[I^-] = \frac{5.1 \times 10^{-12}}{0.010} = 5.1 \times 10^{-10}\ M$$

Using this value of $[I^-]$, we find the silver ion concentration

$$[Ag^+] = \frac{K_{sp}}{[I^-]} = \frac{8.3 \times 10^{-17}}{5.1 \times 10^{-10}} = 1.6 \times 10^{-7}\ M$$

(c) The percent of silver ion remaining in solution is

$$\%\ Ag^+(aq) = \frac{1.6 \times 10^{-7}\ M}{0.010\ M} \times 100\% = 0.0016\% \text{ or } 1.6 \times 10^{-3}\%$$

Is this an effective way to separate silver from copper?

17.19 At what concentrations of hydroxide ion will each metal begin to precipitate? In other words, find the $[OH^-]$ for each metal at which the solution is saturated. Convert the hydroxide ion concentrations to pH values.

Solution: For $Fe(OH)_3$, $K_{sp} = 1.1 \times 10^{-36}$. When $[Fe^{3+}] = 0.010$ M, the $[OH^-]$ value is

$$[OH^-] = \left(\frac{1.1 \times 10^{-36}}{0.010}\right)^{1/3} = 4.8 \times 10^{-12}\ M$$

This $[OH^-]$ corresponds to a pH of 2.68. In other words, $Fe(OH)_3$ will begin to precipitate from this solution at pH of 2.68.

For $Zn(OH)_2$, $K_{sp} = 1.8 \times 10^{-14}$. When $[Zn^{2+}] = 0.010$ M, the $[OH^-]$ value is

$$[OH^-] = \left(\frac{1.8 \times 10^{-14}}{0.010}\right)^{1/2} = 1.3 \times 10^{-6}\ M$$

This corresponds to a pH of 8.11. In other words $Zn(OH)_2$ will begin to precipitate from the solution at pH = 8.11. These results show that $Fe(OH)_3$ will precipitate when pH just exceeds 2.68 and that $Zn(OH)_2$ will precipitate when pH just exceeds 8.11. Therefore, to remove iron as $Fe(OH)_3$, the pH must be greater than 2.68 but less than 8.11.

17.24 The first part of the problem is similar to Example 17.7. Without doing the calculation, using Le Chatelier's principle, should the $PbBr_2$ be more or less soluble in water than in KBr solution? Than in $Pb(NO_3)_2$ solution?

Solution: First, we write $K_{sp} = [Pb^{2+}][Br^-]^2 = 8.9 \times 10^{-6}$

(a) For pure water, let s = $[Pb^{2+}]$ and 2s = $[Br^-]$

$$K_{sp} = (s)(2s)^2 = 4s^3 = 8.9 \times 10^{-6} \quad s = 0.013\ M$$

The molar solubility of $PbBr_2$ is 0.013 M.

(b) In this case $[Br^-] = 0.20$ M (why?). We write

$$[Pb^{2+}] = \frac{K_{sp}}{[Br^-]^2} = \frac{8.9 \times 10^{-6}}{(0.20)^2} = 2.2 \times 10^{-4}\ M$$

Thus the molar solubility is reduced from 0.013 M to 2.2×10^{-4} M as a result of the common ion (Br^-) effect.

(c) In this case $[Pb^{2+}] = 0.20$ M (why?). We write

$$[Br^-] = \sqrt{\frac{8.9 \times 10^{-6}}{0.20}} = 6.7 \times 10^{-3}\ M$$

Since one mole of $PbBr_2$ forms two moles of $Br^-(aq)$, the molar solubility of $PbBr_2$ is half of 6.7×10^{-3} M, or a solubility of 3.4×10^{-3} M.

17.25 First find the molar solubility of $CaCO_3$ in a solution containing 0.050 M calcium ion (see Example 17.7), then convert this to grams of $CaCO_3$ in the given amount of solution.

Solution: First let s be the molar solubility of $CaCO_3$ in this solution.

$$CaCO_3(s) \rightleftharpoons Ca^{2+}(aq) + CO_3^{2-}(aq)$$

		$Ca^{2+}(aq)$	$CO_3^{2-}(aq)$
Initial (M):		0.050	0.000
Change (M):		+s	+s
Equilibrium (M):		(0.050 + s)	s

$$K_{sp} = [Ca^{2+}][CO_3^{2-}] = (0.050 + s)s = 8.7 \times 10^{-9}$$

We can assume 0.050 + s ≅ 0.050, then

$$s = \frac{8.7 \times 10^{-9}}{0.050} = 1.7 \times 10^{-7}\ M$$

The mass of $CaCO_3$ can then be found

$$3.0 \times 10^2 \text{ mL} \left(\frac{1 \text{ L}}{1000 \text{ mL}}\right)\left(\frac{1.7 \times 10^{-7} \text{ mol}}{1 \text{ L}}\right)\left(\frac{100.1 \text{ g } CaCO_3}{1 \text{ mol}}\right) = 5.1 \times 10^{-6} \text{ g } CaCO_3$$

17.26 This problem is like Example 17.7. What is the concentration of chloride ion? Is it the same as the calcium ion concentration?

Solution: We first calculate the concentration of chloride ion in the solution.

$$[Cl^-] = \left(\frac{10.0 \text{ g } CaCl_2}{1 \text{ L soln}}\right)\left(\frac{1 \text{ mol } CaCl_2}{111.0 \text{ g } CaCl_2}\right)\left(\frac{2 \text{ mol } Cl^-}{1 \text{ mol } CaCl_2}\right) = 0.180 \text{ M}$$

	$AgCl(s)$	$\rightleftharpoons$	$Ag^+(aq)$	+	$Cl^-(aq)$
Initial (M):			0.000		0.180
Change (M):			+s		+s
Equilibrium (M):			s		(0.180 + s)

If we assume that (0.180 + s) ≅ 0.180, then

$$K_{sp} = [Ag^+][Cl^-] = 1.6 \times 10^{-10}$$

$$[Ag^+] = \frac{K_{sp}}{[Cl^-]} = \frac{1.6 \times 10^{-10}}{0.180} = 8.9 \times 10^{-10} \text{ M} = s$$

The molar solubility of AgCl is 8.9×10^{-10} M.

17.28 This problem is similar to Problem 17.26 except that it is turned around. Study the solution to the example and try to use it as a model to work backwards to K_{sp}.

Solution: We start with the equilibrium concentration and work toward K_{sp}.

$$[Pb^{2+}] = 2.4 \times 10^{-11} \text{ M} \qquad [IO_3^-] = [2 \times (2.4 \times 10^{-11}) + 0.10] \text{ M} \cong 0.10 \text{ M}$$

$$K_{sp} = [Pb^{2+}][IO_3^-]^2 = (2.4 \times 10^{-11})(0.10)^2 = 2.4 \times 10^{-13}$$

17.29 Study Section 17.4 with the following questions in mind: Why is barium fluoride more soluble in highly acidic solution? Why are the solubilities of the chloride, bromide, and iodide of barium unaffected by acidity?

Solution: When the anion of a salt is a base (see Tables 16.1 and 16.4), the salt will be more soluble in acidic solution because the hydrogen ion decreases the concentration of the anion (Le Chatelier's principle):

$$B^-(aq) + H^+(aq) \rightleftharpoons HB(aq)$$

(a) $BaSO_4$ will be slightly more soluble because SO_4^{2-} is a base (although a weak one).

(b) The solubility of $PbCl_2$ in acid is unchanged over the solubility in pure water because HCl is a strong acid.

(c) $Fe(OH)_3$ will be more soluble in acid because OH^- is a base.

(d) $CaCO_3$ will be more soluble in acidic solution because the H_2CO_3 that is formed decomposes to CO_2 and H_2O. The CO_2 escapes from the solution, shifting the equilibrium. Although it is not important in this case, the carbonate ion is also a base.

17.30 This problem is like Problem 17.29 above.

Solution: (b) $SO_4^{2-}(aq)$ is a weak base; (c) $OH^-(aq)$ is a strong base; (d) $C_2O_4^{2-}(aq)$ is a weak base; (e) $PO_4^{3-}(aq)$ is a weak base. Only (a) is unaffected by the acid solution.

17.32 Find K_{sp} for iron(II) hydroxide in Table 17.1. If you need additional help, refer to Example 17.8.

Solution: From Table 17.1, the value of K_{sp} for iron(II) hydroxide is 1.6×10^{-14}.

(a) At pH = 8.00, pOH = 14.00 - 8.00 = 6.00, and $[OH^-] = 1.0 \times 10^{-6}$ M

$$[Fe^{2+}] = \frac{K_{sp}}{[OH^-]^2} = \frac{1.6 \times 10^{-14}}{(1.0 \times 10^{-6})^2} = 0.016 \text{ M}$$

The molar solubility of iron(II) hydroxide at pH = 8.00 is 0.016 M

(b) At pH = 10.00, pOH = 14.00 - 10.00 = 4.00, and $[OH^-] = 1.0 \times 10^{-4}$ M

$$[Fe^{2+}] = \frac{K_{sp}}{[OH^-]^2} = \frac{1.6 \times 10^{-14}}{(1.0 \times 10^{-4})^2} = 1.6 \times 10^{-6} \text{ M}$$

The molar solubility of iron(II) hydroxide at pH = 10.00 is 1.6×10^{-6} M.

17.33 In other words, find the hydroxide ion concentration when the concentration of magnesium ion is 1.0×10^{-10} M. Example 17.9 in the text is a similar problem.

Solution: The solubility product expression for magnesium hydroxide is

$$K_{sp} = [Mg^{2+}][OH^-]^2 = 1.2 \times 10^{-11}$$

We find the hydroxide ion concentration when $[Mg^{2+}]$ is 1.0×10^{-10} M

$$[OH^-] = \left(\frac{1.2 \times 10^{-11}}{1.0 \times 10^{-10}}\right)^{1/2} = 0.35 \text{ M}$$

Therefore the concentration of OH^- must be slightly greater than 0.35M.

17.34 This is similar in part to Example 17.9. What is the hydroxide ion concentration in the solution after the addition of the ammonia? Does the solution volume change (significant figures)? Which hydroxide of iron will form (see Table 17.1)?

Solution: We first determine the effect of the added ammonia.

$$[NH_3] = 0.60 \text{ M } NH_3 \left(\frac{2.00 \text{ mL}}{1002 \text{ mL}}\right) = 0.0012 \text{ M } NH_3$$

Ammonia is a weak base ($K_b = 1.8 \times 10^{-5}$).

$$NH_3(aq) + H_2O(l) \rightleftharpoons NH_4^+(aq) + OH^-(aq)$$

	NH_3	NH_4^+	OH^-
Initial (M):	0.0012	0.0000	0.0000
Change (M):	-x	+x	+x
Equilibrium (M):	(0.0012 - x)	x	x

$$K_b = \frac{[NH_4^+][OH^-]}{[NH_3]} = \frac{x^2}{(0.0012 - x)} = 1.8 \times 10^{-5}$$

Solving the resulting quadratic equation gives x = 0.00014, or $[OH^-] = 0.00014$ M.

This is a solution of iron(II) sulfate, and we must use K_{sp} for iron(II) hydroxide. We compute the value of Q_c for this solution.

$$Q = [Fe^{2+}]_o[OH^-]_o^2 = (1.0 \times 10^{-3})(1.4 \times 10^{-4})^2 = 2.0 \times 10^{-11}$$

Q is larger than K_{sp} (1.6×10^{-14}); therefore a precipitate of $Fe(OH)_2$ will form.

17.38 Study Section 17.5, Example 17.10, and Table 17.3. What happens to copper(II) ion when it is in the presence of ammonia?

Solution: (a) The equations are as follows:

$$CuI_2(s) \rightleftharpoons Cu^{2+}(aq) + 2I^-(aq)$$

$$Cu^{2+}(aq) + 4NH_3(aq) \rightleftharpoons [Cu(NH_3)_4]^{2+}(aq)$$

The ammonia combines with the Cu^{2+} ions formed in the first step to form the complex ion $[Cu(NH_3)_4]^{2+}$, effectively removing the Cu^{2+} ions, causing the first equilibrium to shift to the right (resulting in more CuI_2 dissovling).

Similarly,

(b) $AgBr(s) \rightleftharpoons Ag^+(aq) + Br^-(aq)$

$Ag^+(aq) + 2CN^- \rightleftharpoons [Ag(CN)_2]^-(aq)$

(c) $HgCl_2(s) \rightleftharpoons Hg^{2+}(aq) + 2Cl^-(aq)$

$Hg^{2+} + 4Cl^- \rightleftharpoons [HgCl_4]^{2-}(aq)$

17.39 This is similar to Example 17.10. Assume the solution volume remains unchanged after addition of the copper(II) sulfate.

Solution: First find the molarity of the copper(II) ion

$$\text{Moles } CuSO_4 = 2.50 \text{ g} \left(\frac{1 \text{ mol}}{159.6 \text{ g}}\right) = 0.0157 \text{ mol}$$

$$[Cu^{2+}] = \frac{0.0157 \text{ mol}}{0.90 \text{ L}} = 0.0174 \text{ M}$$

As in Example 17.10, the position of equilibrium will be far to the right. We assume essentially all the copper ion is complexed with NH_3. The NH_3 consumed is 4×0.0174 M = 0.0696 M. The uncombined NH_3 remaining is (0.30 - 0.0696) M, or 0.23 M. The equilibrium concentrations of $Cu(NH_3)_4{}^{2+}$ and NH_3 are therefore 0.0174 M and 0.23 M, respectively. We find $[Cu^{2+}]$ from the formation constant expression.

$$K_f = \frac{[Cu(NH_3)_4{}^{2+}]}{[Cu^{2+}][NH_3]^4} = 5.0 \times 10^{13} = \frac{0.0174}{[Cu^{2+}][0.23]^4}$$

$$[Cu^{2+}] = 1.2 \times 10^{-13} \text{ M}$$

17.40 This problem is exactly analogous to Problem 17.39 above.

Solution: The concentration of cadmium ion before complex ion formation is

$$[Cd^{2+}] = 0.50 \text{ g } Cd(NO_3)_2 \left(\frac{1 \text{ mol } Cd(NO_3)_2}{236.4 \text{ g } Cd(NO_3)_2}\right)\left(\frac{1}{0.50 \text{ L}}\right) = 4.2 \times 10^{-3} \text{ M}$$

As in Example 17.10, the position of equilibrium will be far to the right. We assume essentially all the cadmium ion is complexed with CN^-. The CN^- consumed is 4×0.0042 M = 0.0168 M. The uncombined CN^- remaining is (0.50 - 0.0168) M, or 0.48 M. The equilibrium concentrations of $[Cd(CN)_4]^{2-}$ and CN^- are therefore 4.2×10^{-3} M and 0.48 M, respectively. We find $[Cd^{2+}]$ from the formation constant expression.

$$K_f = \frac{[Cd(CN)_4{}^{2-}]}{[Cd^{2+}][CN^-]^4} = 7.1 \times 10^{16} = \frac{4.2 \times 10^{-3}}{[Cd^{2+}][0.48]^4}$$

$$[Cd^{2+}] = 1.1 \times 10^{-18} \text{ M}$$

17.41 Calculate and use the equilibrium constant for the reaction of $Al(OH)_3$ with OH^- to form $Al(OH)_4{}^-$.

Solution: The reaction

$$Al(OH)_3(s) + OH^-(aq) \rightleftharpoons Al(OH)_4{}^-(aq)$$

is the sum of the two known reactions

$$Al(OH)_3(s) \rightleftharpoons Al^{3+}(aq) + 3OH^-(aq) \qquad K_{sp} = 1.8 \times 10^{-33}$$

$$Al^{3+}(aq) + 4OH^-(aq) \rightleftharpoons Al(OH)_4^-(aq) \qquad K_f = 2.0 \times 10^{33}$$

The equilibrium constant is

$$K = K_{sp}K_f = (1.8 \times 10^{-33})(2.0 \times 10^{33}) = 3.6 = \frac{[Al(OH)_4^-]}{[OH^-]}$$

When pH = 14.00, $[OH^-]$ = 1.0 M, therefore

$$[Al(OH)_4^-] = K[OH^-] = 3.6 \times 1 = 3.6 \text{ M}$$

This represents the maximum possible concentration of the complex ion at pH 14.00. Since this is much larger than the initial 0.010 M, the complex ion will be the predominant species.

17.42 Use the equations in Table 17.3 as models for the reactions with ammonia. Zinc belongs to the same periodic group as Cd (cadmium). Silver behaves like copper toward ammonia. Why does aluminum hydroxide dissolve in strong base. (See Problem 17.41.)

Solution: The balanced equations are:

$$Ag^+(aq) + 2NH_3(aq) \rightleftharpoons Ag(NH_3)_2^+(aq)$$

$$Zn^{2+}(aq) + 4NH_3(aq) \rightleftharpoons Zn(NH_3)_4^{2+}(aq)$$

Zinc hydroxide forms a complex ion with excess OH^- and silver hydroxide does not.

17.46 Hint: Consider Example 17.11.

Solution: Silver chloride will dissolve in aqueous ammonia because of the formation of a complex ion. Lead chloride will not dissolve; it doesn't form an ammonia complex.

17.48 What happens to each compound when strong aqueous NaOH is added?

Solution: Ammonium chloride is the salt of a weak base (ammonia). It will react with strong aqueous hydroxide to form ammonia (Le Chatelier's principle)

$$NH_4Cl(s) + OH^-(aq) \rightarrow NH_3(g) + H_2O(l) + Cl^-(aq)$$

The human nose is an excellent ammonia detector. Nothing happens between KCl and strong aqueous NaOH.

17.49 Review Section 17.1. A solubility equilibrium is an equilibrium between what and what?

Solution: A solubility equilibrium is an equilibrium between a solid (reactant) and its components (products: ions, neutral molecules, etc.) in solution. Only (d) represents a solubility equilibrium.

Consider part (b). Can you write the equilibrium constant for this reaction in terms of K_{sp} for calcium phosphate?

17.50 Assuming you don't know that copper(II) nitrate is dark blue and that silver nitrate is colorless, what could you do to aqueous solutions of these two compounds to distinguish between them? Study Section 17.6 and Table 17.4.

Solution: Chloride ion will precipitate Ag^+ but not Cu^{2+}. A flame test will also work.

17.51 If necessary, review electron configurations of ions in Section 8.2 and VSEPR structures in Section 10.1.

Solution: The electron configuration of the Al^{3+} ion is $1s^2 2s^2 2p^6$. The $Al(OH)_4^-$ ion will be tetrahedral (AB_4 case).

17.52 Hint: Iron(III) doesn't form a complex ion with OH^-. See Problem 17.41.

Solution: Aqueous hydroxide will precipitate both $Al(OH)_3$ and $Fe(OH)_3$, but only the aluminum compound will redissolve when excess OH^- is added. The reaction is

$$Al(OH)_3(s) + OH^-(aq) \rightleftharpoons Al(OH)_4^-(aq)$$

17.53 See Section 15.8 for a discussion of amphoteric hydroxides. Use Problem 17.41 as a model for writing the complex ion reactions.

Solution: The hydroxides of aluminum, zinc, and copper(II) are amphoteric. The ionic equations are

$$Al(OH)_3(s) + OH^-(aq) \rightleftharpoons Al(OH)_4^-(aq)$$

$$Zn(OH)_2(s) + 2OH^-(aq) \rightleftharpoons Zn(OH)_4^{2-}(aq)$$

$$Cu(OH)_2(s) + 2OH^-(aq) \rightleftharpoons Cu(OH)_4^{2-}(aq)$$

17.55 First, calculate the carbonate concentration just before the magnesium carbonate starts to precipitate. Use that carbonate concentration to calculate the amount of calcium remaining. Compare that amount of calcium to the original amount of calcium.

Solution: Just before magnesium carbonate begins to precipitate,

$$[CO_3^{2-}] = \frac{K_{sp}}{[Mg^{2+}]} = \frac{4.0 \times 10^{-5}}{0.070} = 5.7 \times 10^{-4}\ M$$

If the carbonate concentration is 5.7×10^{-4} M, then the remaining calcium concentration is

$$[Ca^{2+}] = \frac{K_{sp}}{[CO_3^{2-}]} = \frac{8.7 \times 10^{-9}}{5.7 \times 10^{-4}} = 1.5 \times 10^{-5}\ M$$

The percent of calcium remaining in solution is

$$\frac{\text{concentration remaining in solution}}{\text{initial concentration}} \times 100\% = \frac{1.5 \times 10^{-5}\ M}{0.070\ M} \times 100\% = 0.021\%$$

When magnesium carbonate just begins to precipitate, then (100% - 0.021%) = 99.98% of the calcium will be precipitated. Thus, it is possible to precipitate 99.9% of the calcium before the magnesium carbonate will begin to precipitate.

Can you calculate the concentration of carbonate required to precipitate 99.9% of the calcium?

17.56 This problem is of the same general type as Example 17.3. Refer to the Chemistry in Action entitled "pH, Solubility, and Tooth Decay" to find the solubility equations.

Solution: The equations are:

$$Ca_5(PO_4)_3OH(s) \rightleftharpoons Ca_5(PO_4)_3^+(aq) + OH^-(aq)$$
$$Ca_5(PO_4)_3F(s) \rightleftharpoons Ca_5(PO_4)_3^+(aq) + F^-(aq)$$

For $Ca_5(PO_4)_3OH$ as in Example 17.3,

$$K_{sp} = [Ca_5(PO_4)_3^+][OH^-] = s^2 = 7 \times 10^{-37}$$

$$s = 8 \times 10^{-19}\ \text{mol/L}$$

For $Ca_5(PO_4)_3F$,

$$K_{sp} = [Ca_5(PO_4)_3^+][F^-] = s^2 = 1 \times 10^{-60}$$

$$s = 1 \times 10^{-30}\ \text{mol/L}$$

The solubility of fluoroapatite is about 12 orders of magniture less soluble than hydroxyapatite; therefore it should be much more effective against tooth decay.

17.57 Is this problem a K_{sp} problem or is it just a titration problem?

Solution: The number of moles of Ba^{2+} present in the original 50.0 mL of solution is

$$50.0\ \text{mL} \times \frac{1.00\ \text{mol}\ Ba^{2+}}{1\ \text{L soln}} \times \frac{1\ \text{L}}{1000\ \text{mL}} = 0.0500\ \text{mol}\ Ba^{2+}$$

The number of moles of SO_4^{2-} present in the original 86.4 mL solution is

$$86.4\ \text{mL} \times \frac{0.494\ \text{mol}\ SO_4^{2-}}{1\ \text{L soln}} \times \frac{1\ \text{L}}{1000\ \text{mL}} = 0.0427\ \text{mol}\ SO_4^{2-}$$

The overall equation is:	$Ba^{2+}(aq)$	+ $2OH^-(aq)$	+ $2H^+(aq)$	+ $SO_4^{2-}(aq)$	$\rightarrow BaSO_4(s)$	+ $H_2O(l)$
The quantity reacted (mol):	0.0427	0.0854	0.0854	0.0427	0.0427	

Thus the mass of $BaSO_4$ formed is

$$(0.0427\ \text{mol}\ BaSO_4) \times \frac{233.4\ \text{g}\ BaSO_4}{1\ \text{mol}\ BaSO_4} = 9.97\ \text{g}\ BaSO_4$$

$Ba(OH)_2$ was in excess. The quantity of $Ba(OH)_2$ remaining is: (0.0500 - 0.0427) mol = 0.0073 mol

Since $Ba(OH)_2(aq) \rightarrow Ba^{2+}(aq) + 2OH^-(aq)$

then 0.0073 mol $Ba(OH)_2 \rightarrow$ 0.0146 mol OH^-

$$[OH^-] = \frac{0.0146\text{ mol}}{(50.0\text{ mL} + 86.4\text{ mL})} \times \frac{1000\text{ mL}}{1\text{ L}} = 0.107\text{ M}$$

pOH = 0.97 pH = 13.03

17.59 What assumption do we have to make when we fill the container and then empty it to remove dissolved $CaCO_3$?

Solution: $CaCO_3(s) \rightleftharpoons Ca^{2+}(aq) + CO_3^{2-}(aq)$

	Ca^{2+}	CO_3^{2-}
Initial(M):	0.00	0.00
Change (M):	+s	+s
Equilibrium (M):	s	s

$K_{sp} = [Ca^{2+}][CO_3^{2-}] = s^2 = 8.7 \times 10^{-9}$ $s = 9.3 \times 10^{-5}\text{ M} = 9.3 \times 10^{-5}\text{ mol/L}$

Moles of $CaCO_3$ $\frac{116\text{ g}}{100.1\text{ g/mol}} = 1.16$ moles $CaCO_3$

The volume of distilled water to have a concentration of 9.3×10^{-5} mol/L of Ca^{2+} and of CO_3^{2-}

$$\frac{1.16\text{ mol }CaCO_3}{9.3 \times 10^{-5}\text{ mol/L}} = 1.2 \times 10^4\text{ L}$$

The number of times the kettle would have to be filled

$$\frac{1.2 \times 10^4\text{ L}}{2.0\text{ L per filling}} = 6.0 \times 10^3\text{ fillings}$$

Note that the very important assumption is made that each time the kettle is filled, the calcium carbonate is allowed to reach equilibrium before the kettle is emptied.

17.61 The graph indicates that after the precipitate is formed, then addition of more KI causes a redissolution of the solid.

Solution: The original precipitate was HgI_2. $Hg^{2+}(aq) + 2I^-(aq) \rightarrow HgI_2(s)$

With the addition of I^- in the form of KI, a soluble complex is formed and the precipitate redissolves.

$$HgI_2(s) + 2I^-(aq) \rightarrow HgI_4^{2-}(aq)$$

Note: If you have not already done so, please read the "Introduction for the Student" at the beginning of this Student Solutions Manual.

Chapter 18

CHEMISTRY IN THE ATMOSPHERE

18.1 How is mole fraction of gases related to volume fraction? If you need a review of mole fraction, see Section 12.1.

Solution: For ideal gases, mole fraction is the same as volume fraction. From Table 18.1, CO_2 is 0.033% of the composition of dry air, by volume. The value 0.033% means 0.033 volumes (or moles, in this case) out of 100, or

$$X_{CO_2} = \frac{0.033}{100} = 3.3 \times 10^{-4}$$

To change to parts per million (pppm), we multiply the mole fraction by one million.

$$3.3 \times 10^{-4} \times 10^{6} = 330 \text{ ppm}$$

18.2 What is the partial pressure of CO_2 if the atmospheric pressure is one atmosphere?

Solution: Using the information in Table 18.1 and Problem 18.1, 0.033% of the volume (and therefore the pressure) of dry air is due to CO_2. The partial pressure of CO_2 is

$$X_{CO_2}P_T = 3.3 \times 10^{-4} \times (754 \text{ mmHg}) \times \frac{1 \text{ atm}}{760 \text{ mmHg}} = 3.3 \times 10^{-4} \text{ atm}$$

18.5 Would you be able to easily liberate oxygen atoms into the atmosphere on Earth to observe their color as they collided with a surface?

Solution: The experiment would be very difficult because O atoms rapidly combine with themselves to form O_2.

18.6 If we can calulate the volume of the layer of ozone, using the ideal gas law we can easily determine the number of moles. A simple conversion can be made to mass and/or number of molecules.

Solution: The formula for the volume is $4\pi r^2h$, where $r = 6.371 \times 10^6$ m and $h = 3.0 \times 10^{-3}$ m (or 3.0 mm).

$$\text{Volume} = 4\pi(6.371 \times 10^{6}\ \text{m})^2(3.0 \times 10^{-3}\ \text{m})\frac{1 \times 10^{3}\ \text{L}}{1\ \text{m}^3} = 1.5 \times 10^{15}\ \text{L}$$

Using the ideal gas law and remembering that at STP, one mole of a gas occupies 22.4 L,

$$\text{moles } O_3 = (1.5 \times 10^{15}\ \text{L}) \times \frac{1\ \text{mol}}{22.4\ \text{L}} = 6.7 \times 10^{13}\ \text{moles } O_3$$

$$\text{molecules } O_3 = (6.7 \times 10^{13}\ \text{mol } O_3) \times (6.022 \times 10^{23}\ \text{molecules/mol}) = 4.0 \times 10^{37}\ \text{molecules}$$

$$\text{mass } O_3\ (\text{kg}) = (6.7 \times 10^{13}\ \text{mol } O_3) \times \frac{48.00\ \text{g } O_3}{1\ \text{mol } O_3} \times \frac{1\ \text{kg}}{1000\ \text{g}} = 3.2 \times 10^{12}\ \text{kg } O_3$$

18.9 What are the chemical formulas for Freon-11 and Freon-12?

Solution: The formula for Freon-11 is $CFCl_3$ and for Freon-12 is CF_2Cl_2. The equations are:

$$CCl_4 + HF \rightarrow CFCl_3 + HCl$$

$$CFCl_3 + HF \rightarrow CF_2Cl_2 + HCl$$

A catalyst is necessary for both reactions.

18.10 How are compounds decomposed by UV radiation?

Solution: The energy of the photons of UV radiation in the troposphere is insufficient (that is, the wavelength is too long and the frequency is too small) to break the bonds in CFC's.

18.11 How is wavelength related to energy? Example 18.1 in the text is a similar problem.

Solution: Using the relationship

$$\lambda = \frac{hc}{E} = \frac{(6.63 \times 10^{-34}\ \text{J·s})(3.00 \times 10^{8}\text{m/s})}{(460 \times 10^{3}\ \text{J/mol}) / (6.022 \times 10^{23}\ \text{photons/mol})}\left(\frac{10^{9}\ \text{nm}}{1\ \text{m}}\right) = 260\ \text{nm}$$

18.12 Is it possible for for both atoms in chlorine monoxide to have an octet of electrons? Hint: count the number of valence electrons.

Solution: The Lewis structures for chlorine nitrate and chlorine monoxide are

$:\ddot{Cl}-\ddot{O}-\overset{+}{N}(=\ddot{O}:)-\ddot{O}:^{-}$ $\qquad$ $:\ddot{Cl}-\dot{\ddot{O}}:$

18.13 Assume from Problem 18.6 that the amount of ozone is 3.2×10^{12} kg.

Solution: The quantitiy of ozone lost is

$$0.06 \times (3.2 \times 10^{12}\ \text{kg}) = 1.9 \times 10^{11}\ \text{kg of } O_3$$

Assuming no further deterioration

$$\frac{(1.9 \times 10^{11}\ \text{kg of } O_3)}{100\ \text{yr} \times 365\ \text{day/yr}} = 5.2 \times 10^{6}\ \text{kg/day}$$

The standard enthalpy of formation (from Appendix 3) for ozone: $\frac{3}{2}O_2 \rightarrow O_3 \quad \Delta H^{\circ}_f = 142.2$ kJ/mol

The total energy required

$$\frac{1.9 \times 10^{14}\ \text{g of } O_3}{48.00\ \text{g/mol}} \times 142.2\ \text{kJ/mol} = 5.6 \times 10^{14}\ \text{kJ} \quad \text{or} \quad 1.5 \times 10^{10}\ \text{kJ/day}$$

18.16 Refer to Section 18.5.

Solution: Two sources of energy that do not contribute to global warming are nuclear and solar (solar cells).

18.21 What is the shape of ozone? Example 18.2 in the text is a similar problem.

Solution: Since ozone is a polar triatomic molecule with a bent geometry, its normal modes of vibration are exactly like those of water, which are shown in Figure 18.13. Therefore, it is a greenhouse gas.

18.22 As with all stoichiometry problems, the first step in the solution is to have a balanced equation.

Solution: The equation is: $\quad 2ZnS + 3O_2 \rightarrow 2ZnO + 2SO_2$

$$\frac{4.0 \times 10^{4}\ \text{ton ZnS}}{97.46\ \text{ton-mol ZnS}} \times \frac{1\ \text{mol } SO_2}{1\ \text{mol ZnS}} \times 64.07\ \text{ton-mol } SO_2 = 2.6 \times 10^{4}\ \text{tons } SO_2$$

18.23 Refer to Section 18.5 for help.

Solution: Ethane and propane are greenhouse gases.

18.25 This problem is a direct application of the ideal gas law. What does "ppm" mean?

Solution: The quantity 0.42 ppm is 0.42 parts per million = 4.2×10^{-7}. Since for ideal gases volume is directly proportional to moles and pressure,

Mole fraction of ozone $\quad X_{O_3} = 4.2 \times 10^{-7}$

Pressure of ozone $\quad P_{O_3} = (4.2 \times 10^{-7}) \times \dfrac{748\ \text{mmHg}}{760\ \text{mmHg/atm}} = 4.1 \times 10^{-7}\ \text{atm}$

Moles of ozone $\quad n = \dfrac{(4.1 \times 10^{-7}\ \text{atm})(1\ \text{L})}{(0.0821\ \text{L·atm/K·mol})(293\ \text{K})} = 1.7 \times 10^{-8}\ \text{mol}$

Molecules of ozone/L $\quad (1.7 \times 10^{-8}\ \text{mol/L}) \times (6.022 \times 10^{23}\ \text{molecules/mol}) = 1.0 \times 10^{16}\ \text{molecules/L}$

18.27 Refer to Section 14.5 if you have forgotten about the relationship between the magnitude of the equilibrium constant and the thermodynamics of the equilibrium.

Solution: Higher temperature has shifted the equilibrium to the right. Therefore the reaction is endothermic.

18.29 See Chapter 13 for a review of kinetics.

Solution: (a) Since this is an elementary reaction, the rate law is: $\quad \text{Rate} = k[NO]^2[O_2]$

(b) Since $[O_2]$ is very large compared to [NO], then the reaction is a pseudo second-order reaction and the rate law can be simplified to:

$$\text{Rate} = k'[NO]^2 \qquad \text{where } k' = k[O_2]$$

(c) Since for a second-order reaction $\quad t_{1/2} = \dfrac{1}{k[A]_0}$

then $\quad \dfrac{(t_{1/2})_1}{(t_{1/2})_2} = \dfrac{[(A_0)_2]}{[(A_0)_1]}$

$$\frac{6.4 \times 10^3\ \text{min}}{(t_{1/2})_2} = \frac{10\ \text{ppm}}{2\ \text{ppm}}$$

Solving, the new half life is $\quad (t_{1/2})_2 = 1.3 \times 10^3\ \text{min}$

18.30 To work this problem we must first find the volume of the room in units of m^3, which then is converted to L (what is the conversion factor?). For ideal gases the quantity is proportional to the partial pressure of the gas. Therefore we can convert the 800 ppm of the gas to the pressure that it creates. The ideal gas law is used to covert the volume to mass.

Solution: The room volume is: $17.6\ m \times 8.80\ m \times 2.64\ m = 4.09 \times 10^2\ m^3$

Since $1\ m^3 = 1 \times 10^3$ L, then the volume of the container is 4.09×10^5 L. The quantity 8.00×10^2 ppm is

$$\frac{8.00 \times 10^2}{1 \times 10^6} = 8.00 \times 10^{-4} = \text{mole fraction of CO}$$

The pressure of the CO (atm) is $(8.00 \times 10^{-4}) \times \dfrac{756\ \text{mmHg}}{760\ \text{mmHg/atm}} = 7.96 \times 10^{-4}$ atm

The moles of CO is $n = \dfrac{PV}{RT} = \dfrac{(7.96 \times 10^{-4}\ \text{atm})(4.09 \times 10^5\ \text{L})}{(0.0821\ \text{L·atm/K·mol})(293\ \text{K})} = 13.5$ mol

The mass of CO in the room is $\text{mass} = (13.5\ \text{mol}) \times \dfrac{28.01\ \text{g of CO}}{1\ \text{mol CO}} = 378$ g CO

18.31 This problem is a very straightforward example of equilibrium calculations.

Solution: (a) From the balanced equation $K_c = \dfrac{[O_2][HbCO]}{[CO][HbO_2]}$

(b) Using the information provided $212 = \dfrac{[O_2][HbCO]}{[CO][HbO_2]} = \dfrac{[8.6 \times 10^{-3}][HbCO]}{[1.9 \times 10^{-6}][HbO_2]}$

Solving, the ratio of HbCO to HbO_2 is $\dfrac{[HbCO]}{[HbO_2]} = \dfrac{212 \times (1.9 \times 10^{-6})}{(8.6 \times 10^{-3})} = 0.047$

18.33 As with all stoichiometry problems, we must begin with a balanced equation.

Solution: The balanced equation is $Ca(OH)_2 + CO_2 \rightarrow CaCO_3 + H_2O$

The moles of $CaCO_3$ $0.026\ \text{g}\ CaCO_3 \times \dfrac{1\ \text{mol}\ CaCO_3}{100.1\ \text{g}\ CaCO_3/\text{mol}} = 2.6 \times 10^{-4}$ mol $CaCO_3$

The moles of CO_2 (from the balanced equation) is also 2.6×10^{-4} mol.

The moles of air is $n = \dfrac{PV}{RT} = \dfrac{(747\ \text{mmHg}/760\ \text{mmHg})\ \text{atm} \times (5.0\ \text{L})}{(0.0821\ \text{L·atm/K·mol})(291\ \text{K})} = 0.206$ mol

For ideal gases, the mole ratio is the same as the volume ratio. Therefore the % by volume of CO_2 in the air is

$$\frac{2.6 \times 10^{-4}\ \text{mol}\ CO_2}{0.206\ \text{mol air}} \times 100\% = 0.13\%$$

18.34 If you don't remember the names and formulas for the nitrogen oxides, refer to Section 22.4.

Solution: (a)

$$N_2O + O \rightarrow 2NO$$
$$2NO + 2O_3 \rightarrow 2O_2 + 2NO_2$$

Overall $$N_2O + O + 2O_3 \rightarrow 2O_2 + 2NO_2$$

(b) Compounds with a permanent dipole moment such as N_2O are more effective greenhouse gases than nonpolar species such as CO_2 (Section 18.5).

(c) The moles of adipic acid is

$$(2.2 \times 10^9 \text{ kg adipic acid}) \times \frac{1000 \text{ g}}{1 \text{ kg}} \times \frac{1 \text{ mol adipic acid}}{146.1 \text{ g adipic acid}} = 1.5 \times 10^{10} \text{ mol adipic acid}$$

The number of moles of adipic acid is given as being equivalent to the moles of N_2O produced, and from the overall balanced equation, one mole of N_2O will react with two moles of O_3. Thus

$$1.5 \times 10^{10} \text{ mol adipic acid} \rightarrow 1.5 \times 10^{10} \text{ mol } N_2O \text{ which reacts with } 3.0 \times 10^{10} \text{ mol } O_3$$

18.37 To work this problem we first need to calculate the concentration of CO_2 from Henry's law (Section 12.7) and the information in Table 18.1. The H^+ concentration can be calculated from the equilibrium expression for carbonic acid.

Solution: From Henry's law and Problem 18.2 $c = kP$

$$[CO_2] = (0.032 \text{ mol/L·atm}) \times (3.3 \times 10^{-4} \text{ atm}) = 1.06 \times 10^{-5} \text{ mol/L}$$

We assume that all of the dissolved CO_2 is converted to H_2CO_3, thus giving us 1.06×10^{-5} mol/L of H_2CO_3.

The equilibrium expression is $H_2CO_3 \rightleftharpoons H^+ + HCO_3^-$

	H_2CO_3	H^+	HCO_3^-
Initial (M):	1.06×10^{-5}	0	0
Change (M):	$-x$	$+x$	$+x$
Equilibrium (M):	$[(1.06 \times 10^{-5}) - x]$	x	x

$$K \text{ (from Table 16.3)} = 4.2 \times 10^{-7} = \frac{[H^+][HCO_3^-]}{[H_2CO_3]} = \frac{x^2}{(1.06 \times 10^{-5}) - x}$$

Solving for x $x = 1.9 \times 10^{-6} = [H^+]$ pH = 5.72

18.39 If you don't remember what Lewis acids and bases are, review Section 15.7.

Solution: The oxygen in CaO (considered to be ionic) behaves as a Lewis base and the sulfur in SO_2 is the Lewis acid site which can accept a pair of electrons.

$$:\ddot{\underset{..}{O}}:^{2-} + \ddot{O}{=}\ddot{S}{=}\ddot{O} \longrightarrow \left[\ddot{O}{=}\overset{\overset{\displaystyle :\ddot{O}:}{|}}{\underset{..}{S}}{=}\ddot{O} \right]^{2-}$$

Other resonance structures are possible for SO_3^{2-}. What are they?

Note: If you have not already done so, please read the "Introduction for the Student" at the beginning of this Student Solutions Manual.

Chapter 19

ENTROPY, FREE ENERGY AND EQUILIBRIUM

19.4 Do you know how to find $(0.5)^y$ on your calculator?

Solution: (a) The hard way to do this is to multiply 0.5 by itself 6 times. This will give you the correct answer, but it is not the easiest method. Another approach is to take the logarithm of $(0.5)^6$, namely $6 \times \log(0.5)$, and then find the inverse log of that result. The simplest way is to use the x^y function with $x = 0.5$ and $y = 6$. The result is 0.016.

(b) 8.7×10^{-19} (c) $\cong 0$

19.8 In some cases in this problem you must consider the structural complexity of the atoms or molecules being compared. More component particles or less well-ordered structures imply a higher entropy.

Solution: (a) This is easy. The liquid form of any substance always has a greater entropy (molecular level disorder).

(b) This is hard. At first glance there may seem to be no apparent difference between the two substances that might affect the entropy (molecular formulas identical). However, the first has the -O-H structural feature which allows it to participate in hydrogen bonding with other molecules. This allows a more ordered arrangement of molecules in the liquid state. The standard entropy of CH_3OCH_3 is larger.

(c) This is also difficult. Both are monoatomic species. However, the Xe atom has more particles (nucleus plus 54 electrons) than Ar (nucleus plus 18 electrons). Xenon has the higher standard entropy.

(d) Same argument as part (c). Carbon dioxide gas has the higher standard entropy (see Appendix 3).

(e) Graphite has the higher standard entropy. See the discussion in Section 19.2.

(f) Using the same argument as part (c), one mole of N_2O_4 has a larger standard entropy than one mole of NO_2. Compare values in Appendix 3.

Use the data in Appendix 3 to compare the standard entropy of one mole of N_2O_4 with that of two moles of NO_2. In this situation the number of atoms is the same for both. Which is higher and why?

19.10 This problem is like Example 19.2. Obtain the standard entropies from Appendix 3.

Solution: Using Equation (19.7) to calculate ΔS°_{rxn}

(a) ΔS°_{rxn} = S°(SO_2) - [S°(O_2) + S°(S)] = (1 mol)(248.5 J/K · mol) - (1 mol)(205.0 J/K · mol) - (1 mol)(31.88 J/K · mol) = 11.6 J/K

(b) ΔS°_{rxn} = S°(MgO) + S°(CO_2) - S°($MgCO_3$) = (1 mol)(26.78 J/K · mol) + (1 mol)(213.6 J/K · mol) - (1 mol)(65.69 J/K · mol) = 174.7 J/K

(c) ΔS°_{rxn} = S°(Cu) + S°(H_2O) - [S°(H_2) + S°(CuO)] = (1 mol)(33.3 J/K · mol) + (1 mol)(188.7 J/K · mol) - [(1 mol)(131.0 J/K · mol) + (1 mol)(43.5 J/K · mol)] = 47.5 J/K

(d) ΔS°_{rxn} = S°(Al_2O_3) + 3S°(Zn) - [2S°(Al) + 3S°(ZnO)] = (1 mol)(50.99 J/K · mol) + (3 mol)(41.6 J/K · mol) - [(2 mol)(28.3 J/K · mol) + (3 mol)(43.9 J/K · mol)] = -12.5 J/K

(e) ΔS°_{rxn} = S°(CO_2) + 2S°(H_2O) - [S°(CH_4) + 2S°(O_2)] = (1 mol)(213.6 J/K · mol) + (2 mol)(69.9 J/K · mol) - [(1 mol)(186.2 J/K · mol) + (2 mol)(205.0 J/K · mol)] = -242.8 J/K

Why was the entropy value for water different in parts (c) and (e)?

19.11 The predictions in this problem follow from the principles outlined in Section 19.4. Use Example 19.3 as a model.

Solution: All parts of this problem rest on two principles. First, the entropy of a solid is always less than the entropy of a liquid, and the entropy of a liquid is always much smaller than the entropy of a gas. Second, in comparing systems in the same phase, the one with the most complex particles has the higher entropy.

(a) Positive entropy change (increase). One of the products is in the gas phase.

(b) Negative entropy change (decrease). Liquids have lower entropies than gases.

(c) Positive entropy change. Everything is in the gas phase on the product side, and SO_2 is a more complex molecule than O_2.

(d) Positive. Same as (a).

(e) Negative. The product has only one molecule in the gas phase and the reactants have two gas-phase molecules.

(f) Positive. There are two gas-phase species on the product side and only one of the reactant side.

(g) Negative. The reactant side has a gas-phase molecule.

19.12 This problem is exactly like Problem 19.11 above and Example 19.3 in the text..

Solution: (a) $\Delta S < 0$; gas reacting with a liquid to form a solid.

(b) $\Delta S > 0$; solid decomposing to give a liquid and a gas.

(c) $\Delta S > 0$; increase in number of moles of gas.

(d) $\Delta S < 0$; gas reacting with a solid to form a solid.

19.14 Review Section 19.3 and Equation (19.1). Can a reaction that has a decrease in energy be spontaneous?

Solution: The second law states that the entropy of the universe must increase in a spontaneous process. But the entropy of the universe is the sum of two terms: the entropy of the system plus the entropy of the surroundings. One can decrease, but not both. In this case the decrease in system entropy is offset by an increase in the entropy of the surroundings. The reaction in question is exothermic, and the heat released raises the temperature (and the entropy) of the surroundings.

Could this process be spontaneous if the reaction were endothermic?

19.18 Problem 19.19 below and Example 19.4 are similar.

Solution: Using Equation (19.8) to solve for the change in standard free energy,

(a) $\Delta G° = 2\Delta G_f^\circ(NO) - \Delta G_f^\circ(N_2) - \Delta G_f^\circ(O_2) = (2\text{ mol})(86.7\text{ kJ/mol}) - 0 - 0 = 173.4\text{ kJ}$

(b) $\Delta G° = \Delta G_f^\circ(H_2O(g)) - \Delta G_f^\circ(H_2O(l)) = (1\text{ mol})(-228.6\text{ kJ/mol}) - (1\text{ mol})(-237.2\text{ kJ/mol}) = 8.6\text{ kJ}$

(c) $\Delta G° = 4\Delta G_f^\circ(CO_2) + 2\Delta G_f^\circ(H_2O) - 2\Delta G_f^\circ(C_2H_2) - 5\Delta G_f^\circ(O_2) = (4\text{ mol})(-394.4\text{ kJ/mol}) + (2\text{ mol})(-237.2\text{ kJ/mol})$
$- (2\text{ mol})(209.2\text{ kJ/mol}) - (5\text{ mol})(0) = -2470\text{ kJ}$

(d) $\Delta G° = 2\Delta G_f^\circ(MgO) - 2\Delta G_f^\circ(Mg) - \Delta G_f^\circ(O_2) = (2\text{ mol})(-569.6\text{ kJ/mol}) - (2\text{ mol})(0) - (1\text{ mol})(0) = -1139\text{ kJ}$

(e) $\Delta G° = 2\Delta G_f^\circ(SO_3) - 2\Delta G_f^\circ(SO_2) - \Delta G_f^\circ(O_2) = (2\text{ mol})(-370.4\text{ kJ/mol}) - (2\text{ mol})(-300.4\text{ kJ/mol}) - (1\text{ mol})(0)$
$= -140\text{ kJ}$

19.19 Use Example 19.4 as a model for this calculation.

Solution: Using Equation (19.10) as a model we write

$$\Delta G^\circ_{rxn} = 4\Delta G_f^\circ(CO_2(g)) + 6\Delta G_f^\circ(H_2O(l)) - 2\Delta G_f^\circ(C_2H_6(g)) - 7\Delta G_f^\circ(O_2(g))$$

Next substitute the standard free energies of formation from Appendix 3.

$$\Delta G^\circ_{rxn} = (4\text{ mol})(-394.4\text{ kJ/mol}) + (6\text{ mol})(-237.2\text{ kJ/mol}) - (2\text{ mol})(-32.89\text{ kJ/mol}) - (7\text{ mol})(0) = -2935.0\text{ kJ}$$

Is this procedure the same as that for calculating standard enthalpy changes?

19.20 Use Equation (19.7) to compute the free energy change for each reaction. What criterion determines whether the free energy change shows a reaction to be spontaneous? Don't forget to convert kilojoules to joules. See also Example 19.4.

Solution: Reaction A: First apply Equation (19.7) to compute the free energy change at 25°C (298 K)

$$\Delta G = \Delta H - T\Delta S = 10{,}500 \text{ J} - (298 \text{ K} \times 30 \text{ J/K}) = 1560 \text{ J}$$

The +1560 J shows the reaction is not spontaneous at 298 K. The ΔG will change sign (i.e., the reaction will become spontaneous) above the temperature at which $\Delta G = 0$.

$$T = \frac{\Delta H}{\Delta S} = \frac{10500 \text{ J}}{30 \text{ J/K}} = 350 \text{ K}$$

Reaction B: Calculate ΔG

$$\Delta G = \Delta H - T\Delta S = 1800 \text{ J} - (-298 \text{ K} \times 113 \text{ J/K}) = 35{,}500 \text{ J}$$

The free energy change is positive, which shows that the reaction is not spontaneous at 298 K. Since both terms are positive, there is no temperature at which their sum is negative. The reaction is not spontaneous at any temperature.

Reaction C: Calculate ΔG

$$\Delta G = \Delta H - T\Delta S = -126{,}000 \text{ J} - (298 \text{ K} \times 84 \text{ J/K}) = -151{,}000 \text{ J}$$

The free energy change is negative so the reaction is spontaneous at 298 K. Since both terms are negative, the reaction is spontaneous at all temperatures.

Reaction D: Calculate ΔG

$$\Delta G = \Delta H - T\Delta S = -11{,}700 \text{ J} - (-298 \text{ K} \times 105 \text{ J/K}) = 19{,}600 \text{ J}$$

The free energy change is positive at 298 K which means the reaction is not spontaneous at that temperature. The positive sign of ΔG results from the large negative value of ΔS. At lower temperatures the $-T\Delta S$ term will be smaller thus allowing the free energy change to be negative.

$$T = \frac{\Delta H}{\Delta S} = \frac{-11700 \text{ J}}{-105 \text{ J/K}} = 111 \text{ K}$$

19.22 This problem is similar to Example 19.5. Study the part of Section 19.4 on Temperature and Chemical Reactions. At the normal boiling point what is the free energy difference between liquid mercury and gaseous mercury?

Solution: At the temperature of the normal boiling point the free energy difference between the liquid and gaseous forms of mercury (or any other substance) is zero, i.e. the two phases are in equilibrium. We can therefore use Equation (19.7) to find this temperature. For the equilibrium

$$Hg(l) \rightleftharpoons Hg(g)$$

$$\Delta G = \Delta H - T\Delta S = 0$$

$$\Delta H = \Delta H_f^\circ(Hg(g)) - \Delta H_f^\circ(Hg(l)) = 60{,}780 \text{ J} - 0 \text{ J} = 60780 \text{ J}$$

$$\Delta S = S^\circ(Hg(g)) - S^\circ(Hg(l)) = 174.7 \text{ J/K} - 77.4 \text{ J/K} = 97.3 \text{ J/K}$$

$$T_{bp} = \frac{\Delta H}{\Delta S} = \frac{60780 \text{ J}}{97.3 \text{ J/K}} = 625 \text{ K} = 352°\text{C}$$

What assumptions are made? Notice that the given enthalpies and entropies are at standard conditions, namely 25°C and 1.00 atm pressure. In performing this calculation we have tacitly assumed that these quantities don't depend upon temperature. The actual normal boiling point of mercury is 356.58°C. Is the assumption of the temperature independence of these quantities reasonable?

19.23 Does the discussion of free energy in Section 19.4 mention any connection with the rate of a reaction?

Solution: There is no connection between the spontaneity of a reaction predicted by ΔG and the rate at which the reaction occurs. A negative free energy change tells us that a reaction has the potential to happen, but gives no indication of the rate.

Does the fact that a reaction occurs at a measurable rate mean that the free energy difference ΔG is negative?

19.26 This problem involves the use of Equation (19.10) in a manner similar to Example 19.6. Don't forget to convert kilojoules to joules and to use the proper value of R.

Solution: Find the value of K by solving Equation (19.10)

$$K_p = e^{\frac{-\Delta G^\circ}{RT}} = e^{\frac{-2.60 \times 10^3 \text{ J}}{(8.314 \text{ J/K·mol})(298 \text{ K})}} = e^{-1.05} = 0.350$$

19.28 This problem is similar to Problem 19.26 and Example 19.7.

Solution:

$$K_{sp} = [Fe^{2+}][OH^-]^2 = 1.6 \times 10^{-14}$$

$$\Delta G^\circ = -RT\ln K_{sp} = -(8.314 \text{ J/K·mol})(298 \text{ K})\ln(1.6 \times 10^{-14}) = 79 \text{ kJ}$$

19.29 How can you find the standard free energy change for this reaction (see Example 19.4)? Once you know the standard free energy change, how can you compute K (see Problem 19.26)? This problem is similar to Example 19.6.

Solution: We use standard free energies of formation from Appendix 3 to find the standard free energy difference.

$$\Delta G^\circ_{rxn} = 2\Delta G^\circ_f(H_2(g)) + \Delta G^\circ_f(O_2(g)) - 2\Delta G^\circ_f(H_2O(g))$$
$$= (2\ \text{mol})(0) + (1\ \text{mol})(0) - (2\ \text{mol})(-228.6\ \text{kJ/mol}) = 457.2\ \text{kJ}$$

We compute K_p using Equation (19.10)

$$K_p = e^{\frac{-\Delta G^\circ}{RT}} = e^{\frac{-457.2 \times 10^3\ \text{J}}{(8.314\ \text{J/K}\cdot\text{mol})(298\ \text{K})}} = e^{-184.5} = 7 \times 10^{-81}$$

Why only one significant figure?

19.30 The calculations here are very similar to Problem 19.29 above.

Solution: $\Delta G^\circ = 2\Delta G^\circ_f(NO) - \Delta G^\circ_f(N_2) - \Delta G^\circ_f(O_2) = (2\ \text{mol})(86.7\ \text{kJ/mol}) - (1\ \text{mol})(0) - (1\ \text{mol})(0) = 173\ \text{kJ}$

$$\ln K_p = \frac{-\Delta G^\circ}{RT} = \frac{-173 \times 10^3\ \text{J}}{(8.314\ \text{J/K}\cdot\text{mol})(298\ \text{K})} = -69.8 \qquad K_p = 4.9 \times 10^{-31}$$

The calculation shows that because K_p is such a small number, under atmospheric conditions oxygen does not react with nitrogen to produce nitric oxide.

19.32 Write out the expression for K_p for this reaction. How can you calculate the pressure from the data given?

Solution: The expression for K_p is: $\quad K_p = P_{CO_2}$

Thus you can predict the equilibrium pressure directly from the value of the equilibrium constant. The only task at hand is computing the values of K_p using Equations (19.7) and (19.10).

(a) At 25°C, $\quad \Delta G^\circ = \Delta H^\circ - T\Delta S^\circ = (177.8 \times 10^3\text{J}) - (298\ \text{K})(160.5\ \text{J/K}) = 130.0 \times 10^3\ \text{J}$

$$P_{CO_2} = K_p = e^{\frac{-\Delta G^\circ}{RT}} = e^{\frac{-130.0 \times 10^3\ \text{J}}{(8.314\ \text{J/K}\cdot\text{mol})(298\ \text{K})}} = e^{-52.47} = 1.6 \times 10^{-23}\ \text{atm}$$

(b) At 800°C, $\quad \Delta G^\circ = \Delta H^\circ - T\Delta S^\circ = (177.8 \times 10^3\ \text{J}) - (1073\ \text{K})(160.5\ \text{J/K}) = 5.58 \times 10^3\ \text{J}$

$$P_{CO_2} = K_p = e^{\frac{-\Delta G^\circ}{RT}} = e^{\frac{-5.58 \times 10^3\ \text{J}}{(8.314\ \text{J/K}\cdot\text{mol})(1073\ \text{K})}} = e^{-0.625} = 0.535\ \text{atm}$$

What assumptions are made in the second calculation? See Problem 19.22.

19.33 Part (a) is like Problem 19.30 above and Example 19.8. In part (b) the given data are not equilibrium pressures.

Solution: (a) We first find the standard free energy change of the reaction.

$$\Delta G^\circ_{rxn} = \Delta G^\circ_f(PCl_3(g)) + \Delta G^\circ_f(Cl_2(g)) - \Delta G^\circ_f(PCl_5(g))$$
$$= (1\ \text{mol})(-286\ \text{kJ/mol}) + (1\ \text{mol})(0) - (1\ \text{mol})(-325\ \text{kJ/mol}) = 39\ \text{kJ}$$

We can calculate K_p using Equation (19.10)

$$K_p = e^{\frac{-\Delta G^\circ}{RT}} = e^{\frac{-39 \times 10^3\ \text{J}}{(8.314\ \text{J/K}\cdot\text{mol})(298\ \text{K})}} = e^{-16} = 1 \times 10^{-7}$$

(b) We are finding the free energy difference between the reactants and the products at their nonequilibrium values. The result tells us the direction of and the potential for further chemical change. We use the given nonequilibrium pressures to compute Q_P

$$Q_P = \frac{P_{PCl_3} P_{Cl_2}}{P_{PCl_5}} = \frac{0.27 \times 0.40}{0.0029} = 37$$

The value of ΔG (notice that this is not the standard free energy difference) can be found using Equation (19.10) and the result from part (a)

$$\Delta G = \Delta G^\circ + RT\ln Q = (39 \times 10^3\ \text{J}) + (8.314\ \text{J/K}\cdot\text{mol})(298\ \text{K})\ln(37) = 48\ \text{kJ}$$

Which way is the direction of spontaneous change for this system? What would be the value of ΔG if the given data were equilibrium pressures? What would be the value of Q_P in that case?

19.34 Part (a) is like Example 19.8 and part (b) is like the corresponding part of Problem 19.33 above. Refer to those problems if you need more detail in how to work the problem.

Solution: (a) $\Delta G^\circ = -RT\ln K_P = -(8.314\ \text{J/K}\cdot\text{mol})(2000\ \text{K})\ln(4.4) = -24.6\ \text{kJ}$

(b) $$\Delta G = \Delta G^\circ + RT\ln Q = (-24.6\ \text{kJ}) + \frac{(8.314\ \text{J/K}\cdot\text{mol})(2000\ \text{K})}{(1000\ \text{J/kJ})}\ln\frac{(0.66)(1.20)}{(0.25)(0.78)} = -1.3\ \text{kJ}$$

19.35 Find the standard free energy of formation of CO in Appendix 3. Use Example 19.8 as a model to find the standard free energy change for the reaction.

Solution: We use the given K_P to find the standard free energy change by means of Equation (19.8).

$$\Delta G^\circ = -RT\ln K = -(8.314\ \text{J/K}\cdot\text{mol})(298\ \text{K})\ln(5.62 \times 10^{35}) = -204\ \text{kJ}$$

The standard free energy of formation of one mole of $COCl_2$ can now be found using Equation (19.8)

$$(1\ \text{mol})\Delta G^\circ_f(COCl_2(g)) = \Delta G^\circ_{rxn} + \Delta G^\circ_f(CO(g)) + \Delta G^\circ_f(Cl_2(g))$$
$$= -204\ \text{kJ} + (1\ \text{mol})(-137.3\ \text{kJ/mol}) + (1\ \text{mol})(0) = -341\ \text{kJ}$$

Therefore $\Delta G^\circ_f(COCl_2(g)) = -341$ kJ/mol. Why is the standard free energy of formation of Cl_2 equal to zero?

19.36 Use Equation (19.10) to find the equilibrium constant.

Solution: The equilibrium constant expression is: $K_P = P_{H_2O}$

We are actually finding the equilibrium vapor pressure of water (compare to Problem 19.32). We use Equation (19.10).

$$P_{H_2O} = K_P = e^{\frac{-\Delta G^\circ}{RT}} = e^{\frac{-8.6 \times 10^3\ \text{J}}{(8.314\ \text{J/K}\cdot\text{mol})(298\ \text{K})}} = e^{-3.47} = 3.1 \times 10^{-2}\ \text{atm}$$

The positive value of ΔG° implies this is an unfavorable process at 25°C. Why? (Hint: What is the pressure of $H_2O(g)$ under standard-state conditions?)

19.38 This problem is similar to Problem 19.36 above.

Solution: Since $\Delta G^\circ = -RT\ln K_P$, we have

$$\Delta G^\circ = 212 \times 10^3\ \text{J} = -(8.314\ \text{J/K}\cdot\text{mol})(298\ \text{K})\ln K_P \qquad K_P = 6.9 \times 10^{-38}$$

$$\text{Since } K_P = \sqrt{P_{O_2}} \qquad P_{O_2} = 4.8 \times 10^{-75}\ \text{atm}$$

This pressure is far too small to measure.

19.40 This is like Example 19.9 and part (b) of Problem 19.33 above. For which set of conditions will ΔG be zero?

Solution: In each part of this problem we use Equation (19.9).

$$\Delta G = \Delta G^\circ + RT\ln Q = \Delta G^\circ + RT\ln\,[H^+][OH^-]$$

(a) In this case the given concentrations are equilibrium concentrations for 25°C and we can state that $\Delta G = 0$.

$$\Delta G^\circ = -RT\ln K_w = -(8.314\ \text{J/K}\cdot\text{mol})(298\ \text{K})\ln\,(1.0 \times 10^{-14}) = 8.0 \times 10^4\ \text{J}$$

(b) $\Delta G = \Delta G^\circ + RT\ln Q = \Delta G^\circ + RT\ln\,[H^+][OH^-]$

$$= 8.0 \times 10^4\ \text{J} + (8.314\ \text{J/K}\cdot\text{mol})(298\ \text{K})\ln\,(1.0 \times 10^{-3})(1.0 \times 10^{-4}) = 4.0 \times 10^4\ \text{J}$$

(c) $\Delta G = \Delta G^\circ + RT\ln Q = \Delta G^\circ + RT\ln\,[H^+][OH^-]$

$$= 8.0 \times 10^4\ \text{J} + (8.314\ \text{J/K}\cdot\text{mol})(298\ \text{K})\ln\,(1.0 \times 10^{-12})(2.0 \times 10^{-8}) = -3.2 \times 10^4\ \text{J}$$

(d) $\Delta G = \Delta G^\circ + RT\ln Q = \Delta G^\circ + RT\ln\,[H^+][OH^-]$

$$= 8.0 \times 10^4\ \text{J} + (8.314\ \text{J/K}\cdot\text{mol})(298\ \text{K})\ln\,(3.5)(4.8 \times 10^{-4}) = 6.4 \times 10^4\ \text{J}$$

19.42 In thinking about your answers, consider Problem 19.23 above.

Solution: One possible explanation is simply that no reaction is possible, namely that there is an unfavorable free energy difference between products and reactants ($\Delta G > 0$).

A second possibility is that the potential for spontaneous change is there ($\Delta G < 0$), but that the reaction is extremely slow (very large activation energy).

A remote third choice is that the student accidentally prepared a mixture in which the components were already at their equilibrium concentrations.

Which of the above situations would be altered by the addition of a catalyst?

19.43 What is the sign of ΔS for a solid to liquid phase transition? What is the sign of ΔH for any solid to liquid phase transition?

Solution: For a solid to liquid phase transition (melting) the entropy always increases ($\Delta S > 0$) and the reaction is always endothermic ($\Delta H > 0$). Melting is not spontaneous below the melting point (-100°C case) so $\Delta G > 0$. At the melting point (-77.7°C) solid and liquid are in equilibrium so $\Delta G = 0$. Melting is always spontaneous above the melting point (-60°C case) so $\Delta G < 0$.

19.44 See the discussions of these processes in this chapter.

Solution: (a) An ice cube melting in a glass of water at 20°C. The value of ΔG for this process is negative so it must be spontaneous.

(b) A "perpetual motion" machine. In one version, a model has a flywheel which, once started up, drives a generator which drives a motor which keeps the flywheel running at a constant speed and also lifts a weight.

(c) A perfect air conditioner; it extracts heat energy from the room and warms the outside air without using any energy to do so. (Note: this process does not violate the first law of thermodynamics.)

(d) Same example as (a).

(e) A closed flask at 25°C containing $NO_2(g)$ and $N_2O_4(g)$ at equilibrium.

19.45 If the process is spontaneous and the enthalpy change is essentially zero, what can you say about the sign of ΔS?

Solution: When ΔH is zero and ΔG is negative (spontaneous process), the sign of ΔS must be positive. Gaseous diffusion is a process in which the molecular disorder increases.

19.46 If the process is spontaneous and the enthalpy change is positive, what can you say about the sign of ΔS?

Solution: If the process is spontaneous as well as endothermic, the signs of ΔG and ΔH must be negative and positive, respectively. Since $\Delta G = \Delta H - T\Delta S$, the sign of ΔS must be positive if ΔG is to be negative.

19.48 Is it unusual for a spontaneous reaction to lead to a decrease in entropy?

Solution: For a reaction to be spontaneous, ΔG must be negative. If ΔS is negative, as it is in this case, then the reaction must be exothermic (why?). When water freezes, it gives off heat (exothermic). Consequently, the entropy of the surroundings increases and $\Delta S_{universe} > 0$.

19.50 This problem involves the mathematical manipulation of Equations (19.7) and (19.10).

Solution: (a) At two different temperatures T_1 and T_2,

$$\Delta G_1^\circ = \Delta H^\circ - T_1\Delta S^\circ = -RT\ln K_1 \qquad (1)$$

$$\Delta G_2^\circ = \Delta H^\circ - T_2\Delta S^\circ = -RT\ln K_2 \qquad (2)$$

$$\ln K_1 = -\frac{\Delta H^\circ}{RT_1} + \frac{\Delta S^\circ}{R} \qquad (3)$$

$$\ln K_2 = -\frac{\Delta H^\circ}{RT_2} + \frac{\Delta S^\circ}{R} \qquad (4)$$

Subtracting equation (3) above from equation (4),

$$\ln\frac{K_2}{K_1} = \frac{\Delta H^\circ}{R}\left(\frac{1}{T_1} - \frac{1}{T_2}\right) = \frac{\Delta H^\circ}{R}\left(\frac{T_2 - T_1}{T_1T_2}\right)$$

(b) Using the equation that we just derived,

$$\ln\frac{K_2}{4.63 \times 10^{-3}} = \frac{58.0 \times 10^3\ \text{J}}{8.314\ \text{J/K}\cdot\text{mol}}\left(\frac{338\ \text{K} - 298\ \text{K}}{338\ \text{K} \times 298\ \text{K}}\right) = 2.77$$

$$\frac{K_2}{4.63 \times 10^{-3}} = 15.96 \qquad K_2 = 0.0739$$

$K_2 > K_1$, as predicted by the positive ΔH°.

19.51 Don't forget to convert the temperature to kelvins.

Solution: Using the relationship: $\dfrac{\Delta H_{vap}}{T_{b.p.}} = \Delta S_{vap} \cong 90\ \text{kJ/K}\cdot\text{mol}$

benzene	$\Delta S_{vap} = 87.8\ \text{kJ/K}\cdot\text{mol}$	hexane	$\Delta S_{vap} = 90.1\ \text{kJ/K}\cdot\text{mol}$
mercury	$\Delta S_{vap} = 93.7\ \text{kJ/K}\cdot\text{mol}$	toluene	$\Delta S_{vap} = 91.8\ \text{kJ/K}\cdot\text{mol}$

Most liquids have ΔS_{vap} approximately equal to a constant value because they have similar structures in the liquid state. Thus, ΔS for (liquid $\rightarrow$ vapor) is also similar. (The "structure" of most vapors is totally random.)

(b) Using the data in Table 11.6, we find: ethanol $\Delta S_{vap} = 111.9$ kJ/K · mol
water $\Delta S_{vap} = 109.4$ kJ/K · mol

Both water and ethanol have a larger ΔS_{vap} because the liquids are more ordered due to hydrogen bonding.

19.52 Is HF a species that is hydrogen-bonded in the liquid phase?

Solution: Evidence shows that HF, which is strongly hydrogen-bonded in the liquid phase, is still considerably hydrogen-bonded in the vapor state such that its ΔS_{vap} is smaller than most other substances.

19.54 The second part of the problem is analogous to Problem 19.36.

Solution: Using data in Appendix 3, we solve for $\Delta G°$.

$$\Delta G° = \Delta G_f^{\circ}(CuSO_4) + 5\Delta G_f^{\circ}(H_2O) - \Delta G_f^{\circ}(CuSO_4{\cdot}5H_2O)$$

$$\Delta G° = (1 \text{ mol})(-661.9 \text{ kJ/mol}) + (5 \text{ mol})(-228.6 \text{ kJ/mol}) - (1 \text{ mol})(-1879.7 \text{ kJ/mol})$$

$$\Delta G° = 74.8 \text{ kJ}$$

Using Equation (19.10): $K_p = e^{\frac{-\Delta G°}{RT}} = e^{\frac{-74.8 \times 10^3 \text{ J}}{(8.314 \text{ J/K·mol})(298 \text{ K})}} = 7.7 \times 10^{-14}$

$$K_p = \left(P_{H_2O}\right)^5 \qquad P_{H_2O} = (K_p)^{1/5} = (7.7 \times 10^{-14})^{1/5} = 0.002 \text{ atm} = 1.5 \text{ mmHg}$$

With a pressure of only 1.5 mmHg, you would have to have a sensitive method of measuring pressure.

19.56 Refer to the Chemistry in Action: The Efficiency of Heat Engines at the last of the chapter.

Solution: Using the equation:

$$\text{Efficiency} = \frac{T_h - T_c}{T_h} = \frac{773 \text{ K} - 288 \text{ K}}{773 \text{ K}} \times 100\% = 62.7\%$$

Friction and other defects cause the actual efficiency to be much lower than the thermodynamic efficiency.

19.58 Use the data in Appendix 3 for the required enthalpies and entropies.

Solution: Assuming that both $\Delta H°$ and $\Delta S°$ are temperture independent, we calculate both $\Delta H°$ and $\Delta S°$.

$$\Delta H° = [\Delta H_f^{\circ}(CO) + \Delta H_f^{\circ}(H_2)] - [\Delta H_f^{\circ}(H_2O) + \Delta H_f^{\circ}(C)]$$

$$\Delta H° = [(1 \text{ mol})(-110.5 \text{ kJ/mol}) + (1 \text{ mol})(0 \text{ kJ/mol})] - [(1 \text{ mol})(-241.8 \text{ kJ/mol}) + (1 \text{ mol})(0 \text{ kJ/mol})]$$

$\Delta H° = 131.3 \text{ kJ}$

$\Delta S° = [S°(CO) + S°(H_2)] - [S°(H_2O) + S°(C)]$

$\Delta S° = [(1 \text{ mol})(197.9 \text{ J/K} \cdot \text{mol})) + (1 \text{ mol})(131.0 \text{ J/K} \cdot \text{mol})] - [(1 \text{ mol})(188.7 \text{ J/K} \cdot \text{mol}) + (1 \text{ mol})(5.69 \text{ J/K} \cdot \text{mol})]$

$\Delta S° = 134.5 \text{ J/K}$

It is obvious from the given conditions that the reaction must take place at a fairly high temperature (in order to have red-hot coke). Setting $\Delta G° = 0$

$$0 = \Delta H° - T\Delta S°$$

$$T = \frac{\Delta H°}{\Delta S°} = \frac{(131.3 \text{ kJ})(1000 \text{ J/1 kJ})}{134.5 \text{ J/K}} = 976 \text{ K} = 703°\text{C}$$

19.59 What is the Third Law of Thermodynamics? Review Section 19.4 if needed.

Solution: According to the Third Law, a perfect crystalline substance at zero kelvin has an entropy of zero. A substance can never have a negative entropy.

19.60 What do we know about the relative acidity of HF versus HCl (see Table 15.2).

Solution: (a) We know that HCl is a strong acid and HF is a weak acid. Thus the equilibrium constant $K < 1$.

(b) The number of particles on each side of the equation is the same, so $\Delta S° \approx 0$. Therefore $\Delta H°$ will dominate.

(c) HCl is a weaker bond than HF (see Table 9.4), therefore $\Delta H° > 0$.

19.62 Use the hint and refer to the equation derived in Problem 19.50.

Solution: For the reaction $CaCO_3(s) \rightleftharpoons CaO(s) + CO_2(g)$ $\quad K_P = P_{CO_2}$

Using the equation from Problem 19.50

$$\ln\frac{K_2}{K_1} = \frac{\Delta H°}{R}\left(\frac{1}{T_1} - \frac{1}{T_2}\right) = \frac{\Delta H°}{R}\left(\frac{T_2 - T_1}{T_1T_2}\right)$$

Substituting,

$$\ln\frac{1829}{22.6} = \frac{\Delta H°}{8.314 \text{ J/K·mol}}\left(\frac{1223 \text{ K} - 973 \text{ K}}{(973 \text{ K})(1223 \text{ K})}\right)$$

Solving ,

$$\Delta H° = 174 \text{ kJ/mol}$$

19.63 This problem is a direct application of the Gibbs free energy equation 19.7

Solution: For the reaction to be spontaneous $\Delta G < 0$. $\Delta G = \Delta H - T\Delta S$

Given that $\Delta H = 19$ kJ = 19,000 J, then $\Delta G = 19{,}000 - (273\text{ K} + 72\text{ K})(\Delta S)$

Solving the equation with the value of $\Delta G = 0$, $\Delta S = 55$ J/K

This value of ΔS which we solved for is the value needed to produce a ΔG value of zero. The *minimum* value of ΔS that will produce a spontaneous reaction will be any value of entropy *greater than* 55 J/K.

Note: If you have not already done so, please read the "Introduction for the Student" at the beginning of this Student Solutions Manual.

Chapter 20

ELECTROCHEMISTRY

20.8 This problem is similar to Example 20.3 and Problem 20.9 below. Use Figure 20.3 as a model for your cell diagram.

Solution: Using the standard reduction potentials found in Table 20.1

$$Mg^{2+}(aq) + 2e^- \rightarrow Mg(s) \qquad E° = -2.37\ V$$
$$Cd^{2+}(aq) + 2e^- \rightarrow Cd(s) \qquad E° = -0.40\ V$$

The diagonal rule shows that Cd^{2+} will oxidize Mg so that the magnesium half-reaction occurs at the anode.

$$Mg(s) + Cd^{2+}(aq) \rightarrow Mg^{2+}(aq) + Cd(s)$$

$$E° = -0.40\ V - (-2.37\ V) = 1.97\ V$$

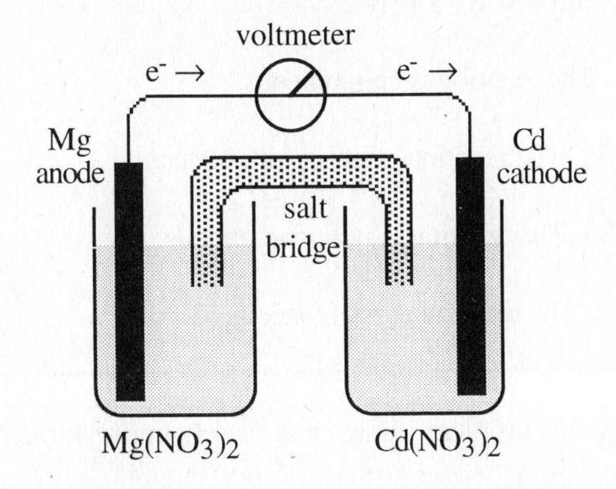

20.9 Study Example 20.3 before working this problem. Identify the cathode and anode half-reactions.

Solution: We use the standard reduction potentials found in Table 20.1

$$Al^{3+}(aq) + 3e^- \rightarrow Al(s) \qquad E° = -1.66\ V$$
$$Ag^{+}(aq) + e^- \rightarrow Ag(s) \qquad E° = 0.80\ V$$

The diagonal rule shows that Ag^+ will oxidize Al so that the aluminum half-reaction occurs at the anode.

$$3Ag^+(aq) + Al(s) \rightarrow 3Ag(s) + Al^{3+}(aq)$$

We can find the standard emf by using Equation (20.1)

$$E^\circ_{cell} = E^\circ_{ox} + E^\circ_{red} = 1.66\ V + 0.80\ V = 2.46\ V$$

20.11 Study Examples 20.2 and 20.4 before working this problem (iron(III) will be reduced to iron(II)). How can you tell from the standard emf whether the reaction will be spontaneous?

Solution: The appropriate half-reactions from Table 20.1 are

$$I_2(s) + 2e^- \rightarrow 2I^-(aq) \qquad E^\circ = 0.53\ V$$
$$Fe^{3+}(aq) + e^- \rightarrow Fe^{2+}(aq) \qquad E^\circ = 0.77\ V$$

The diagonal rule shows that iron(III) should oxidize iodide ion to iodine. This makes the iodide ion/iodine half-reaction the anode. The standard emf can be found using Equation (20.1)

$$E^\circ_{cell} = E^\circ_{ox} + E^\circ_{red} = -0.53\ V + 0.77\ V = 0.24\ V$$

(The emf was not required in this problem, but the fact that it is positive confirms that the reaction should be spontaneous.)

20.12 This is like Problem 20.11 above. The standard reduction potentials are obtained from Table 20.1.

Solution: In each case we write $E^\circ_{cell} = E^\circ_{ox} + E^\circ_{red}$

(a) $E^\circ = 2.87\ V - 0.40\ V = 2.47\ V$. The reaction is spontaneous.

(b) $E^\circ = -\ 1.07\ V - 0.14\ V = -1.21\ V$. The reaction is not spontaneous.

(c) $E^\circ = -\ 0.80\ V - 0.25\ V = -1.05\ V$. The reaction is not spontaneous.

(d) $E^\circ = -\ 0.15\ V + 0.77\ V = 0.62\ V$. The reaction is spontaneous.

20.14 Study Example 20.2 before starting this problem. In (a) and (c) assume that the metal ion, not sulfate, is the reactant.

Solution: In each part find the appropriate half-reactions in Table 20.1 and either apply the diagonal rule or calculate the standard emf of the cell reaction.

(a) In order for Zn^{2+} to react with silver metal, Ag(*s*) must appear on the right side of Table 20.1 above Zn^{2+}. Find these half-reactions in the table and you will see that this is not the case.

(b) This is just like Example 20.2. There will be no reaction.

(c) We will work this part by calculating the standard emf. The relative positions of the Zn^{2+}/Zn and Ni^{2+}/Ni half-reactions in the table show that nickel(II) will oxidize zinc metal. This makes the Zn^{2+}/Zn half-reaction the anode. Using Equation (20.1),

$$E°_{cell} = E°_{ox} + E°_{red} = 0.76 \text{ V} - 0.25 \text{ V} = 0.51 \text{ V}$$

Since the emf is positive, the reaction is spontaneous (we could see this using the diagonal rule).

(d) This is just like Example 20.2. Cl_2 will oxidize I^-, forming Cl^- and I_2.

20.15 If necessary, review balancing redox reactions by the half-reaction method.

Solution: (a) The half-reactions are:

$$H_2(g) \rightarrow 2H^+(aq) + 2e^-$$
$$Ni^{2+}(aq) + 2e^- \rightarrow Ni(s)$$

The complete balanced equation is

$$Ni^{2+}(aq) + H_2(g) \rightarrow Ni(s) + 2H^+(aq)$$

To find the direction of spontaneous change we can apply the diagonal rule. Ni(*s*) is above and to the right of $H^+(aq)$ in Table 20.1 (see the half-reactions at -0.25 and 0.00 V). Therefore, the spontaneous reaction is the reverse of the above reaction, that is:

$$Ni(s) + 2H^+(aq) \rightarrow Ni^{2+}(aq) + H_2(g)$$

(b) The half-reactions are:

$$Ni^{2+}(aq) + 2e^- \rightarrow Ni(s)$$
$$Cd(s) \rightarrow Cd^{2+}(aq) + 2e^-$$

The complete balanced equation is:

$$Ni^{2+}(aq) + Cd(s) \rightarrow Ni(s) + Cd^{2+}(aq)$$

In Table 20.1 Cd(*s*) is above and to the right of $Ni^{2+}(aq)$; therefore the spontaneous reaction is as written.

(c) The half-reactions are:

$$MnO_4^-(aq) + 8H^+(aq) + 5e^- \rightarrow Mn^{2+}(aq) + 4H_2O$$
$$2Cl^-(aq) \rightarrow Cl_2(g) + 2e^-$$

The complete balanced equation is:

$$2MnO_4^-(aq) + 16H^+(aq) + 10Cl^-(aq) \rightarrow 2Mn^{2+}(aq) + 8H_2O + 5Cl_2(g)$$

In Table 20.1, $Cl^-(aq)$ is above and to the right of $MnO_4^-(aq)$; therefore the spontaneous reaction is as written.

(d) The half reactions are:

$$Ce^{3+}aq) \rightarrow Ce^{4+}(aq) + e^-$$
$$2H^+(aq) + 2e^- \rightarrow H_2(g)$$

The complete balanced equation is:

$$2Ce^{3+}aq) + 2H^+(aq) \rightarrow 2Ce^{4+}(aq) + H_2(g)$$

In Table 20.1, $H_2(g)$ is above and to the right of $Ce^{4+}(aq)$; therefore the spontaneous reaction is the reverse of the reaction as written.

(e) The half-reactions are:

$$Cr(s) \rightarrow Cr^{3+}(aq) + 3e^-$$
$$Zn^{2+}(aq) + 2e^- \rightarrow Zn(s)$$

The complete balanced equation is:

$$2Cr(s) + 3Zn^{2+}(aq) \rightarrow 2Cr^{3+}(aq) + 3Zn(s)$$

In Table 20.1, Zn(*s*) is above and to the right of $Cr^{3+}(aq)$; therefore the spontaneous reaction is the reverse of the reaction as written.

20.16 Find the given half-reactions in Table 20.1. Where are the oxidizing and reducing agents located in this table? Where are the strongest of each found? Parts (a) and (b) are similar to Example 20.1. Part (c) is like Problem 20.9. Use Figure 20.1 as a model for part (d).

Solution: (a) In Table 20.1 the oxidizing agents are listed on the left with the strongest at the bottom. If a particular species can't be found in this column, it isn't an oxidizing agent. In the given list $Cl_2(g)$ is in the lowest position on the left; it is the strongest oxidizing agent.

(b) In Table 20.1 the reducing agents are listed on the right with the strongest at the top. If a particular species can't be found in this column, it isn't a reducing agent. In the given list $Ca(s)$ is in the highest position on the right; it is the strongest reducing agent.

(c) The half-reactions are:

$$Fe^{2+}(aq) + 2e^- \rightarrow Fe(s) \qquad E° = -0.44\ V$$
$$Ag^+(aq) + e^- \rightarrow Ag(s) \qquad E° = 0.80\ V$$

By the diagonal rule $Ag^+(aq)$ will oxidize $Fe(s)$; the Fe^{2+}/Fe half-reaction will be the anode. We calculate the standard emf for the reaction below using Equation (20.1)

$$E°_{cell} = E°_{ox} + E°_{red} = 0.44\ V + 0.80\ V = 1.24\ V$$

(d)

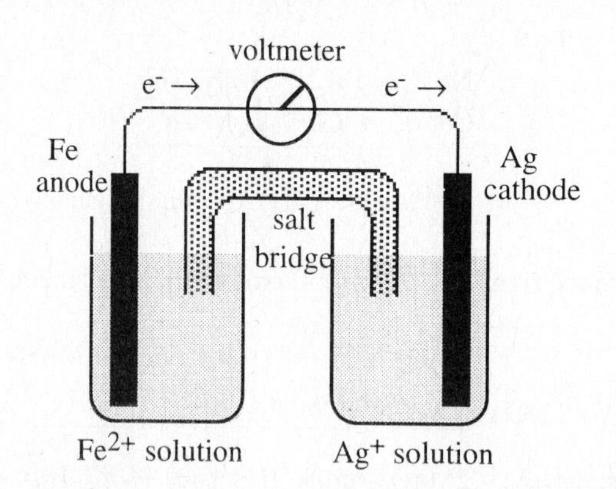

20.18 A discussion of oxidizing agents and reducing agents is presented in Problem 20.16 (a) above.

Solution: From Table 20.1 we compare the standard reduction potentials for the half-reactions. The more positive the potential, the better the substance as an oxidizing agent.

(a) Au^{3+}, (b) Ag^+, (c) Cd^{2+}, (d) O_2 in acidic solution.

20.22 This problem is like Example 20.5 in the text. How many electrons are involved in the overall balanced equation? Use Equation (20.5) for your calculations.

Solution: Equation (20.5) is:

$$\log K = \frac{nE°}{0.0591\ V}$$

Rearranging, $$E° = \frac{(0.0591\text{ V})\log K}{n} = \frac{(0.0591\text{ V})\log(2.69 \times 10^{12})}{2} = 0.367\text{ V}$$

Why do we use n = 2?

20.23 Study Examples 20.5 and 20.6 before working this problem.

Solution: (a) We break the equation into two half-reactions:

$$Mg(s) \rightarrow Mg^{2+}(aq) + 2e^-$$
$$Pb^{2+}(aq) + 2e^- \rightarrow Pb(s)$$

Notice that the Mg/Mg^{2+} half-reaction must be the anode; this follows from writing the original equation with $Mg(s)$ on the reactant side. We calculate the standard emf using Equation (20.1)

$$E°_{cell} = E°_{ox} + E°_{red} = 2.37\text{ V} - 0.13\text{ V} = 2.24\text{ V}$$

We find $\Delta G°$ and K_c using Equations (20.3) and (20.5)

$$\Delta G° = -nFE°_{cell} = -(2\text{ mol})(96500\text{ J/V·mol})(2.24\text{ V}) = -432\text{ kJ}$$

$$K_c = 10^{nE°/0.0591} = 10^{(2)(2.24)/(0.0591)} = 10^{75.8} = 6 \times 10^{75}$$

Note: K_c can also be found using Equation (19.10).

(b) We break the equation into two half-reactions:

$$Br_2(l) + 2e^- \rightarrow 2Br^-(aq)$$
$$2I^-(aq) \rightarrow I_2(s) + 2e^-$$

Notice that the I^-/I_2 half-reaction must be the anode; this follows from writing the original equation with I^- on the reactant side. We calculate the standard emf using Equation (20.1)

$$E°_{cell} = E°_{ox} + E°_{red} = -0.53\text{ V} + 1.07\text{ V} = 0.54\text{ V}$$

We find $\Delta G°$ and K_c using Equations (20.3) and (20.5)

$$\Delta G° = -nFE°_{cell} = -(2\text{ mol})(96500\text{ J/V·mol})(0.54\text{ V}) = -104\text{ kJ}$$

$$K_c = 10^{nE°/0.0591} = 10^{(2)(0.54)/(0.0591)} = 10^{18.3} = 2 \times 10^{18}$$

(c) This is worked in an exactly analogous manner. $E°_{cell} = 0.46\text{ V}$

$$\Delta G° = -nFE°_{cell} = -(4\text{ mol})(96500\text{ J/V·mol})(0.46\text{ V}) = -178\text{ kJ}$$

$$K_c = 10^{nE°/0.0591} = 10^{(4)(0.46)/(0.0591)} = 10^{31.1} = 1 \times 10^{31}$$

(d) This is worked in an exactly analogous manner. $E°_{cell} = 2.19\text{ V}$

$$\Delta G° = -nFE°_{cell} = -(6\ mol)(96500\ J/V{\cdot}mol)(2.19\ V) = -1.27 \times 10^3\ kJ$$

$$K_c = 10^{nE°/0.0591} = 10^{(6)(2.19)/(0.0591)} = 10^{222} = 1 \times 10^{222}$$

20.24 This similar to Example 20.5 and Problem 20.23 above.

Solution: We find the standard cell voltage using Table 20.1.

$$E°_{cell} = E°_{ox} + E°_{red} = 2.37\ V - 0.76\ V = 1.61\ V$$

$$\log K = \frac{(2)(1.61\ V)}{(0.0591\ V)} = 54.5 \qquad K = 3 \times 10^{54}$$

20.25 Find the listed species in Table 20.1. See Problem 20.9 for an example of deciding on the correct spontaneous reaction. The last part is like Problem 20.23 above.

Solution: The half-reactions are:

$$Fe^{3+}(aq) + e^- \rightarrow Fe^{2+}(aq) \qquad E° = 0.77\ V$$
$$Ce^{4+}(aq) + e^- \rightarrow Ce^{3+}(aq) \qquad E° = 1.61\ V$$

The diagonal rule shows that Ce^{4+} will oxidize Fe^{2+} to Fe^{3+}; this makes the Fe^{2+}/Fe^{3+} half-reaction the anode. The standard cell emf is found using Equation (20.1)

$$E°_{cell} = E°_{ox} + E°_{red} = -0.77\ V + 1.61\ V = 0.84\ V$$

The values of $\Delta G°$ and K_c are found using Equations (20.3) and (20.5)

$$\Delta G° = -nFE°_{cell} = -(1\ mol)(96500\ J/V{\cdot}mol)(0.84\ V) = -81\ kJ$$

$$K_c = 10^{nE°/0.0591} = 10^{(1)(0.84)/(0.0591)} = 2 \times 10^{14}$$

20.26 Notice that the given Cu^+/Cu half-reaction is not in Table 20.1. Locate its appropriate position in the Table. Where is it in relation to the Cu^{2+}/Cu^+ half-reaction? What is the relationship between these two half-reactions and the reaction given in this problem? Which one is the anode reaction?

Solution: The overall reaction is the sum of the two half-reactions

$$Cu^+(aq) \rightarrow Cu^{2+}(aq) + e^- \qquad E°_{ox} = -0.15\ V$$
$$Cu^+(aq) + e^- \rightarrow Cu(s) \qquad E°_{red} = 0.52\ V$$

$$2Cu^+(aq) \rightarrow Cu^{2+}(aq) + Cu(s)$$

The Cu^{2+}/Cu^+ half-reaction is the anode process. The standard cell emf can be found using Equation (20.1).

$$E°_{cell} = E°_{ox} + E°_{red} = -0.15\ V + 0.52\ V = 0.37\ V$$

The standard free energy and the equilibrium constant can be calculated with Equations (20.3) and (20.5).

$$\Delta G° = -nFE°_{cell} = -(1 \text{ mol})(96500 \text{ J/V·mol})(0.37 \text{ V}) = -36 \text{ kJ}$$

$$K_c = 10^{nE°/0.0591} = 10^{(1)(0.37)/(0.0591)} = 2 \times 10^6$$

Is this a spontaneous reaction? Can copper(I) salts exist in aqueous solution?

20.28 Is the cell described a standard cell? Why not? Study Section 20.5 and Example 20.7 if you don't know.

Solution: If this were a standard cell, the concentrations would all be 1.00 M, and the voltage would just be the standard emf calculated from Table 20.1. Since cell emf's depend on the concentrations of the reactants and products, we must use the Nernst equation (Equation (20.7)) to find the emf of a nonstandard cell.

$$E = E° - \frac{0.0591 \text{ V}}{n} \log Q = 1.10 \text{ V} - \frac{0.0591 \text{ V}}{2} \log \frac{[Zn^{2+}]}{[Cu^{2+}]} = 1.10 \text{ V} - \frac{0.0591 \text{ V}}{2} \log \frac{0.25}{0.15} = 1.09 \text{ V}$$

How did we find the value of 1.10 V for E°?

20.30 This problem is similar to Example 20.8. How do we incorporate the concentration of hydrogen gas into the equilibrium expression?

Solution: The overall cell reaction is: $2H^+(aq) + Pb(s) \rightarrow H_2(g) + Pb^{2+}(aq)$

$$E°_{cell} = E°_{ox} + E°_{red} = 0.13 \text{ V} + 0.00 \text{ V} = 0.13 \text{ V}$$

$$E = E° - \frac{0.0591 \text{ V}}{n} \log \frac{[Pb^{2+}]P_{H_2}}{[H^+]^2} = 0.13 \text{ V} - \frac{0.0591 \text{ V}}{2} \log \frac{(1.0)(0.10)}{(0.050)^2} = 0.083 \text{ V}$$

20.31 Finding the emf's is like Example 20.7 and Problem 20.28 above. How do you find the free energy difference under non-standard conditions?

Solution: (a) We use Table 20.1 to find the standard cell emf

$$E°_{cell} = E°_{ox} + E°_{red} = 2.37 \text{ V} - 0.14 \text{ V} = 2.23 \text{ V}$$

We use the Nernst equation to compute the emf at the given nonstandard concentrations

$$E = E° - \frac{0.0591 \text{ V}}{n} \log Q = 2.23 \text{ V} - \frac{0.0591 \text{ V}}{2} \log \frac{0.045}{0.035} = 2.23 \text{ V}$$

We can find the free energy difference at the given concentrations using Equation (20.2)

$$\Delta G = -nFE_{cell} = -(2 \text{ mol})(96500 \text{ J/V·mol})(2.23 \text{ V}) = -430 \text{ kJ}$$

What is another way to compute ΔG°?

(b) We use Table 20.1 to find the standard cell emf

$$E^\circ_{cell} = E^\circ_{ox} + E^\circ_{red} = 0.76 \text{ V} - 0.74 \text{ V} = 0.02 \text{ V}$$

We use the Nernst equation to compute the emf at the given nonstandard concentrations

$$E = E^\circ - \frac{0.0591 \text{ V}}{n} \log Q = 0.02 \text{ V} - \frac{0.0591 \text{ V}}{6} \log \frac{(0.0085)^3}{(0.010)^2} = 0.04 \text{ V}$$

We can find the free energy difference at the given concentrations using Equation (20.2)

$$\Delta G = -nFE_{cell} = -(6 \text{ mol})(96500 \text{ J/V·mol})(0.04 \text{ V}) = -23 \text{ kJ}$$

20.32 Write out the Nernst equation for this reaction. Is the reaction spontaneous under standard conditions? How can you tell? At what voltage will the reaction change from nonspontaneous to spontaneous?

Solution: As written, the reaction is not spontaneous under standard conditions; the cell emf is negative.

$$E^\circ_{cell} = E^\circ_{ox} + E^\circ_{red} = -0.34 \text{ V} - 0.76 \text{ V} = -1.10 \text{ V}$$

The reaction will become spontaneous when the concentrations of zinc(II) and copper(II) ions are such as to make the emf positive. The turning point is when the emf is zero. We solve the Nernst equation for the $[Cu^{2+}]/[Zn^{2+}]$ ratio at this point

$$E_{cell} = 0 = E^\circ - \frac{0.0591 \text{ V}}{n} \log Q = -1.10 \text{ V} - \frac{0.0591 \text{ V}}{2} \log \frac{[Cu^{2+}]}{[Zn^{2+}]}$$

$$\frac{[Cu^{2+}]}{[Zn^{2+}]} = 10^{nE^\circ/0.0591} = 10^{(2)(-1.10)/(0.0591)} = 6.0 \times 10^{-38}$$

In other words for the reaction to be spontaneous, the $[Cu^{2+}]/[Zn^{2+}]$ ratio must be less than 6.0×10^{-38}. Is the reduction of zinc(II) by copper metal a practical use of copper?

20.33 Read about concentration cells in Section 20.5. What is the standard emf of a concentration cell?

Solution: All concentration cells have the same standard emf: zero volts. (Why?) We use the Nernst equation to compute the emf.

$$E = E^\circ - \frac{0.0591 \text{ V}}{n} \log Q = 0 \text{ V} - \frac{0.0591 \text{ V}}{2} \log \frac{0.24}{0.53} = 0.010 \text{ V}$$

What is the direction of spontaneous change in all concentration cells?

20.37 Study the discussion of fuel cells in Section 20.6. Given that a current of one ampere represents a flow of one coulomb per second, how many coulombs pass through the conductor in 3.0 h at a current of 8.5 amperes? How does this relate to the amount of hydrogen (see Section 20.4)? How can you find the volume of the hydrogen gas? What amount of oxygen is required every minute?

Solution: (a) The total charge passing through the circuit is

$$3.0\ \text{h}\left(\frac{8.5\ \text{C}}{1\ \text{s}}\right)\left(\frac{3600\ \text{s}}{1\ \text{h}}\right) = 9.2 \times 10^4\ \text{C}$$

From the anode half-reaction we can find the amount of hydrogen

$$9.2 \times 10^4\ \text{C}\left(\frac{2\ \text{mol}\ H_2}{4\ \text{F}}\right)\left(\frac{1\ \text{F}}{96500\ \text{C}}\right) = 0.48\ \text{mol}\ H_2$$

The volume can be computed using the ideal gas equation

$$V = \frac{nRT}{P} = \frac{(0.48\ \text{mol})(0.0821\ \text{L·atm/K·mol})(298\ \text{K})}{155\ \text{atm}} = 0.076\ \text{L}$$

(b) The charge passing through the circuit in one minute is

$$\left(\frac{8.5\ \text{C}}{1\ \text{s}}\right)\left(\frac{60\ \text{s}}{1\ \text{min}}\right) = 510\ \text{C/min}$$

We can find the amount of oxygen from the cathode half-reaction and the ideal gas equation

$$\left(\frac{510\ \text{C}}{1\ \text{min}}\right)\left(\frac{1\ \text{F}}{96500\ \text{C}}\right)\left(\frac{1\ \text{mol}\ O_2}{4\ \text{F}}\right) = 1.3 \times 10^{-3}\ \text{mol}\ O_2\text{/min}$$

$$\text{Volume} = \left(\frac{1.3 \times 10^{-3}\ \text{mol}\ O_2}{1\ \text{min}}\right)\left(\frac{(0.0821\ \text{L·atm/K·mol})(298\ \text{K})}{1\ \text{atm}}\right)\left(\frac{1.0\ \text{L air}}{0.20\ \text{L}\ O_2}\right) = 0.16\ \text{L air/min}$$

20.38 Use the thermodynamic data in Appendix 3 to compute the standard free energy change for the combustion of propane (see the balanced equation in the discussion of fuel cells). What is the relationship between standard free energy change and the standard cell emf?

Solution: The standard free energy change is+

$$\Delta G^\circ_{rxn} = 3\Delta G^\circ_f(CO_2(g)) + 4\Delta G^\circ_f(H_2O(l)) - \Delta G^\circ_f(C_3H_8(g)) - 5\Delta G^\circ_f(O_2(g))$$

$$\Delta G^\circ_{rxn} = (3\ \text{mol})(-394.4\ \text{kJ/mol}) + (4\ \text{mol})(-237.2\ \text{kJ/mol}) - (1\ \text{mol})(-23.5\ \text{kJ/mol}) - (5\ \text{mol})(0) = -2108.5\ \text{kJ}$$

If we know the standard free energy difference for a reaction, we can calculate the standard emf using Equation (20.3).

$$E^\circ_{cell} = \frac{-\Delta G^\circ}{nF} = \frac{-(-2108.5 \times 10^3\text{J})}{(20\ \text{mol})(96500\ \text{J/V·mol})} = 1.09\ \text{V}$$

Does this suggest that, in theory, it should be possible to construct a galvanic cell (battery) based on any conceivable spontaneous reaction?

20.46 Study Example 20.10 before trying this problem.

Solution: (a) The only ions present in molten $CaCl_2$ are Ca^{2+} and Cl^-. The electrode reactions are:

anode: $2Cl^-(aq) \rightarrow Cl_2(g) + 2e^-$

cathode: $Ca^{2+}(aq) + 2e^- \rightarrow Ca(s)$

(b) The amount of charge passing through the cell is:

$$0.50\ \text{A}\left(\frac{1\ \text{C/s}}{1\ \text{A}}\right)\left(\frac{60\ \text{s}}{1\ \text{min}}\right)\left(\frac{1\ \text{F}}{96500\ \text{C}}\right)30\ \text{min} = 9.3 \times 10^{-3}\ \text{F}$$

The amount of calcium metal formed is:

$$9.3 \times 10^{-3}\ \text{F}\left(\frac{1\ \text{mol Ca}}{2\ \text{F}}\right)\left(\frac{40.08\ \text{g Ca}}{1\ \text{mol Ca}}\right) = 0.19\ \text{g Ca}$$

20.47 How many moles of oxygen gas are formed?

Solution: Find the amount of oxygen using the ideal gas equation.

$$n = \frac{PV}{RT} = \frac{(755\ \text{mmHg})\left(\frac{1\ \text{atm}}{760\ \text{mmHg}}\right)(0.076\ \text{L})}{(0.0821\ \text{L·atm/K·mol})(298\ \text{K})} = 3.1 \times 10^{-3}\ \text{mol}\ O_2$$

Since the half-reaction shows that one mole of oxygen requires four faradays of electric charge, we write

$$3.1 \times 10^{-3}\ \text{mol}\ O_2\left(\frac{4\ \text{F}}{1\ \text{mol}\ O_2}\right) = 0.012\ \text{F}$$

20.48 This is like the electrolysis of aqueous sodium chloride; see the discussion of this topic in Section 20.8. How does the electrolysis of aqueous sodium chloride differ from the electrolysis of molten sodium chloride? This problem is similar to Example 20.9.

Solution: The electrolysis of molten zinc chloride follows the expected half-reactions (see Section 20.8 and Example 20.9)

anode: $2Cl^-(aq) \rightarrow Cl_2(g) + 2e^-$

cathode: $Zn^{2+}(aq) + 2e^- \rightarrow Zn(s)$

The electrolysis of aqueous zinc chloride is similar to the case of aqueous sodium chloride. Water is reduced more readily than zinc ion.

$$2H_2O(l) + 2e^- \rightarrow H_2(g) + 2OH^-(aq)$$

The anode reaction remains the same.

20.50 Use Problem 20.47 above as a model to work this problem.

Solution: (a) The half-reaction is: $2H_2O(l) \rightarrow O_2(g) + 4H^+(aq) + 4e^-$

We find the number of moles of oxygen produced:

$$n_{O_2} = \frac{PV}{RT}$$

The number of faradays required is

$$n_{O_2}\left(\frac{4\ F}{1\ mol\ O_2}\right) = \frac{PV}{RT}\left(\frac{4\ F}{1\ mol\ O_2}\right) = \frac{(1.0\ atm)(0.84\ L)}{(0.0821\ L{\cdot}atm/K{\cdot}mol)(298\ K)}\left(\frac{4\ F}{1\ mol\ O_2}\right) = 0.14\ F$$

(b) The half-reaction is:

$$2Cl^-(aq) \rightarrow Cl_2(g) + 2e^-$$

The number of moles of chlorine produced is:

$$n_{Cl_2} = \frac{PV}{RT}$$

The number of faradays required is:

$$n_{Cl_2}\left(\frac{2\ F}{1\ mol\ Cl_2}\right) = \frac{(750/760)atm(1.50\ L)}{(0.0821\ L{\cdot}atm/K{\cdot}mol)(293\ K)}\left(\frac{2\ F}{1\ mol\ Cl_2}\right) = 0.123\ F$$

(c) The half-reaction is:

$$Sn^{2+}(aq) + 2e^- \rightarrow Sn(s)$$

The number of faradays required is:

$$6.0\ g\ Sn\left(\frac{1\ mol\ Sn}{118.7\ g}\right)\left(\frac{2\ F}{1\ mol\ Sn}\right) = 0.10\ F$$

20.52 Hint: Read the discussion of the electrolysis of aqueous NaCl in Section 20.8.

Solution: (a) The half-reaction is: $Ag^+(aq) + e^- \rightarrow Ag(s)$

(b) The probable oxidation is: $2H_2O(l) \rightarrow O_2(g) + 4H^+(aq) + 4e^-$

(c) The quantity of electricity used is:

$$0.67\ g\ Ag\left(\frac{1\ mol\ Ag}{107.9\ g\ Ag}\right)\left(\frac{1\ F}{1\ mol\ Ag}\right)\left(\frac{96500\ C}{1\ F}\right) = 6.0 \times 10^2\ C$$

20.54 Study the discussion of the quantitative aspects of electrolysis in Section 20.8. The central idea in this problem is that the same quantity of electric charge passes through both cells; you neither gain nor lose electrons in an electrolysis. Remember that the number of amperes is the number of coulombs flowing per second.

Solution: (a) First find the amount of charge needed to produce 2.00 g of silver according to the half-reaction:

$$Ag^+(aq) + e^- \rightarrow Ag(s)$$

$$2.00 \text{ g Ag}\left(\frac{1 \text{ mol Ag}}{107.9 \text{ g Ag}}\right)\left(\frac{96500 \text{ C}}{1 \text{ F}}\right)\left(\frac{1 \text{ F}}{1 \text{ mol Ag}}\right) = 1.79 \times 10^3 \text{ C}$$

The half-reaction for the reduction of copper(II) is:

$$Cu^{2+}(aq) + 2e^- \rightarrow Cu(s)$$

We can write for the amount of copper formed:

$$1.79 \times 10^3 \text{ C}\left(\frac{1 \text{ F}}{96500 \text{ C}}\right)\left(\frac{1 \text{ mol Cu}}{2 \text{ F}}\right)\left(\frac{63.55 \text{ g Cu}}{1 \text{ mol Cu}}\right) = 0.589 \text{ g Cu}$$

(b) The current flowing through the cells is:

$$1.79 \times 10^3 \text{ C}\left(\frac{1}{3.75 \text{ h}}\right)\left(\frac{1 \text{ h}}{3600 \text{ s}}\right)\left(\frac{1 \text{ A}}{1 \text{ C/s}}\right) = 0.133 \text{ A}$$

20.55 Study Example 20.10 as a model for the computation of amount of electrolysis product from a given electric current. The anode efficiency simply means that only 93.0% of the current produces chlorine gas (the other 7.0% presumably oxidizes water to oxygen gas).

Solution: The half-reaction for the oxidation of chloride ion is:

$$2Cl^-(aq) \rightarrow Cl_2(g) + 2e^- \qquad E^\circ_{ox} = -1.36 \text{ V}$$

The amount of chlorine formed in one hour will be:

$$1500 \text{ A}\left(\frac{1 \text{ C/s}}{1 \text{ A}}\right)\left(\frac{3600 \text{ s}}{1 \text{ h}}\right)\left(\frac{1 \text{ F}}{96500 \text{ C}}\right)\left(\frac{1 \text{ mol } Cl_2}{2 \text{ F}}\right)\left(\frac{0.07090 \text{ kg } Cl_2}{1 \text{ mol } Cl_2}\right)\left(\frac{93.0\%}{100\%}\right) = 1.84 \text{ kg } Cl_2\text{/h}$$

20.56 Balance the half-reaction, compute the volume of the chromium metal, convert this to mass, and calculate the time required to produce this quantity of chromium with a current of 25.0 amperes.

Solution: The balanced half-reaction is:

$$Cr_2O_7^{2-}(aq) + 14H^+(aq) + 12e^- \rightarrow 2Cr(s) + 7H_2O(l)$$

We find the quantity of chromium metal by calculating the volume and converting this to mass using the given density.

$$1.0 \times 10^{-2} \text{ mm}\left(\frac{1 \text{ cm}}{10 \text{ mm}}\right)\left(\frac{100 \text{ cm}}{1 \text{ m}}\right)^2 0.25 \text{ m}^2 = 2.5 \text{ cm}^3$$

$$\text{Mass Cr} = 2.5 \text{ cm}^3\left(\frac{7.19 \text{ g}}{1 \text{ cm}^3}\right) = 18 \text{ g Cr}$$

The time required to produce this amount of chromium is:

$$18 \text{ g Cr} \left(\frac{1 \text{ mol Cr}}{52.00 \text{ g Cr}}\right)\left(\frac{12 \text{ F}}{2 \text{ mol Cr}}\right)\left(\frac{96500 \text{ C}}{1 \text{ F}}\right)\left(\frac{1 \text{ s}}{25.0 \text{ C}}\right)\left(\frac{1 \text{ h}}{3600 \text{ s}}\right) = 2.2 \text{ h}$$

Would any time be saved by connecting several bumpers together in a series?

20.57 What is the charge of copper ion in $CuSO_4$? What quantity of electric charge passes through the solution in 304 s at a current of 3.00 amperes? What are the units of molar mass?

Solution: The quantity of charge passing through the solution is:

$$0.750 \text{ A} \left(\frac{1 \text{ C/s}}{1 \text{ A}}\right)\left(\frac{60 \text{ s}}{1 \text{ min}}\right)\left(\frac{1 \text{ F}}{96500 \text{ C}}\right) 25.0 \text{ min} = 1.17 \times 10^{-2} \text{ F}$$

Since the charge of the copper ion is +2, the number of moles of copper formed must be:

$$1.17 \times 10^{-2} \text{ F} \left(\frac{1 \text{ mol Cu}}{2 \text{ F}}\right) = 5.85 \times 10^{-3} \text{ mol Cu}$$

The units of molar mass are grams per mole. The molar mass of copper is:

$$\left(\frac{0.369 \text{ g}}{5.85 \times 10^{-3} \text{ mol}}\right) = 63.1 \text{ g/mol}$$

20.58 This problem is similar in principle to Problem 20.57 above and to Example 20.11.

Solution: One faraday will produce half a mole of copper (63.55/2 g). The number of coulombs required to generate this amount is

$$3.00 \text{ A} \times 304 \text{ s} \left(\frac{1 \text{ C}}{1 \text{ A·s}}\right)\left(\frac{(63.55/2) \text{ g Cu}}{0.300 \text{ g Cu}}\right) = 9.66 \times 10^{4} \text{ C}$$

20.60 This is similar to Problem 20.49 above.

Solution: The costs for producing various metals can be calculated by setting up the ratios of electrons needed to reduce each type of ion. The reductions are:

$Mg^{2+} + 2e^- \rightarrow Mg$ $\quad Al^{3+} + 3e^- \rightarrow Al$ $\quad Na^+ + e^- \rightarrow Na$ $\quad Ca^{2+} + 2e^- \rightarrow Ca$

(a) For aluminum: $\left(\frac{\$155}{1 \text{ ton}}\right)\left(\frac{3 \text{ e}^-}{2 \text{ e}^-}\right)\left(\frac{24.31 \text{ g Mg}}{26.98 \text{ g Al}}\right) \times 10.0 \text{ tons} = \$2.09 \times 10^3$

(b) For sodium: $\left(\frac{\$155}{1 \text{ ton}}\right)\left(\frac{1 \text{ e}^-}{2 \text{ e}^-}\right)\left(\frac{24.31 \text{ g Mg}}{22.99 \text{ g Na}}\right) \times 30.0 \text{ tons} = \$2.46 \times 10^3$

(c) For calcium: $\left(\frac{\$155}{1\ \text{ton}}\right)\left(\frac{2\ e^-}{2\ e^-}\right)\left(\frac{24.31\ \text{g Mg}}{40.08\ \text{g Ca}}\right) \times 50.0\ \text{tons} = \$4.70 \times 10^3$

20.62 What quantity of charge (in coulombs) is needed to produce the given amount of oxygen gas (compare to Problem 20.47)? How many oxygen molecules does this amount represent? How many electrons are needed to form one oxygen molecule?

Solution: The half-reaction for the oxidation of water to oxygen is:

$$2H_2O(l) \rightarrow O_2(g) + 4H^+(aq) + 4e^- \qquad E^\circ_{ox} = -1.23\ V$$

Knowing that one mole of any gas at STP occupies a volume of 22.4 L, we find the number of moles of oxygen.

$$4.26\ L\ O_2 \left(\frac{1\ mol}{22.4\ L}\right) = 0.190\ mol\ O_2$$

Since four electrons are required to form one oxygen molecule, the number of electrons must be:

$$0.190\ mol\ O_2 \left(\frac{4\ mol\ e^-}{1\ mol\ O_2}\right)\left(\frac{6.022 \times 10^{23}}{1\ mol}\right) = 4.58 \times 10^{23}\ e^-$$

The amount of charge passing through the solution is:

$$6.00\ A \left(\frac{1\ C/s}{1\ A}\right)\left(\frac{3600\ s}{1\ h}\right) 3.40\ hr = 7.34 \times 10^4\ C$$

We find the electron charge by dividing the amount of charge by the number of electrons.

$$\frac{7.34 \times 10^4\ C}{4.58 \times 10^{23}\ e^-} = 1.60 \times 10^{-19}\ C/e^-$$

In actual fact this sort of calculation can be used to find Avogadro's number, not the electron charge. The latter can be measured independently, and one can use this charge together with electrolytic data like the above to calculate the number of objects in one mole. See also Problem 20.78.

20.65 Parts (a)-(c) are like Problems 20.9 and 20.28 above. A pH meter measures a cell voltage that is dependent upon hydrogen ion concentration.

Solution: (a)The half-reactions are:

$$2H^+(aq) + 2e^- \rightarrow H_2(g) \qquad E^\circ = 0.00\ V$$

$$Ag^+(aq) + e^- \rightarrow Ag(s) \qquad E^\circ = 0.80\ V$$

$$E^\circ_{cell} = E^\circ_{ox} + E^\circ_{red} = 0.00\ V + 0.80\ V = 0.80\ V$$

(b) The spontaneous cell reaction under standard-state conditions is:

$$2Ag^+(aq) + H_2(g) \rightarrow 2Ag(s) + 2H^+(aq)$$

(c) Using the Nernst equation: $E = E° - \frac{0.0591\ V}{n} \log \frac{[H^+]^2}{[Ag^+]^2 P_{H_2}}$

(i) The potential is: $E = 0.80\ V - \frac{0.0591\ V}{2} \log \frac{(1.0 \times 10^{-2})^2}{(1.0)^2(1.0)} = 0.92\ V$

(ii) The potential is: $E = 0.80\ V - \frac{0.0591\ V}{2} \log \frac{(1.0 \times 10^{-5})^2}{(1.0)^2(1.0)} = 1.10\ V$

(d) From the results in (c) we deduce that this cell is a pH meter; its potential is a sensitive function of the hydrogen ion concentration.

20.67 Silver oxalate is a sparingly soluble ionic compound (oxalate ion is $C_2O_4^{2-}$). Write out the balanced equation for the solubility equilibrium of silver oxalate. If you know the silver ion concentration, can you determine the oxalate ion concentration (see Section 17.1)? Can you determine the silver ion concentration from the given potential difference? What is the SHE (see Section 20.3)? What are the conditions required for the operation of the SHE?

Solution: The overall cell reaction is:

$$2Ag^+(aq) + H_2(g) \rightarrow 2Ag(s) + 2H^+(aq)$$

We write the Nernst equation for this system.

$$E = E° - \frac{0.0591\ V}{n} \log Q = 0.80\ V - \frac{0.0591\ V}{2} \log \frac{[H^+]^2}{[Ag^+]^2 P_{H_2}}$$

The measured voltage is 0.589 V, and we can find the silver ion concentration as follows:

$$0.589\ V = 0.80\ V - \frac{0.0591\ V}{2} \log \frac{1}{[Ag^+]^2}$$

$\log \frac{1}{[Ag^+]^2} = 7.14$ $\qquad \frac{1}{[Ag^+]^2} = 1.4 \times 10^7$ $\qquad [Ag^+] = 2.7 \times 10^{-4}\ M$

Knowing the silver ion concentration, we can calculate the oxalate ion concentration and the solubility product constant.

$$[C_2O_4^{2-}] = \frac{1}{2}[Ag^+] = 1.35 \times 10^{-4}\ M$$

$$K_{sp} = [Ag^+]^2[C_2O_4^{2-}] = (2.7 \times 10^{-4})^2(1.35 \times 10^{-4}) = 9.8 \times 10^{-12}$$

20.68 Part (a) is similar to Problem 20.28 above. For the second part you must recalculate the concentrations of silver and magnesium ions after the deposition of the 1.20 g of silver metal.

Solution: (a) If this were a standard cell, the concentrations would all be 1.00 M, and the voltage would just be the standard emf calculated from Table 20.1. Since cell emf's depend on the concentrations of the reactants and products, we must use the Nernst equation (Equation (20.7)) to find the emf of a nonstandard cell.

$$E = E^\circ - \frac{0.0591\ V}{n}\log Q = 3.17\ V - \frac{0.0591\ V}{2}\log\frac{[Mg^{2+}]}{[Ag^+]^2} = 3.17\ V - \frac{0.0591\ V}{2}\log\frac{0.10}{(0.10)^2} = 3.14\ V$$

(b) First we calculate the concentration of silver ion remaining in solution after the deposition of 1.20 g of silver metal

Ag originally in solution: $\frac{0.100\ mol\ Ag^+}{1\ L} \times 0.346\ L = 3.46 \times 10^{-2}\ mol\ Ag^+$

Ag deposited: $1.20\ g\ Ag \times \frac{1\ mol}{107.9\ g} = 1.11 \times 10^{-2}\ mol\ Ag$

Ag remaining in solution: $3.46 \times 10^{-2}\ mol\ Ag - 1.11 \times 10^{-2}\ mol\ Ag = 2.35 \times 10^{-2}\ mol\ Ag$

$$[Ag^+] = \frac{2.35 \times 10^{-2}\ mol}{0.346\ L} = 6.79 \times 10^{-2}\ M$$

The overall reaction is: $Mg(s) + 2Ag^+(aq) \rightarrow Mg^{2+}(aq) + 2Ag(s)$

We use the balanced equation to find the amount of magnesium metal suffering oxidation and dissolving.

$$1.11 \times 10^{-2}\ mol\ Ag\left(\frac{1\ mol\ Mg}{2\ mol\ Ag}\right) = 5.55 \times 10^{-3}\ mol\ Mg$$

The amount of magnesium originally in solution was

$$0.288\ L\left(\frac{0.100\ mol}{1\ L}\right) = 2.88 \times 10^{-2}\ mol$$

The new magnesium ion concentration is:

$$\frac{(5.55 \times 10^{-3} + 2.88 \times 10^{-2})\ mol}{0.288\ L} = 0.119\ M$$

The new cell emf is:

$$E = E^\circ - \frac{0.0591\ V}{n}\log Q = 3.17\ V - \frac{0.0591\ V}{2}\log\frac{0.119}{(6.79 \times 10^{-2})^2} = 3.13\ V$$

20.69 This is similar to Problem 20.11 above.

Solution: In hydrochloric acid the oxidizing species is $H^+(aq)$; Cl^- cannot accept another electron because it already has a complete octet. By using the diagonal rule, we see that copper metal is located above and to the right side of nitrate ion in acid solution (i.e., nitric acid at +0.96 V). Copper will react spontaneously with nitric acid. On the other hand, we see that copper metal is located below and to the right of hydrogen ion. Copper does not react spontaneously with hydrochloric acid.

20.70 Write out the balanced redox equation. Set up the Nernst equation for this system; notice that you are given all the concentrations and that the emf for the Pb^{2+}/Pb half-reaction can be found in Table 20.1. Which term in the Nernst equation don't you know? What is the relationship between standard cell emf and the equilibrium constant for the cell reaction?

Solution: (a) The balanced redox equation is:

$$Pb(s) + 2VO^{2+}(aq) + 4H^+(aq) \rightarrow Pb^{2+}(aq) + 2V^{3+}(aq) + 2H_2O(l)$$

The Nernst equation takes the form

$$E_{cell} = E^\circ_{cell} - \frac{0.0591\text{ V}}{n}\log Q = E^\circ_{cell} - \frac{0.0591\text{ V}}{2}\log\frac{[V^{3+}]^2[Pb^{2+}]}{[VO^{2+}]^2[H^+]^4} = 0.67\text{ V}$$

The standard emf is:

$$E^\circ_{cell} = 0.67\text{ V} + \frac{0.0591\text{ V}}{2}\log\frac{(1.0\times10^{-5})^2(0.10)}{(0.10)^2(0.10)^4} = 0.52\text{ V} = E^\circ_{ox} + E^\circ_{red} = 0.13 + E^\circ_{VO^{2+}/V^{3+}}$$

$$E^\circ_{VO^{2+}/V^{3+}} = 0.39\text{ V}$$

(b) The standard cell emf is 0.52 V. The equilibrium constant can be found from Equation (20.5).

$$K = 10^{nE^\circ/0.0591} = 10^{(2)(0.52)/(0.0591)} = 4\times10^{17}$$

20.71 Study the discussion of the electrolysis of aqueous sodium chloride in Section 20.8, and compare the standard reduction potentials for Cl_2 and F_2 (Table 20.1).

Solution: The overvoltage of oxygen is not large enough to prevent its formation at the anode. Water is oxidized before fluoride ion.

This fact was one of the major obstacles preventing the discovery of fluorine for many years. HF was usually chosen as the substance for electrolysis, but two problems interfered with the experiment. First, any water in the HF was oxidized before the fluoride ion. Second, pure HF without any water in it is a nonconductor of electricity (HF is a weak acid!). The problem was finally solved by dissolving KF in liquid HF to give a conducting solution.

20.72 Look at the anode and cathode half-reactions. The apparatus is designed to keep the products formed at each electrode separated so they cannot react. What are the relative amounts of gases formed at each electrode? What happens to the pH of the solution at each electrode?

Solution: The two electrode processes are:

anode: $2H_2O(l) \rightarrow O_2(g) + 4H^+(aq) + 4e^-$

cathode: $4H_2O(l) + 4e^- \rightarrow 2H_2(g) + 4OH^-(aq)$

The amount of hydrogen formed is twice the amount of oxygen. Notice that the solution at the anode will become acidic and that the solution at the cathode will become basic (test with litmus paper). What are the relative amounts of H^+ and OH^- formed in this process? Would the solutions surrounding the two electrodes neutralize each other exactly? If not, would the resulting solution be acidic or basic?

20.73 If you don't remember how to work this problem, review the section "Concentration Cells" in Section 20.5. A similar problem is Problem 20.33 above.

Solution: The cell voltage is given by:

$$E = E° - \frac{0.0591\ V}{2} \log \frac{[Cu^{2+}]_{dilute}}{[Cu^{2+}]_{concentrated}} = 0 - \frac{0.0591\ V}{2} \log \frac{0.080}{1.2} = 0.035\ V$$

20.74 According to the cathode reaction given, how many coulombs of electric charge can be produced from 4.0 g of MnO_2? What is the definition of the ampere (Section 20.8)?

Solution: The 4.0 g of MnO_2 can produce:

$$4.0\ g\ MnO_2 \left(\frac{1\ mol}{86.94\ g}\right)\left(\frac{2\ F}{2\ mol\ MnO_2}\right)\left(\frac{96500\ C}{1\ F}\right) = 4.44 \times 10^3\ C$$

Since a current of one ampere represents a flow of one coulomb per second, we can find the time it takes for this amount of charge to pass.

$$4.44 \times 10^3\ C \left(\frac{1\ s}{0.0050\ C}\right)\left(\frac{1\ h}{3600\ s}\right) = 2.5 \times 10^2\ h$$

20.75 Study the discussion of concentration cells in Section 20.5. Use Equation (20.6) to set up the expressions for the concentration cell emf (notice that the temperature is not 25°C) for the cases of Hg^+ and Hg_2^{2+}. What must be the concentration ratio in both cases? What factor in the equation is different in the two cases?

Solution: Since this is a concentration cell, the standard emf is zero. (Why?) The two cell voltage equations are:

$$E_{cell} = -\frac{2.303RT}{nF} \log Q = -\frac{2.303RT}{2F} \log \frac{[Hg_2^{2+}]_{soln\ A}}{[Hg_2^{2+}]_{soln\ B}}$$

$$E_{cell} = -\frac{2.303RT}{nF} \log Q = -\frac{2.303RT}{1F} \log \frac{[Hg^+]_{soln\ A}}{[Hg^+]_{soln\ B}}$$

In the first case two electrons are transferred per mercury ion (n = 2), while in the second only one is transferred (n = 1). Note that the concentration ratio will be 1:10 in both cases. The voltages calculated at 18°C are:

$$E_{cell} = \frac{-2.303 \times 8.314\ J/K \cdot mol \times 291\ K}{2 \times 96500\ J \cdot V^{-1} mol^{-1}} \log 10^{-1} = 0.0289\ V$$

$$E_{cell} = \frac{-2.303 \times 8.314\ J/K \cdot mol \times 291\ K}{1 \times 96500\ J \cdot V^{-1} mol^{-1}} \log 10^{-1} = 0.0577\ V$$

We conclude that the mercury(I) ion is really Hg_2^{2+}.

20.76 What reactions will occur in this experiment? Study the discussion of the electrolysis of aqueous NaCl in Section 20.8. Is iodide ion easier to oxidize than water (see Table 20.1)? What will happen at the cathode?

Solution: Iodide ion is much easier to oxidize than water so the anode reaction will be:

$$2I^-(aq) \rightarrow I_2(s) + 2e^- \qquad E°_{ox} = -0.53 \text{ V}$$

The solution surrounding the anode will become brown because of the formation of I_3^-.

The cathode reaction will be the same as in the NaCl electrolysis. (Why?) Since OH^- is a product, the solution around the cathode will become basic which will cause the phenolphthalein indicator to turn red.

20.77 First of all, verify that Ag^+ is the reactant in short supply, i.e. the limiting reagent (see Section 3.5). Write the balanced equation for the reaction and calculate the amount of magnesium remaining (if any). Compute the standard cell emf for the reaction and convert this to the equilibrium constant (Section 20.4). Can the silver ion concentration be zero?

Solution: We begin by treating this like an ordinary stoichiometry problem. The number of moles of magnesium is:

$$1.56 \text{ g Mg} \left(\frac{1 \text{ mol Mg}}{24.31 \text{ g Mg}}\right) = 0.0642 \text{ mol Mg}$$

The number of moles of silver ion in the solution is:

$$\left(\frac{0.100 \text{ mol Ag}^+}{1 \text{ L}}\right) 0.1000 \text{ L} = 0.0100 \text{ mol Ag}^+$$

The balanced equation for the reaction is:

$$2Ag^+(aq) + Mg(s) \rightarrow 2Ag(s) + Mg^{2+}(aq)$$

The limiting reagent is silver ion. The amount of Mg consumed is:

$$0.0100 \text{ mol Ag}^+ \left(\frac{1 \text{ mol Mg}}{2 \text{ mol Ag}}\right) = 0.00500 \text{ mol Mg}$$

Assuming 100% reaction, the Mg^{2+} concentration is:

$$[Mg^{2+}]_o = \frac{0.00500 \text{ mol}}{0.100 \text{ L}} = 0.0500 \text{ M}$$

The amount of magnesium remaining is:

$$(0.0642 - 0.00500) \text{ mol Mg} \times 24.31 \text{ g/mol} = 1.44 \text{ g Mg}$$

The standard emf for this reaction is found by applying Equation (20.1).

$$E°_{cell} = E°_{ox} + E°_{red} = 2.37 \text{ V} + 0.80 \text{ V} = 3.17 \text{ V}$$

We can then compute the equilibrium constant.

$$K = 10^{nE°_{cell}/0.0591} = 10^{(2)(3.17)/(0.0591)} = 1 \times 10^{107}$$

To find the equilibrium concentrations of Mg^{2+} and Ag^+, let 2x be the concentration of Ag^+ at equilibrium.

$$2Ag^+(aq) + Mg(s) \rightleftharpoons 2Ag(s) + Mg^{2+}(aq)$$

	Ag^+	Mg^{2+}
Initial (M):	0.0000	0.0500
Change (M):	+2x	-x
Equilibrium (M):	2x	(0.0500 - x)

$$K = \frac{[Mg^{2+}]}{[Ag^+]^2} = \frac{(0.0500 - x)}{(2x)^2} = 1 \times 10^{107}$$

We can assume $0.0500 - x \cong 0.0500$.

$$2x = [Ag^+] = 2 \times 10^{-55}\ M$$

How many silver ions will be present in 100 mL of this solution? The concentration of magnesium ion will be 0.0500 M (three significant figures).

20.78 Before starting this, study Problem 20.62. For this problem assume that the only species in solution capable of reduction is hydrogen ion. How many electrons must have passed through the circuit under the given conditions? How many moles of copper dissolved?

Solution: (a) The gas must be hydrogen from the reduction of hydrogen ion. The two electrode reactions and the overall cell reaction are:

anode: $Cu(s) \rightarrow Cu^{2+}(aq) + 2e^-$

cathode: $2H^+(aq) + 2e^- \rightarrow H_2(g)$

Since 0.584 g of copper was consumed, the amount of hydrogen gas produced was:

$$0.584\text{ g Cu}\left(\frac{1\text{ mol Cu}}{63.55\text{ g Cu}}\right)\left(\frac{1\text{ mol }H_2}{1\text{ mol Cu}}\right) = 9.20 \times 10^{-3}\text{ mol }H_2$$

Volume at STP: $(9.20 \times 10^{-3}\text{ mol})(22.41\text{ L/mol}) = 0.206\text{ L}$

(b) From the current and the time we can calculate the amount of charge:

$$1.18\text{ A}\left(\frac{1\text{ C/s}}{1\text{ A}}\right) 1.52 \times 10^3\text{ s} = 1.79 \times 10^3\text{ C}$$

Since we know the charge of the electron, we can compute the number of electrons.

$$1.79 \times 10^3\text{ C}\left(\frac{1\ e^-}{1.6022 \times 10^{-19}\text{ C}}\right) = 1.12 \times 10^{22}\ e^-$$

Using the amount of copper consumed in the reaction, we can calculate Avogadro's number.

$$\frac{1.12 \times 10^{22}\ e^-}{9.20 \times 10^{-3}\text{ mol Cu}} \times \left(\frac{1\text{ mol Cu}}{2\text{ mol }e^-}\right) = 6.09 \times 10^{23}\text{ mol}^{-1}$$

In practice Avogadro's number can be determined by electrochemical experiments like this. The charge of the electron can be found independently by Millikan's experiment.

20.79 What happens to the electrodes when a galvanic cell is operating?

Solution: Weigh the zinc and copper electrodes before operating the cell and re-weigh afterwards. The anode (Zn) should lose mass and the cathode (Cu) should gain mass.

20.81 Use Table 20.1 for cell potentials.

Solution: The half reactions are:

$$H_2O_2(aq) \rightarrow O_2(g) + 2H^+(aq) + 2e^- \qquad E° = -0.68\ V$$
$$H_2O_2(aq) + 2H^+(aq) + 2e^- \rightarrow 2H_2O(l) \qquad E° = 1.77\ V$$
$$2H_2O_2(aq) \rightarrow 2H_2O(l) + O_2(g) \qquad E° = 1.09\ V$$

Thus, this is a spontaneous reaction. H_2O_2 is not stable (it disproportionates).

20.83 What is a formation constant?

Solution: The half reactions are:

$$Zn(s) + 4OH^-(aq) \rightarrow Zn(OH)_4^{2-}(aq) + 2e^- \qquad E° = 1.36\ V$$
$$Zn^{2+}(aq) + 2e^- \rightarrow Zn(s) \qquad E° = -0.76\ V$$
$$Zn^{2+}(aq) + 4OH^-(aq) \rightarrow Zn(OH)_4^{2-}(aq) \qquad E° = 0.60\ V$$

$$K_f = 10^{nE°/0.0591} = 10^{(2)(0.60)/(0.0591)} = 2 \times 10^{20}$$

20.85 Obtain the free energies from Appendix 3.

Solution: (a) Calculating ΔG°

$$\Delta G° = 2\Delta G_f^\circ(N_2) + 6\Delta G_f^\circ(H_2O) - 4\Delta G_f^\circ(NH_3) - 3\Delta G_f^\circ(O_2)$$
$$\Delta G° = 0 + (6\ mol)(-237.2\ kJ/mol) - (4\ mol)(-16.6\ kJ/mol) - 0 = -1356.8\ kJ$$

(b) The half-reactions are:

$$4NH_3(g) \rightarrow 2\ N_2(g) + 12H^+(aq) + 12e^-$$
$$3O_2(g) + 12H^+(aq) + 12e^- \rightarrow 6H_2O(l)$$

The overall reaction is a 12-electron process.

$$E°_{cell} = \frac{-\Delta G°}{nF} = \frac{-(-1356.8 \text{ x } 1000)}{(12)(96500)} = 1.17\ V$$

20.87 This problem is a Nernst-type calculation that is combined with Le Chatelier's principle.

Solution: (a) The reaction is: $Zn + Cu^{2+} \rightarrow Zn^{2+} + Cu$

Using the Nernst equation: $$E = E° - \frac{0.0591}{2} \log \frac{0.20}{0.20} = 1.10\ V$$

If NH_3 is added to the $CuSO_4$ solution: $Cu^{2+} + 4NH_3 \rightarrow Cu(NH_3)_4^{2+}$

The concentration of copper ions $[Cu^{2+}]$ decreases, so the log term becomes more positive and E decreases. If NH_3 is added to the $ZnSO_4$ solution:

$$Zn^{2+} + 4NH_3 \rightarrow Zn(NH_3)_4^{2+}$$

The concentration of zinc ions $[Zn^{2+}]$ decreases, so the log term becomes less positive and E increases.

(b) After addition of 25.0 mL of 3.0 M NH_3,

$$Cu^{2+} + 4NH_3 \rightarrow Cu(NH_3)_4^{2+}$$

Assume that all Cu^{2+} becomes $Cu(NH_3)_4^{2+}$: $[Cu(NH_3)_4^{2+}] = 0.10\ M$

$$[NH_3] = \frac{3.0\ M}{2} - 0.40\ M = 1.10\ M$$

$$E = E° - \frac{0.0591}{2} \log \frac{[Zn^{2+}]}{[Cu^{2+}]}$$

$$0.68\ V = 1.10\ V - \frac{0.0591}{2} \log \frac{0.20}{[Cu^{2+}]}$$

$$[Cu^{2+}] = 1.2 \times 10^{-15}\ M$$

$$K_f = \frac{[Cu(NH_3)_4^{2+}]}{[Cu^{2+}][NH_3]^4} = \frac{0.10}{(1.2 \times 10^{-15})(1.1)^4} = 5.7 \times 10^{13}$$

Note: this value differs somewhat from that listed in Table 17.3.

20.88 What are the common oxidation states for silver? For gold? Refer to Table 20.1.

Solution: The silver reaction is: $Ag^+(aq) + e^- \rightarrow Ag$

The moles of Ag^+ is: $$\frac{2.64\ g\ Ag}{107.9\ g\ Ag/mol\ Ag} = 2.45 \times 10^{-2}\ mol\ Ag$$

The moles of Au^{n+} is: $$\frac{1.61\ g\ Au}{197.0\ g\ Au/mol\ Au} = 8.17 \times 10^{-3}\ mol\ Au$$

The relative moles of Ag^+ to Au^{n+} $$\frac{2.45 \times 10^{-2}\ mol\ Ag}{8.17 \times 10^{-3}\ mol\ Au} = 3$$

That is, the same number of electrons that reduced the Ag^+ to Ag reduced only one-third the number of moles of the Au^{n+} to Au. Thus each Au^{n+} required three electrons per ion for every one electron for Ag^+. The oxidation state for the gold ion is +3; the ion is Au^{3+}.

$$Au^{3+}(aq) + 3e^- \rightarrow Au$$

Could we determine how many faradays of electricity was involved?

20.90 Part (c) of the problem is very much like Problem 20.47 turned around. It is a direct application of Coulomb's law coupled with the ideal gas law.

Solution: (a)

Anode	$2F^- \rightarrow F_2(g) + 2e^-$
Cathode	$2H^+ + 2e^- \rightarrow H_2(g)$
Overall	$2H^+ + 2F^- \rightarrow H_2(g) + F_2(g)$

(b) KF increases the electrical conductivity (what type of electrolyte is $HF(l)$)? The K^+ is not reduced.

(c) Calculating the moles of F_2

$$(15\ h \times 502\ A)\left(\frac{60\ min}{1\ h}\right)\left(\frac{60\ s}{1\ min}\right)\left(\frac{1\ C}{1\ A\cdot s}\right)\left(\frac{1\ F}{96500\ C}\right)\left(\frac{1\ mol\ F_2}{2\ F}\right) = 140\ mol\ F_2$$

Using the ideal gas law $\quad V = \frac{nRT}{P} = \frac{(140\ mol)(0.0821\ L\cdot atm/K\cdot mol)(297\ K)}{1.2\ atm} = 2.8 \times 10^3\ L$

20.91 This problem is similar to Problem 20.23.

Solution: We reverse the first half-raection and add it to the second:

$$Hg_2^{2+} \rightarrow 2Hg^{2+} + 2e^- \qquad E° = -0.92\ V$$
$$Hg_2^{2+} + 2e^- \rightarrow 2Hg \qquad E° = 0.85\ V$$

$$2Hg_2^{2+} \rightarrow 2Hg^{2+} + 2Hg \qquad E° = -0.07\ V$$

Since the cell potential is an intensive property, $\quad Hg_2^{2+} \rightarrow Hg^{2+} + Hg, \quad E° = -0.07\ V$

$$\Delta G° = -nFE° = -(2\ mol)(96500\ J/V\cdot mol)(-0.07\ V) = 14\ kJ$$

The corresponding equilibrium constant is $\quad K = \frac{[Hg^{2+}]}{[Hg_2^{2+}]}$

$$\Delta G° = -RT\ln K \qquad \ln K = \frac{-14 \times 10^3\ J}{(8.314\ J/K\cdot mol)(298\ K)}$$

$$K = 4 \times 10^{-3}$$

20.92 Does metallic mercury react with hydrochloric acid? Consult Example 20.4.

Solution: It is mercury ion in solution that is extremely hazardous. Since mercury metal does not react with hydrochloric acid (the acid in gastric juice), it does not dissolve and passes through the human body unchanged. Nitric acid (not part of human gastric juices) dissolves mercury metal (see Example 20.4); if nitric acid were secreted by the stomach, ingestion of mercury metal would be fatal.

20.94 This problem is a Coulomb's law problem. It is somewhat unusual in that we need to caluculate a concentration of OH^- via pH; we then calculate moles of OH^-; then knowing the time of the electrolysis, we can caluculate the average current.

Solution: Electroylis of NaCl*(aq)*

Anode: $2Cl^-(aq) \rightarrow Cl_2(g) + 2e^-$

Cathode: $2H_2O(l) + 2e^- \rightarrow H_2(g) + 2OH^-(aq)$

Overall: $2H_2O(l) + 2Cl^-(aq) \rightarrow H_2(g) + Cl_2(g) + 2OH^-(aq)$

pH = 12.24 pOH = 1.76 $[OH^-] = 1.74 \times 10^{-2}$ M

Moles of $OH^- = (1.74 \times 10^{-2}$ mol/L$) \times (0.300$ L$) = 5.22 \times 10^{-3}$ mol OH^-

From the balanced equation 1 mole OH^- is equivalent to 1 mole e^-,

Coulombs = A·s

$$5.22 \times 10^{-3} \text{ mol} \times \frac{1 \text{ F}}{1 \text{ mol}} \times \frac{96500 \text{ C}}{1 \text{ F}} = \text{A} \times 6.00 \text{ min} \times \frac{60 \text{ s}}{1 \text{ min}}$$

A = 1.4 amps

20.96 This problem is similar to Problem 20.58.

Solution: The number of coulombs:

$$? \text{ C} = 2.00 \text{ h} \times \frac{3600 \text{ s}}{1 \text{ h}} \times \frac{2.50 \text{ C}}{1 \text{ s}} = 1.80 \times 10^4 \text{ C}$$

$$? \text{ F} = (1.80 \times 10^4 \text{ C}) \times \frac{1 \text{ F}}{96500 \text{ C}} = 0.187 \text{ F}$$

The reaction is: $Pt^{n+} + ne^- \rightarrow Pt$ Thus, n mol of e^- are required per mol Pt.

The mass of Pt is:

$$9.09 \text{ g Pt} = 0.187 \text{ F} \times \frac{1 \text{ mol Pt}}{n \text{ F}} \times \frac{195.1 \text{ g Pt}}{1 \text{ mol Pt}}$$

Solving for n, we get: n = 4 Thus the platinum ion is Pt^{4+}.

Note: If you have not already done so, please read the "Introduction for the Student" at the beginning of this Student Solutions Manual.

Chapter 21

METALLURGY AND THE CHEMISTRY OF METALS

21.14 Assume the anode is made of slightly impure copper. The electrolytic purification of copper is described in Section 21.2. Review20 19.8 if you don't remember how to do electrolysis calculations. This is similar to Problem 20.56.

Solution: The anode reaction is: $Cu(s) \rightarrow Cu^{2+}(aq) + 2e^-$

We see that one mole of copper is stoichiometrically equivalent to two faradays. The time required is:

$$5.00 \text{ kg Cu} \left(\frac{1000 \text{ g}}{1 \text{ kg}}\right)\left(\frac{1 \text{ mol Cu}}{63.55 \text{ g Cu}}\right)\left(\frac{2 \text{ F}}{1 \text{ mol Cu}}\right)\left(\frac{96500 \text{ C}}{1 \text{ F}}\right)\left(\frac{1 \text{ A·s}}{1 \text{ C}}\right)\left(\frac{1}{37.8 \text{ A}}\right)$$

$$= 4.02 \times 10^5 \text{ s} = 112 \text{ h}$$

21.15 Review Section 20.3. Locate the standard reduction potentials of these metals in Table 20.1. Which metals are easier to oxidize than copper?

Solution: Table 20.1 shows that Pb, Fe, Co, Zn are more easily oxidized (stronger reducing agents) than copper. The Ag, Au, and Pt are harder to oxidize and will not dissolve.

Would you throw away the sludge if you were in charge of the copper refining plant? Why is it still profitable to manufacture copper even though the market price is very low?

21.17 Use the information contained in the hint.

Solution: The trick in this process centers on the fact that $TiCl_4$ is a liquid with a boiling point (136.4°C) a little higher than that of water. The tetrachloride can be formed by treating the oxide (rutile) with chlorine gas at high temperature. The balanced equation is

$$TiO_2(s) + 2Cl_2(g) \rightarrow TiCl_4(l) + O_2(g)$$

The liquid tetrachloride can be isolated and purified by simple distillation. Purified $TiCl_4$ is then reduced with magnesium (a stronger reducing agent than Ti) at high temperature.

$$TiCl_4(g) + 2\,Mg(l) \rightarrow Ti(s) + 2MgCl_2(l)$$

The other product, $MgCl_2$, can be separated easily from titanium metal by dissolving in water.

21.18 Note that you are given the mass of the copper metal, not the mass of the ore. You must find this from the mass percent of the copper.

Solution: (a) We first find the mass of ore containing 2.0×10^8 kg of copper.

$$2.0 \times 10^8 \text{ kg} \left(\frac{100\%}{0.80\%}\right) = 2.5 \times 10^{10} \text{ kg}$$

We can then compute the volume from the density of the ore.

$$2.5 \times 10^{10} \text{ kg} \left(\frac{1000 \text{ g}}{1 \text{ kg}}\right)\left(\frac{1 \text{ cm}^3}{2.8 \text{ g}}\right) = 8.9 \times 10^{12} \text{ cm}^3$$

(b) From the formula of chalcopyrite it is clear that two moles of sulfur dioxide will be formed per mole of copper. The mass of sulfur dioxide formed will be:

$$2.0 \times 10^8 \text{ kg Cu} \left(\frac{1 \text{ mol Cu}}{0.06355 \text{ kg Cu}}\right)\left(\frac{2 \text{ mol } SO_2}{1 \text{ mol Cu}}\right)\left(\frac{0.06407 \text{ kg } SO_2}{1 \text{ mol } SO_2}\right) = 4.0 \times 10^8 \text{ kg } SO_2$$

21.19 What determines whether electrolysis will be the method of choice for producing a metal from its compounds? See Section 21.2.

Solution: Very electropositive metals (i.e., very strong reducing agents) can only be isolated from their compounds by electrolysis. No chemical reducing agent is strong enough. In the given list $CaCl_2$, NaCl, and Al_2O_3 would require electrolysis.

21.20 How are iron and aluminum produced? See Sections 21.2 and 21.7.

Solution: Iron can be produced by reduction with coke in a blast furnace, whereas aluminum is usually produced electrolytically which is a much more expensive process.

21.28 Section 21.5 has a detailed discussion of the purification of potassium.

Solution: Metallic potassium cannot be easily prepared by the electrolysis of molten KCl because it is too soluble in the molten KCl to float on top of the Downs cell for collection. Moreover, it vaporizes readily at the operating temperatures, creating hazardous conditions.

21.30 Consult Section 21.5 for additional information.

Solution: Sodium and potassium metals will combine with almost any reducible substance to form sodium and potassium cations. These metals also form anions in liquid ammonia solution; under special conditions (see Section 21.5) it is possible to isolate stable crystalline salts of these anions.

21.31 Section 21.5 contains the descriptive chemistry of the alkali metals.

Solution: All of these reactions are discussed in Section 21.5 of the text.

(a) $2K(s) + 2H_2O(l) \rightarrow 2KOH(aq) + H_2(g)$

(b) $NaH(s) + H_2O(l) \rightarrow NaOH(aq) + H_2(g)$

(c) $2Na(s) + O_2(g) \rightarrow Na_2O_2(s)$

(d) $K(s) + O_2(g) \rightarrow KO_2(s)$

21.33 If necessary, review the ideal gas law.

Solution: The balanced equation is: $Na_2CO_3(aq) + 2HCl(aq) \rightarrow 2NaCl(aq) + CO_2(g) + H_2O(l)$

$$n = 25.0 \text{ g } Na_2CO_3 \left(\frac{1 \text{ mol } Na_2CO_3}{106.0 \text{ g } Na_2CO_3}\right)\left(\frac{1 \text{ mol } CO_2}{1 \text{ mol } Na_2CO_3}\right) = 0.236 \text{ mol } CO_2$$

$$V = \frac{nRT}{P} = \frac{(0.236 \text{ mol})(0.0821 \text{ L·atm/K·mol})(283 \text{ K})}{(746/760) \text{ atm}} = 5.59 \text{ L}$$

21.35 Refer to Appendix for required thermodynamic data.

Solution: (a) $\Delta H° = \Delta H_f^o(MgO) + \Delta H_f^o(CO_2) - \Delta H_f^o(MgCO_3)$

$$\Delta H° = (1 \text{ mol})(-601.8 \text{ kJ/mol}) + (1 \text{ mol})(-393.5 \text{ kJ/mol}) - (1 \text{ mol})(-1112.9 \text{ kJ/mol}) = 117.6 \text{ kJ}$$

(b) $\Delta H° = \Delta H_f^o(CaO) + \Delta H_f^o(CO_2) - \Delta H_f^o(CaCO_3)$

$$\Delta H° = (1 \text{ mol})(-635.6 \text{ kJ/mol}) + (1 \text{ mol})(-393.5 \text{ kJ/mol}) - (1 \text{ mol})(-1206.9 \text{ kJ/mol}) = 177.8 \text{ kJ}$$

$\Delta H°$ is less for $MgCO_3$; therefore, it is more easily decomposed by heat.

21.36 Write out the balanced equations implied by the hint. See Section 21.6 for the concentrated nitric acid reaction. The other products of the thermal decomposition of magnesium nitrate are NO_2 and O_2.

Solution: First magnesium is treated with concentrated nitric acid (redox reaction) to obtain magnesium nitrate:

$$3Mg(s) + 8HNO_3(aq) \rightarrow 3Mg(NO_3)_2(aq) + 4H_2O(l) + 2NO(g)$$

The magnesium nitrate is recovered from solution by evaporation, dried, and heated in air to obtain magnesium oxide:

$$2Mg(NO_3)_2(s) \rightarrow 2MgO(s) + 4NO_2(g) + O_2(g)$$

21.37 Does magnesium metal react with acids? Study Section 21.6.

Solution: As described in Section 21.6, magnesium metal will combine with chlorine.

$$Mg(s) + Cl_2(g) \rightarrow MgCl_2(s)$$

Magnesium will also react with HCl.

$$Mg(s) + 2HCl(aq) \rightarrow MgCl_2(aq) + H_2(g)$$

Neither of the above methods are really practical because magnesium metal is expensive to produce (electrolysis of magnesium chloride!). Can you suggest a method starting with a magnesium compound like $MgCO_3$?

21.39 Why isn't helium grouped with the alkaline earth metals? Why aren't the Group II metals grouped with the noble gases?

Solution: Even though helium and the Group 2A metals have ns^2 outer electron configurations, helium has a closed shell noble gas configuration and the Group 2A metals do not. There is no 1p subshell. (Remember why?)

21.40 Look up the solubilities of the sulfates of Be, Mg, Ca, Sr, Ba, and Ra.

Solution: The water solubilities of the sulfates increase in the order Ra < Ba < Sr < Ca < Mg. The trend in this series is clearly in the sense of smaller ionic radius favoring greater solubility. Probably the smaller ion size results in much greater hydration energy (Section 6.6). Which sulfate in this series should have the largest lattice energy (Section 9.3)? Which is the more important factor in determining solubility in this series: hydration energy or lattice energy?

According to the Handbook of Chemistry and Physics, $BeSO_4$ reacts with water to form "$BeSO_4 \cdot 4H_2O$". In this sense it is not strictly comparable with the other sulfates of the Group 2A metals. However, this compound is really comprised of a sulfate ion and a $Be(H_2O)_4^{2+}$ complex ion. The latter is just a very strongly hydrated Be^{2+} ion. The solubility of the "$BeSO_4 \cdot 4H_2O$" is higher than any of other Group 2A sulfates, so it really does fit at the high solubility end of the series.

21.41 Study the section on the chemistry of calcium in Section 21.6.

Solution: The formation of calcium oxide is: $2Ca(s) + O_2(g) \rightarrow 2CaO(s)$

The conversion of calcium oxide to calcium hydroxide is: $CaO(s) + H_2O(l) \rightarrow Ca(OH)_2(s)$

The reaction of calcium hydroxide with carbon dioxide is: $Ca(OH)_2(s) + CO_2(g) \rightarrow CaCO_3(s) + H_2O(l)$

If calcium metal were exposed to extremely humid air, do you think that the oxide would still form?

21.43 An outline of the Hall process is presented in Section 21.7.

Solution: Anhydrous aluminum oxide is reduced to aluminum by the Hall process. A Hall electrolytic cell contains a series of carbon anodes. The cathode is also made of carbon and constitutes the lining inside the cell. The key to the Hall process is the use of cryolite, or Na_3AlF_6 (m.p. 1000°C), as solvent for aluminum oxide (m.p. 2045°C). The mixture is electrolyzed to produce aluminum and oxygen gas:

Anode (oxidation):	$3[2O^{2-} \rightarrow O_2(g) + 4e^-]$
Cathode (reduction):	$4[Al^{3+} + 3e^- \rightarrow Al(l)]$
Overall:	$2Al_2O_3 \rightarrow 4Al(l) + 3O_2(g)$

Oxygen gas reacts with the carbon anodes (at elevated temperatures) to form carbon monoxide, which escapes as a gas. The liquid aluminum metal (m.p. 660.2°C) sinks to the bottom of the vessel, from which it can be drained from time to time during the procedure.

21.45 Check the standard reduction potentials in Table 20.1. Which metals are stronger reducing agents than aluminum?

Solution: According to Table 20.1, the following metals can reduce aluminum ion to aluminum:

Be, Mg, Na, Ca, Sr, Ba, K, Li

Today (1989) the cheapest of these metals (magnesium) costs $12.00 per lb. The current cost of aluminum is $0.70 per lb. Is the Hall process an improvement?

21.46 What is the relationship between free energy and cell voltage? If you don't remember, study Section 20.4 and Equation (20.3). Use the same relationship for part (b).

Solution: (a) The relationship between cell voltage and free energy difference is:

$$\Delta G = -nFE$$

In the given reaction n = 6. (Why?) We write:

$$E = \frac{-\Delta G}{nF} = \frac{-594 \times 10^3 \text{ J}}{6 \times 96500 \text{ C}} = -1.03 \text{ V}$$

The balanced equation shows <u>two</u> moles of aluminum. Is this the voltage required to produce <u>one</u> mole of aluminum? If we divide everything in the equation by two, we obtain:

$$\frac{1}{2}Al_2O_3(s) + \frac{3}{2}C(s) \rightarrow Al(l) + \frac{3}{2}CO(g)$$

For the new equation n = 3 and ΔG is $\frac{1}{2} \times 594$ kJ = 297 kJ. We write:

$$E = \frac{-\Delta G}{nF} = \frac{-297 \times 10^3 \text{ J}}{3 \times 96500 \text{ C}} = -1.03 \text{ V}$$

The minimum voltage required to produce one mole or one thousand moles of aluminum is the same; the amount of <u>current</u> will be different in each case. Why is the voltage <u>negative</u>?

(b) First we convert 1.00 kg of Al to moles.

$$1.00 \text{ kg Al} \left(\frac{1 \text{ mol Al}}{0.02698 \text{ kg Al}}\right) = 37.1 \text{ mol Al}$$

The reaction equation in part (a) shows us that three faradays of charge are required to produce one mole of aluminum. The voltage is three times the minimum calculated above (namely, -3.09 V or -3.09 J/C). We can find the electrical energy by using the same equation with the higher voltage

$$\Delta G = -nFE = -\left(\frac{3 \text{ F}}{1 \text{ mol Al}}\right)\left(\frac{96500 \text{ C}}{1 \text{ F}}\right)\left(\frac{-3.09 \text{ J}}{1 \text{ C}}\right)\left(\frac{37.1 \text{ mol Al}}{1 \text{ kg Al}}\right) = 3.32 \times 10^4 \text{ kJ/kg Al}$$

This equation can be used because electrical work can be calculated by multiplying the voltage by the amount of charge transported through the circuit (joules = volts × coulombs). The nF term in Equation (20.3) represents the amount of charge.

What is the significance of the positive sign of the free energy change? Would the manufacturing of aluminum be a different process if the free energy difference were negative?

21.47 Everybody knows that aluminum doesn't rust, but consult Table 20.1. Which is a stronger reducing agent, iron or aluminum? Based on the information in Table 20.1, should aluminum react with oxygen? Study Section 21.7.

Solution: A coating of aluminum oxide rapidly forms on the surface of a fresh piece of aluminum metal when exposed to air. Unlike iron and many other metals, the aluminum oxide coat seals the surface from the atmosphere and prevents further oxidation.

Aluminum oxide is soluble in both aqueous acid and base (amphoteric oxide, Section 15.8). What happens to aluminum metal when it is exposed to acidic or basic water?

21.48 Use VSEPR theory (Section 10.1) to predict the complex ion shapes.

Solution: The two complex ions can be classified as AB_4 and AB_6 structures (no unshared electron pairs on Al and 4 or 6 attached atoms, respectively). Their VSEPR geometries are tetrahedral and octahedral.

The accepted explanation for the nonexistence of $AlCl_6^{3-}$ is that the chloride ion is too big to form an octahedral cluster around a very small Al^{3+} ion. What is your guess for the formulas of complex ions formed between Al^{3+} and bromide or iodide ions? What about Ga^{3+} and chloride ion?

21.49 Work out the appropriate basic solution half-reactions for each case before trying to write a balanced equation. If you don't remember how to write a balanced half-reaction in basic solution, review Section 3.6.

Solution: The half-reaction for the oxidation of Al to AlO_2^- in basic solution is:

$$Al(s) + 4OH^-(aq) \rightarrow AlO_2^-(aq) + 2H_2O(l) + 3e^-$$

(a) The nitrate-ammonia half-reaction is: $NO_3^-(aq) + 6H_2O(l) + 8e^- \rightarrow NH_3(aq) + 9OH^-(aq)$

Combining the equations: $8Al(s) + 5OH^-(aq) + NO_3^-(aq) + 2H_2O(l) \rightarrow 8AlO_2^-(aq) + 3NH_3(aq)$

(b) The water-hydrogen half-reaction is: $H_2O(l) + e^- \rightarrow OH^-(aq) + \frac{1}{2}H_2(g)$

Combining the equations: $Al(s) + OH^-(aq) + H_2O(l) \rightarrow AlO_2^-(aq) + \frac{3}{2}H_2(g)$

(c) The SnO_3^{2-}-Sn half-reaction is: $SnO_3^{2-}(aq) + 3H_2O(l) + 4e^- \rightarrow Sn(s) + 6OH^-(aq)$

Combining the equations: $4Al(s) + 3SnO_3^{2-}(aq) + H_2O(l) \rightarrow 4AlO_2^-(aq) + 3Sn(s) + 2OH^-(aq)$

21.51 The properties of aluminum are discussed in detail in Section 21.7.

Solution: Some of aluminum's useful properties are: low density (light weight), high tensile strength, high electrical conductivity, high thermal conductivity, inert protective oxide surface coating.

21.52 Study the chemical reactions of aluminum in Section 21.7. Note that $AlCl_3(aq)$ is not the same thing as Al_2Cl_6. The compound in part (d) is an alum.

Solution: (a) $2Al(s) + 3Cl_2(g) \rightarrow Al_2Cl_6(s)$

(b) $4Al(s) + 3O_2(g) \rightarrow 2Al_2O_3(s)$

(c) $2Al(s) + 3H_2SO_4(aq) \rightarrow Al_2(SO_4)_3(aq) + 3H_2(g)$

(d) $Al_2(SO_4)_3(aq) + (NH_4)_2SO_4(aq) \rightarrow 2NH_4AlSO_4)_2 \cdot 12H_2O(s)$, followed by evaporation of the solution.

21.53 The "bridge" Al-Cl-Al linkages in Al_2Cl_6 are relatively weak bonds. What would happen to these bonds at high temperature?

Solution: The "bridge" bonds in Al_2Cl_6 break at high temperature: $Al_2Cl_6(g) \rightleftharpoons 2AlCl_3(g)$

This increases the number of molecules in the gas phase and causes the pressure to be higher than expected for pure Al_2Cl_6.

If you knew the equilibrium constants for the above reaction at higher temperatures, could you calculate the expected pressure of the $AlCl_3$-Al_2Cl_6 mixture?

21.54 See Section 21.7 for a description of the bonding in Al_2Cl_6. Use VSEPR theory (Section 10.1) to predict the geometry of $AlCl_3$. Correlate the VSEPR geometry with the hybrid orbitals expected for the Al atom.

Solution: VSEPR analysis shows $AlCl_3$ to be an AB_3-type molecule (no lone pairs on the central atom). The geometry should be trigonal planar, and the aluminum atom should therefore be sp^2 hybridized.

21.55 Given the oxidation number of the metal, what are the formulas of the metal sulfide and oxide? Write a balanced equation for the reaction.

Solution: The formulas of the metal oxide and sulfide are MO and MS (why?). The balanced equation must therefore be:

$$2MS(s) + 3O_2(g) \rightarrow 2MO(s) + 2SO_2(g)$$

The number of moles of MO and MS are equal. We let x be the molar mass of the metal. The number of moles of metal oxide is:

$$0.972 \text{ g} \left(\frac{1 \text{ mol}}{(x + 16.00) \text{ g}}\right)$$

The number of moles of metal sulfide is: $1.164 \text{ g} \left(\frac{1 \text{ mol}}{(x + 32.07) \text{ g}}\right)$

The two amounts are equal. $\frac{0.972}{(x + 16.00)} = \frac{1.164}{(x + 32.07)}$

We solve for x. $0.972(x + 32.07) = 1.164(x + 16.00)$

$$x = 65.4 \text{ g/mol}$$

21.56 Consult Table 20.1. Is copper(II) ion more easily reduced than H^+ or water? Is copper metal more easily oxidized than water?

Solution: Copper(II) ion is more easily reduced than either water or hydrogen ion (How can you tell? See Section 19.3.). Copper metal is more easily oxidized than water. Water should not be affected by the copper purification process.

21.58 This problem is similar to Problem 19.19.

Solution: Using Equation (19.8),

(a) $\Delta G^\circ_{rxn} = 4\Delta G^\circ_f(Fe) + 3\Delta G^\circ_f(O_2) - 2\Delta G^\circ_f(Fe_2O_3) = (4 \text{ mol})(0) + (3 \text{ mol})(0) - (2 \text{ mol})(-741.0 \text{ kJ/mol}) = 1482 \text{ kJ}$

(b) $\Delta G^\circ_{rxn} = 4\Delta G^\circ_f(Al) + 3\Delta G^\circ_f(O_2(g) - 2\Delta G^\circ_f(Al_2O_3) = (4\ mol)(0) + (3\ mol)(0) - (2\ mol)(-1576.4\ kJ/mol)$
$= 3152.8\ kJ$

21.59 What reaction occurs between nitrogen gas and magnesium metal at high temperature? See Section 21.6.

Solution: At high temperature magnesium metal reacts with nitrogen gas to form magnesium nitride.

$$3Mg(s) + N_2(g) \rightarrow Mg_3N_2(s)$$

Can you think of any gas other than a noble gas that could provide an inert atmosphere for processes involving magnesium at high temperatures?

21.61 What causes metal ions to alter the pH of aqueous solutions (Section 16.5) ? What is amphoterism (Section 15.5)? Why is diamond harder than gray tin (both have the same structure)?

Solution: (a) In water the aluminum(III) ion causes an increase in the concentration of hydrogen ion (lower pH). This results from the effect of the small diameter and high charge (3+) of the aluminum ion on surrounding water molecules. The aluminum ion draws electrons in the O-H bonds to itself, thus allowing easy formation of H^+ ions.

(b) $Al(OH)_3$ is an amphoteric hydroxide (Section 15.5). It will dissolve in strong base with the formation of a complex ion

$$Al(OH)_3(s) + 3OH^-(aq) \rightarrow Al(OH)_6^{3-}(aq)$$

The concentration of OH^- in aqueous ammonia is too low for this reaction to occur.

21.63 (a) What are the usual products in the thermal decomposition of a metal carbonate? (b) Is potassium metal an oxidizing agent or a reducing agent? (c) What insoluble compound will form?

Solution: The reactions are:

(a) $Al_2(CO_3)_3(s) \rightarrow Al_2O_3(s) + 3CO_2(g)$

(b) $AlCl_3(s) + 3K(s) \rightarrow Al(s) + 3KCl(s)$

(c) $Ca(OH)_2(aq) + Na_2CO_3(aq) \rightarrow CaCO_3(s) + 2NaOH(aq)$

21.64 The reaction involves the acid-base properties of CaO and NH_4^+. Which one is an acid? Which one is a base?

Solution: Calcium oxide is a base. The reaction is a neutralization.

$$CaO(s) + 2HCl(aq) \rightarrow CaCl_2(aq) + H_2O(l)$$

Note: If you have not already done so, please read the "Introduction for the Student" at the beginning of this Student Solutions Manual.

Chapter 22

NONMETALLIC ELEMENTS AND THEIR COMPOUNDS

22.5 How many electrons must a hydrogen atom gain to achieve a closed shell configuration? How many electrons must a hydrogen atom lose to form a stable positive ion? Can any other element make such a statement?

> ***Solution:*** The fact that the n = 1 subshell consists of just a single 1s orbital makes hydrogen unique. In one sense it needs only one more electron to acquire a noble gas structure (helium), so it could be classed with the halogens. On the other hand hydrogen only has to lose one electron to form a stable positive ion, so it could also be grouped with the alkali metals. In terms of its electronegativity hydrogen falls between boron and carbon in the middle of the periodic table.

22.6 The synthesis of hydrogen is discussed in detail in Section 22.2.

> ***Solution:*** (a) Laboratory preparations: $Zn(s) + 2HCl(aq) \rightarrow ZnCl_2(aq) + H_2(g)$
>
> $$2H_2O(l) \rightarrow 2H_2(g) + O_2(g)$$
>
> (b) Industrial preparations: $C_3H_8(g) + 3H_2O(g) \rightarrow 3CO(g) + 7H_2(g)$
>
> $$C(s) + H_2O(g) \rightarrow CO(g) + H_2(g)$$
>
> The products in the last reaction are referred to as *water gas.*

22.8 Locate the elements with atomic numbers 17 and 20 in the periodic table. What type of hydrogen compound will each form (see Problem 22.7)?

> ***Solution:*** Element number 17 is the halogen chlorine. Since it is a nonmetal, chlorine will form the molecular compound HCl. Element 20 is the alkaline earth metal calcium which will form an ionic hydride CaH_2. A water solution of HCl is called hydrochloric acid. Calcium hydride will react according to the equation (see Section 22.2)
>
> $$CaH_2(s) + 2H_2O(l) \rightarrow Ca(OH)_2(aq) + 2H_2(g)$$

22.10 Refer to the discussions of these types of hydrides in Section 22.2.

Solution: Ionic hydrides are formed when molecular hydrogen combines directly with any alkali metal or with some of the alkaline earth metals.

$$2Li(s) + H_2(g) \rightarrow 2LiH(s)$$

LiH is an ionic compound with a melting point of 680°C.

A typical covalent hydride is ammonia. It is a discrete covalently-bonded molecular unit (distorted tetrahedron).

$$N_2(g) + 3H_2(g) \rightarrow 2NH_3(g)$$

22.12 A discussion of interstitial hydrides can be found in Section 22.2.

Solution: Molecular hydrogen forms a number of hydrides with transition metals. In some of these compounds, the ratio of hydrogen atoms to metal atoms is *not* a constant. Such compounds are called *interstitial hydrides*.

22.13 Read the discussion of interstitial hydrides in Section 22.2. What would happen to a mixture of hydrogen and neon gases if it were heated in a container of metallic palladium?

Solution: Hydrogen forms an interstitial hydride with palladium that behaves almost like a solution of hydrogen atoms in the metal. At elevated temperatures hydrogen atoms can pass through solid palladium; other substances cannot.

22.14 The first part is like Problem 22.8. Use this equation as the basis for the stoichiometry calculation.

Solution: The equation is: $CaH_2(s) + 2H_2O(l) \rightarrow Ca(OH)_2(aq) + 2H_2(g)$

Using the ideal gas law and the stoichiometry in the equation:

$$\text{g } CaH_2 = \frac{\left(\frac{746}{760}\text{ atm}\right)(26.4\text{ L})}{(0.0821\text{ L·atm/mol·K})(293\text{ K})} \times \frac{1\text{ mol } CaH_2}{2\text{ mol } H_2} \times \frac{42.10\text{ g}}{1\text{ mol } CaH_2} = 22.7\text{ g}$$

22.15 Use the ideal gas equation to find the number of moles of D_2 gas.

Solution: The number of moles of deuterium gas is:

$$n = \frac{PV}{RT} = \frac{0.90\text{ atm} \times 2.0\text{ L}}{0.0821\text{ L·atm/K·mol} \times 298\text{ K}} = 0.074\text{ mol}$$

If the abundance of deuterium is 0.015%, the number of moles of water must be:

$$0.074 \text{ mol} \times \frac{100\%}{0.015\%} = 4.9 \times 10^2 \text{ mol}$$

At a recovery of 80% the amount of water needed is:

$$4.9 \times 10^2 \text{ mol } H_2O \times \frac{100\%}{80\%} \times \frac{0.018 \text{ kg}}{1.0 \text{ mol}} = 11 \text{ kg } H_2O$$

22.16 Both reactions would be reductions if they occurred. Can H_2 reduce Cu^{2+}? Can H_2 reduce Na^+? See Table 20.1.

Solution: According to Table 20.1, H_2 can reduce copper(II), but not Na^+. (How can you tell?) The reaction is:

$$CuO(s) + H_2(g) \rightarrow Cu(s) + H_2O(l)$$

22.18 Study the appropriate part of Section 22.2.

Solution: The term "hydrogen economy" simply means using hydrogen gas as a major alternate fuel source. An extensive discussion of the reactions and methods is presented in Section 22.2.

22.19 The answer to this problem can be found by reading Section 22.3.

Solution: Examples of carbides are CaC_2 and Be_2C. Examples of cyanides are NaCN and HCN.

22.21 A thorough discussion of the oxides of carbon is in Section 22.3.

Solution: Carbon monoxide is prepared industrially by passing steam over heated coke. It is not an acidic oxide and is only slightly soluble in water. It is relatively unreactive. Carbon dioxide is produced when any form of carbon or carbon-containing compounds are burned in an excess of oxygen. It is a colorless and odorless gas. Unlike carbon monoxide, carbon dioxide is nontoxic and is acidic.

22.23 Refer to the Chemistry in Action article entitled "Synthetic Gas From Coal" at the end of Section 22.3.

Solution: The conversion of coal to a gaseous form such as CH_4 is referred to as coal gasification.

22.25 Draw Lewis structures of carbon dioxide and hydroxide ion and include the polarities of the O-H and C=O bonds. In terms of simple electrical forces, what part of the carbon dioxide molecule will attract the oxygen end of the hydroxide ion?

Solution: The reaction can be represented:

$$H-\ddot{O}:^- \rightarrow C(=O)=O \longrightarrow H-\ddot{O}-C(-O^-)=O$$

The lone pair on the hydroxide oxygen becomes a new carbon-oxygen bond. The octet rule requires that one of the electron pairs in the double bond be changed to a lone pair. What is the other resonance form of the product ion?

22.26 How many valence electrons are there in the ion?

Solution: The Lewis structure is:

$$:C\equiv C:^{2-}$$

22.27 The products of the reactions of these carbides with water are shown in Section 22.3. Are carbide ions good Brønsted bases?

Solution: The reactions are: (a) $Be_2C(s) + 4H_2O(l) \rightarrow 2Be(OH)_2(aq) + CH_4(g)$

(b) $CaC_2(s) + 2H_2O(l) \rightarrow Ca(OH)_2(aq) + C_2H_2(g)$

22.28 In other words, write a balanced equation for the decomposition of $NaHCO_3$. The other product is water.

Solution: (a) The reaction is: $2NaHCO_3(s) \rightarrow Na_2CO_3(s) + H_2O(g) + CO_2(g)$

Is this an endo- or an exothermic process?

(b) The hint is generous. The reaction is:

$$Ca(OH)_2(aq) + CO_2(g) \rightarrow CaCO_3(s) + H_2O(l)$$

The visual proof is the formation of a white precipitate of $CaCO_3$. Why would a water solution of NaOH be unsuitable to qualitatively test for carbon dioxide?

22.29 Based on the solubility properties of $CaCO_3$, what is your guess for $MgCO_3$?

Solution: Magnesium and calcium carbonates are insoluble; the bicarbonates are soluble. Formation of a precipitate after addition of $MgCl_2$ solution would show the presence of Na_2CO_3.

Assuming similar concentrations, which of the two sodium salt solutions would have a higher pH?

22.30 Heat causes the decomposition of most bicarbonate salts; see Problem 22.28 for the reactions.

Solution: Heat causes bicarbonates to decompose according to the reaction:

$$2HCO_3^- \rightarrow CO_3^{2-} + H_2O + CO_2$$

Generation of carbonate ion causes precipitation of the insoluble $MgCO_3$.

Do you think there is much chance of finding natural mineral deposits of calcium or magnesium bicarbonates?

22.31 Write equations for the reactions of bicarbonate ion with hydrogen ion and with hydroxide ion. Which happens in this case?

Solution: Bicarbonate ion can react with either H^+ or OH^-.

$$HCO_3^-(aq) + H^+(aq) \rightarrow H_2CO_3(aq)$$

$$HCO_3^-(aq) + OH^-(aq) \rightarrow CO_3^{2-}(aq) + H_2O(l)$$

Since ammonia is a base, carbonate ion is formed which causes precipitation of $CaCO_3$.

22.33 The term "burn" does not always imply reaction with atmospheric oxygen; more generally, it can refer to any highly exothermic oxidation process; in this case the oxygen source is CO_2.

Solution: Table 20.1 shows that magnesium metal has the potential to be an extremely powerful reducing agent. It appears inert at room temperature, but at high temperatures it can react with almost any source of oxygen atoms (including water!) to form MgO. In this case carbon dioxide is reduced to carbon.

$$2Mg(s) + CO_2(g) \rightarrow 2MgO(s) + C(s)$$

How does one extinguish a magnesium fire?

22.34 If you need help refer to Chapter 8, Secion 8.2 for a discussion of the meaning of isoelectronic.

Solution: Carbon monoxide and molecular nitrogen are isoelectronic. Both have 14 electrons. What other diatomic molecules discussed in these problems are isoelectronic with CO?

22.36 In part (a) the products are water and copper metal. In part (b) water is also a product.

Solution: The preparations are:

$$2NH_3(g) + 3CuO(s) \rightarrow N_2(g) + 3Cu(s) + 3H_2O(g)$$

$$(NH_4)_2Cr_2O_7(s) \rightarrow N_2(g) + Cr_2O_3(s) + 4H_2O(g)$$

22.38 Review the Brønsted system of acids and bases in Section 15.1. What is the difference between a base and an acid?

Solution: The balanced equation is: $NH_2^-(aq) + H_2O(l) \rightarrow NH_3(aq) + OH^-(aq)$

In this system the acid is H_2O (proton donor) and the base is NH_2^- (proton acceptor). What are the conjugate acid and the conjugate base?

22.39 The other product is water; urea is a solid at room temperature. Review the discussion of the Le Chatelier principle in Section 14.5 if you are unsure of how to proceed in this problem.

Solution: The balanced equation is: $2NH_3(g) + CO_2(g) \rightarrow (NH_2)_2CO(s) + H_2O(l)$

If pressure increases, the position of equilibrium will shift in the direction with the smallest number of molecules in the gas phase, i.e., to the right. The reaction is best run at high pressure.

Write the expression for Q_p for this reaction (Section 14.2). Does increasing pressure cause Q_p to increase or decrease? Is this consistent with the above prediction?

22.40 What happens to atmospheric N_2 and O_2 under the influence of lightning (see Section 22.4)?

Solution: Lightning can cause N_2 and O_2 to react: $N_2(g) + O_2(g) \rightarrow 2NO(g)$

The NO formed naturally in this manner eventually suffers oxidation to nitric acid (see the Ostwald process in Section 22.4 and Problem 22.42 below) which precipitates as rain. The nitrate ion is a natural source of nitrogen for growing plants.

22.41 Is there a relationship between the density and the molar mass of a gas? Review Section 5.4 if you don't remember.

Solution: The density of a gas depends on temperature, pressure, and the molar mass of the substance. When two gases are at the same pressure and temperature, the ratio of their densities should be the same as the ratio of their molar masses. The molar mass of ammonium chloride is 53.5 g/mol, and the ratio of this to the molar mass of molecular hydrogen (2.02 g/mol) is 26.8. The experimental value of 14.5 is roughly half this amount. Such results usually indicate breakup or dissociation into smaller molecules in the gas phase (note the temperature). The measured molar mass is the average of all the molecules in equilibrium.

$$NH_4Cl(g) \rightleftharpoons NH_3(g) + HCl(g)$$

Knowing that ammonium chloride is a stable substance at 298 K, is the above reaction exo- or endothermic?

22.42 The Ostwald synthesis of nitric acid is discusssed in detail in Section 13.5.

Solution: A brief summary of the Ostwald process involves the following steps: (1) ammonia is converted with molecular oxygen to water and NO, (b) NO readily oxidizes to NO_2, (c) NO_2 when dissolved in water forms both nitric and nitrous acid, and (d) on heating, nitrous acid is converted to nitric acid and NO (which can be recycled through step (2).

Can you write balanced equations for all of the reactions?

22.44 Section 22.4 describes the process of nitrogen fixation.

Solution: Nitrogen fixation is the conversion of molecular nitrogen into nitrogen compounds.

22.45 What is the oxidation state of nitrogen in nitric acid?

Solution: The highest oxidation state possible for a Group 5A element is +5. This is the oxidation state of nitrogen in nitric acid.

22.46 What is the oxidation number of nitrogen in nitrous acid? Are there nitrogen compounds with lower oxidation numbers? Are there nitrogen compounds with higher oxidation numbers? See Table 22.2.

Solution: The oxidation number of nitrogen in nitrous acid is +3. Since this value is between the extremes of +5 and -3 for nitrogen, nitrous acid can be either oxidized or reduced. Nitrous acid can oxidize HI to I_2 (in other words HI acts as a reducing agent).

$$2HI(g) + 2HNO_2(aq) \rightarrow I_2(s) + 2NO(g) + 2H_2O(l)$$

A strong oxidizing agent can oxidize nitrous acid to nitric acid (oxidation number of nitrogen +5).

$$2Ce^{4+}(aq) + HNO_2(aq) + H_2O(l) \rightarrow 2Ce^{3+}(aq) + HNO_3(aq) + 2H^+(aq)$$

22.48 Study the discussion of the properties of nitric acid in Section 22.4.

Solution: Nitric acid is a strong oxidizing agent in addition to being a strong acid (Table 20.1). The primary action on a good reducing agent like zinc is reduction of nitrate ion to ammonium ion.

$$4Zn(s) + NO_3^-(aq) + 10H^+(aq) \rightarrow 4Zn^{2+}(aq) + NH_4^+(aq) + 3H_2O(l)$$

22.49 The second product of the potassium nitrate-carbon reaction is carbon dioxide.

Solution: The balanced equation is: $2KNO_3(s) + C(s) \rightarrow 2KNO_2(s) + CO_2(g)$

The maximum amount of potassium nitrite (theoretical yield) is:

$$57.0 \text{ g KNO}_3 \left(\frac{1 \text{ mol}}{101.1 \text{ g}}\right)\left(\frac{1 \text{ mol KNO}_2}{1 \text{ mol KNO}_3}\right)\left(\frac{85.11 \text{ g}}{1 \text{ mol}}\right) = 48.0 \text{ g KNO}_2$$

22.50 This problem is similar to Problem 19.34.

Solution: (a) If $\Delta G°$ for the reaction is 173.4 kJ, then $\Delta G^{\circ}_{f} = \frac{173.4 \text{ kJ}}{2 \text{ mol}} = 86.7$ kJ/mol (why?)

(b) From Equation (19.10): $\Delta G° = -RT \ln K$

$$173.4 \times 10^{3} \text{ J} = -(8.314 \text{ J/K·mol})(298 \text{ K}) \ln K$$

$$K = 4 \times 10^{-31}$$

(c) Since using Equation (14.5) $\Delta n = 0$, then $K_p = K_c = K = 4 \times 10^{-31}$

22.51 Review Sections 9.8 and 10.1 if you aren't sure how to work this problem.

Solution: One of the best Lewis structures for nitrous oxide is:

$$:\overset{-}{\underset{..}{N}}=\overset{+}{N}=\underset{..}{\ddot{O}}$$

There are no lone pairs on the central nitrogen, making this an AB_2 VSEPR case. All such molecules are linear. Other resonance forms are:

$$:N\equiv\overset{+}{N}-\underset{..}{\overset{..-}{O}}: \qquad {}^{2-}:\underset{..}{\ddot{N}}-\overset{+}{N}\equiv\overset{+}{O}:$$

Are all the resonance forms consistent with a linear geometry?

22.52 The required thermodynamic data is tabulated in Appendix 3.

Solution:

$$\Delta H° = 4\Delta H^{\circ}_{f}(NO) + 6\Delta H^{\circ}_{f}(H_2O) - [4\Delta H^{\circ}_{f}(NH_3) + 5\Delta H^{\circ}_{f}(O_2)]$$

$$\Delta H° = (4 \text{ mol})(90.4 \text{ kJ/mol}) + (6 \text{ mol})(-285.8 \text{ kJ/mol}) - (4 \text{ mol})(-46.3 \text{ kJ/mol}) = -1168 \text{ kJ}$$

22.54 What is the difference in the electronic structure between nitrogen and phosphorus?

Solution: The atomic radius of P (128 pm) is considerably larger than that of N (92 pm); consequently, the 3*p* orbital on a P atom cannot overlap effectively with a 3*p* orbital on a neighboring P atom to form a pi bond. Simply stated, the phosphorus is too large to allow effective overlap of the 3*p* orbitals to form π bonds.

22.55 What are the P-P-P bond angles in the P_4 molecule? How do these compare with bond angles usually found in trigonal pyramidal molecules (Section 10.1)?

Solution: Since the P_4 molecule has a tetrahedral structure, the P-P-P bond angles must be 60°, which is extremely acute. Most trigonal pyramidal molecules have bond angles close to 109°. This distortion is believed to result in weaker bonds and lower molecular stability.

22.56 Often the synthesis of a particular substance takes several steps. The best strategy for this type of problem is to work backwards from the goal, phosphoric acid, to the beginning, elemental phosphorus. Study the discussion of phosphorus chemistry in Section 22.4 and find a reaction in which the product is phosphoric acid. Look at the reactants and try to see how they might be formed from elemental phosphorus.

Solution: You won't find a reaction that starts with elemental phosphorus and ends with phosphoric acid. However there is more than one reaction having phosphoric acid as a product. One possibility is the reaction of P_4O_{10} with water.

$$P_4O_{10}(s) + 6H_2O(l) \rightarrow 4H_3PO_4(aq)$$

Can P_4O_{10} be formed from elemental phosphorus? Study of Section 22.4 shows that P_4 combines with oxygen to form P_4O_{10}.

$$P_4(s) + 5O_2(g) \rightarrow P_4O_{10}(s)$$

The synthesis of phosphoric acid is the result of these two steps in sequence.

Can you come up with an alternative synthesis starting with elemental phosphorus and chlorine gas?

22.58 The only products are N_2O_5 and HPO_3.

Solution: The balanced equation is:

$$P_4O_{10}(s) + 4HNO_3(aq) \rightarrow 2N_2O_5(g) + 4HPO_3(l)$$

The theoretical yield of N_2O_5 is:

$$79.4 \text{ g } P_4O_{10} \left(\frac{1 \text{ mol}}{283.9 \text{ g}}\right)\left(\frac{2 \text{ mol } N_2O_5}{1 \text{ mol } P_4O_{10}}\right)\left(\frac{108.0 \text{ g}}{1 \text{ mol}}\right) = 60.4 \text{ g } N_2O_5$$

22.60 Part (a) is similar to Problem 15.67. In part (b) what elements are capable of forming hydrogen bonds, and in part (c) what elements can have expanded octets? For part (d) study 22.55.

Solution: (a) See Problem 15.67.

(b) The electronegativity of nitrogen is greater than that of phosphorus. The N-H bond is much more polar than the P-H bond and can participate in hydrogen bonding. This increases intermolecular attractions and results in a higher boiling point.

(c) Elements in the second period never expand their octets. A common explanation is the absence of 2d atomic orbitals.

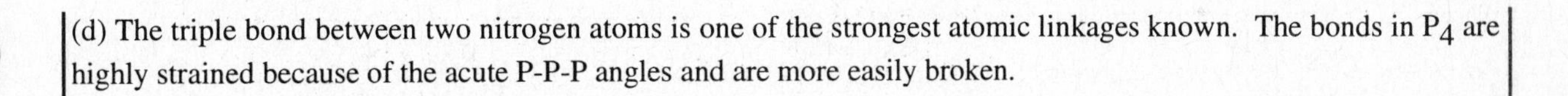

(d) The triple bond between two nitrogen atoms is one of the strongest atomic linkages known. The bonds in P_4 are highly strained because of the acute P-P-P angles and are more easily broken.

22.62 Study the introductory paragraphs of Section 22.5.

Solution: The industrial preparation of oxygen is the fractional distillation of liquified air. In the laboratory, oxygen is prepared by heating potassium chlorate.

$$2KClO_3(s) \rightarrow 2KCl(s) + 3O_2(g)$$

22.64 Find the oxides in appropriate parts of Sections 22.3, 22.4, and 22.5.

Solution: (a) Carbon dioxide is a simple triatomic molecule; silicon dioxide has a complex three-dimensional network structure of alternating Si-O bonds. The principal reason for the profound difference is the inability of representative elements beyond the second period to form multiple bonds. See Problems 22.54 and 22.60 (c).

(b) The two structures are (see also Figure 22.14)

```
     O            O—O
    / \          /   \
   H   H        H     H
```

The O-O single bond is weak, making hydrogen peroxide a very reactive, unstable substance.

(c) MgO is a typical metal-nonmetal ionic solid. NO is a simple diatomic molecular covalent (nonmetal-nonmetal) compound. Can you draw a Lewis structure for NO?

22.66 Consult the appropriate parts of Section 22.5.

Solution: Hydrolysis means reaction with water, usually in the context of breaking water into H^+ and OH^-. Hydration means incorporating one or more molecules of water into a compound such as $CuSO_4 \cdot 5H_2O$.

22.67 Use the basic diatomic molecular orbital energy-level diagram in Section 10.7.

Solution: The molecular orbital energy level diagrams are

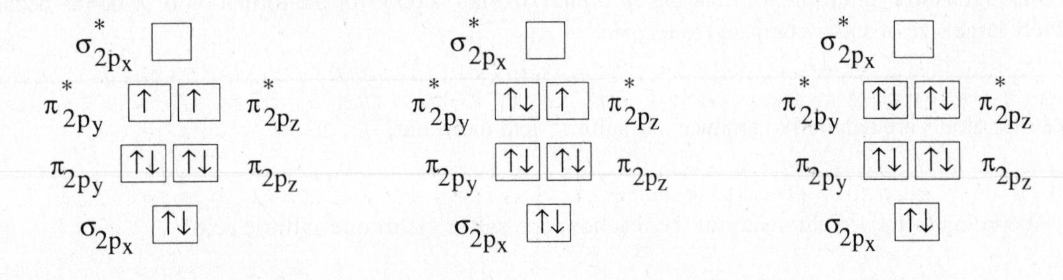

O_2	O_2^-	O_2^{2-}
σ^*_{2s} ↑↓	σ^*_{2s} ↑↓	σ^*_{2s} ↑↓
σ_{2s} ↑↓	σ_{2s} ↑↓	σ_{2s} ↑↓
σ^*_{1s} ↑↓	σ^*_{1s} ↑↓	σ^*_{1s} ↑↓
σ_{1s} ↑↓	σ_{1s} ↑↓	σ_{1s} ↑↓

Which of the three has the strongest bonding?

22.68 Study Section 22.5 to find the appropriate equation. Would this preparation still work if barium sulfate were soluble in water?

Solution: The equation is: $BaO_2 \cdot 8H_2O(s) + H_2SO_4(aq) \rightarrow BaSO_4(s) + H_2O_2(aq) + 8H_2O(l)$

22.70 Percent by mass and STP gas volume are discussed in Sections 2.7 and 5.5, respectively.

Solution: (a) As stated in the problem, the decomposition of hydrogen peroxide is accelerated by light. Storing solutions of the substance in dark-colored bottles helps to prevent this form of decomposition.

(b) The STP volume of oxygen gas formed is:

$$15.0 \text{ g} \left(\frac{7.50\%}{100\%}\right)\left(\frac{1 \text{ mol } H_2O_2}{34.02 \text{ g}}\right)\left(\frac{1 \text{ mol } O_2}{2 \text{ mol } H_2O_2}\right)\left(\frac{22.41 \text{ L}}{1 \text{ mol}}\right) = 0.371 \text{ L } O_2$$

22.71 Use the rules for assigning oxidation numbers given in Section 3.5.

Solution: Following the rules given in Section 3.5, we assign hydrogen an oxidation number of +1 and fluorine an oxidation number of -1. Since HFO is a neutral molecule, the oxidation number of oxygen is zero. Can you think of other compounds in which oxygen has this oxidation number?

22.72 This is similar to Problem 22.54. Is this a general trend?

Solution: Analogous to P in Problem 22.54, the 3*p* orbital overlap is poor for the formation of π bonds because of the relatively large size of sulfur compared to oxygen.

22.74 How many sulfur atoms are required to produce one sulfuric acid molecule?

Solution: According to the stoichiometry of the reaction, one sulfur yields one sulfuric acid.

$$41 \times 10^6 \text{ tons } H_2SO_4 \times \frac{2000 \text{ lb}}{1 \text{ ton}} \times \frac{453.6 \text{ g}}{1 \text{ lb}} \times \frac{32.07 \text{ g S}}{98.09 \text{ g } H_2SO_4} = 1.2 \times 10^{13} \text{ g}$$

Converting to moles: $\frac{1.2 \times 10^{13} \text{ g}}{32.07 \text{ g/mol}} = 3.7 \times 10^{11}$ mol

22.75 The chemical effect of a dehydrating agent is to remove the elements of water (H_2O) from the formula of the substance acted upon. It is OK to write an equation with two molecules of the substance suffering dehydration. Compare to Problem 22.58; P_4O_{10} is also a dehydrating agent.

Solution: Each reaction uses H_2SO_4 as a reagent.

(a) $HCOOH(l) \rightleftharpoons CO(g) + H_2O(l)$

(b) $4H_3PO_4(l) \rightleftharpoons P_4O_{10}(s) + 6H_2O(l)$

(c) $2HNO_3(l) \rightleftharpoons N_2O_5(g) + H_2O(l)$

(d) $2HClO_3(l) \rightleftharpoons Cl_2O_5(l) + H_2O(l)$

Notice that these processes do not involve oxidation or reduction. The reaction between sulfuric acid and water (hydration) is extremely exothermic.

22.76 The chemical reactions of sulfuric acid are discussed in Section 22.5.

Solution: Copper reacts with hot concentrated sulfuric acid to yield copper(II) sulfate, water, and sulfur dioxide. Concentrated sulfuric acid also reacts with carbon to produce carbon dioxide, water, and sulfur dioxide.

Can you write balanced equations for these processes?

22.78 Explanations of the nonexistence of chemical substances have a way of being embarrassing if and when someone finally succeeds in making the compound. Try to fashion a rationalization in terms of relative atomic sizes.

Solution: The usual explanation for the fact that no chemist has yet succeeded in making SCl_6, SBr_6 or SI_6 is based on the idea of excessive crowding of the six chlorine, bromine, or iodine atoms around the sulfur. Others suggest that sulfur in the +6 oxidation state would oxidize chlorine, bromine, or iodine in the 1- oxidation state to the free elements. In any case none of these substances has been made as of the date of this writing.

It is of interest to point out that thirty years ago all textbooks confidently stated that compounds like ClF_5 could not be prepared (see Section 22.6).

Note that PCl_6^- is a known species (See Problem 22.61). How different are the sizes of S and P?

22.80 One mole of calcium carbonate combines with one mole of sulfur dioxide.

Solution: There are actually several steps involved in removing sulfur dioxide from industrial emissions with calcium carbonate. First calcium carbonate is heated to form carbon dioxide and calcium oxide.

$$CaCO_3(s) \rightleftharpoons CaO(s) + CO_2(g)$$

The CaO combines with sulfur dioxide to form calcium sulfite.

$$CaO(s) + SO_2(g) \rightarrow CaSO_3(s)$$

Alternatively, calcium sulfate forms if enough oxygen is present.

$$2CaSO_3(s) + O_2(g) \rightarrow 2CaSO_4(s)$$

The amount of calcium carbonate (limestone) needed in this problem is:

$$50.6 \text{ g } SO_2 \left(\frac{1 \text{ mol } SO_2}{64.07 \text{ g } SO_2}\right)\left(\frac{1 \text{ mol } CaCO_3}{1 \text{ mol } SO_2}\right)\left(\frac{100.1 \text{ g } CaCO_3}{1 \text{ mol } CaCO_3}\right) = 79.1 \text{ g } CaCO_3$$

The calcium oxide-sulfur dioxide reaction is an example of a Lewis acid-base reaction (Section 15.7) between oxide ion and sulfur dioxide. Can you make Lewis structures showing this process? Which substance is the Lewis acid and which is the Lewis base?

22.82 This is an oxidation-reduction problem. Try to work it using the ion-electron method (Section 3.6). Do the sodium ions participate in the redox process? Is the reaction in acidic or basic solution?

Solution: A check of Table 20.1 shows that sodium ion cannot be reduced by any of the substances mentioned in this problem. It is a "spectator ion." We focus on the substances that are actually undergoing oxidation or reduction and write half-reactions for each.

$$2I^-(aq) \rightarrow I_2(s)$$

$$H_2SO_4(aq) \rightarrow H_2S(g)$$

Balancing the oxygen, hydrogen, and charge gives:

$$2I^-(aq) \rightarrow I_2(s) + 2e^-$$

$$H_2SO_4(aq) + 8H^+(aq) + 8e^- \rightarrow H_2S(g) + 4H_2O(l)$$

Multiplying the iodine half-reaction by four and combining gives the balanced redox equation.

$$H_2SO_4(aq) + 8I^-(aq) + 8H^+(aq) \rightarrow H_2S(g) + 4I_2(s) + 4H_2O(l)$$

The hydrogen ions come from extra sulfuric acid. We add one sodium ion for each iodide ion to obtain the final equation.

$$9H_2SO_4(aq) + 8NaI(aq) \rightarrow H_2S(g) + 4I_2(s) + 4H_2O(l) + 8NaHSO_4(aq)$$

22.84 Consult the descriptive parts of Section 22.6.

Solution: A number of methods are available for the preparation of metal chlorides.

(a) $Na(s) + Cl_2(g) \rightarrow 2NaCl(s)$

(b) $2HCl(aq) + Mg(s) \rightarrow MgCl_2(aq) + H_2(g)$

(c) $HCl(aq) + NaOH(aq) \rightarrow NaCl(aq) + H_2O(l)$

(d) $CaCO_3(s) + 2HCl(aq) \rightarrow CaCl_2(aq) + CO_2(g) + H_2O(l)$

(e) $AgNO_3(aq) + NaCl(aq) \rightarrow AgCl(s) + NaNO_3(aq)$

22.86 Can copper(II) ion oxidize fluoride or chloride ions? Consult Table 20.1. Can this be a redox reaction? How about formation of a complex ion?

Solution: Fluoride, chloride, bromide, and iodide form complex ions (Section 17.5) with many transition metals. With chloride the green $CuCl_4^{2-}$ forms

$$Cu^{2+}(aq) + 4Cl^-(aq) \rightarrow CuCl_4^{2-}(aq)$$

With fluoride, copper(II) forms an insoluble green salt, CuF_2. Copper(II) cannot oxidize fluoride or chloride.

22.87 These substances are held together by strong hydrogen bonds (Section 10.2). The $(HF)_2$ is polar; the HF_2^- is not.

Solution: The structures are:

$$H-F\cdots H-F \qquad [F\cdots H\cdots F]^-$$

The HF_2^- ion has the strongest known hydrogen bond. More complex hydrogen bonded HF clusters are also known. $(HF)_6$ is not polar. Can you guess a possible structure?

22.88 Read the problem statement carefully. Is sulfuric acid added to a water solution of NaCl or to pure, solid NaCl? What do you know about the boiling points of hydrogen chloride and sulfuric acid?

Solution: Sulfuric acid is added to solid sodium chloride, not aqueous sodium chloride. Hydrogen chloride is a gas at room temperature and can escape from the reacting mixture.

$$H_2SO_4(l) + NaCl(s) \rightarrow HCl(g) + NaHSO_4(s)$$

The reaction is driven to the right by the continuous loss of $HCl(g)$ (Le Chatelier's principle).

What happens when sulfuric acid is added to a water solution of NaCl? Could you tell the difference between this solution and the one formed by adding hydrochloric acid to aqueous sodium sulfate?

22.90 Study Problem 22.82 before answering this one.

Solution: As with iodide salts, a redox reaction occurs between sulfuric acid and sodium bromide.

$$2H_2SO_4(aq) + 2NaBr(aq) \rightarrow SO_2(g) + 2Br_2(l) + 2H_2O(l) + Na_2SO_4(aq)$$

22.91 This is a redox reaction. Write down the balanced equation. How many moles of bromine are formed from one mole of chlorine?

Solution: The balanced equation is:

$$Cl_2(g) + 2Br^-(aq) \rightarrow 2Cl^-(aq) + Br_2(g)$$

The number of moles of bromine is the same as the number of moles of chlorine, so this problem is essentially a gas law exercise in which P and T are changed for some given amount of gas

$$V_2 = \frac{P_1V_1}{T_1} \times \frac{T_2}{P_2} = \frac{760\text{ mmHg} \times 2.00\text{ L}}{288\text{ K}} \times \frac{373\text{ K}}{700\text{ mmHg}} = 2.81\text{ L}$$

22.92 Consult Section 10.1 to review VSEPR principles.

Solution:

Molecule or Ion	Class	Structure or Shape
(a) IF_7	AB_7	pentagonal bipyramid
(b) I_3^-	AB_2E_3	linear
(c) $SiCl_4$	AB_4	tetrahedral
(d) PF_5	AB_5	trigonal bipyramidal
(e) SF_4	AB_4E	T-shaped
(f) ClO_2^-	AB_2E_2	bent

22.93 If you don't remember how to balance redox equations, see Problem 22.82 above or review Section 3.6.

Solution: The balanced equation is:

$$I_2O_5(s) + 5CO(g) \rightarrow I_2(s) + 5CO_2(g)$$

The oxidation numbers of iodine and carbon change from +5 to 0 and from +2 to +4, respectively. Iodine suffers reduction; carbon is oxidized.

22.94 In part (c) water is also a product.

Solution: The balanced equations are:

(a) $2H_3PO_3(aq) \rightarrow H_3PO_4(aq) + PH_3(g)$

(b) $Li_4C(s) + 4HCl(aq) \rightarrow 4LiCl(aq) + CH_4(g)$

(c) $2HI(g) + 2HNO_2(aq) \rightarrow I_2(s) + 2NO(g) + 2H_2O(l)$

(d) $H_2S(g) + 2Cl_2(g) \rightarrow 2HCl(g) + SCl_2(l)$

22.96 You will not find equations in the text for these conversions. Isotopic synthesis is special. Remember that deuterium oxide behaves chemically just like ordinary water (only reaction rates are different). Find reactions in this chapter for the syntheses of ammonia and acetylene. Do any of them use water? Can the reactants in any of them be made from water?

> ***Solution:*** (a) Ammonia can be made by treating a metal nitride with water (Section 22.4).
>
> $$Li_3N(s) + 3D_2O(l) \rightarrow ND_3(g) + 3LiOD(s)$$
>
> It can also be made by the Haber process; the D_2 can be prepared from D_2O by electrolysis (Section 20.8).
>
> $$2D_2O(l) \rightarrow 2D_2(g) + O_2(g)$$
>
> (b) Acetylene can be made from calcium carbide and water (Section 22.3).
>
> $$CaC_2(s) + 2D_2O(l) \rightarrow C_2D_2(g) + Ca(OD)_2(s)$$
>
> Neither ND_3 nor C_2D_2 can be made by combining NH_3 or C_2H_2 with D_2O. You get a mixture of products containing varying amounts of H and D.

22.97 Find chemical reactions in which O_2 and N_2O differ. How do they differ in physical properties?

> ***Solution:*** (a) Physical properties: the substances have different molar masses; a molar mass or gas density measurement will distinguish between them. Nitrous oxide is polar while molecular oxygen is not; a measurement of dipole moment will distinguish between the two. Can you think of other differences? How about magnetic properties?
>
> (b) Chemical properties: Oxygen reacts with NO to form the red-brown gas NO_2; nitrous oxide doesn't react with NO. Can you think of other differences? What happens when you breathe nitrous oxide?

22.98 Consult Section 3.5 for a discussion of the rules for assigning oxidation numbers.

> ***Solution:*** There is no change in oxidation number; it is zero for both compounds.

22.100 The hint should provide sufficient clues for you to make a good guess at the answer.

> ***Solution:*** The green color is due to iron(II) sulfide.

Note: If you have not already done so, please read the "Introduction for the Student" at the beginning of this Student Solutions Manual.

Chapter 23

TRANSITION METAL CHEMISTRY AND COORDINATION COMPOUNDS

23.14 Study Section 23.2 to review the meanings of the terms in this problem.

Solution: (a) En is the abbreviation for ethylenediamine ($H_2NCH_2CH_2NH_2$).

(b) The oxidation number of Co is +3. (Why?)

(c) The coordination number of Co is six. (Why isn't this the same as the number of ligands?)

(d) Ethylenediamine (en) is a bidentate ligand. Could cyanide ion be a bidentate ligand? Ask your instructor.

23.16 Study Example 23.1 before working this problem.

Solution: (a) The net charge of the complex ion is the sum of the charges of the ligands and the central metal ion. In this case the complex ion has a -3 charge. (Potassium is always +1. Why?) Since the six cyanides are -1 each, the Fe must be +3.

(b) The complex ion has a -3 charge. Each oxalate ion has a -2 charge (Table 23.3), therefore the Cr must be +3.

(c) Since cyanide ion has a -1 charge, Ni must have a +2 charge to make the complex ion carry a -2 net charge.

(d) Since sodium is always +1 and the oxygens are -2, then Mo must have an oxidation number of +6.

(e) Magnesium is +2 and oxygen -2; therefore W is +6.

23.17 Study Example 23.2 and the discussion of the Naming of Coordination Compounds in Section 23.2 before working this problem.

Solution:

(a) tetraamminedichlorocobalt(III)

(b) triamminetrichlorochromium(III)

(f) *cis*-dichlorobis(ethylenediamine)cobalt(III)

(g) pentaamminechloroplatinum(IV) chloride

(c) dibromobis(ethylenediamine)cobalt(III)

(d) pentacarbonyliron(0)

(e) *trans*-diamminedichloroplatinum(II)

(h) hexaamminecobalt(III) chloride

(i) pentaamminechlorocobalt(III) chloride

23.18 Before working this problem, study Example 23.3 and the discussion of the Naming of Coordination Compounds in Section 23.2.

Solution: The formulas are:

(a) $[Zn(OH)_4]^{2-}$ (b) $[CrCl(H_2O)_5]Cl_2$ (c) $[CuBr_4]^{2-}$ (d) $[Fe(EDTA)]^{2-}$

(e) $[Cr(en)_2Cl_2]^+$ (f) $Fe(CO)_5$ (g) $K_2[Cu(CN)_4]$ (h) $[Co(NH_3)_4(H_2O)Cl]Cl_2$

In (b), why two chloride ions at the end of the formula? In (d), does the "(II)" following ferrate refer to the -2 charge of the complex ion or to the +2 charge of the iron atom? In (g), why are there two potassium ions?

23.25 Examine Figure 23.10 carefully. How many different structures are shown? Are some just views of the same structure from a different angle? For part (b) is it possible to have all three Cl atoms cis? How about all three trans? Are other configurations possible?

Solution: (a) In general for any MA_2B_4 octahedral molecule, only two geometric isomers are possible. The only real distinction is whether the two A-ligands are cis or trans. In Figure 23.10 (a) and (c) are the same compound (Cl atoms cis in both), and (b) and (d) are identical (Cl atoms trans in both).

(b) A model or a careful drawing is very helpful to understand the MA_3B_3 octahedral structure. There are only two possible geometric isomers. The first has all A's (and all B's) cis; this is called the facial isomer. The second has two A's (and two B's) at opposite ends of the molecule (trans). Try to make or draw other possibilities. What happens?

23.26 Before attempting this problem, study Section 23.3 and Figures 23.10 and 23.12. What is the coordination number of cobalt in each of these complexes? Note: bidentate ligands like ethylenediamine and oxalate $C_2O_4^{2-}$ can only bond to the metal in a cis configuration; they are not big enough for a trans structure.

Solution: (a) All six ligands are identical in this octahedral complex. There are no geometric or optical isomers.

NH_3
H_3N NH_3
Co
H_3N NH_3
NH_3

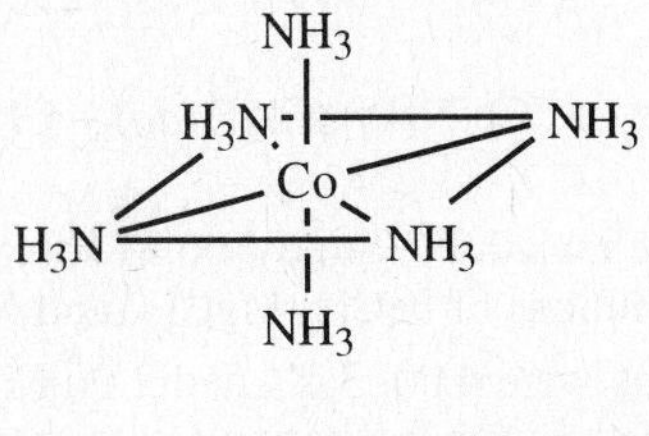

(b) Again there are no geometric or optical isomers. To have cis and trans isomers there would have to be two chlorine ligands.

Cl
H_3N — Co — NH_3
H_3N — NH_3
NH_3

(c) There are cis and trans geometric isomers (see Problem 23.25). No optical isomers.

Cl
H_3N — Co — NH_3
H_3N — NH_3
Cl

trans isomer

Cl
H_3N — Co — Cl
H_3N — NH_3
NH_3

cis isomer

(d) There are two optical isomers. They are like Figure 23.12(a) with the two chlorine atoms replaced by one more bidentate ligand. The three bidentate en ligands are represented by the curved lines.

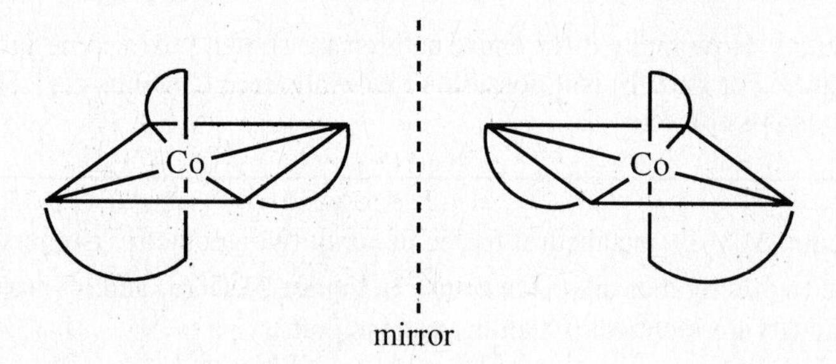

(e) This is exactly like (d) directly above. The oxalate ligand is bidentate as is ethylenediamine.

23.28 The electrolytic conductance of a solution depends directly on the number of ions in the solution. Imagine that you have a 1.00 M solution of each of the possible cobalt complexes. How do the numbers of ions differ in each of the solutions? The strong electrolyte salts could be used as reference solutions.

Solution: The three cobalt compounds would dissociate as follow:

$$[Co(NH_3)_6]Cl_3(aq) \rightarrow [Co(NH_3)_6]^{3+}(aq) + 3Cl^-(aq)$$

$$[Co(NH_3)_5Cl]Cl_2(aq) \rightarrow [Co(NH_3)_5Cl]^{2+}(aq) + 2Cl^-(aq)$$

$$[Co(NH_3)_4Cl_2]Cl(aq) \rightarrow [Co(NH_3)_4Cl_2]^{+}(aq) + Cl^-(aq)$$

In other words, the concentration of free ions in the three 1.00 M solutions would be 4.00 M, 3.00 M, and 2.00 M, respectively. If you made up 1.00 M solutions of $FeCl_3$, $MgCl_2$, and NaCl, these would serve as reference solutions in which the ion concentrations were 4.00, 3.00, and 2.00 M, respectively. A 1.00 M solution of $[Co(NH_3)_5Cl]Cl_2$ would have an electrolytic conductivity close to that of the $MgCl_2$ solution, etc.

23.34 Consult the discussion of crystal field splitting near the end of Section 23.4.

Solution: There are six ligands in an octahedral complex and only four in a tetrahedral complex. Consequently, the electrostatic interaction is greater in an octahedral complex, resulting in a larger crystal field splitting

23.35 When a material appears yellow to the eye, what colors are being absorbed? What about something that looks blue? Which color of absorbed light has more energy?

Solution: When a substance appears to be yellow, it is absorbing light from the blue-violet, high energy end of the visible spectrum. Often this absorption is just the tail of a strong absorption in the ultraviolet. Substances that appear green or blue to the eye are absorbing light from the lower energy red or orange part of the spectrum.

Cyanide ion is a very strong field ligand. It causes a larger crystal field splitting than water, resulting in the absorption of higher energy (shorter wavelength) radiation.

23.36 Consult Figures 23.21 and 23.22. Do both nickel atoms have the same number of 3d electrons? How are these electrons distributed in the two splitting patterns? If the metal ion had nine 3d electrons in the two complexes, would there be a difference in magnetic properties?

Solution:

$[Ni(CN)_4]^{2-}$:

[ ] $d_{x^2-y^2}$

[↑↓] d_{xy}

[↑↓] d_{z^2}

d_{xz} [↑↓] [↑↓] d_{yz}

$[NiCl_4]^{2-}$:

[↑↓] d_{xy} [↑] d_{xz} [↑] d_{yz}

[↑↓] $d_{x^2-y^2}$ [↑↓] d_{z^2}

23.37 Assume both complexes are octahedral and use the crystal field splitting pattern in Figure 23.18. How many 3d electrons are associated with the chromium ion? Are the ligands strong field or weak field ligands? Example 22.4 in the text is a similar problem.

Solution: (a) Cyanide is a strong field ligand (see Problem 23.35).

$[Cr(CN)_6]^{4-}$:

[ ] $d_{x^2-y^2}$ [ ] d_{z^2}

[↑↓] d_{xy} [↑] d_{xz} [↑] d_{yz}

$[Cr(H_2O)_6]^{2+}$:

[↑] $d_{x^2-y^2}$ [ ] d_{z^2}

[↑] d_{xy} [↑] d_{xz} [↑] d_{yz}

The four 3d electrons should occupy the three lower orbitals as shown; there should be two unpaired electrons.

(b) Water is a weak field ligand. The four 3d electrons should occupy the five orbitals as shown. There should be four unpaired electrons.

If these were chromium(III) complexes, how many unpaired electrons would there be?

23.38 Consult Table 23.6. How do you convert wavelength to photon energy (see Section 7.1)?

Solution: (a) Wavelengths of 470 nm fall between blue and blue-green (Table 23.6), corresponding to an observed color in the orange part of the spectrum.

(b) We convert wavelength to photon energy using the Planck relationship

$$\Delta E = \frac{hc}{\lambda} = \frac{(6.63 \times 10^{-34}\ \text{J·s})(3.00 \times 10^{8}\ \text{m/s})}{470 \times 10^{-9}\ \text{m}} = 4.23 \times 10^{-19}\ \text{J}$$

$$(4.23 \times 10^{-19}\ \text{J})(6.022 \times 10^{23}\ \text{mol}^{-1})\left(\frac{10^{-3}\ \text{kJ}}{1\ \text{J}}\right) = 255\ \text{kJ/mol}$$

23.39 Does longer wavelength mean lower or higher energy? How does this correlate with ligand position in the spectrochemical series?

Solution: A lower position in the spectrochemical series means less crystal field splitting and light absorption at longer wavelengths.

(a) The aquo complex will absorb at longer wavelengths.

(b) The fluoro complex will absorb at longer wavelengths.

(c) The chloro complex will absorb at longer wavelengths.

23.40 Use the freezing point data to compute the average molar mass of the complex. Into how many ions has the complex split?

Solution: First compute the average molar mass:

$$\text{Molality} = \frac{\Delta T_f}{K_f} = \frac{0.56°\text{C}}{1.86°\text{C/m}} = 0.30\ \text{mol/kg}\ H_2O$$

$$\text{Moles} = \frac{0.30\ \text{mol}}{1\ \text{kg}\ H_2O} \times 0.025\ \text{kg}\ H_2O = 0.0075\ \text{mol}$$

$$\text{Molar mass} = \frac{0.875\ \text{g}}{0.0075\ \text{mol}} = 117\ \text{g/mol}$$

The molar mass of $Co(NH_3)_4Cl_3$ is 233.4 g/mol, which is twice the computed molar mass. This implies dissociation into two ions in solution. The formula must be:

$$[Co(NH_3)_4Cl_2]Cl$$

Refer to Problem 23.26 (c) for a diagram of the structure of the complex ion.

23.44 Removing rust stains is a case of trying to dissolve insoluble iron hydroxides. Study the part of Section 23.6 dealing with plant growth and iron(III)-EDTA complexes. Note that oxalic acid is a chelating agent (bidentate ligand).

Solution: Rust stain removal involves forming a water soluble oxalate ion complex of iron like $[Fe(C_2O_4)_3]^{3-}$. The overall reaction is:

$$Fe_2O_3(s) + 6H_2C_2O_4(aq) \rightarrow 2Fe(C_2O_4)_3^{3-}(aq) + 3H_2O(l) + 6H^+(aq)$$

Does this reaction depend on pH?

23.46 This problem is like Problem 22.86.

Solution: The green precipitate is CuF_2. When KCl is added, the bright green solution is due to the formation of $CuCl_4^{2-}$:

$$Cu^{2+}(aq) + 4Cl^-(aq) \rightleftharpoons CuCl_4^{2-}(aq)$$

23.47 The described events sound very much like initial precipitation of a cyanide salt followed by formation of a water soluble cyanide complex. Try to write balanced equations for these steps. When no cyanide ion is present, what normally happens when hydrogen sulfide is bubbled through a solution containing copper(II)?

Solution: The white precipitate is copper(II) cyanide.

$$Cu^{2+}(aq) + 2CN^-(aq) \rightarrow Cu(CN)_2(s)$$

This forms a soluble complex with excess cyanide.

$$Cu(CN)_2(s) + 2CN^-(aq) \rightarrow Cu(CN)_4^{2-}(aq)$$

Copper(II) sulfide is normally a very insoluble substance. In the presence of excess cyanide ion the concentration of free copper(II) ion is so low that CuS precipitation cannot occur. In other words, the cyanide complex of copper has a very large formation constant.

23.48 Hint: the tetrachlorocuprate(II) ion is green; the hexaaquocopper(II) ion is blue.

Solution: The overall reaction is: $CuCl_4^{2-}(aq) + 6H_2O(l) \rightleftharpoons Cu(H_2O)_6^{2+} + 4Cl^-(aq)$

Addition of excess water (dilution) shifts the equilibrium to the right (Le Chatelier's principle).

23.49 Write out the formation constant expression for the complex and make appropriate substitutions.

Solution: The formation constant expression is:

$$K_f = \frac{[Fe(H_2O)_5NCS^{2+}]}{[Fe(H_2O)_6^{3+}][NCS^-]}$$

Notice that the original volumes of the Fe(III) and SCN^- solutions were both 1.0 mL and that the final volume is 10.0 mL. This represents a tenfold dilution, and the concentrations of Fe(III) and SCN^- become 0.020 M and 1.0×10^{-4} M, respectively. We make a table:

	$Fe(H_2O)_6^{3+}$ +	NCS^- ⇌	$Fe(H_2O)_5NCS^{2+}$
Initial (M):	0.020	1.0×10^{-4}	0
Change (M):	7.3×10^{-5}	-7.3×10^{-5}	$+7.3 \times 10^{-5}$
Equilibrium (M):	0.020	2.7×10^{-5}	7.3×10^{-5}

$$K_f = \frac{7.3 \times 10^{-5}}{(0.020)(2.7 \times 10^{-5})} = 1.4 \times 10^2$$

23.50 What mass of iron is present in 100.00 g of hemoglobin? What mass of hemoglobin contains one mole of iron?

Solution: A 100.00 g sample of hemoglobin contains 0.34 g of iron. In moles this is:

$$0.34 \text{ g Fe} \left(\frac{1 \text{ mol}}{55.85 \text{ g}}\right) = 6.1 \times 10^{-3} \text{ mol Fe}$$

The amount of hemoglobin that contains one mole of iron must be:

$$\frac{100.00 \text{ g hemoglobin}}{6.1 \times 10^{-3} \text{ mol Fe}} = 1.6 \times 10^4 \text{ g hemoglobin/mol Fe}$$

We compare this to the actual molar mass of hemoglobin:

$$\frac{6.5 \times 10^4 \text{ g hemoglobin}}{1 \text{ mol hemoglobin}} \left(\frac{1 \text{ mol Fe}}{1.6 \times 10^4 \text{ g hemoglobin}}\right) = 4 \text{ Fe/hemoglobin}$$

The hemoglobin molecule contains four iron atoms.

23.51 Review Section 23.1. How does the electron configuration of zinc(II) differ from that of other transition metal ions?

Solution: (a) Zinc(II) has a completely filled 3d subshell giving the ion greater stability.

(b) Normally the colors of transition metal ions result from transitions within incompletely filled d subshells. The 3d subshell of zinc(II) ion is filled.

23.52 How do you find a equilibrium constant for an overall reaction that is the sum of other reactions, for each of which you know the equilibrium constant?

Solution: Reversing the first equation: $Ag(NH_3)_2^+(aq) \rightleftharpoons Ag^+(aq) + 2NH_3(aq)$

$$K_1 = \frac{1}{1.5 \times 10^7} = 6.7 \times 10^{-8}$$

$$Ag^+(aq) + 2CN^-(aq) \rightleftharpoons Ag(CN)_2^-(aq)$$

$$K_2 = 1.0 \times 10^{21}$$

For the equilibrium: $Ag(NH_3)_2^+(aq) + 2CN^-(aq) \rightleftharpoons Ag(CN)_2^-(aq) + 2NH_3(aq)$

$$K = K_1K_2 = (6.7 \times 10^{-8})(1.0 \times 10^{21}) = 6.7 \times 10^{13}$$

$$\Delta G° = -RT\ln K = -(8.314\ J/K{\cdot}mol)(298\ K)\ln(6.7 \times 10^{13}) = -79\ kJ$$

23.54 Do you think of silver or platinum as being more "reactive"?

Solution: (a) The half-reactions are:

$Pt^{2+}(aq) + 2e^- \rightarrow Pt(s)$ $\quad E° = 1.20\ V$

$Ag^+(aq) + e^- \rightarrow Ag(s)$ $\quad E° = 0.80\ V$

$$2Ag(s) + Pt^{2+}(aq) \rightarrow 2Ag^+(aq) + Pt(s)$$

$$E°_{cell} = E°_{ox} + E°_{red} = -0.80\ V + 1.20\ V = 0.40\ V$$

Since the cell voltage is positive, the reaction is spontaneous.

$$\ln K = \frac{nFE°}{RT} = \frac{(2\ mol)(96500\ C/V{\cdot}mol)(0.40\ V)}{(8.314\ J/K{\cdot}mol)(298\ K)} \qquad K = 3 \times 10^{13}$$

23.56 The role of chelating agents in medicine is discussed in Section 23.7.

Solution: BAL and EDTA are used in the treatment of mercury and lead poisoning. Recent studies indicate that some platinum complexes inhibit the growth of cancerous cells *via* a chelation process.

23.58 The differences between geometric and optical isomers are discussed in Section 23.4.

Solution: *Geometric isomers* are compounds with the same type and number of atoms and the same chemical bonds but different spatial arrangements; such isomers cannot be interconverted without breaking a chemical bond.

Optical isomers are compounds that are nonsuperimposable mirror images.

23.59 Consult the "Chemistry in Action" segment in Section 23.7. How do the magnetic properties of the two compounds differ? Does this correlate with the light absorption properties?

Solution: Oxyhemoglobin absorbs higher energy light than deoxyhemoglobin (Table 23.6). Oxyhemoglobin is diamagnetic (low spin), while deoxyhemoglobin is paramagnetic (high spin). These differences occur because oxygen (O_2) is a strong-field ligand. The crystal field splitting diagrams are:

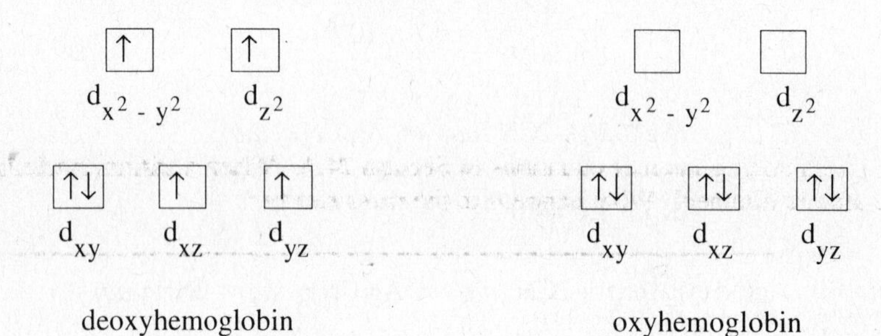

23.60 Draw the crystal field splitting diagram for the $Mn(H_2O)_6^{2+}$ complex (remember, water is a weak-field ligand). If an electron jumps from the lower to the upper set of 3d orbitals, will there be a change in the number of unpaired electrons?

Solution: The orbital splitting diagram below shows five unpaired electrons. The Pauli exclusion principle requires that an electron jumping to the higher 3d orbital would have to change its spin.

↑ ↑

$d_{x^2-y^2}$ d_{z^2}

↑ ↑ ↑

d_{xy} d_{xz} d_{yz}

$Mn(H_2O)_6^{2+}$

23.61 When do we expect cation complexes to be colored? Colorless?

Solution: Complexes are expected to be colored when the highest occupied orbitals have between one and nine d electrons. Such complexes can therefore have d → d transitions (that are usually in the visible part of the electromagnetic radiation spectrum). The ions V^{5+}, Ca^{2+}, and Sc^{3+} have d^0 electron configurations and Cu^+, Zn^{2+}, and Pb^{2+} have d^{10} electron configurations: these complexes are colorless. The other complexes have outer electron configurations of d^1 to d^9 and are therefore colored.

Note: If you have not already done so, please read the "Introduction for the Student" at the beginning of this Student Solutions Manual.

Chapter 24

NUCLEAR CHEMISTRY

24.5 Study the discussion on balancing nuclear equations in Section 24.1. When a tritium nucleus emits a beta particle, what happens to the atomic number? What happens to the mass number?

Solution: The balanced nuclear equations are:

(a) ${}^{3}_{1}H \rightarrow {}^{3}_{2}He + {}^{0}_{-1}\beta$

(b) ${}^{242}_{94}Pu \rightarrow {}^{4}_{2}He + {}^{238}_{92}U$

(c) ${}^{131}_{53}I \rightarrow {}^{131}_{54}Xe + {}^{0}_{-1}\beta$

(d) ${}^{251}_{98}Cf \rightarrow {}^{247}_{96}Cm + {}^{4}_{2}\alpha$

24.6 This like Example 24.1. What quantities must balance on both sides of a nuclear equation?

Solution: (a) The atomic number sum and the mass number sum must remain the same on both sides of a nuclear equation. On the left side of this equation the atomic number sum is 13 (12 + 1) and the mass number sum is 27 (26 + 1). These sums must be the same on the right side. The atomic number of X is therefore 11 (13 - 2) and the mass number is 23 (27 - 4). X is sodium-23 (${}^{23}_{11}Na$).

(b) X is ${}^{1}_{1}H$ or ${}^{1}_{1}p$

(c) X is ${}^{1}_{0}n$

(d) X is ${}^{1}_{0}n$

(e) X is ${}^{56}_{26}Fe$

(f) X is ${}^{0}_{-1}e$ or ${}^{0}_{-1}\beta$

(g) X is ${}^{0}_{-1}e$ or ${}^{0}_{-1}\beta$

(h) X is ${}^{40}_{20}Ca$

(i) X is ${}^{4}_{2}He$ or ${}^{4}_{2}\alpha$

24.7 The shorthand notation in this problem is explained in Section 24.4 and is illustrated in Example 24.3.

Solution: In the shorthand notation for nuclear reactions, the first symbol inside the parentheses is the "bombarding" particle (reactant) and the second symbol is the "ejected" particle (product).

(a) ${}^{15}_{7}N + {}^{1}_{1}p \rightarrow {}^{12}_{6}C + {}^{4}_{2}He$

(b) ${}^{27}_{13}Al + {}^{2}_{1}d \rightarrow {}^{25}_{12}Mg + {}^{4}_{2}He$

(c) ${}^{55}_{25}Mn + {}^{1}_{0}n \rightarrow {}^{56}_{25}Mn + \gamma$

(d) ${}^{80}_{34}Se + {}^{2}_{1}d \rightarrow {}^{81}_{34}Se + {}^{1}_{1}p$

24.8 Write out the full nuclear symbols for these isotopes showing the atomic numbers. How many protons are needed to convert bismuth to astatine? Hint: astatine-213 and -212 are neutron emitters.

Solution: All you need is a high-intensity alpha particle emitter. Any heavy element like plutonium or curium will do. Place the bismuth-209 sample next to the alpha emitter and wait. The reaction is:

$$^{209}_{83}\text{Bi} + {}^{4}_{2}\text{He} \rightarrow {}^{213}_{85}\text{At} \rightarrow {}^{212}_{85}\text{At} + {}^{1}_{0}\text{n} \rightarrow {}^{211}_{85}\text{At} + {}^{1}_{0}\text{n}$$

24.16 Do you remember how to convert amu to grams? Assume the nucleus to be spherical. See Section 24.2.

Solution: We assume the nucleus to be spherical. The mass is:

$$235 \text{ amu}\left(\frac{1 \text{ g}}{6.022 \times 10^{23} \text{ amu}}\right) = 3.90 \times 10^{-22} \text{ g}$$

The volume is:

$$\left(\frac{4\pi}{3}\right)\left(7.0 \times 10^{-3} \text{ pm} \times \left(\frac{1 \text{ cm}}{1 \times 10^{10} \text{ pm}}\right)\right)^3 = 1.4 \times 10^{-36} \text{ cm}^3$$

The density is:

$$\frac{3.90 \times 10^{-22} \text{ g}}{1.4 \times 10^{-36} \text{ cm}^3} = 2.8 \times 10^{14} \text{ g/cm}^3$$

24.17 Study the nuclear stability rules in Section 24.2.

Solution: (a) Lithium-9 should be less stable. The neutron-to-proton ratio is too high.

(b) Actually, both sodium-22 and sodium-25 are unstable. The first has an odd number of protons and an odd number of neutrons (very rare). For the second the neutron-to-proton ratio is probably too high.

(c) Scandium-48 is less stable because of odd numbers of protons and neutrons (verify).

24.18 Study Section 24.2 and Table 24.2.

Solution: Nickel, selenium, and cadmium have more stable isotopes. All three have even atomic numbers (see Table 24.2).

24.19 What can you say about the neutron/proton ratio when an isotope is above the belt of stability? How do isotopes of this type alter the situation?

Solution: When an isotope is above the belt of stability, the neutron/proton ratio is too high. The only mechanism to correct this situation is beta emission; the process turns a neutron into a proton. Direct neutron emission doesn't happen.

$$^{18}_{7}N \rightarrow ^{18}_{8}O + ^{0}_{-1}\beta$$

Oxygen-18 is a stable isotope.

24.20 Study Section 24.2. One isotope of each pair violates a stability rule.

Solution: (a) Neon-17 should be radioactive. It falls below the belt of stability.

(b) Calcium-45 should be radioactive. It falls above the belt of stability.

(c) All technetium isotopes are radioactive.

(d) Mercury-195 should be radioactive. Mercury-196 has an even number of both neutrons and protons.

(e) All curium isotopes are unstable.

24.21 Use the Einstein mass-energy equivalence relationship to convert to mass. Note that the equation requires energy in joules, mass in kilograms, and the speed of light in meters per second. Assume ΔE to be the same as ΔH.

Solution: The mass change is:

$$\Delta m = \frac{\Delta E}{c^2} = \frac{-436400 \text{ J}}{(3.00 \times 10^8 \text{ m/s})^2} = -4.85 \times 10^{-12} \text{ kg/mol } H_2$$

Is this mass measurable with ordinary laboratory analytical balances?

24.22 Use the Einstein mass-energy equivalence relationship to convert joules to kilograms.

Solution: We use the procedure shown in Example 24.2.

(a) There are 4 neutrons and 3 proton in a Li-7 nucleus. The predicted mass is:

$$(3 \text{ x mass of proton}) + (4 \text{ x mass of neutron}) = (3 \text{ x } 1.007825 \text{ amu}) + (4 \text{ x } 1.008665 \text{ amu})$$

$$\text{predicted mass} = 7.058135 \text{ amu}$$

The mass defect, that is the difference between the predicted mass and the measured mass is:

$$\Delta m = 7.01600 \text{ amu} - 7.058135 \text{ amu} = -0.042135 \text{ amu}$$

The mass that is converted in energy, that is the energy released is:

$$\Delta E = \Delta mc^2 = (-0.042135 \text{ amu})\left(\frac{1 \text{ kg}}{6.022 \times 10^{26} \text{ amu}}\right)(3.00 \times 10^8 \text{ m/s})^2 = -6.30 \times 10^{-12} \text{ J}$$

The nuclear binding energy is 6.30×10^{-12} J. The binding energy per nucleon is:

$$(6.30 \times 10^{-12} \text{ J})/7 \text{ nucleons} = 9.00 \times 10^{-13} \text{ J/nucleon}$$

Using the same procedure as in (a), using 1.007825 amu for $^{1}_{1}H$ and 1.008665 amu for $^{1}_{0}n$, we can show that:

(b) For chlorine-35: Nuclear binding energy $= 4.92 \times 10^{-11}$ J

Nuclear binding energy per nucleon $= 1.41 \times 10^{-12}$ J/nucleon.

(c) For helium-4: Nuclear binding energy $= 4.54 \times 10^{-12}$ J

Nuclear binding energy per nucleon $= 1.14 \times 10^{-12}$ J/nucleon.

24.24 What happens to the atomic number of an atom when an alpha particle is lost? What happens to the mass number? What about loss of a beta particle?

Solution: Alpha emission decreases the atomic number by two and the mass number by four. Beta emission increases the atomic number by one and has no effect on the mass number.

(a) $^{232}_{90}Th \xrightarrow{\alpha} {}^{228}_{88}Ra \xrightarrow{\beta} {}^{228}_{89}Ac \xrightarrow{\beta} {}^{228}_{90}Th$

(b) $^{235}_{92}U \xrightarrow{\alpha} {}^{231}_{90}Th \xrightarrow{\beta} {}^{231}_{91}Pa \xrightarrow{\alpha} {}^{227}_{89}Ac$

(c) $^{237}_{93}Np \xrightarrow{\alpha} {}^{233}_{91}Pa \xrightarrow{\beta} {}^{233}_{92}U \xrightarrow{\alpha} {}^{229}_{90}Th$

24.26 What fraction of the original sample remains after one half-life? After two half-lives? How many half-lives are represented by 42.0 min? This problem is similar to Example 24.3.

Solution: The number of atoms decreases by half for each half-life. For ten half-lives we have:

$$5.00 \times 10^{22} \text{ atoms} \left(\frac{1}{2}\right)^{10} = 4.89 \times 10^{19} \text{ atoms}$$

24.27 Notice that you are given the decay rates and the overall time interval. If you knew the amount of yttrium-90 at the start of the experiment and at the end of the experiment, you could use Equation (13.3) to calculate k. What is proportional to the amount of yttrium-90 at any time t?

Solution: Since all radioactive decay processes have first-order rate laws, the decay rate is proportional to the amount of radioisotope at any time:

If we combine equations 13.3 and 13.5:

$$\ln\frac{N_o}{N_t} = kt \qquad 0.693 = kt_{1/2}$$

$$t_{1/2} = 0.693t \times \left(\ln\frac{N_o}{N_t}\right)^{-1} \qquad (1)$$

The time interval is: (2:15 pm, 12/17/82) - (1:00 pm, 12/3/82) = 14 days + 1 hour + 15 min = 20235 min

Substituting into equation (1):

$$t_{1/2} = 0.693 \times 20235 \text{ min} \times \left(\ln \frac{9.8 \times 10^5 \text{ dis}\cdot\text{min}^{-1}}{2.6 \times 10^4 \text{ dis}\cdot\text{min}^{-1}} \right)^{-1}$$

$$t_{1/2} = 3.9 \times 10^3 \text{ min or } 2.7 \text{ day}$$

24.28 What happens to the atomic and mass numbers in alpha emission processes? What about beta emission? This is similar to Problem 24.24.

Solution: The equation for the overall process is:

$$^{232}_{90}\text{Th} \rightarrow 6\,^{4}_{2}\text{He} + 4\,^{0}_{-1}\beta + \text{X}$$

The final product isotope must be $^{208}_{82}\text{Pb}$.

24.29 What does the order of a rate law imply about the mechanism of the rate process?

Solution: A truly first-order rate law implies that the mechanism is unimolecular; in other words the rate is determined only by the properties of the decaying atom or molecule and does not depend on collisions or interactions with other objects. This is why radioactive dating is so reliable.

24.30 This problem is like Problem 24.27 and Example 24.3.

Solution: Using Equation (13.3): $\ln \frac{[A]_0}{[A]} = \ln \frac{1.00}{0.200} = kt$

Using Equation (13.5): $0.693 = kt_{1/2}$ $\quad k = \frac{0.693}{28.1 \text{ yr}} = 0.0247 \text{ yr}^{-1}$

Substituting: $t = \ln \frac{1.00}{0.200} \times \frac{1}{0.0247 \text{ yr}^{-1}} = 65.2 \text{ yr}$

24.32 Part (a) is like Problem 24.6. The second part is similar to Problems 24.27 and 24.30 above.

Solution: (a) The balanced equation is: $^{40}_{19}\text{K} \rightarrow ^{40}_{18}\text{Ar} + ^{0}_{+1}\beta$

(b) Using Equation (13.5): $k = \frac{0.693}{t_{1/2}} = \frac{0.693}{1.2 \times 10^9 \text{ yr}}$

Using Equation (13.3): $t = \frac{1}{k}\ln\frac{[A]_o}{[A]_t} = \frac{1.2 \times 10^9\ yr}{0.693}\ln\frac{1.00}{0.18} = 3.0 \times 10^9\ yr$

24.45 Study the experimental strategies for using radioactive tracers in Section 24.7.

Solution: The easiest experiment would be to add a small amount of aqueous iodide containing some radioactive iodine to a saturated solution of lead(II) iodide. If the equilibrium is dynamic, radioactive iodine will eventually be detected in the solid lead(II) iodide.

Could this technique be used to investigate the forward and reverse rates of this reaction?

24.46 Conceivably, the iodine atoms in the I_2 product could come from the I^- or the IO_4^-. Does the experimental result allow for any conclusion about the source of the iodine atoms?

Solution: The fact that the radioisotope appears only in the I_2 shows that the IO_3^- is formed only from the IO_4^-. Does this result rule out the possibility that I_2 could be formed from IO_4^- as well? Can you suggest an experiment to answer the question?

24.47 In other words can you suggest an experiment to detect the diffusion of labeled ions throughout a solid crystal?

Solution: On paper this is a simple experiment. If one were to dope part of a crystal with a radioactive tracer, one could demonstrate diffusion in the solid state by detecting the tracer in a different part of the crystal at a later time. This actually happens with many substances. In fact in some compounds one type of ion migrates easily while the other remains in fixed position!

24.48 Study Problems 24.45 and 24.47 above and Section 24.7.

Solution: Allow a few days for the iron-59 isotope to be incorporated into a person's body. Isolate red blood cells from a blood sample and monitor radioactivity from the hemoglobin molecules present in the red blood cells.

24.50 Do isotopes of the same element behave differently in chemical reactions? Are there differences in physical properties?

Solution: Chlorine-39 should undergo exactly the same chemical reactions as chlorine-35. There should be some slight differences in reaction rates (the kinetic isotope effect, Section 22.2), but products will be the same. With the passage of time the chlorine-39 containing compounds will suffer decomposition because of the decay of the unstable isotope, and of course the radioactivity will be detectable. The chemistry of the product of the decay of chlorine-39 will be different than that of chlorine-35.

$$^{39}_{17}Cl \rightarrow\ ^{39}_{18}Ar +\ ^{0}_{-1}\beta$$

Since no such compound such as HAr exists, the compound $^{39}_{19}Cl–H$, after undergoing radioactive decay would become $^{39}_{18}Ar\text{–}H$ that would decompose to give H and Ar atoms.

24.52 See Section 24.4.

Solution: Using the system explained in the subsection entitled "The Transuranium Elements" in Section 24.4:

(a) element 112 (Uub) (b) element 138 (Uto) (c) element 176 (Ush) (d) element 204 (Bnq)

What are the atomic numbers of One, Hot, Tub, Sot, and Pot?

24.53 Hint: even-even nuclei have no net spin (like the electron, protons and neutrons have half-integral spins).

Solution: Apparently there is a sort of Pauli principle for nucleons as well as for electrons. When neutrons pair with neutrons and when protons pair with protons, their spins cancel. Even-even nuclei are the only ones with no net spin.

24.54 Part (a) is like Example 24.1. How many tritium atoms are in a 1.00 kg sample of water? What is the value of the specific rate constant for the decay of tritium? Can you express k in min^{-1}?

Solution: (a) The balanced equation is:

$$^{3}_{1}H \rightarrow \, ^{3}_{2}He + \, ^{0}_{-1}\beta$$

(b) The number of tritium (T) atoms in 1.00 kg of water is:

$$1.00 \times 10^3 \text{ g } H_2O \left(\frac{1 \text{ mol}}{18.02 \text{ g}}\right)\left(\frac{6.022 \times 10^{23}}{1 \text{ mol}}\right)\left(\frac{2 \text{ H atoms}}{1 \text{ } H_2O}\right)\left(\frac{1 \text{ T atom}}{1.0 \times 10^{17} \text{ H atoms}}\right) = 6.68 \times 10^8 \text{ T atoms}$$

The number of disintegrations per minute will be: rate = k(number of T atoms) = $kN = \frac{0.693}{t_{1/2}} N$

$$\text{rate} = \left(\frac{0.693}{12.5 \text{ yr}}\right)\left(\frac{1 \text{ yr}}{365 \text{ days}}\right)\left(\frac{1 \text{ day}}{24 \text{ h}}\right)\left(\frac{1 \text{ h}}{60 \text{ min}}\right) 6.68 \times 10^8 \text{ T atoms}$$

rate = 70.5 T atoms/min = 70.5 disintegrations/min

24.56 What is a curie? See Section 24.8. Study Problem 24.54 to see how to find the number of disintegrations per second in the neptunium-237 sample. Part (b) is like Example 24.1.

Solution: (a) One millicurie represents 3.70×10^7 disintegtrations/s. The rate of decay of the isotope is given by the rate law: rate = kN, where N is the number of atoms in the sample. We find the value of k in units of s^{-1}:

$$k = \frac{0.693}{t_{1/2}} = \left(\frac{0.693}{2.20 \times 10^6 \text{ yr}}\right)\left(\frac{1 \text{ yr}}{365 \text{ day}}\right)\left(\frac{1 \text{ day}}{24 \text{ hr}}\right)\left(\frac{1 \text{ hr}}{3600 \text{ s}}\right) = 9.99 \times 10^{-15} \text{ s}^{-1}$$

The number of atoms in a 0.500 g sample of neptunium-237 is:

$$0.500\ \text{g}\left(\frac{1\ \text{mol}}{237.0\ \text{g}}\right)\left(\frac{6.022 \times 10^{23}\ \text{atoms}}{1\ \text{mol}}\right) = 1.27 \times 10^{21}\ \text{atoms}$$

$$\text{rate of decay} = kN = (9.99 \times 10^{-15}\ \text{s}^{-1})(1.27 \times 10^{21}\ \text{atoms}) = 1.27 \times 10^{7}\ \text{atoms/s}$$

We can also say that: rate of decay = 1.27×10^7 disintegrations/s

The activity is:

$$1.27 \times 10^{7}\ \text{disintegrations/s}\left(\frac{1\ \text{millicurie}}{3.70 \times 10^{7}\ \text{disintegrations/s}}\right) = 0.343\ \text{millicurie}$$

(b) The decay equation is: $^{237}_{93}Np \rightarrow {}^{4}_{2}He + {}^{233}_{91}Pa$

24.58 Hint: Strontium is very similar to calcium in its chemical behavior.

Solution: Because both Ca and Sr belong to Group 2A, radioactive strontium that has been ingested into the human body becomes concentrated in bones (replacing Ca) and can damage blood cell production.

24.60 Radioactive iodine is a common nuclear reactor byproduct. Where does the human body concentrate iodine? Study Section 24.7.

Solution: Normally the human body concentrates iodine in the thyroid gland. The purpose of the large doses of KI is to displace radioactive iodine from the thyroid and allow its excretion from the body.

24.62 How do X-rays work in detectors?

Solution: (a) The nuclear equation is: $^{14}_{7}N + {}^{1}_{0}n \rightarrow {}^{15}_{7}N + \gamma$

(b) X-ray analysis only detects shapes, particularly of metal objects. Bombs can be made in a variety of shapes and sizes and can be constructed of "plastic" explosives. Thermal neutron analysis is much more specific than X-ray analysis. However, articles that are high in nitrogen other than explosives (such as silk, wool, and polyurethane) will give "false positive" test results.

24.63 How does the gravitational force on earth compare with that of the sun?

Solution: Because of the relative masses, the force of gravity on the sun is much greater than it is on earth. Thus the nuclear particles on the sun are already held much closer together than the equivalent nuclear particles on earth. Less energy (lower temperature) is required on the sun to force fusion collisions between the nuclear particles.

Note: If you have not already done so, please read the "Introduction for the Student" at the beginning of this Student Solutions Manual.

Chapter 25

ORGANIC CHEMISTRY

25.10 Is this compound an alkane? Is it an alkene? Is it a cycloalkane? How can you tell?

> ***Solution:*** The molecular formula shows the compound is either an alkene or a cycloalkane. (Why?) You can't tell which from the formula. The possible isomers are:
>
> ```
>                                                         H  H
>    H  H  H  H          H  H  H  H                       |  |
>    |  |  |  |          |  |  |  |                    H—C—C—H
> H—C—C—C=C          H—C—C=C—C—H                      |  |
>    |  |     |          |        |                    H—C—C—H
>    H  H     H          H        H                       |  |
>                                                         H  H
> ```
>
> The structure in the middle (2-butene) can exist as cis or trans isomers. There are two more isomers. Can you find and draw them? Can you have an isomer with a double bond and a ring? What would the molecular formula be like in that case?

25.12 What are the expected C-C-C bond angles in benzene and in cyclohexane? What are the expected bond angles in cyclobutadiene?

> ***Solution:*** In benzene the C-C-C angles should be 120° because of the sp^2 hybridization (trigonal planar carbon). These angles can be accommodated perfectly by a planar hexagonal structure. In cyclohexane the C-C-C angles should be 109.5° (sp^3 hybridized, tetrahedral carbon). These angles are not possible in a planar structure.
>
> If cyclobutadiene were square or rectangular, the C-C-C angles must be 90°. If the molecule is diamond-shaped, two of the C-C-C angles must be less than 90°. Both of these situations result in a great deal of distortion and strain in the molecule. Cyclobutadiene is very unstable for these and other reasons.

25.13 What type of hydrocarbon is each compound? What are characteristic chemical reactions of each type?

> ***Solution:*** One compound is an alkane; the other is an alkene. Alkenes characteristically undergo addition reactions with halogens (Cl_2, Br_2, I_2) and with hydrogen halides (HCl, HBr, HI). Alkanes do not react with these substances under ordinary conditions.

25.14 Consult Section 25.1 for a discussion of geometric isomers of alkenes. Which of the two structures would be more spatially crowded (i.e., less stable)?

Solution: The two isomers are:

trans: $H(H_3C)C{=}C(CH_3)H$ — cis: $(H_3C)HC{=}CH(CH_3)$

A simplified method of presenting the structures is:

trans — cis

The cis structure is more crowded and a little less stable. As a result, slightly more heat (energy) would be released when the alkene adds a molecule of hydrogen to form butane, C_4H_{10}. Note that butane is the product when either alkene is hydrogenated.

25.16 Study the discussion of the light-initiated reaction between methane and chlorine in Section 25.1. The fact that bromine vapor is deep red implies that it absorbs light in the visible region of the spectrum. What might happen to the Br_2 molecule when it absorbs light energy?

Solution: The red bromine vapor absorbs photons of blue light and dissociates to form bromine atoms:

$$Br_2 \rightarrow 2Br\cdot$$

The bromine atoms collide with methane molecules and abstract hydrogen atoms.

$$Br\cdot + CH_4 \rightarrow HBr + \cdot CH_3$$

The methyl radical then reacts with Br_2, giving the observed product and regenerating a bromine atom to start the process over again:

$$\cdot CH_3 + Br_2 \rightarrow CH_3Br + Br\cdot$$

$$Br\cdot + CH_4 \rightarrow HBr + \cdot CH_3 \qquad \text{and so on...}$$

25.17 Study Markovnikov's rule for alkene addition reactions in Section 25.1.

Solution: (a) Ethylene is symmetrical; there is no preference in the addition.

$$CH_3{-}CH_2{-}OSO_3H$$

(b) The positive part of the polar reagent adds to the carbon atom that already has the most hydrogen atoms.

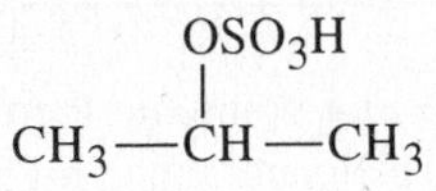

$$\begin{array}{c} OSO_3H \\ | \\ CH_3{-}CH{-}CH_3 \end{array}$$

(c) The positive part of the polar reagent adds to the carbon atom that already has the most hydrogen atoms.

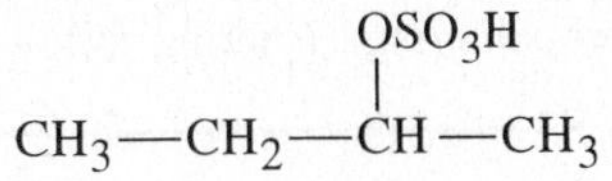

$$CH_3{-}CH_2{-}\underset{}{\overset{OSO_3H}{\overset{|}{C}H}}{-}CH_3$$

(d) 2-butene is symmetrical like ethylene. The product is the same as part (c).

25.18 If the bromine atom were replaced by a hydrogen atom, what type of hydrocarbon would this be? Try replacing a hydrogen atom by a bromine atom in the possible structures. How many different ways can this be done?

Solution: C_3H_6 could be an alkene or a cycloalkane.

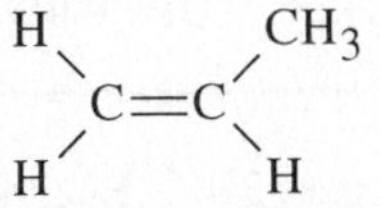

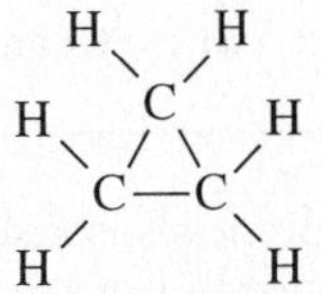

$$(H)(H)C{=}C(CH_3)(H) \qquad \text{cyclo-}C_3H_6$$

Three ways to replace a hydrogen by a bromine are:

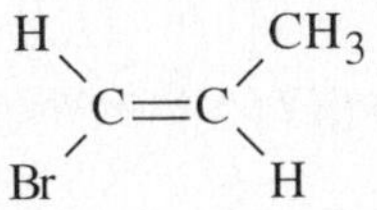

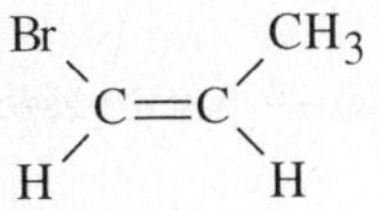

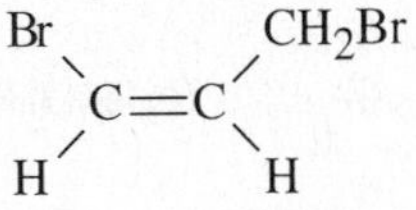

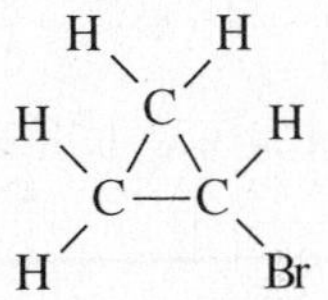

$$(H)(Br)C{=}C(CH_3)(H) \qquad (Br)(H)C{=}C(CH_3)(H) \qquad (Br)(H)C{=}C(CH_2Br)(H) \qquad \text{cyclo-}C_3H_5Br$$

There is only one isomer for the cycloalkane. Note that all three carbons are equivalent in this structure.

25.20 Study the reaction patterns of alkanes, cycloalkanes (different from alkanes?), alkenes, and aromatic hydrocarbons in Section 25.1.

Solution: (a) Cyclopropane is more reactive than propane. Both are saturated compounds, but the acute C-C-C bond angles in cyclopropane cause a great deal of strain.

(b) Ethylene is more reactive than methane. The double bond easily undergoes addition reactions (Problems 25.13 and 25.17).

(c) Benzene is more reactive than cyclohexane. Aromatic compounds undergo substitution reactions; cyclohexane is essentially a saturated hydrocarbon.

25.22 Which type of structure has the lowest boiling point? Which type has the highest? Higher boiling point correlates with stronger intermolecular attractions. Can you see a difference?

Solution: The straight chain molecules have the highest boiling point and therefore the strongest intermolecular attractions. These chains can pack together more closely and efficiently than highly branched, cluster structures. This allows intermolecular forces to operate more effectively and cause stronger attractions.

25.24 The triangular figure is meant to represent a cyclopropane ring. Naming follows the standard rules.

Solution: The names are: (a) *cis*-1,2-dichlorocyclopropane; and (b) *trans*-1,2-dichlorocyclopropane.

Are any other dichlorocyclopropane isomers possible?

25.27 Go back to Review Question 25.26 before working this problem.

Solution: (a) ether (b) amine (c) aldehyde (d) ketone (e) carboxylic acid (f) alcohol (g) amine and carboxylic acid (amino acid)

25.28 The reaction of alcohols with alkali metals is discussed in Section 25.2. Reactions of alcohols with hydrogen chloride usually result in the formation of water as one product. What is the other product?

Solution: The reactions are:

(a) $C_2H_5OH(l) + HCl(g) \rightarrow C_2H_5Cl(l) + H_2O(l)$

(b) $C_2H_5OH(l) + Na(s) \rightarrow NaOC_2H_5(s) + \frac{1}{2}H_2(g)$

25.29 Review Section 25.2. What product forms from the reaction of an alcohol with a carboxylic acid?

Solution: Alcohols react with carboxylic acids to form esters. The reaction is:

$$HCOOH + CH_3OH \rightarrow HCOOCH_3 + H_2O$$

The structure of the product is:

$$\mathrm{H{-}\overset{\overset{\large O}{\|}}{C}{-}O{-}CH_3}$$

25.30 What products form when aldehydes are oxidized? What about ketones? Is there a difference in the two reactions?

Solution: Aldehydes can be oxidized easily to carboxylic acids. The oxidation reaction is:

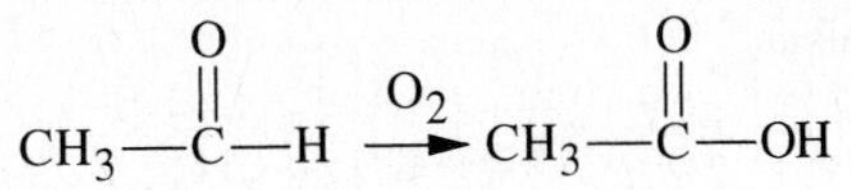

$$CH_3-\overset{\overset{\displaystyle O}{\|}}{C}-H \xrightarrow{O_2} CH_3-\overset{\overset{\displaystyle O}{\|}}{C}-OH$$

Oxidation of a ketone requires that the carbon chain be broken:

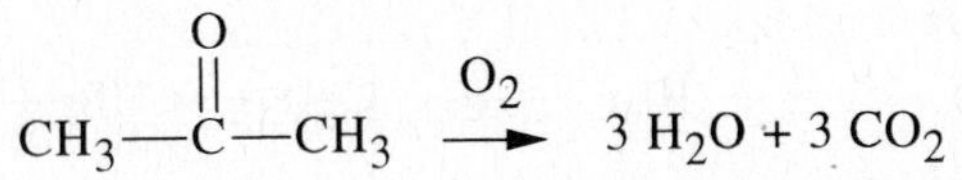

$$CH_3-\overset{\overset{\displaystyle O}{\|}}{C}-CH_3 \xrightarrow{O_2} 3\ H_2O + 3\ CO_2$$

25.32 What types of compounds are possible for this formula? What does the lack of reaction with sodium imply about the possible functional groups? See Problem 25.28.

Solution: The fact that the compound does not react with sodium metal eliminates the possibility that the substance is an alcohol. The only other possibility is the ether functional group. There are three ethers possible with this molecular formula:

$CH_3\text{-}CH_2\text{-}O\text{-}CH_2\text{-}CH_3$ $\quad$ $CH_3\text{-}CH_2\text{-}CH_2\text{-}O\text{-}CH_3$ $\quad$ $(CH_3)_2CH\text{-}O\text{-}CH_3$

Light-induced reaction with chlorine results in substitution of a chlorine atom for a hydrogen atom (the other product is HCl). For the first ether there are only two possible chloro derivatives:

$ClCH_2\text{-}CH_2\text{-}O\text{-}CH_2\text{-}CH_3$ $\quad$ $CH_3\text{-}CHCl\text{-}O\text{-}CH_2\text{-}CH_3$

For the second there are four possible chloro derivatives Three are shown below. Can you draw the fourth?

$CH_3\text{-}CHCl\text{-}CH_2\text{-}O\text{-}CH_3$ $\quad$ $CH_3\text{-}CH_2\text{-}CHCl\text{-}O\text{-}CH_3$ $\quad$ $CH_2Cl\text{-}CH_2\text{-}CH_2\text{-}O\text{-}CH_3$

For the third there are three possible chloro derivatives:

$$CH_3-\overset{\overset{\displaystyle CH_2Cl}{|}}{CH}-O-CH_3 \qquad CH_3-\overset{\overset{\displaystyle CH_3}{|}}{CH}-O-CH_2Cl \qquad (CH_3)_2-\overset{\overset{\displaystyle Cl}{|}}{C}-O-CH_3$$

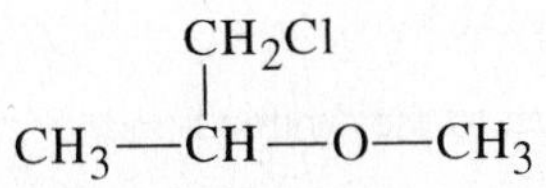

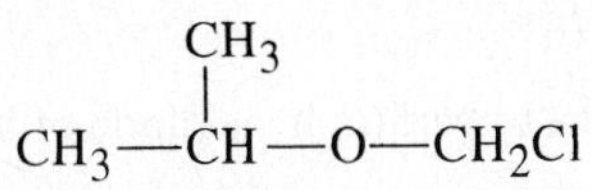

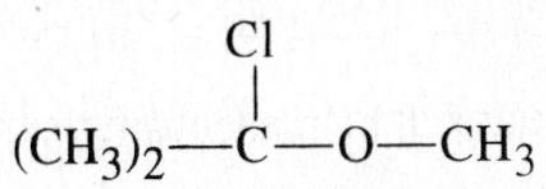

The $(CH_3)_2CH\text{-}O\text{-}CH_3$ choice is the original compound.

25.34 Part (a) is like Problem 25.29 above. For parts (b) and (c) study the characteristic reactions of alkenes and alkynes in Section 25.1. This problem is similar to Example 25.3 in the text.

Solution: (a) The product is similar to that in Problem 25.29.

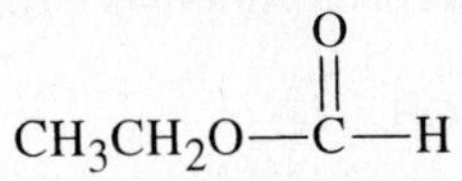

$$CH_3CH_2O-\overset{\overset{\displaystyle O}{\|}}{C}-H$$

(b) Addition of hydrogen to an alkyne gives an alkene.

$$H-C\equiv C-CH_3 + H_2 \rightarrow H_2C{=}CH-CH_3$$

The alkene can also add hydrogen to form an alkane.

$$H_2C{=}CH{-}CH_3 + H_2 \rightarrow CH_3{-}CH_2{-}CH_3$$

(c) HBr will add to the alkene as shown (Note: the carbon atoms at the double bond have been omited for simplicity).

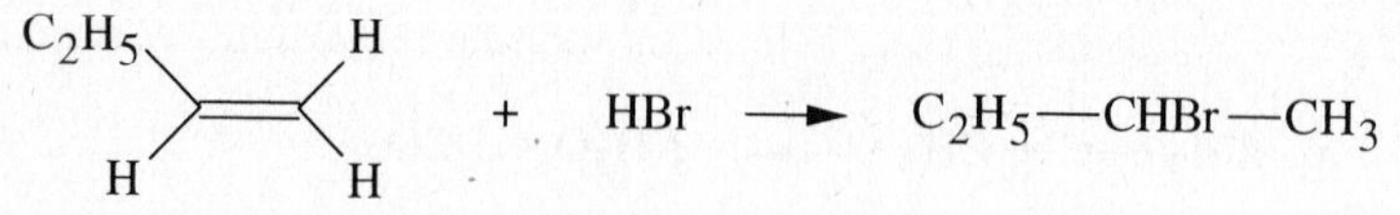

$$(C_2H_5)(H)C{=}C(H)(H) + HBr \longrightarrow C_2H_5{-}CHBr{-}CH_3$$

How do you know that the hydrogen adds to the CH_2 end of the alkene?

25.36 Start by drawing the structure of 3-methyl-1-butyne. A common strategy in synthesis problems is to work backwards from the products to the starting compound. In part (a) what has been added to the starting molecule? What sort of reaction gives this product? Can the product from part (a) be used to make the compounds in parts (b) and (c)?

Solution: In comparing the compound in part (a) with the starting alkyne, it is clear that a molecule of HBr has been added to the triple bond. Is the position of the bromine atom in agreement with the Markovnikov rule? The reaction for (a) is:

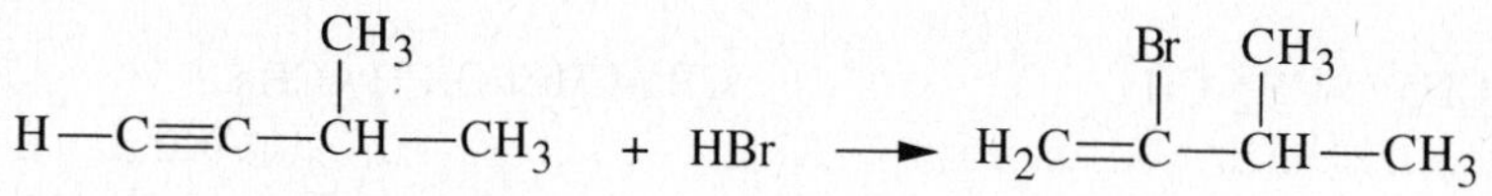

$$H{-}C{\equiv}C{-}CH(CH_3){-}CH_3 + HBr \longrightarrow H_2C{=}CBr{-}CH(CH_3){-}CH_3$$

(b) This compound can be made from the product formed in part (a) by addition of bromine to the double bond.

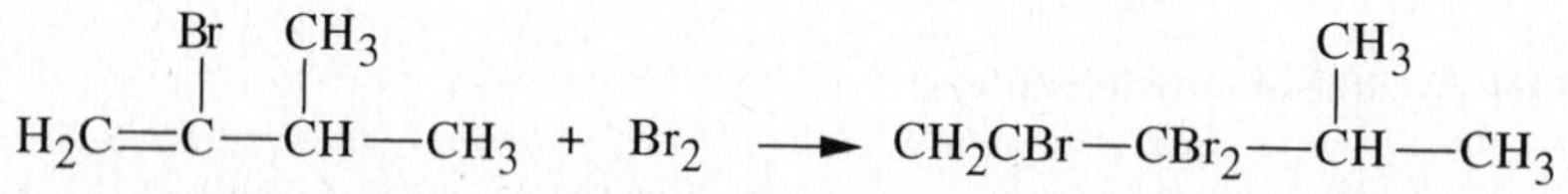

$$H_2C{=}CBr{-}CH(CH_3){-}CH_3 + Br_2 \longrightarrow CH_2CBr{-}CBr_2{-}CH(CH_3){-}CH_3$$

(c) This compound can be made from the product of part (a) by addition of hydrogen to the double bond.

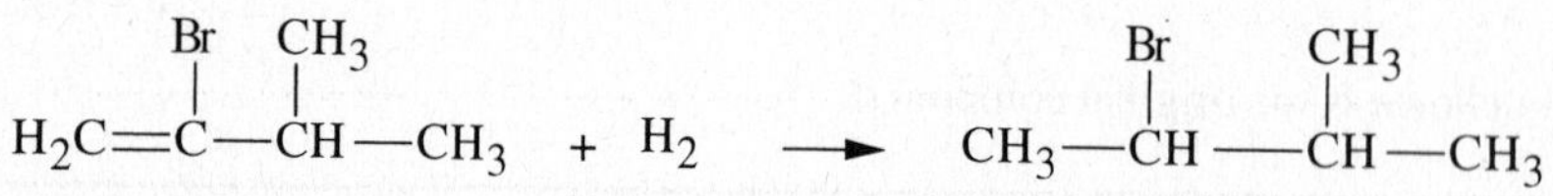

$$H_2C{=}CBr{-}CH(CH_3){-}CH_3 + H_2 \longrightarrow CH_3{-}CHBr{-}CH(CH_3){-}CH_3$$

25.37 What functional groups containing oxygen can be oxidized to ketones? There is more than one possible original compound.

Solution: Alcohols can be oxidized to ketones under controlled conditions. The possible starting compounds are:

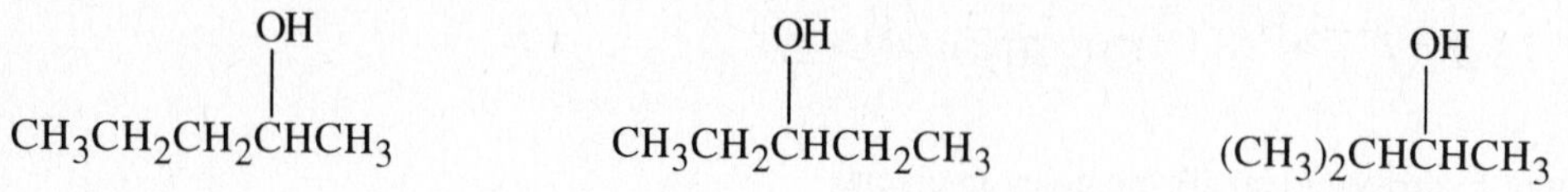

$$CH_3CH_2CH_2CH(OH)CH_3 \qquad CH_3CH_2CH(OH)CH_2CH_3 \qquad (CH_3)_2CHCH(OH)CH_3$$

The corresponding products are:

$$CH_3CH_2CH_2\overset{O}{\overset{\|}{C}}CH_3 \qquad CH_3CH_2\overset{O}{\overset{\|}{C}}CH_2CH_3 \qquad (CH_3)_2CH\overset{O}{\overset{\|}{C}}CH_3$$

Why isn't the alcohol $CH_3CH_2CH_2CH_2CH_2OH$ a possible starting compound?

25.38 Study the rules of nomenclature in Section 25.1.

Solution: (a) This is a branched hydrocarbon. The name is based on the longest carbon chain. It is 2-methylpentane.

(b) This is also a branched hydrocarbon. The longest chain includes the C_2H_5 group; the name is based on hexane, not pentane. This is an old trick. Carbon chains are flexible and don't have to lie in a straight line. The name is 2,3,4-trimethylhexane. Why not 3,4,5-trimethylhexane?

(c) How many carbons in the longest chain? It doesn't have to be straight! The name is 3-ethylhexane.

(d) An alkene with two double bonds is called a diene. The name is 3-methyl-1,4-pentadiene.

(e) The name is 2-pentyne.

(f) The name is 3-phenyl-1-pentene.

25.39 Review the rules for naming aromatic compounds in Section 25.1.

Solution: (a) This compound is 1,3-dichloro-4-methylbenzene.

(b) 1,4-dinitro-2-ethylbenzene. Note that the numbers should be kept as low as possible. Starting the numbering with the top nitro group to give 1,4-dinitro-3-ethylbenzene would be incorrect.

(c) Again keep the numbers low. 1,2,4,5-tetramethylbenzene is correct, not 1,3,4,6-tetramethylbenzene.

25.40 Review the rules for hydrocarbon derivative nomenclature in Section 25.1 if you are not sure about how to answer this problem.

Solution: The structures are:

(a)

$$CH_3—CH_2—\underset{}{\overset{CH_3}{\overset{|}{C}H}}—CH_2—CH_2—CH_3$$

(b)

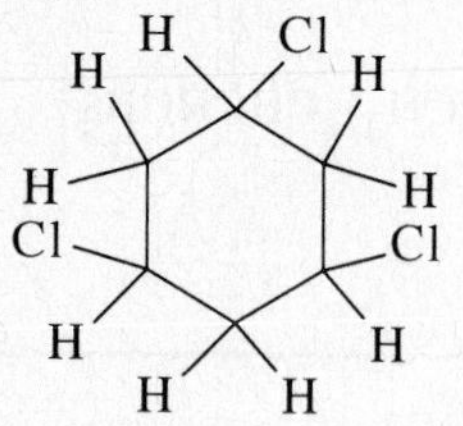

H H Cl H

H H

Cl Cl

H H

H H

Note: the carbon atoms in the ring have been omitted for simplicity.

(c)

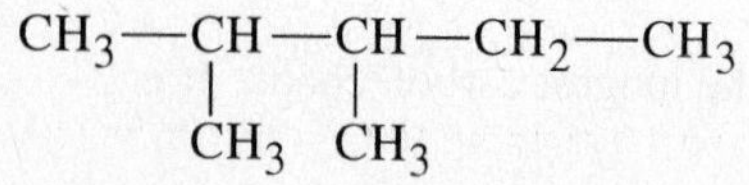

$CH_3—CH—CH—CH_2—CH_3$ (with CH_3 on the second and third carbons)

(d)

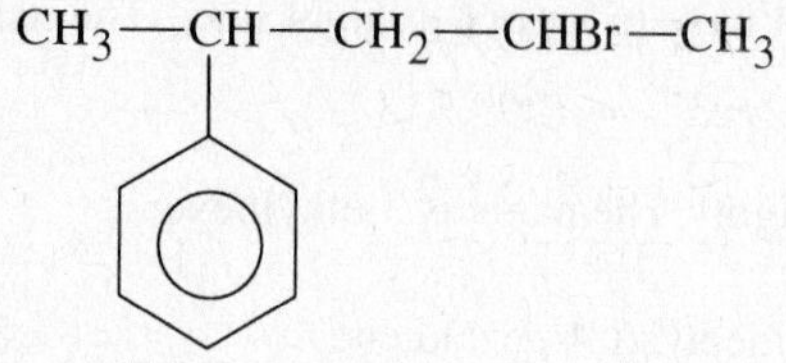

$CH_3—CH—CH_2—CHBr—CH_3$ (with a benzene ring on the second carbon)

(e)

$CH_3—CH_2—CH—CH—CH—CH_2—CH_2—CH_3$ (with CH_3 on the third, fourth and fifth carbons)

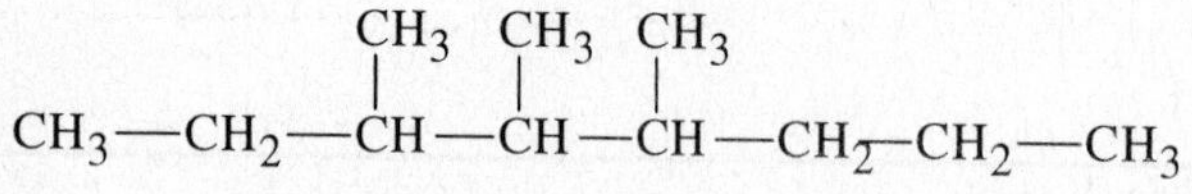

25.42 This is similar to Problems 25.40 and 25.41 above.

Solution: The structures are:

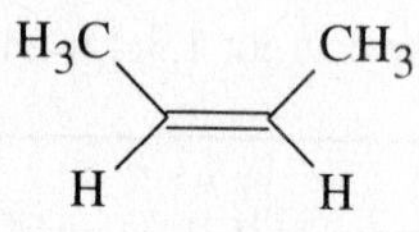

(a) CH_2, H_2C, CH_2, $H_2C—CH_2$ (a five-membered ring)

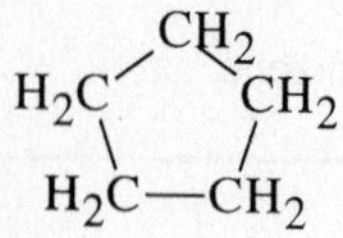

(b) H_3C and CH_3, H and H on a C=C double bond

(c) $CH_3—CH—CH_2—CH_2—CH_2—CH_3$ (with OH on the second carbon)

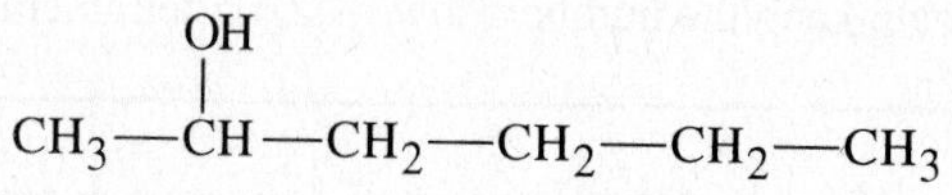

(d) Br—(benzene ring)—Br

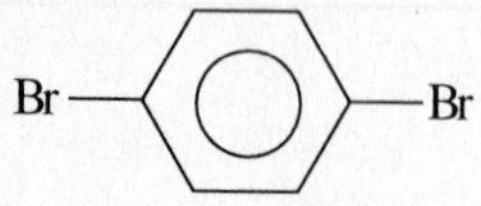

(e) $CH_3—C\equiv C—CH_3$

25.43 The formula indicates the compound is a chlorine substituted toluene (methylbenzene).

Solution: The four isomers are:

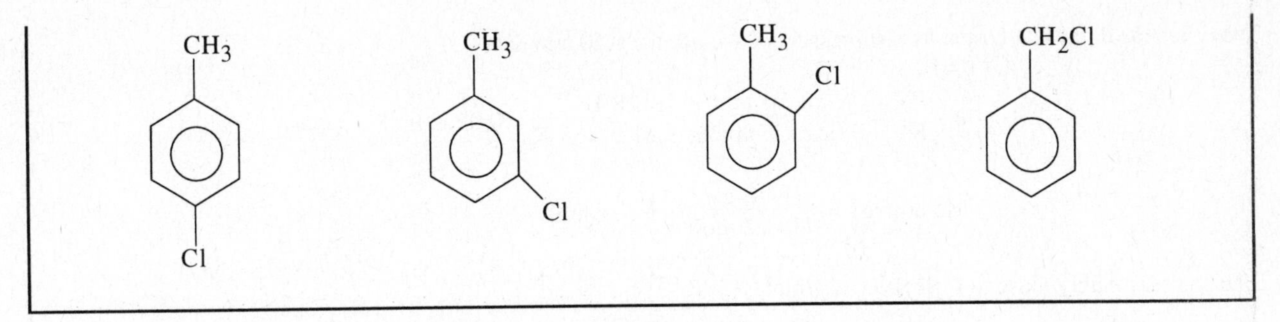

25.44 This is a Hess's law problem. If you need review, refer to Section 6.5 of the textbook.

Solution: If we rearrange the equations given and multiply times the necessary factors, we have:

$$2CO_2(g) + 2H_2O(l) \rightarrow C_2H_4(g) + 3O_2(g) \qquad \Delta H° = 1411\ kJ$$
$$C_2H_2(g) + \tfrac{5}{2}O_2(g) \rightarrow 2CO_2(g) + H_2O(l) \qquad \Delta H° = -1299.5\ kJ$$
$$H_2(g) + \tfrac{1}{2}O_2(g) \rightarrow H_2O(l) \qquad \Delta H° = -285.8\ kJ$$

$$C_2H_2(g) + H_2(g) \rightarrow C_2H_4(g) \qquad \Delta H° = -174\ kJ$$

The heat of hydrogenation for acetylene equals -174 kJ.

25.45 Problem 2.89 is an example of finding an empirical formula from percent by mass data. When temperature and amount of gas are constant, what is the relationship between pressure and volume for an ideal gas. How can you find molar mass from PVT data.

Solution: (a) The empirical formula is:

H: $\frac{3.2}{1.008} = 3.17$

C: $\frac{37.5}{12.01} = 3.12$

F: $\frac{59.3}{19.00} = 3.12$

The empirical is $H_{3.17}C_{3.12}F_{3.12}$. This reduces to an empirical formula of HCF.

(b) When temperature and amount of gas are constant, the product of pressure times volume is constant (Boyle's law).

$$2.00\ atm \times 0.332\ L = 0.664\ atm{\cdot}L$$
$$1.50\ atm \times 0.409\ L = 0.614\ atm{\cdot}L$$
$$1.00\ atm \times 0.564\ L = 0.564\ atm{\cdot}L$$
$$0.50\ atm \times 1.028\ L = 0.514\ atm{\cdot}L$$

The substance does not obey the ideal gas law.

(c) Since the gas does not obey the ideal gas equation exactly, the molar mass will only be approximate. Gases obey the ideal gas law best at lowest pressures. We use the 0.50 atm data

$$n = \frac{PV}{RT} = \frac{0.50 \text{ atm} \times 1.028 \text{ L}}{0.0821 \text{ L·atm/K·mol} \times 363 \text{ K}} = 0.0172 \text{ mol}$$

$$\text{Molar mass} = \frac{1.00 \text{ g}}{0.0172 \text{ mol}} = 58.1 \text{ g/mol}$$

This is reasonably close to $C_2H_2F_2$ (64 g/mol)

(d) The $C_2H_2F_2$ formula is that of a difluoroethylene. Three isomers are possible. The carbon atoms are omitted for simplicity (see Problem 25.14 above).

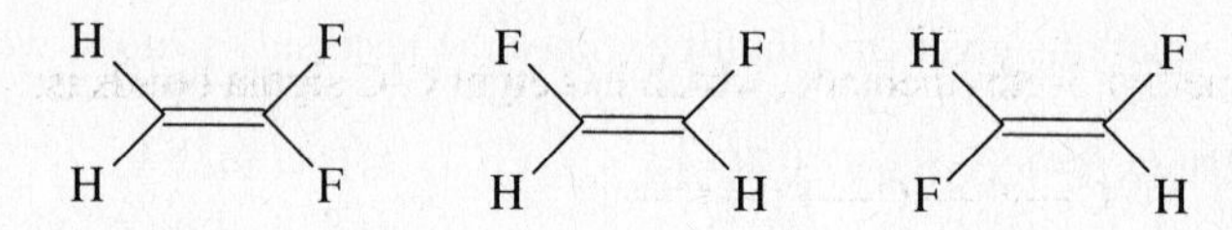

Only the third isomer has no dipole moment.

(e) The name is *trans*-difluoroethylene.

25.46 What structural features are necessary for hydrogen bond formation? See Section 10.2.

Solution: To form a hydrogen bond with water a molecule must have at least one H-F, H-O, or H-N bond, or must contain an O, N, or F atom. The following can form hydrogen bonds with water:

(a) carboxylic acids (c) ethers (d) aldehydes (f) amines

25.48 What are the products of the combustion of octane?

Solution: The combustion equation is:

$$2C_8H_{18}(l) + 25O_2(g) \rightarrow 16CO_2(g) + 18H_2O(l)$$

The number of moles of octane in one liter is:

$$1.0 \text{ L}\left(\frac{1000 \text{ mL}}{1 \text{ L}}\right)\left(\frac{0.70 \text{ g } C_8H_{18}}{1 \text{ mL } C_8H_{18}}\right)\left(\frac{1 \text{ mol } C_8H_{18}}{114.2 \text{ g } C_8H_{18}}\right) = 6.13 \text{ mol } C_8H_{18}$$

By stoichiometry, the number of moles of oxygen used is:

$$6.13 \text{ mol } C_8H_{18}\left(\frac{25 \text{ mol } O_2}{2 \text{ mol } C_8H_{18}}\right) = 76.6 \text{ mol } O_2$$

The volume of oxygen is:

$$V = \frac{nRT}{P} = \frac{(76.6 \text{ mol})(0.0821 \text{ L·atm/K·mol})(293 \text{ K})}{(1.00 \text{ atm})} = 1.84 \times 10^3 \text{ L}$$

The volume of air is:

$$1.84 \times 10^3 \text{ L} \left(\frac{100}{22}\right) = 8.4 \times 10^3 \text{ L}$$

25.49 What is the difference between a sigma bond and a pi bond? How many sigma bonds are in a single bond? In a double bond? In a triple bond?

Solution: (a) Benzene has six C-C sigma bonds.

(b) Cyclobutane has four C-C sigma bonds.

(c) The carbon skeleton of 2-methyl-3-ethylpentane, which has eight C-C sigma bonds is:

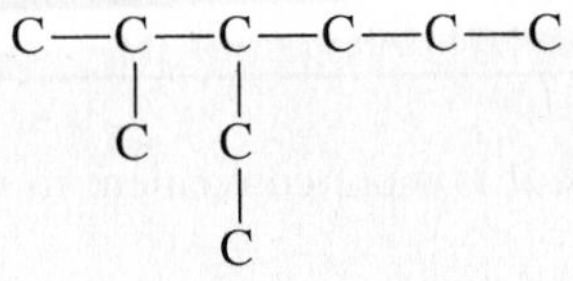

```
C—C—C—C—C—C
  |  |
  C  C
     |
     C
```

(d) 2-butyne (See Problem 25.42(e)) has three C-C sigma bonds.

(e) Anthracene is:

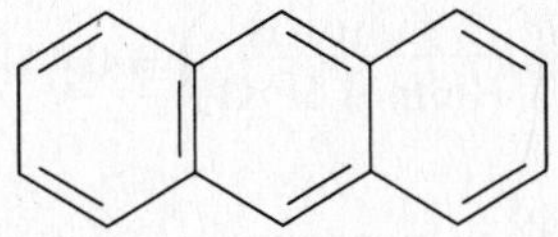

There are sixteen C-C sigma bonds.

25.50 This is similar to Problems 25.18 and 25.19.

Solution: The structural isomers are:

1,2-dichlorobutane

$CH_3—CH_2—\overset{*}{C}HCl—CH_2Cl$

1,3-dichlorobutane

$CH_3—\overset{*}{C}HCl—CH_2—CH_2Cl$

2,3-dichlorobutane

$CH_3—\overset{*}{C}HCl—\overset{*}{C}HCl—CH_3$

1,4-dichlorobutane

$CH_2Cl—CH_2—CH_2—CH_2Cl$

1,1-dichlorobutane

$CH_3—CH_2—CH_2—CHCl_2$

2,2-dichlorobutane

$CH_3—CH_2—CCl_2—CH_3$

1,3-dichloro-2-methylpropane

$$CH_2Cl-\underset{\displaystyle CH_3}{\underset{|}{CH}}-CH_2Cl$$

1,2-dichloro-2-methylpropane

$$CH_3-\underset{\displaystyle CH_3}{\underset{|}{CCl}}-CH_2Cl$$

1,1-dichloro-2-methylpropane

$$CH_3-\underset{\displaystyle CH_3}{\underset{|}{CH}}-CHCl_2$$

The asterisk identifies the asymmetric carbon atom.

25.52 This is similar to Problems 25.51 and 25.45 above.

Solution: Since the masses are given in milligrams, it is more convenient to work with millimoles (mmol). The number of mmoles of carbon is:

$$\text{C:} \quad 9.708 \text{ mg } CO_2 \left(\frac{1 \text{ mmol } CO_2}{44.01 \text{ mg } CO_2}\right)\left(\frac{1 \text{mmol C}}{1 \text{ mmol } CO_2}\right) = 0.2206 \text{ mmol C}$$

$$\text{H:} \quad 3.969 \text{ mg } H_2O \left(\frac{1 \text{ mmol } H_2O}{18.02 \text{ mg } H_2O}\right)\left(\frac{2 \text{ mmol H}}{1 \text{ mmol } H_2O}\right) = 0.441 \text{ mmol H}$$

The mass of oxygen is found by difference: 3.795 mg compound - (2.649 mg C + 0.445 mg H) = 0.701 mg O

$$\text{O:} \quad 0.701 \text{ mg} \left(\frac{1 \text{ mmol O}}{16.00 \text{ mg O}}\right) = 0.0438 \text{ mmol O}$$

The empirical formula is $C_{0.2206}H_{0.441}O_{0.0438}$ or $C_5H_{10}O$.

The molar mass is found using the ideal gas equation:

$$M = \frac{mRT}{PV} = \frac{(0.205 \text{ g})(0.0821 \text{ L·atm/K·mol})(473 \text{ K})}{(1.00 \text{ atm})(0.0898 \text{ L})} = 88.7 \text{ g/mol}$$

The formula mass of $C_5H_{10}O$ is 86.13 g, so this is also the molecular formula. Three possible structures are:

(ring: CH_2, H_2C, CH_2, H_2C, CH_2, O) (ring: H_2C—CH_2, H_2C, CH, O, with CH_3 on CH)

$$CH_2{=}CH-CH_2-O-CH_2-CH_3$$

25.54 In other words pairs of adjacent carbon atoms connected by a single bond would be different from pairs connected by a double bond.

Solution: The isomers are:

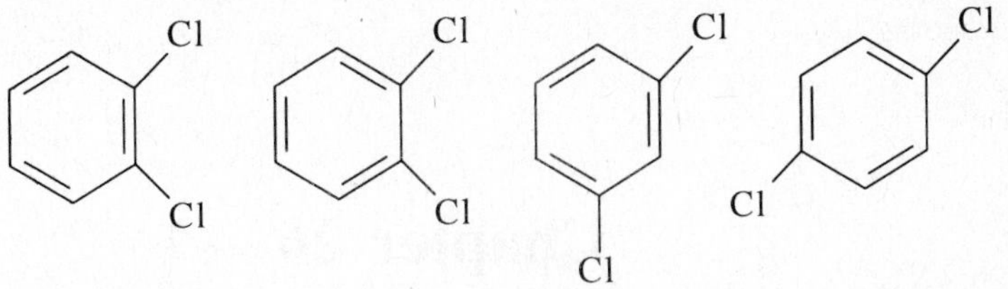

Did you have more isomers? Remember that benzene is a planar molecule; "turning over" a structure does not create a new isomer.

25.55 Consult an appropriate handbook for the needed data. Ask your instructor if you don't know where to look.

Solution: Ethanol has a melting point of -117.3°C, a boiling point of +78.5°C, and is miscible with water. Dimethyl ether has a melting point of -138.5°C, a boiling point of -25°C (It is a gas at room temperature), and dissolves in water to the extent of 37 volumes of gas to one volume of water.

25.56 In this problem you are asked to think about several different ways in which the hydrogen atoms can attach themselves to the alkyne. Chemists can often make conclusions about the mechanism of a reaction from the distribution of possible products in experiments of this type.

Solution: A mixture of cis and trans isomers would imply some sort of random addition mechanism in which one hydrogen atom at a time adds to the molecule.

The formation of pure cis or pure trans isomer indicates a more specific mechanism. For example, a pure cis product suggests simultaneous addition of both hydrogen atoms in the form of a hydrogen molecule to one side of the alkyne. In practice we get the cis isomer.

Note: If you have not already done so, please read the "Introduction for the Student" at the beginning of this Student Solutions Manual.

Chapter 26

ORGANIC POLYMERS: SYNTHETIC AND NATURAL

26.7 Use the description of the polymerization of ethylene in Section 26.2 as a model for this reaction.

Solution: The reaction is initiated by a radical R·

$$R\cdot + CF_2{=}CF_2 \rightarrow R\text{-}CF_2\text{-}CF_2\cdot$$

The product is also a radical, and the reaction continues

$$R\text{-}CF_2\text{-}CF_2\cdot + CF_2{=}CF_2 \rightarrow R\text{-}CF_2\text{-}CF_2\text{-}CF_2\text{-}CF_2\cdot \quad \text{etc...}$$

26.8 Assume that the two monomers alternate in the chain. Study the polymer structures shown in Section 26.2. What happens to the double bond in ethylene and its derivatives?

Solution: The repeating structural unit of the polymer is:

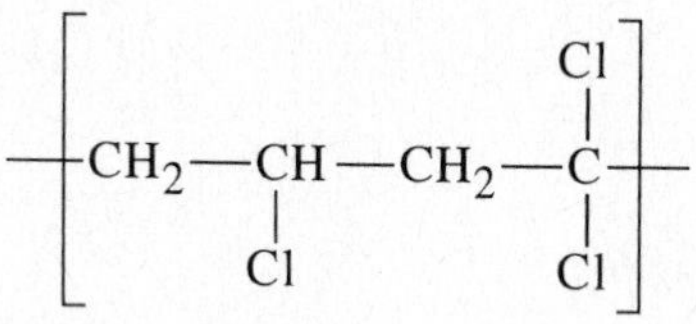

$$\left[-CH_2-\underset{\underset{Cl}{|}}{CH}-CH_2-\overset{\overset{Cl}{|}}{\underset{\underset{Cl}{|}}{C}}- \right]$$

Does each carbon atom still obey the octet rule?

26.10 Consult Section 26.2 and Table 26.1.

Solution: Polystyrene is formed by an addition polymerization reaction with the monomer, styrene, which is a phenyl-substituted ethylene. The structures of styrene and polystyrene are shown in Table 26.1.

26.12 Hint: the connections in nylon are amide linkages, which can be split by the action of acid. This reverses the polymerization reaction (see Figure 26.6).

Solution: The cleavage reaction is:

$$-(CH_2)_4-\overset{\displaystyle O}{\overset{\|}{C}}-NH-(CH_2)_6-NH-\overset{\displaystyle O}{\overset{\|}{C}}- \xrightarrow{H^+} HCOOH-(CH_2)_4-COOH + \overset{+}{H_3N}-(CH_2)_4-\overset{+}{NH_3}$$

26.23 This is the same as Problem 23.50.

Solution: The details of this problem are presented in this book in Problem 23.50. Hemoglobin must contain four Fe atoms per molecule for the actual molar mass to be four times the empirical formula.

26.24 Find the amino acids in Table 26.2. Study Figures 26.9-26.12.

Solution: (a) alanylglycine and glycylalanine are shown in Figure 26.10.

(b) glycyllysine

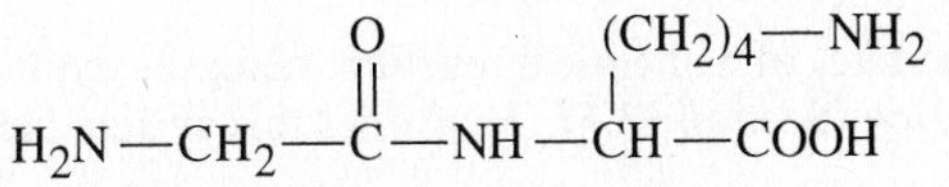

$$H_2N-CH_2-\overset{\displaystyle O}{\overset{\|}{C}}-NH-\overset{\displaystyle (CH_2)_4-NH_2}{\overset{|}{CH}}-COOH$$

lysylglycine

$$H_2N-\overset{\displaystyle H_2N-(H_2C)_4}{\overset{|}{CH}}-\overset{\displaystyle O}{\overset{\|}{C}}-NH-CH_2-COOH$$

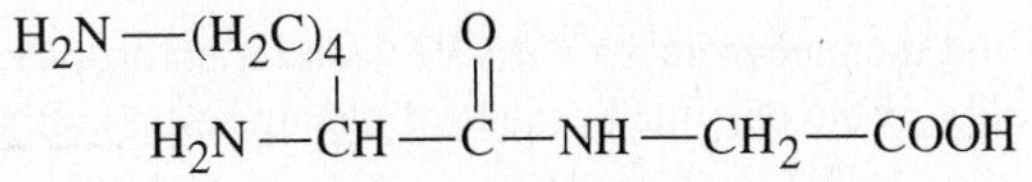

26.26 In other words all the monomers are the same. This is like Problems 26.24 and 26.25 above.

Solution: The structure of the polymer is:

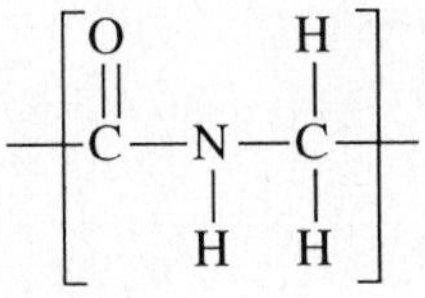

$$\left[-\overset{\displaystyle O}{\overset{\|}{C}}-\underset{\displaystyle H}{\underset{|}{N}}-\underset{\displaystyle H}{\underset{|}{\overset{\displaystyle H}{\overset{|}{C}}}}-\right]$$

26.27 This is like Problem 26.25 except that only two amino acids are available. How many ways can the two monomers be arranged into three-member chains? Remember that you can use a given monomer more than once. Don't draw structures; just write the possible chain sequences using the shorthand in Problem 26.28 below.

Solution: The best way to attack this type of problem is with a systematic approach. Start with all the possible tripeptides with three lysines (one), then all possible tripeptides with two lysines and one alanine (three), one lysine and two alanines (three also--Why the same number?), and finally three alanines (one).

Lys-Lys-Lys

Lys-Lys-Ala   Lys-Ala-Lys   Ala-Lys-Lys

Lys-Ala-Ala   Ala-Lys-Ala   Ala-Ala-Lys

Ala-Ala-Ala

Any other possibilities?

26.28 This is as much a puzzle as it is a chemistry problem. The puzzle involves breaking up a nine-link chain in various ways and trying to deduce the original chain sequence from the various pieces. Examine the pieces and look for patterns. Remember that depending on how the chain is cut, the same link (amino acid) can show up in more than one fragment. How many different amino acids are represented in the pieces? Is it necessary for one or more amino acids to appear more than once?

Solution: Since there are only seven different amino acids represented in the fragments, at least one must appear more than once. The nonapeptide is:

Gly-Ala-Phe-Glu-His-Gly-Ala-Leu-Val

Do you see where all the pieces come from?

26.30 Normally the rate constant (and therefore the rate) of a chemical reaction increases exponentially with increasing temperature according to the Arrhenius equation (Section 13.4). How does this system behave? What could cause this behavior?

Solution: The rate increases in an expected manner from 10°C to 30°C and then drops rapidly. The probable reason for this is the loss of catalytic activity of the enzyme because of denaturation at high temperature.

26.31 Draw the crystal field orbital splitting diagrams (Section 23.5 and Problem 23.59) for the two iron(II) complexes. Assume octahedral structure. Which orbitals are empty in the oxyhemoglobin complex? What is the location of these orbitals relative to the ligand donor atoms?

Solution: The two splitting diagrams are:

	deoxyhemoglobin	oxyhemoglobin
$d_{x^2-y^2}$	↑	(empty)
d_{z^2}	↑	(empty)
d_{xy}	↑↓	↑↓
d_{xz}	↑	↑↓
d_{yz}	↑	↑↓

In the oxyhemoglobin low-spin complex the two orbitals pointing directly toward the ligand donor atoms are unoccupied, thus allowing closer approach of the ligand and a smaller effective size of the iron atom. Alternatively, in deoxyhemoglobin the electrons are farther apart (occupying all five d orbitals); therefore, Fe^{2+} is larger.

26.36 Study Sections 26.3 and 26.4 and Figures 26.13 and 26.23.

Solution: There are two comon structures for protein molecules, an α helix and a β-pleated sheet. The α-heical structure is stabilized by intramolecular hydrogen bonds between the NH and CO groups of the main chain, giving rise to an overall rodlike shape. The CO group of each amino acid is hydrogen-bonded to the NH group of the amino acid that is four residues away in the sequence. In this manner all the main-chain CO and NH groups take part in hydrogen bonding. The β-pleated structure is like a sheet rather than a rod. The polypeptide chain is almost fully extended, and each chain forms many intermolecular hydrogen bonds with adjacent chains. In general, then, the hydogen bonding is responsible for the three dimensional geometry of the protein molecules.

In nucleic acids, the key to the double-helical structure is the formation of hydrogen bonds between bases in the two strands. Although hydrogen bonds can fom between any two bases, called base pairs, the most favorable couplings are between adenine and thymine and between cytosine and guanine.

Additional information concerning the importance of hydrogen bonding in biological systems is in Sections 26.3 and 26.4.

26.38 Study Section 26.3 and Problem 26.30.

Solution: When proteins are heated above body temperature they can lose some or all of their secondary and tertiary structure and become denatured. The denatured proteins no longer exhibit normal biological activity.

26.39 Refer to Figure 26.23(a) in Section 26.4. What is the most noticeable difference between C-G base pairs and A-T base pairs?

Solution: The sample that has the higher percentage of C-G base pairs has a higher melting point because C-G base pairs are held together by three hydrogen bonds. The A-T base pair interaction is relatively weaker because they have only two hydrogen bonds.

26.40 What is the effect of heat or acids or bases on proteins?

Solution: As is described in Section 26.3, acids *denature* enzymes. The citric acid in lemon juice denatures the enzyme that catalyzes the oxidation so as to inhibit the oxidation (browning).

26.42 Using the hint provided in the question, what is the purpose of hemoglobin molecules in human blood?

Solution: Insects have blood that contains no hemoglobin. Thus they rely on simple diffusion to supply oxygen. It is unlikely that an human-sized insect could obtain sufficient oxygen by diffusion alone to sustain its metabolic requirements.